Climate and Weather of the Sun-Earth System (CAWSES)

Selected Papers from the 2007 Kyoto Symposium

Cover illustration: The logo of CAWSES promoted by SCOSTEP

Climate and Weather of the Sun-Earth System (CAWSES)

Selected Papers from the 2007 Kyoto Symposium

Edited by

Toshitaka Tsuda

Research Institute for Sustainable Humanosphere, Kyoto University, Japan

Ryoichi Fujii

Solar-Terrestrial Environment Laboratory, Nagoya University, Japan

Kazunari Shibata

Kwasan and Hida Observatories, Kyoto University, Japan

Marvin A. Geller

School of Marine and Atmospheric Sciences, Stony Brook University, USA

TERRAPUB, Tokyo

Climate and Weather of the Sun-Earth System (CAWSES)
Selected Papers from the 2007 Kyoto Symposium
Edited by Toshitaka Tsuda, Ryoichi Fujii, Kazunari Shibata, and Marvin A. Geller
ISBN 978-4-88704-147-9

Published by TERRAPUB, 2003 Sansei Jiyuugaoka Haimu, 27-19 Okusawa 5-chome, Setagaya-ku, Tokyo 158-0083, Japan.
Tel: +81-3-3718-7500 Fax: +81-3-3718-4406
URL http://www.terrapub.co.jp/

Printed in Japan

Preface

SCOSTEP (Scientific Committee on Solar-Terrestrial Physics) began implementing its CAWSES (Climate and Weather of the Sun-Earth System) program in 2004, and it formally ended at the end of 2008. An international symposium on CAWSES was held at Kyoto University, Kyoto, Japan during October 23–27, 2007, to examine the progress that was made during CAWSES as well as to look forward to the future.

This symposium included tutorial lectures by Prof. E. Parker, on "Hydrodynamics, Magnetohydrodynamics, and Astrophysical Plasmas"; by Prof. A. Nishida on "Early Japanese Contributions to Space Weather Research—1945–1960"; and by Prof. M. A. Geller on "The 1960s—A Decade of Remarkable Advances in Middle Atmosphere Research." They were given with the intent of introducing all participants to the various aspects of solar-terrestrial physics no matter what their specialty area. There were also 16 keynote papers, again with the intent to familiarize all participants to research on various aspects of the solar-terrestrial system.

This book contains written versions of all three of the tutorial talks (Geller, Nishida, Parker) and 11 of the keynote talks (Beer, Dikpati, Forbes, Fröhlich, Gopalswamy, Haigh, Labitzke, Fukao, Tsuneta, Vincent, Zelenyi).

The editors hope that this book gives a good sense of both the historical context for present day solar-terrestrial research as well as its status during the latter stages of the CAWSES program.

The editors wish to acknowledge the excellent work of the international program committee for this conference, which was chaired by Prof. Y. Omura, and the local organizing committee for this symposium, which was chaired by Prof. T. Ogino. All success of this book is a consequence of the excellence of the symposium. We also wish to acknowledge that two of the most prominent Japanese scientists in SCOSTEP and CAWSES research retired near the time of the Symposium, Professors S. Fukao and Y. Kamide. We thank them for their years of valuable contributions.

Toshitaka Tsuda
Ryoichi Fujii
Kazunari Shibata
Marvin A. Geller

International CAWSES Symposium, October 23–27, 2007, Kyoto University, Kyoto, Japan

Contents

Climate and Weather of the Sun-Earth System (CAWSES): Selected Papers from the 2007 Kyoto Symposium,
Edited by T. Tsuda, R. Fujii, K. Shibata, and M. A. Geller, pp. 1–22.

Early Japanese contributions to space weather research —1945–1960—

A. Nishida

Graduate University for Advanced Studies,
Hayama, Kanagawa 240-0193, Japan
E-mail: hirosoph@dj.mbn.or.jp

Major contributions by Japanese scientists in the period of 1945 to 1960 are reviewed. This was the period when the groundwork for the space weather research was laid by ground-based observations and theoretical research. Important contributions were made on such subjects as; equatorial ionosphere in quiet times, tidal wind system in the ionosphere, formation of the F_2 layer, VLF propagation above the ionosphere, and precursory phenomena (type IV radio outburst and polar cap absorption) to storms. At the IGY (1957–1958), research efforts were intensified and new programs in space and Antarctica were initiated. Japanese scientists in this discipline held a tight network for communication and collaboration that has been kept to this day.

1 Introduction

On the occasion of the International CAWSES Symposium held at Kyoto in October, 2007, I presented a short history of research in this field that was made in Japan over the period of 15 years from 1945 to 1960. As the designation of the research discipline, "Space Weather" followed "Solar Terrestrial Physics" which had developed from the basic disciplines of ionospheric physics, geomagnetism, solar physics and cosmic ray physics. The year 1945 was when the World War II ended and basic research was reactivated, and the year 1960 was roughly the time when in-situ observations by space vehicles became a common tool of research and the frontier of the research rapidly expanded.

In these 15 years the research depended on remote sensing observations from the ground. Structures and dynamics of the ionosphere and above were inferred from the analyses of such data and new observations were designed to advance the understanding. Many of the basic concepts of the solar terrestrial physics that developed out of these efforts stand today as the basis of the Space Weather research. This paper reviews important contributions made by Japanese scientists to this development.

Before highlighting individual contributions in the following chapters, we should note that the Japanese scientists in this discipline held a tight network of communication and collaboration. Already in 1946 the Special Committee on the Ionosphere was established in the Science Council of Japan. Receiving the grants from the government to support travels, scientists nationwide met together periodically to exchange the latest information and to discuss collaborative programs. "*Report of Ionosphere Research in Japan*" that was inaugurated in 1947 was the publication from this Committee. The Committee is still active (though the title has changed) and being run by the Solar Terrestrial Environment Laboratory of the Nagoya University. While the Committee on the Ionosphere was run by public support, the Society of Terrestrial Magnetism and Electricity was founded in 1947 as an organization of individual scientists in this field. Its publication, "*Journal of Geomagnetism and Geoelectricity*," was inaugurated in 1949. Society of Geomagnetism, Earth and Planetary Sciences (SGEPSS) is the current name of the same Society, and "*Earth, Planets and Space*" is the successor of JGG.

The author himself who enrolled in the graduate school in 1958 was a late comer to the generation of scientists who were active in the 1945–1960 period. But I remember how exciting it was to attend the meetings of the above Committee and Society and listen to fresh outcomes of the latest research efforts. For writing this article I went through the literatures of the time and was impressed over again by the world-class contributions made by our predecessors.

2 Prologue

Ionospheric research had been conducted intensively from earlier times in Japan because it was necessary to specify and forecast ionospheric conditions in these days of the short wave communications. The ionospheric warning was particularly important for Japan because the radio waves to/from North America and Europe traverse high latitudes and are liable to ionospheric disturbances. In addition to this, during the wartime when the Japanese forces were deployed to South-East Asia the conditions of the equatorial ionosphere had to be monitored in order to select the optimum frequency for radio communications. For the latter purpose, fourteen stations were built, spanning the equatorial region and low latitudes.

A major outcome from this low-latitude network of stations was the discovery of the equatorial anomaly in f_0F_2 that is the measure of the peak electron density at the F_2 layer. Figure 1 shows the f_0F_2 versus local time at eight stations. While the diurnal variation of f_0F_2 peaked in the daytime at most of these stations, there were exceptions around the geomagnetic equator. At Singapore (geomagnetic latitude: -8.9°) and Penang (-4.7°), f_0F_2 was flat in the daytime and were lower than the values that were observed further away from the geomagnetic equator. This feature was named later as Equatorial Anomaly by Appleton.

In geomagnetic research, electric current systems in the ionosphere were studied by using geomagnetic data obtained during the second polar year. Daily variations

"Report of Japanese Research on Radio Wave Propagation," published from General Headquarters, United States Army Forces, Pacific Office of the Chief Signal Officer, May 1946, gives a fair account of this activity. It is said that D. K. Bailey was instrumental in compiling this report and helping the resumption of ionospheric research in the post-war Japan. See also a review article on the equatorial F-layer by Rishbeth (2000).

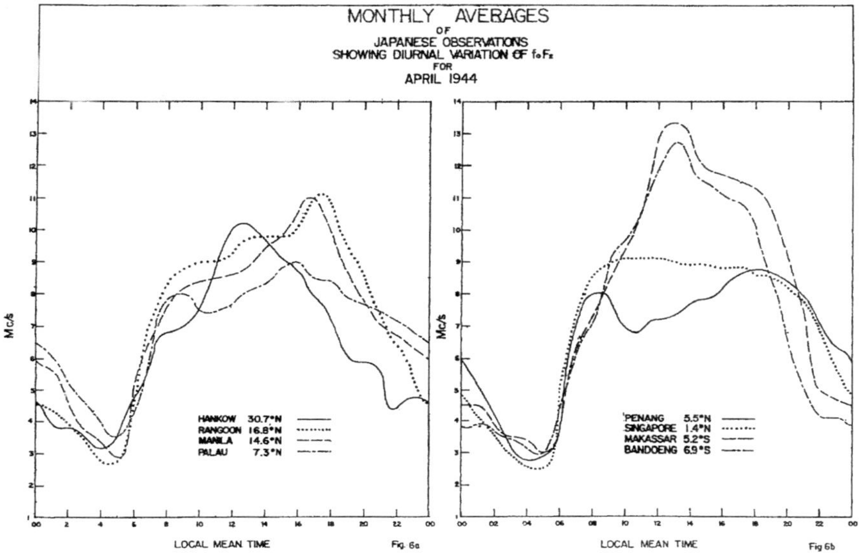

Fig. 1. Daily variations of f_oF_2 observed at eight stations at or near the geomagnetic equator in 1944. Noon time peak is absent or weak at stations around the geomagnetic equator. (Geomagnetic latitudes are, Hnakou: 20.4°, Rangoon: 6.8°, Manila: 4.6°, Palau, −1.9°, Penang: −4.7°, Singapore: −8.9°, Makassar: −15.2°, Bandoeng: −17.2°). This diagram is taken from "Report of Japanese Research on Radio Wave Propagation," published by Pacific Office of the Chief Signal Officer, General Headquarters, United States Army Forces, May 1946.

in geomagnetically quiet days, S_q, were studied extensively by Hasegawa who was attracted to this phenomenon while he was assigned to monitoring the magnetic disturbance accompanying the volcanic activity. He found that the focus of this current vortex was not fixed in latitude changing its position from day to day (Hasegawa, 1936). He was also aware that S_q had an anomalously large amplitude at the geomagnetic equator and noted this phenomenon (that is, the equatorial enhancement), as one of the important questions to solve. Concerning the geomagnetic disturbances, Hatakeyama made a pioneering study of the magnetic bays. He showed that global distribution of the bay disturbance was essentially the same as for the diurnal part of magnetic storms, and interpreted the disturbance in terms of the intensification of the ionospheric current due to enhancement of the electric conductivity by the charged corpuscles from the sun. The electromotive force provided by the dynamo effect of the atmospheric motion was thought to be the same for both quiet and disturbed times (Hatakeyama, 1938).

3 Electrodynamics of the Ionosphere in Quiet Times

The equatorial anomaly of f_oF_2 and the equatorial enhancement of S_q were explained using the concepts of plasma drift and Hall conductivity.

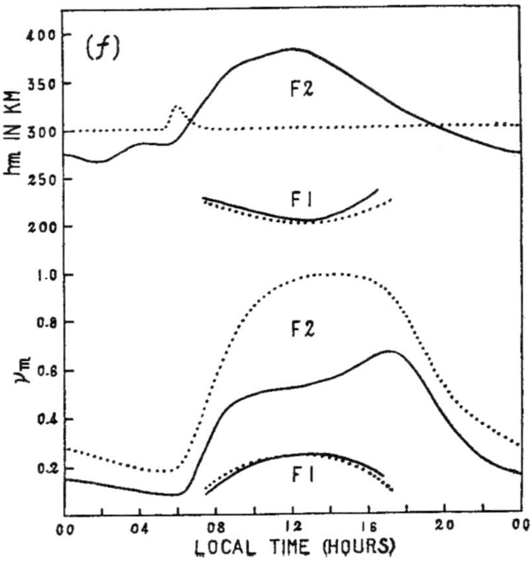

Fig. 2. Theoretical models of the daily variation of the F_2 layer at the geomagnetic equator. In the model where the vertical drift is taken into account (solid curve) the depression in density that was observed as "equatorial anomaly" was reproduced, while the Chapman layer (dashed curve) was peaked at noon. Top: virtual height and bottom: electron density at the F_2 peak (Hirono and Maeda, 1954).

3.1 Equatorial anomaly of f_oF_2

The geomagnetic daily variation S_q had been considered to result from the current generated in the ionospheric E region by the dynamo action of the atmospheric tide. When the atmosphere is in motion with velocity $\mathbf{v}$ relative to the Earth the dynamo electric field $\mathbf{v} \times \mathbf{B}$ is generated where $\mathbf{B}$ is the magnetic field. The total electric field $\mathbf{E}$ is the sum of $\mathbf{v} \times \mathbf{B}$ with the electrostatic field $\mathbf{E}_s$ which is produced to balance the polarization produced by the current due to $\mathbf{v} \times \mathbf{B}$, namely, to satisfy $\mathrm{div}[\Sigma](\mathbf{E}_s + \mathbf{v} \times \mathbf{B}) = 0$ where $[\Sigma]$ is the electric conductivity tensor integrated over the thickness of the E region.

This electric field $\mathbf{E} = \mathbf{E}_s + \mathbf{v} \times \mathbf{B}$ causes the charged particles to drift in the direction of $\mathbf{E} \times \mathbf{B}$ (Hirono, 1950). In the F region where the collision frequency is lower than the gyrofrequency for both ions and electrons, the drift motion is shared by ions and electrons. Since the electric field $\mathbf{E}$ of S_q is directed eastward in low latitudes, the drift is directed upward and transports ions and electrons away from the peak of the F_2 layer. This acts to reduce the electron density at the peak of the F_2 layer. The effect is most pronounced in the geomagnetic equatorial region where the magnetic field $\mathbf{B}$ is in the horizontal direction. Hence a minimum is expected to form at the geomagnetic equator in the latitudinal distribution of f_oF_2. The equatorial anomaly of f_oF_2 was explained in this way (Hirono and Maeda, 1954).

This was demonstrated in Fig. 2 where daily variations of height (top) and electron density (bottom) at the peak of the equatorial F_2 layer were compared between the above model (solid lines) and the Chapman model (dashed lines). The above model gave higher height and lower density at the F_2 peak in the daytime than the Chapman model in agreement with observations. (Horizontal diffusion was not incorporated in their model.)

3.2 Equatorial enhancement of magnetic variations

In the E region where the drift motion of ions is impeded by collisions and only electrons can drift, the $\mathbf{E} \times \mathbf{B}$ drift produces a downward current. In other words, the Hall current flows downward under the eastward electric field and the northward magnetic field. The downward current builds up negative charges at the top of the E region and the polarization produces an upward electric field. The Hall current due to the upward electric field adds to the eastward Pedersen current generated by the eastward electric field that is originally imposed, so that the eastward current $\mathbf{J}_{\mathrm{East}}$ is given by $\mathbf{J}_{\mathrm{East}} = (\Sigma_{\mathrm{P}} + \Sigma_{\mathrm{H}}^2/\Sigma_{\mathrm{P}})\mathbf{E}_{\mathrm{East}}$, where Σ_{P} and Σ_{H} represent Pedersen and Hall conductivities, respectively. Figure 3 shows the ratio between $(\Sigma_{\mathrm{P}} + \Sigma_{\mathrm{H}}^2/\Sigma_{\mathrm{P}})$ and Σ_{P} against latitude. Case A is for the case where electrons carry negative charges while case C is for a hypothetical case where negative ions do so. For the relevant case A, a pronounced peak is seen at the equator. The equatorial enhancement was attributed to this effect (Hirono, 1952).

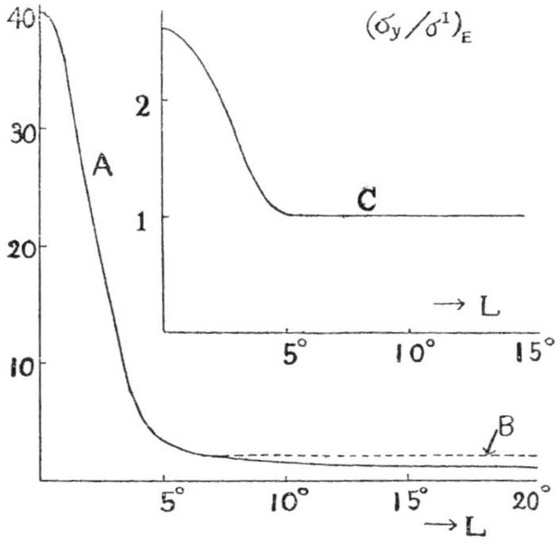

Fig. 3. Enhancement of the electric conductivity along the equator when the vertical polarization field produced by Hall current is taken into account. The model A is the relevant case; C is a reference model where carriers of the negative charge are assumed to be negative ion (Hirono, 1950).

3.3 Tidal wind system in the E region

On the basis of the dynamo theory, the wind system in the ionosphere was derived from the observation of the S_{q} current system. Using the current density $\mathbf{J}$ obtained from S_{q} and the conductivity model, $\mathbf{J} = [\Sigma](\mathbf{E}_{\mathrm{s}} + \mathbf{v} \times \mathbf{B})$ was solved for $\mathbf{E}_{\mathrm{s}}$ and $\mathbf{v}$

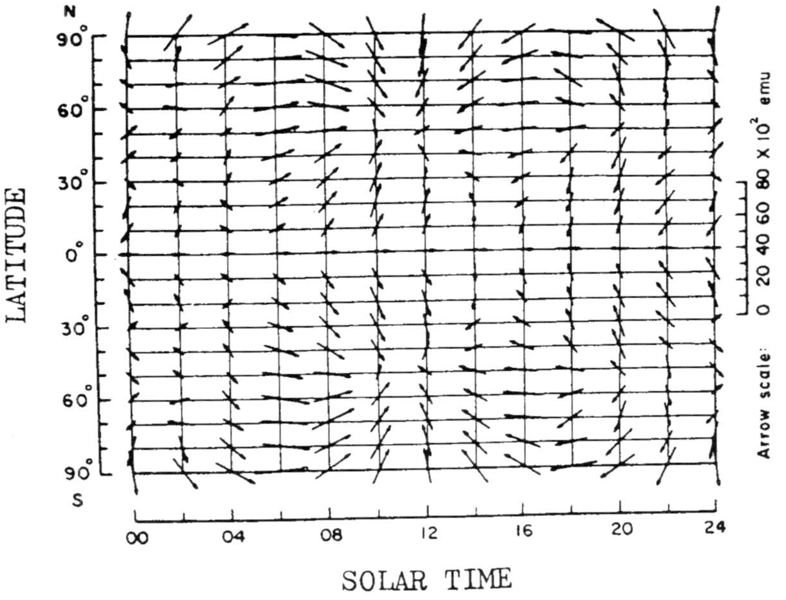

Fig. 4. Tidal wind system in the E region derived from the geomagnetic S_q variation (Maeda, 1955).

using the conditions div $\mathbf{J} = 0$ and rot $\mathbf{E} = 0$. In an earlier calculation $\mathbf{v}$ was assumed to be derivable from a potential (Maeda, 1955) but both rotational and irrotational components of $\mathbf{v}$ were considered later (Kato, 1957). Figure 4 shows the distribution of the tidal wind velocity obtained by the former author. In both cases, the tidal motion at the ionospheric height was found to be predominantly diurnal. This made a marked contrast to the atmospheric tidal oscillation observed on the ground where the semi-diurnal component was dominant. The difference in predominant modes of the tidal oscillation between ground and ionospheric heights posed a serious question.

4 Disturbances

4.1 Ionospheric disturbance in the F_2 layer

Disturbances in the F_2 layer had been studied extensively (Uyeda, 1949). World-wide morphology of the disturbances was constructed by superposing the observations of ten typical ionospheric storms during 1950–1951 (Fig. 5). The disturbance in f_oF_2 is characterized by bifurcation between positive and negative changes. The positive and the negative regions in the main phase cover the morning and afternoon sides, respectively, and then the area of the negative region expands to the evening side, spreading the decreases of f_oF_2 on the global scale (Obayashi, 1954).

Attempts were made to interpret the disturbance observed in the F_2 region in terms of the drift motion. The drift velocity was derived from the electric field ob-

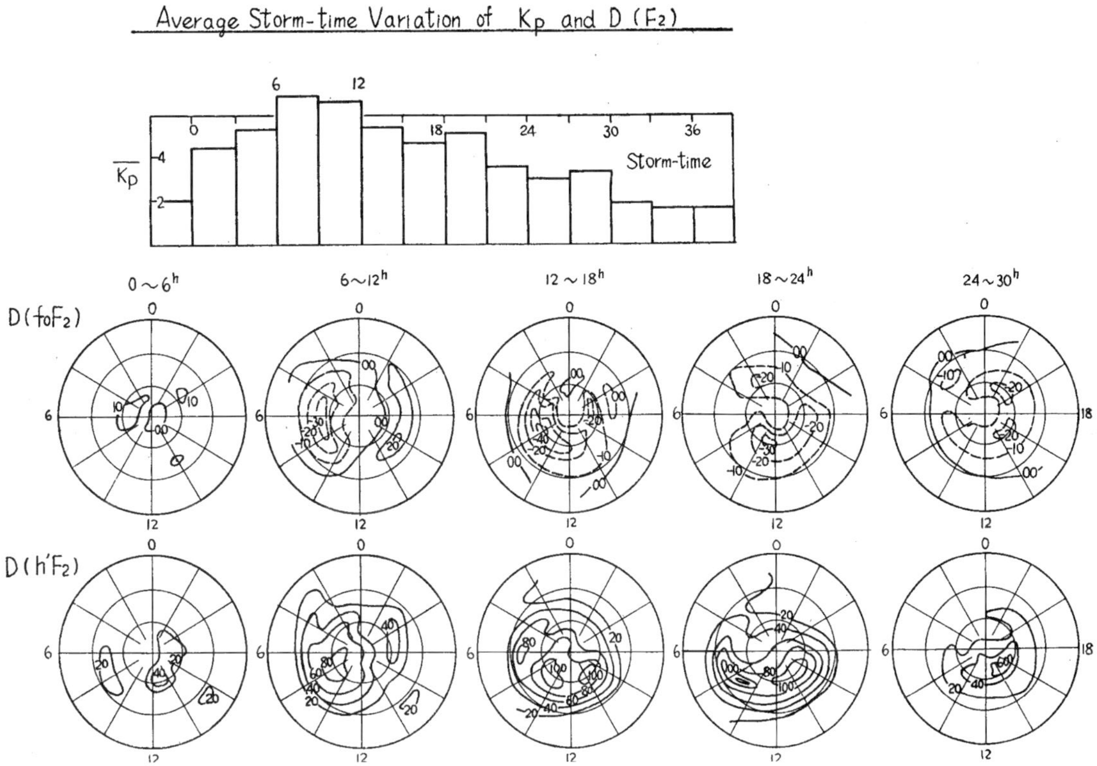

Fig. 5. Ionospheric disturbance in the F_2 layer obtained by averaging ten storms in 1950–1951 every six hours after the beginning. Top: K_p index, middle: percentage deviations of f_oF_2, and bottom: deviations of $h'F_2$ in km (Obayashi, 1954).

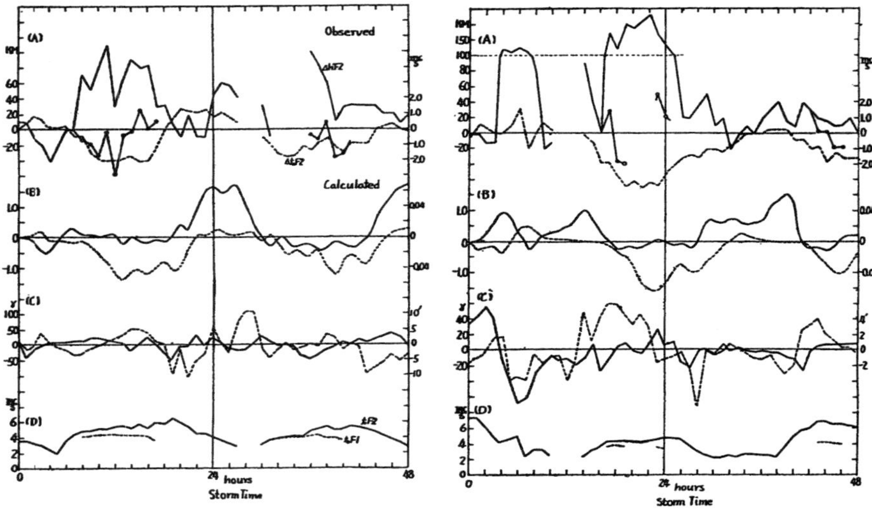

(a) **Washington, 0ʰ, Sep. 7, 1952** (b) **Washington, 16ʰ, Feb. 23, 1952**

Fig. 6. Comparison between (A) the observed and (B) the calculated variations of the F_2 layer at Washington during ionospheric disturbances. The bottom panels show observations of (C) the magnetic field and (D) the ionosphere (Sato, 1957).

tained from the simultaneous measurement of the magnetic field. In Fig. 6, calculated values of $h'F_2$ and f_oF_2 (B) are compared with the observed deviations from quiet times (A) at Washington D.C. for two intervals of 48 hours. Dashed curves that represent f_oF_2 seem to show fair agreement between (A) and (B) in the cases of negative changes in f_oF_2. Solid curves representing $h'F_2$ seem to differ but when the effect of retardation of the sounding radio wave is taken into account (yielding solid curves with small circles) the agreement seems improved. However, the calculated and observed values of f_oF_2 do not agree in the cases of positive changes of f_oF_2 (Sato, 1957). In the light of the findings in later years that F_2 storms involve several mechanisms, it has to be considered reasonable that the drift alone was not able to fully explain the observations.

4.2 Magnetic disturbances and the dynamo theory

Geomagnetic bays and elementary disturbances that constituted magnetic storms were compared with the popular model where the electromotive force was the same dynamo field as in quiet times but the electric conductivity was locally enhanced along the auroral zone. But the model did not agree with the observation; there was a very significant difference in phase (that means the direction of the current system in the frame of local time) amounting to 110°–150° (Nagata and Fukushima, 1952; Fukushima and Oguti, 1953).

Viewed from the vintage point of the present, it is amazing how broadly and deeply the success of the dynamo theory in interpreting the quiet-time features of the ionosphere influenced the research in these days. It was even suggested that winds in the D/lower E and upper E layers have different configurations so that the transition of the current system from S_q to S_D is caused by the switch of the current layer from one to the other (Matsushita, 1949; Nagata *et al.*, 1950). Combination of the wind-driven dynamo and a localized enhancement in the ionospheric conductivity was invoked also to explain the preliminary impulse of SC* (Nagata and Abe, 1955).

The enhancement in the electric conductivity in high latitudes was attributed to the ionizing effect of the impinging solar corpuscles. The "corpuscular stream" in these days corresponded broadly to what we call the solar wind today, but the properties attributed to them were not entirely the same. (1) It was supposed that the corpuscular stream was emitted from the sun for some limited intervals or from some limited locations. (2) As a typical speed, values of 1,000–1,500 km/s were often quoted. This was partly because the sudden commencement of magnetic storms was attributed to the encounter with the corpuscular stream itself rather than to the impact of the shock wave that propagated thereupon. (3) Some of the corpuscles were supposed to penetrate deep into the geomagnetic field. This should have been taken as a very serious problem since Störmer's theory of the motion of charged particles in the geomagnetic field was well established. (4) Dynamical effect of the momentum of the corpuscular flow was not considered.

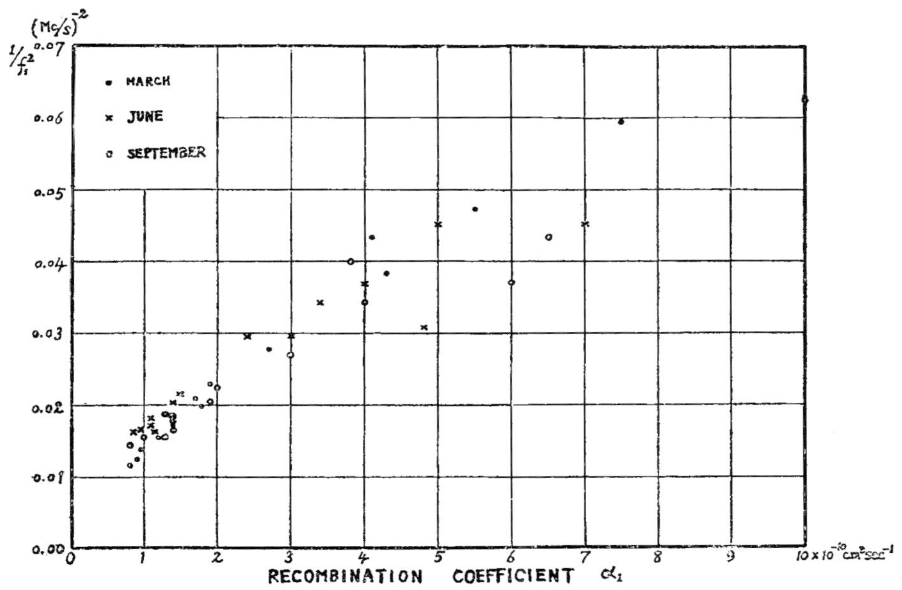

Fig. 7. Horizontal axis is the coefficient that is obtained by fitting the recombination model and vertical axis is the inverse of $(f_oF_2)^2$. There should not be a linear relation if the recombination model were applicable (Yonezawa, 1951).

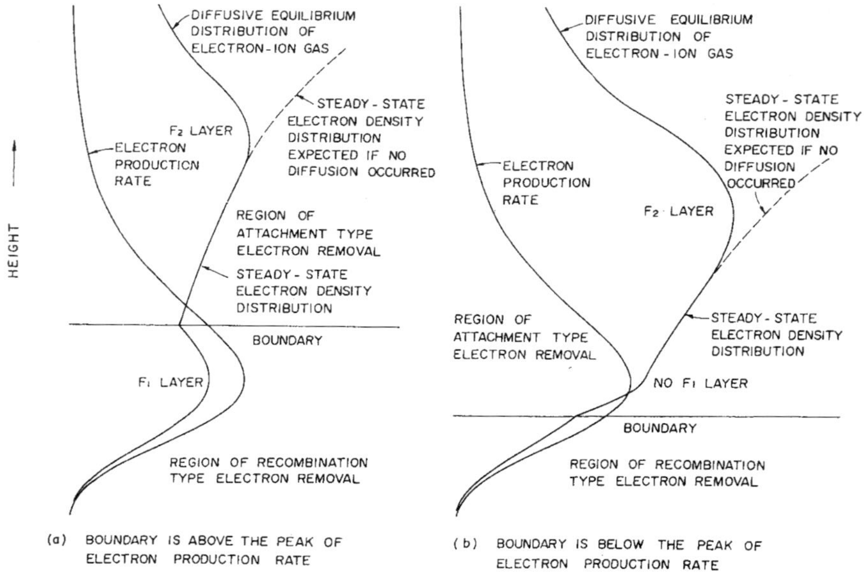

Fig. 8. Schematic illustrations of the height distribution of the electron density. See text. (With kind permission from Springer Science+Business Media: *Space Sci. Rev.*, Theory of formation of the ionosphere, **5**, 1966, 3–56, Yonezawa, T., figure 5.)

5 Formation of the F_2 Layer

According to the classical theory of formation of the ionospheric F_2 layer by Chapman, electrons are supposed to be lost by recombination. This means that the rate of the loss is proportional to the square of the electron density. However, when the apparent recombination coefficient was derived from observations at sunset, this was found to depend on the maximum electron density N_{max} of the F_2 layer (Fig. 7). The proportionality of the presumed recombination coefficient to $1/N_{max}$ suggested that the electrons are lost by attachment, rather than by recombination (Yonezawa, 1951).

Based on this finding, a new theory of the F_2 layer formation was developed. It was argued that the electron loss proceeds in two steps. The first is the charge transfer between atomic ions of oxygen and neutral molecules of oxygen or excited nitrogen molecules. The second step is the dissociative recombination between the resulting molecular ions and electrons (Yonezawa, 1955). At higher altitudes in the F region the first step, that is, attachment, is the rate-determining, while at lower altitudes the second step is so, because of rapid decreases in the number densities of N_2 and O_2 with height.

This reaction alone cannot produce the density peak of the F_2 layer. The loss rate of electrons by attachment is proportional to the density of molecules (O_2 or N_2)

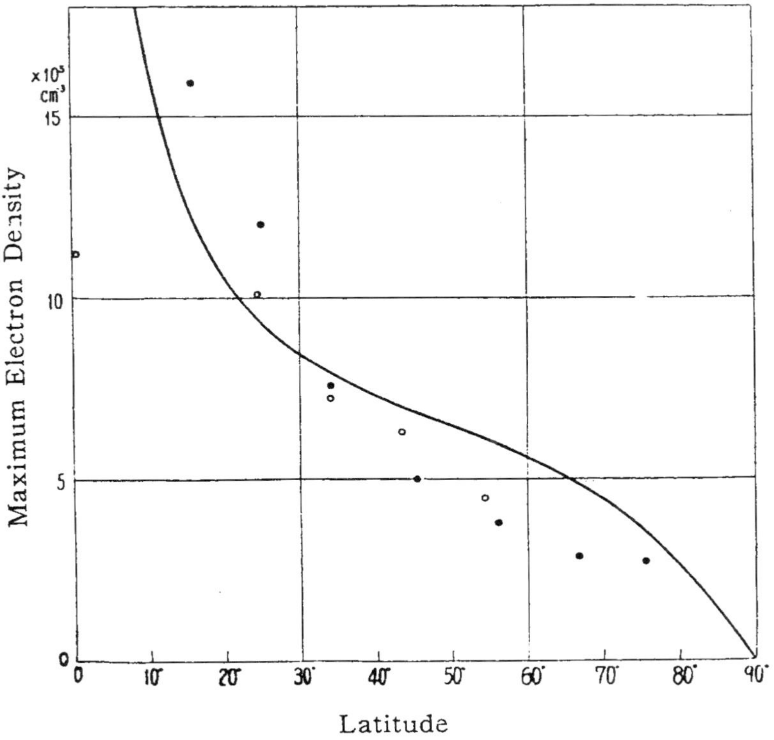

Fig. 9. Comparison between the observed and the calculated variations of the electron density at the peak of F_2 layer with latitudes (Yonezawa, 1956).

while their production rate is proportional to the density of atoms (O or N), and the electron density is proportional to the ratio of the latter to the former. Since density of atoms decreases more gradually with height as compared to molecules, the above ratio increases monotonically with height and no peak would result. The decrease of the attachment coefficient with height enhances this tendency.

This can be seen in Fig. 8 which schematically illustrates the distribution of the electron density with height. Two panels correspond to different cases of the transition height of the loss mechanism from recombination to attachment; it occurs at a relatively high altitude in the left-hand case and at a lower altitude in the right-hand case. In both cases, the dashed curves that extend to high altitudes represent the density distribution that may be expected when the loss is by attachment.

However, Yonezawa (1956) proposed that the peak of the electron density at F_2 could still be produced by the effect of diffusion in the earth's gravitational field. Electron-ion pairs generated in the upper portion of the F_2 layer come down quickly by diffusion, so that the distribution takes a form of the diffusive equilibrium. When this effect is incorporated, the electron distributions have a peak as drawn by solid curves in Fig. 8. The model is compared with observations in Fig. 9 where variations of the F_2 maximum electron density with latitude are compared between the model (solid curve) and the observations (dots). They agree very well except in the equatorial region where the diffusion in the horizontal direction need be incorporated. Thus, a new theory of the formation of the F_2 layer was established (Yonezawa, 1956).

6 Plasma Waves

6.1 Ray paths of VLF waves

While the ionospheric sounding used high frequency radio waves in the HF range, radio emissions of natural origin were observed in the VLF range. In this frequency range the whistler atmospherics are produced by lightening discharges. Extraordinary ray component of these waves penetrate the ionosphere and propagate through the outer region along geomagnetic field lines. In a classical paper published in 1953

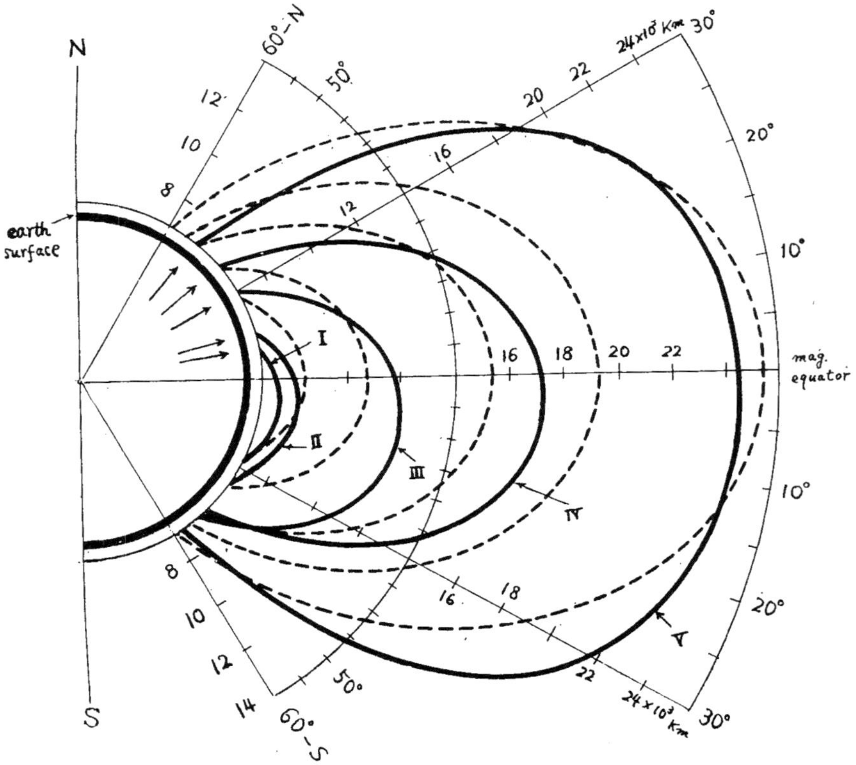

Fig. 10. Calculated ray paths of VLF waves (solid curves) and geomagnetic field lines (dashed curves) (Maeda and Kimura, 1956).

Storey found that the plasma extended above the F_2 layer and derived the electron density of 400 cm^{-3} at a height of two earth-radii by analyzing the frequency dispersion of the whistler atmospherics.

His work demonstrated that VLF waves of natural origin could be a powerful tool for investigating the physics of the outer ionosphere. (The term magnetosphere not yet existed at the time.) In order to provide the basis for interpreting the whistler observations Maeda and Kimura (1956) performed an exact calculation of ray paths of whistlers. Rays were emitted from several locations in the ionosphere; the highest L value of the emitting site was 4.5. Tracing was performed graphically using Fermat's principle.

The result of their ray tracing is shown in Fig. 10. They found that for the sources that are located at relatively low geomagnetic latitudes the rays arrive at a latitude higher than the source latitude. For the sources located at high latitudes, on the other hand, the rays are more symmetric but deviate appreciably from geomagnetic field lines. This paper has played a pioneering role to later studies of VLF waves as it became known that there are many other kinds of wave sources that are generated by wave-particle interactions in the magnetosphere.

6.2 Resonance model of magnetic pulsations

Waves in the ULF range found in magnetograms are called magnetic pulsations. They are classified into several types according to frequency, regularity and spatial distribution. Giant pulsations (which are called Pc5 according to the current nomenclature) are characterized by amplitudes of 15–30 nT, periods of several minutes, and occurrence in auroral latitudes. Kato and Watanabe (1954) found that the spectra of these pulsations had harmonics at $1/n$ of the fundamental period (Fig. 11), and

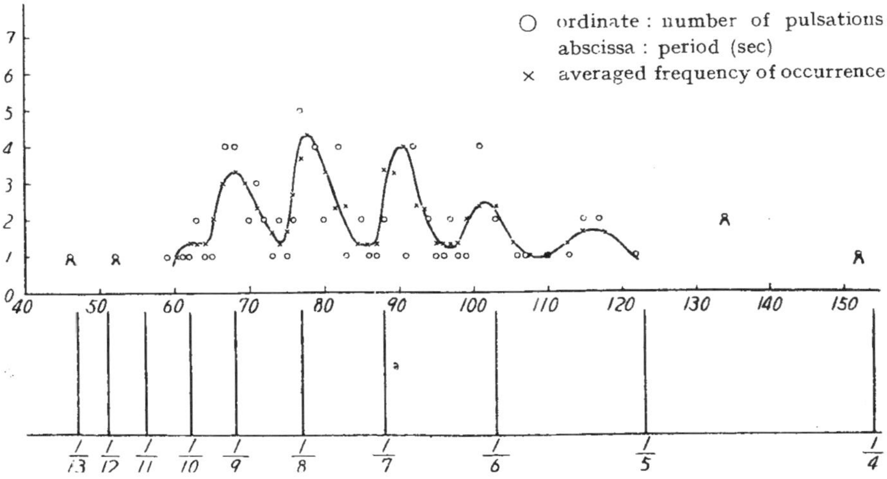

Fig. 11. Harmonic structure of the spectrum of giant pulsations (Kato and Watanabe, 1954).

suggested that they were resonant oscillations of magnetohydrodynamic waves excited by external disturbances. This suggestion was published in the same year as Dungey's well known report from the Pennsylvania State University. It was unfortunate that Kato and Watanabe did not seem to be aware of Storey's paper on the plasma density above the ionosphere. Later Obayashi and Jacobs (1958) calculated the resonance period of pulsations by using a realistic model of the density distribution and found that the latitude dependence of the period can be well explained by this model.

7 Solar Terrestrial Relation

From the intensive efforts during the IGY to forecast geomagnetic and ionospheric storms, two important phenomena were identified that preceded the storms. One was the type-IV solar radio outburst and the other was the blackout in the ionospheric radio propagation across the polar cap.

When solar radio outbursts from 200 MHz to 9400 MHz were compared with the ionospheric disturbances, it was found that the spectral property of the outbursts was related to the character of the disturbance they produced. As shown in Fig. 12, outbursts whose spectral intensity increased toward lower frequencies (left panel) tended to cause storms but not SWFs. (Short Wave Fadeout (SWF) means sudden increase in the absorption in the D and E regions due to enhancement in the ionization.) In contrast, the outbursts whose intensity increased toward higher frequencies (right panel) almost always caused SWFs but not storms. There were also cases where two types were combined (Sinno and Hakura, 1958). Outbursts with the low frequency component were identified with the synchrotron radiation emitted by energetic electrons and were called Type-IV. The source location of Type-IV radio outburst moved outward through the corona. Hence the occurrence of solar eruptions that eject corpuscular clouds, that is, "coronal mass ejections" according to the current nomenclature, could be monitored from the ground by looking out for this type of the outburst.

The polar cap blackout is caused by the abnormal enhancement in the radio wave absorption in the polar cap. It was found that the polar cap blackouts began several hours after solar flares accompanied by major Type-IV radio bursts. An example is shown in Fig. 13. The absorption was caused by precipitation of energetic particles having energies of the order of 10–100 MeV which were high enough for entering the geomagnetic field but not enough for crossing the ionosphere. Since it began well ahead of a storm, the polar cap blackout served as a very useful tool for forecasting disturbances in the radio propagation. At the onset of magnetic storms the region of the blackout spread to auroral latitudes and took a spiral-like shape centered on the morning side (Hakura, Takenoshita, and Otsuki, 1958).

The delay time between the Type-IV radio outburst and the beginning of the polar cap blackout, that is, the propagation time of energetic particles having energies of 10–100 MeV, depended on the solar longitude of the outburst site and was shorter for outbursts that occurred on the western side of the solar central meridian than those on the eastern side: see Fig. 14. This was interpreted in terms of the spiral structure of the interplanetary magnetic field where the field lines that reach the Earth originate from the western side of the central meridian. Particles from the west could stream

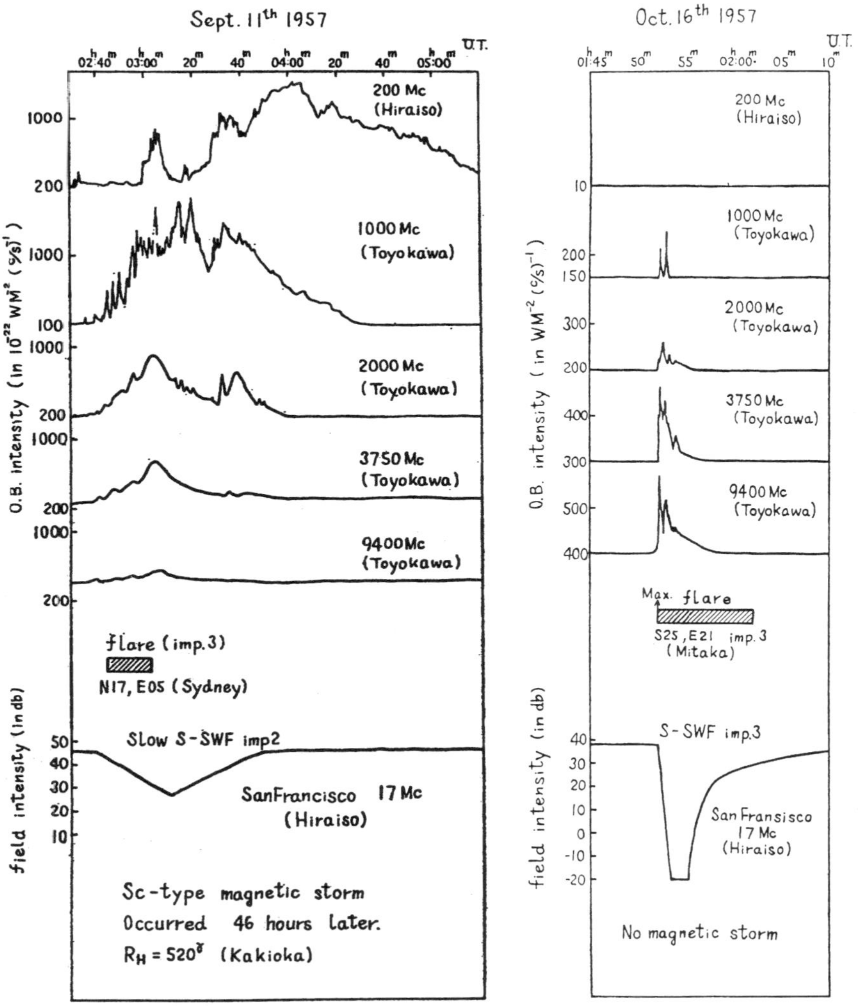

Fig. 12. Comparison of spectrum of solar radio bursts (five top curves), interval of flare (shaded box), field intensity of intercontinental radio transmission indicating the strength of SWF (bottom curve) and property of magnetic disturbance (handwritten at the bottom). Left: a case of low frequency burst and magnetic storm, and right: a case of high frequency burst and no magnetic storm (Sinno and Hakura, 1958).

directly along the field lines while those from the east had to diffuse across field lines. No such asymmetry was found for the delay between the Type-IV outburst and the onset of storms (Obayashi and Hakura, 1960b).

The above sequence of events, namely, (1) solar flare accompanied by Type-IV

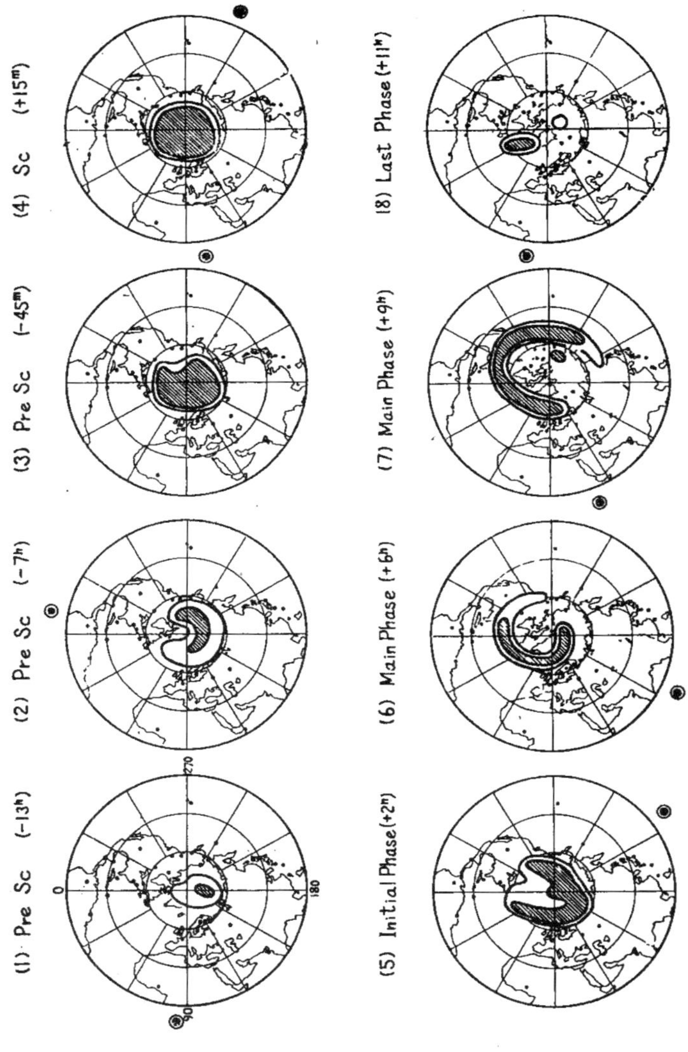

Fig. 13. Sequence of the ionospheric absorption following a solar radio burst. (2) Absorption begins several hours before the storm onset, (3) fills the polar cap, and (5 to 8) moves to the auroral zone after the onset (Obayashi and Hakura, 1960a).

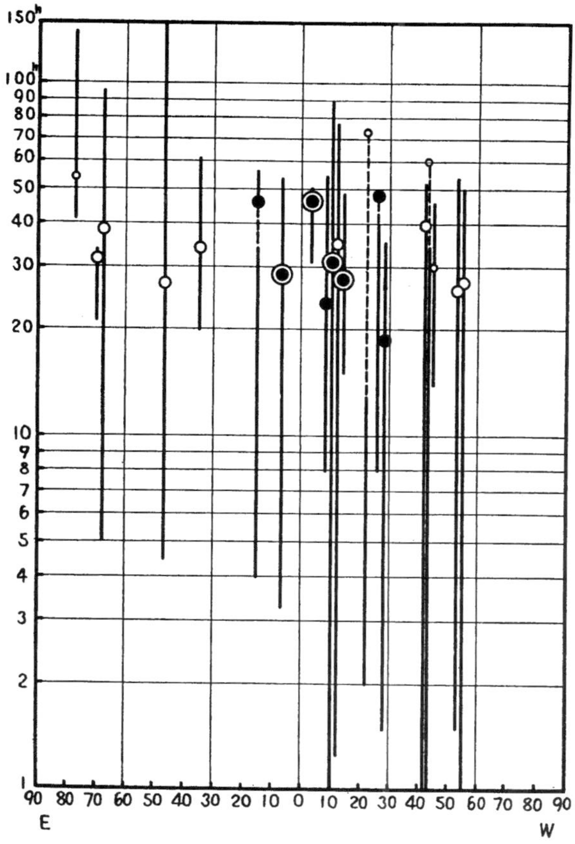

Fig. 14. Heliographic distribution of time delays to the polar cap absorptions (where solid lines are drawn throughout their durations) and of geomagnetic storms (where dots indicate the storm intensity) (Obayashi and Hakura, 1960b).

burst, (2) polar cap absorption, and (3) onset of magnetic and ionospheric storm at the arrival of the solar corpuscles, was demonstrated by numerous case studies, and it presented a powerful tool for warning the disturbance in the radio propagation. This study may be symbolic of a happy marriage between the curiosity-driven basic research and the applied research to improve the social infrastructure.

It was also recognized that there were exceptions to the sequence of events (1), (2) and (3). There were events where (1) type-IV burst was followed by (2) polar cap absorption, but magnetic storm did not follow and absorption did not spread to auroral latitudes. The only signature of the magnetic disturbance was a sudden impulse (Obayashi and Hakura, 1960a). There were also events where magnetic disturbance was produced several hours before a sudden impulse (Hakura, 1961). They suggested that some of the key factors in the solar terrestrial relation remained unidentified at the time.

The auroral zone absorption was also attributed to the direct entry of solar particles and the shift to lower latitudes was interpreted to be due to decrease in the magnetic field by the ring current. This interpretation had to be changed when more became known about the internal dynamics of the magnetosphere and the auroral zone absorption was found to result from the precipitation from inside the magnetosphere.

8 IGY Programs of Sounding Rocket and Antarctic Research

The International Geophysical Year (IGY, 1957–1958) gave a great impetus to the geophysical research worldwide. Japan was no exception. Most significantly, programs of the sounding rockets and the Antarctic research were initiated on this occasion.

The rocket program started in 1955 with the development of the rockets themselves. The first observation of the ionosphere was made in September 1960 with a Langmuir probe. Its record (Fig. 15) shows the electron density profile at middle latitude up to the height of ~200 km in the afternoon. Maximum electron densities of E, E_s and F layers obtained by the vertical incidence sounder at a nearby ground station are also shown for reference. It is not difficult to imagine the excitement at seeing the topside structure for the first time that had been hidden to traditional sounders.

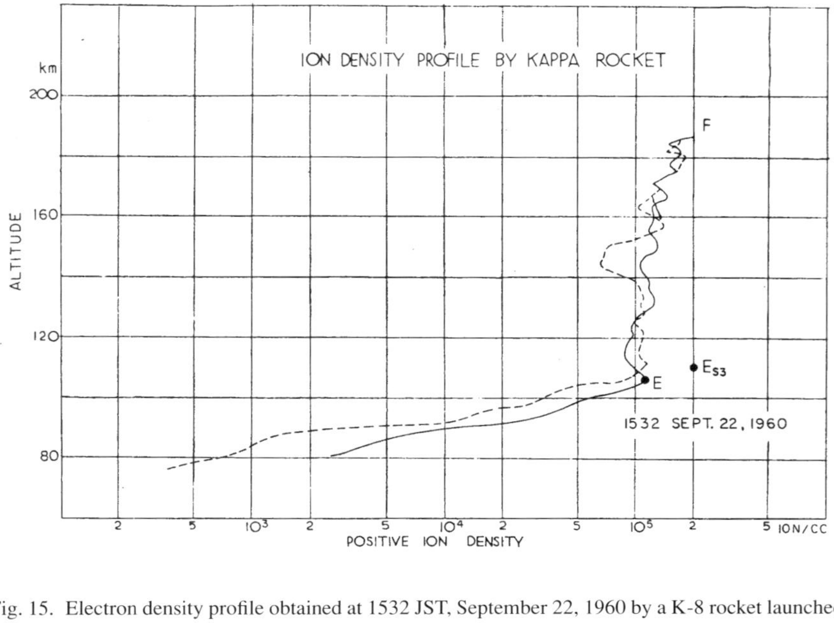

Fig. 15. Electron density profile obtained at 1532 JST, September 22, 1960 by a K-8 rocket launched from Michikawa, Japan. (Reprinted from *Planet. Space Sci.*, **9**, Maeda, K. and K. Hirao, A review of upper atmosphere rocket research in Japan, 355–369, Copyright 1962, with permission from Elsevier.)

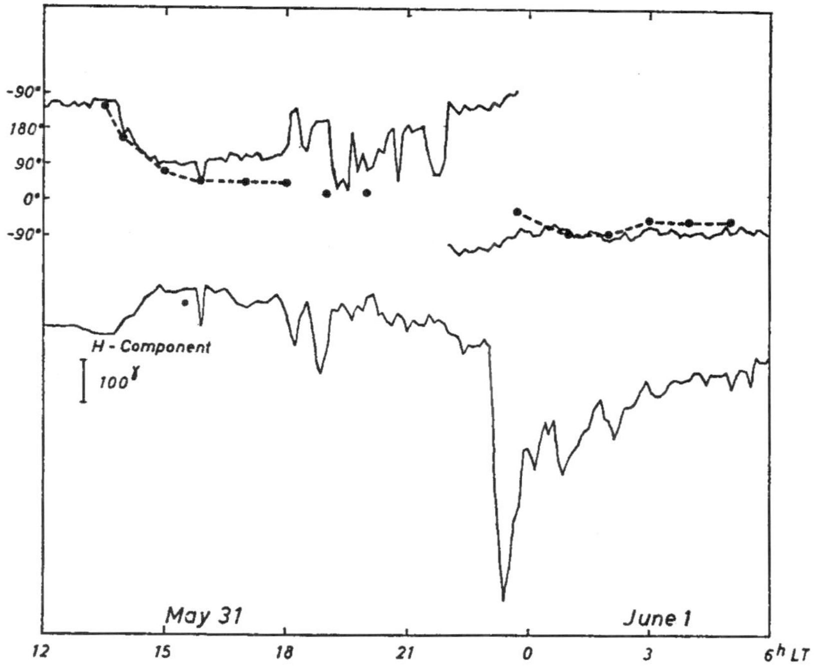

Fig. 16. Comparison between directions of ionospheric current and lower boundary of aurora (Oguti, 1960).

In the Antarctic, the Japanese Syowa Base was established in January, 1957. Observations of magnetic field, ionosphere, aurora and airglow were basic elements of the program. Since Syowa is located under the auroral zone it represented an ideal platform for studying aurora and its related phenomena. A full set of instruments were deployed during the wintering in 1959. An example of the observed relation between aurora and geomagnetic disturbance is shown in Fig. 16. The solid line in the upper part of the figure is the direction of the overhead current flow and the dots show the direction of the lower boundary of aurora. Their excellent agreement demonstrated that the ionospheric current was flowing along the latter (Oguti, 1960).

The Japanese community in geomagnetism and ionospheric physics has benefited from having enthusiastic and competent leaders. Figure 17 presents pictures of three prominent leaders at the time of the IGY. Prof. Ken-ichi Maeda (left), an ionospheric physicist, promoted experiments in space by organizing developments of rocket-borne scientific instruments. Prof. Takesi Nagata (right), a geomagnetician, led the Antarctic Research Expedition and established the Syowa Base. Prof. Mankichi Hasegawa (center), a geomagnetician who had founded the Japan Society of Geomagnetism and Geoelectrity, chaired the Japanese IGY Committee. We have

Fig. 17. Japanese leaders at the time of IGY: K. Maeda (1909–1995), M. Hasegawa (1894–1970) and T. Nagata (1913–1991). All the pictures were dated between 1955 and 1958.

a wealth of research programs that have been developed upon the heritage built by them. The scientific satellite programs at ISAS/JAXA, Antarctic Research conducted by the National Institute of Polar Research and the atmospheric radar program at the Institute of Sustainable Humanosphere at Kyoto University are the outstanding examples. CAWSES is also along the line of international programs in which the Japanese community has been deeply involved.

9 Epilogue

In the preceding Chapters I have noted that three major questions were raised from the space weather or solar terrestrial research before 1960. The first is why the tidal wind in the ionospheric *E* region is predominantly diurnal while the semi-diurnal tide prevails in the lower atmosphere. The second is how the electric field can be generated in the magnetosphere and ionosphere by mechanisms other than the atmospheric dynamo. The third is what other factors govern the solar wind-magnetosphere interaction than the properties of the solar corpuscles themselves.

The diurnal mode of the tidal oscillation which is essentially confined in the ionosphere was found in the solution of the tidal equation when the eigen value (representing the thickness of the layer) is allowed to be negative. Such a mode does not propagate but decreases or increases exponentially in the vertical direction, that is, the oscillation is confined to a region where the energy is supplied. The solution of the first negative mode agreed very well with the pressure variation deduced from the geomagnetic S_{q} variation (Kato, 1966).

A new type of geomagnetic daily variations, $S_{\mathrm{q}}^{\mathrm{p}}$, was discovered by Kokubun (1964) in the sunlit polar cap on geomagnetically quiet days. The current density corresponding to $S_{\mathrm{q}}^{\mathrm{p}}$ was too high to be generated by the combination of the dynamo electric field and the reasonable estimate of conductivity in the polar cap. Hence it was suggested that the electric field of S_q^p was the counterpart of the magnetospheric convection that is generated by the force exerted from the solar wind as proposed by Axford and Hines; see also Nagata and Kokubun (1962).

The geomagnetic field was found to respond globally (that is, from the equator to the pole) to the north-south component of the interplanetary magnetic field (IMF). The magnetic variation which showed this feature seemed to be the same as S_q^p when smoothed. Its correlation with IMF was so distinct that the peaks of respective variables matched with a time delay of several minutes. It was considered that the magnetic variation of this type, named DP2, reflects the time variability of the magnetospheric convection, and the excellent correlation with IMF suggests that the convection is driven by magnetic reconnection between the geomagnetic and interplanetary magnetic field lines; convection is stronger when IMF is closer to being southward and so is the magnetic disturbances produced in the magnetosphere as expected by Dungey's theory (Nishida, 1968). Thus the additional factor which exerts dominant role in the solar wind-magnetosphere interaction was identified to be the interplanetary magnetic field.

Once it was recognized that the ionosphere is coupled with the magnetosphere, it became untenable to use a model where the current was practically confined in the ionosphere. The field-aligned current that transports energy between the two regions had to be explicitly incorporated. An inversion method was developed to derive the electric field and the three-dimensional current system from observations of the magnetic disturbance field on the ground coupled with a realistic model of the conductivity distribution (Kamide *et al.*, 1981). The method has become a common tool for the magnetospheric research that employ ground magnetograms.

Acknowledgments. I thank Dr. Kiyoshi Igarashi for providing me with valuable records of the ionospheric research in the wartime and the immediate postwar period. I also thank Prof. Iwane Kimura and Hiroshi Maeda for sharing their personal recollections with me. "M. Hasegawa and geomagnetism and Geoelectricity" a monograph in Japanese by Drs. Hiroshi Nagano and Yasuharu Sanoo was a very useful summary of the research conducted by Prof. M. Hasegawa and his group at Kyoto University.

References

Fukushima, N. and T. Oguti, Polar magnetic storms and geomagnetic bays, Appendix I. A theory of S_D field, *Rept. Ionosph. Res. Jpn.*, **7**, 137, 1953.

Hakura, Y., Development of ionospheric and geomagnetic storms caused by solar corpuscular emissions, II. Polar blackouts, storm Es, and geomagnetic storms, *Rept. Ionosph. Space Res. Jpn.*, **15**, 14, 1961.

Hakura, Y., Y. Takenoshita, and T. Otsuki, Polar blackouts associated with severe geomagnetic storms on Sept. 13th, 1957 and Feb. 11th, 1958, *Rept. Ionosph. Res. Jpn.*, **12**, 459, 1958.

Hasegawa, M., On the type of the diurnal variations of the terrestrial magnetism on quiet days, *Proc. Imp. Acad. Tokyo*, **12**, 185, 1936.

Hatakeyama, H., On the bay-disturbance in the terrestrial magnetic field, *Geophys. Monogr. Cent. Meteorol. Obs., Tokyo*, **12**, 1938.

Hirono, K., On the influence of the Hall current to the electrical conductivity of the ionosphere, *J. Geomag. Geoelectr.*, **2**, 1, 1950.

Hirono, K., A theory of diurnal magnetic variations in equatorial region and conductivity of the ionospheric *E* region, *J. Geomag. Geoelectr.*, **4**, 7, 1952.

Hirono, M. and H. Maeda, Geomagnetic distortion of the $F2$ region on the magnetic equator, *J. Geomag. Geoelectr.*, **6**, 127, 1954.

Kamide, Y., A. D. Richmond, and S. Matsushita, Estimation of ionospheric electric fields, ionospheric currents, and field aligned currents from ground magnetic records, *J. Geophys. Res.*, **86**, 801, 1981.

Kato, S., Horizontal wind systems in the ionospheric E region deduced from the dynamo theory of geomagnetic S_q variation, Part IV, *J. Geomag. Geoelectr.*, **9**, 107, 1957.

Kato, S., Diurnal atmospheric oscillation, 1. Eigenvalues and Hough functions, *J. Geophys. Res.*, **71**, 3201, 1966.

Kato, Y. and T. Watanabe, A possible explanation of the cause of giant pulsations, *Sci. Rept. Tohoku Univ., Ser. 5*, **6**, 95, 1954.

Kokubun, S., Polar magnetic variations and their conjugacy, Ph.D. Thesis, University of Tokyo, 1964.

Maeda, H., Horizontal wind systems in the ionospheric E region deduced from the dynamo theory of geomagnetic S_q variation, Part I. Non-rotating Earth, *J. Geomag. Geoelectr.*, **7**, 121, 1955.

Maeda, K. and I. Kimura, A theoretical investigation on the propagation path of the whistling atmospherics, *Rept. Ionosph. Res. Jpn.*, **10**, 105, 1956.

Maeda, K. and K. Hirao, A review of upper atmosphere rocket research in Japan, *Planet. Space Sci.*, **9**, 355, 1962.

Matsushita, S., Circulatory motion in the ionospheric atmosphere and their relation to the S field of the terrestrial magnetism. II, *J. Geomag. Geoelectr.*, **1**, 41, 1949.

Nagata, T. and N. Fukushima, Constitution of polar magnetic storms, *Rept. Ionosph. Res. Jpn.*, **6**, 85, 1952.

Nagata, T. and S. Abe, Notes on the distributions of SC* in high latitudes, *Rept. Ionosph. Res. Jpn.*, **9**, 39, 1955.

Nagata, T. and S. Kokubun, An additional geomagnetic daily variation field (S_q^p-field) in the polar region on geomagnetically quiet day, *Rept. Ionosph. Space Res. Jpn.*, **16**, 256, 1962.

Nagata, T., N. Fukushima, and M. Sugiura, Electro-dynamical behavior of the ionosphere region viewed from geomagnetic variation, *J. Geomag. Geoelectr.*, **2**, 35, 1950.

Nishida, A., Coherence of geomagnetic DP 2 fluctuations with interplanetary magnetic variations, *J. Geophys. Res.*, **73**, 5549, 1968.

Obayashi, T., On the world-wide disturbances of the ionosphere, *Rept. Ionosph. Res. Jpn.*, **8**, 135, 1954.

Obayashi, T. and J. A. Jacobs, Geomagnetic pulsations and the Earth's outer atmosphere, *Geophys. J.*, **1**, 53, 1958

Obayashi, T. and Y. Hakura, Enhanced ionization in the polar ionosphere caused by solar corpuscular emissions, *Rept. Ionosph. Space Res. Jpn.*, **14**, 1, 1960a.

Obayashi, T. and Y. Hakura, Propagation of solar cosmic rays through the interplanetary magnetic field, *Rept. Ionosph. Space Res. Jpn.*, **14**, 427, 1960b.

Oguti, T., On the Earth storms V, Inter relations among the upper atmosphere disturbance phenomena over the polar regions, *Rept. Ionosph. Space Res. Jpn.*, **14**, 291, 1960.

Rishbeth, H., The equatorial F-layer: progress and puzzles, *Ann. Geophys.*, **18**, 730, 2000.

Sato, T., Disturbances in the ionospheric $F2$ region associated with geomagnetic storms, II. Middle latitudes, *J. Geomag. Geoelectr.*, **9**, 1, 1957.

Sinno, K. and Y. Hakura, On the relation of solar eruptions to geomagnetic and ionospheric disturbances, I. On the power spectrum of solar radio outbursts, *Rept. Ionosph. Res. Jpn.*, **12**, 285, 1958.

Uyeda, H., World-wide distribution and time variation of ordinary component critical frequency of$F2$-layer, *Rept. Ionosph. Res. Jpn.*, **3**, 1949.

Yonezawa, T., An analysis of electron density variations in the F2 layer after sunset, *Rept. Ionosph. Res. Jpn.*, **5**, 1, 1951.

Yonezawa, T., A consideration of the mechanism of electron removal in the F2 layer of the ionosphere, II, *Rept. Ionosph. Res. Jpn.*, **9**, 17, 1955.

Yonezawa, T., A new theory of formation of the $F2$ layer, *J. Radio Res. Lab.*, **3**, 1, 1956.

Yonezawa, T., Theory of formation of the ionosphere, *Space Sci. Rev.*, **5**, 3, 1966.

Climate and Weather of the Sun-Earth System (CAWSES): Selected Papers from the 2007 Kyoto Symposium,
Edited by T. Tsuda, R. Fujii, K. Shibata, and M. A. Geller, pp. 23–40.

Hydrodynamics, magnetohydrodynamics, and astrophysical plasmas

E. N. Parker

Dept. of Physics, University of Chicago
E-mail: parker@oddjob.uchicago.edu

1 Introduction

It is not uncommon to encounter the assertion that the dynamical evolution of the large-scale bulk velocity **v** of a collisionless gas is not described by the hydrodynamic (HD) equations. A favorite "reason" is the statement that pressure cannot be defined in the absence of collisions. Then in some quarters it is believed that the large-scale bulk dynamics of a swirling magnetized ionized gas must be described in terms of the electric current density **j** and the electric field **E** rather than **B** and **v**, thereby rejecting magnetohydrodynamics (MHD). This point of view arises from the conviction that **j** is the cause of **B**, and, therefore, the more fundamental field variable.

As a matter of fact, HD is based on the concepts of conservation of particles, momentum and energy, so HD cannot be avoided whether there are interparticle collisions (which conserve particles, momentum and energy) or not. So the basic HD equations cannot be avoided.

The dynamics of a swirling magnetized plasma follows from the equations of Newton, Maxwell, Lorentz, etc., from which one discovers that the dynamical theory cannot be formulated in terms of **j** and **E** in tractable form. The dynamical equations become nonlinear global integro-differential equations, of little use in a time varying system. Unfortunately the response often is to maintain the supremacy of **j** and **E** and turn away from Newton and Maxwell to seemingly plausible, but largely false, concepts.

The standard textbooks of HD and MHD are generally correct in their derivation of the dynamical equations from Newton, Maxwell, and Lorentz, but they do not address the popular misunderstandings and the rejection of HD and MHD. In fact, the basic equations of HD and MHD are simply derived and cannot be avoided in the large-scale bulk dynamics of a collisionless gas, with or without a magnetic field. The purpose of this article is to show that Newton, Maxwell, and Lorentz constrain the theory to HD and MHD and exclude the aforementioned misconceptions.

2 Hydrodynamics

Consider a collisionless gas in swirling motion in the absence of any applied forces, e.g. gravity, magnetic stresses, etc. In this simple case each particle moves in a straight line with constant speed regardless of the presence or absence of the other particles, and it is not immediately obvious that the sum of all these free particle motions has the overall large-scale nature of HD. So we must proceed formally from the principles of conservation of particles, momentum and energy.

The first question is how many particles are necessary to justify the fluid concept that we call HD? It is evident there must be *enough particles to provide a statistically well defined local mean number density* N. For if there are not enough particles, then neither the mean density nor the local bulk velocity can be defined with precision, and the fluid concept described by HD is not useful. Consider, then, a system with characteristic scale L. That implies a statistically well defined local mean density on some much smaller scale l. Experience with numerical simulations on a grid with spacing l suggests that l as small as $10^{-3}L$ works very well, allowing spatial derivatives to be computed with adequate accuracy. Using this as our criterion, it follows that there are Nl^3 particles in the basic cell, with volume $V = l^3$. The statistical fluctuation ΔN in the number Nl^3 of particles in the cell is of the order of $(Nl^3)^{1/2}$, so that $\Delta N/N$ is of the order of $1/(Nl^3)^{1/2}$. This condition is quite generally satisfied in astrophysical settings. For instance, the flow of the solar wind around the terrestrial magnetosphere is on a scale of $L \geq 10^8$ cm, so we might need l as small as 10^5 cm. The solar wind density is typically 5 ions/cm^3, so that the number of particles is $Nl^3 = 5 \times 10^{15}$. The statistical fluctuations are hardly more than one part in 10^8, and quite sufficient for our purposes. Needless to say, the internal small-scale structure of the standing shock upstream from Earth cannot be treated with HD, and we content ourselves there with application of the Rankine-Hugoniot relations.

Now, if the local mean density N is statistically well defined, then so is the local mean bulk velocity $\mathbf{v}$, or v_i. Denoting the velocity of an individual particle by u_i, it follows that

$$N = \frac{1}{V}\Sigma, \; Nv_i = \frac{1}{V}\Sigma u_i, \tag{1}$$

where Σ means the sum over all the particles in the volume $V = l^3$. Let w_i represent the thermal velocity of an individual particle relative to the bulk motion v_i, so that

$$u_i = v_i + w_i \tag{2}$$

and

$$\frac{1}{V}\Sigma w_i = 0. \tag{3}$$

It follows that the momentum density of the particles of mass M is

$$\frac{1}{V}\Sigma M u_i = NMv_i, \tag{4}$$

and the flux of momentum density is

$$\frac{1}{V}\Sigma M u_i u_j = NMv_iv_j + p_{ij}, \tag{5}$$

where

$$p_{ij} = \frac{1}{V} \Sigma M w_i w_j \tag{6}$$

is the pressure tensor, representing the flux of thermal momentum transported by the thermal motions. A momentum flux is equivalent to a force per unit area, i.e. pressure, regardless of the presence or absence of interparticle collisions. The first term on the right hand side of Eq. (5) is the Reynolds stress tensor,

$$R_{ij} = N M v_i v_j \tag{7}$$

representing the momentum of the bulk flow transported by the bulk flow.

Note that the diagonal terms of $\frac{1}{V}\Sigma M u_i u_j$ represent twice the kinetic energy density, so the flux of kinetic energy is contained in the tensor

$$\begin{aligned} \frac{1}{V} \Sigma M u_i u_j u_k &= \frac{1}{V} \Sigma M (v_i + w_i)(v_j + w_j)(v_k + w_k) \\ &= N M v_i v_j v_k + v_i p_{jk} + v_j p_{ki} + v_k p_{ij} + T_{ijk}, \end{aligned} \tag{8}$$

where

$$T_{ijk} = \frac{1}{V} \Sigma M w_i w_j w_k \tag{9}$$

represents the thermal transport of thermal energy, i.e. the heat flow tensor.

Now particles, momentum, and kinetic energy are conserved quantities. Suppose, then, that Q represents the density of a conserved quantity. The time rate of change of that quantity in a fixed volume U is equal to the rate at which Q flows inward across the surface S of the volume U,

$$\begin{aligned} \frac{\partial}{\partial t} \int_U d^3 r\, Q &= - \int_S dS_j (Q_{\text{flux}})_j \\ &= - \int_U d^3 r \frac{\partial}{\partial x_j} (Q_{\text{flux}})_j, \end{aligned}$$

upon applying Gauss's theorem. The equality holds for every volume U, requiring that

$$\frac{\partial Q}{\partial t} = - \frac{\partial}{\partial x_j} (Q_{\text{flux}})_j \tag{10}$$

at every point. So the time rate of change of the density Q of a conserved quantity is equal to the negative divergence of the flux of Q.

It follows that conservation of particles requires

$$\frac{\partial N}{\partial t} + \frac{\partial}{\partial x_j} N v_j = 0. \tag{11}$$

Conservation of momentum requires

$$\frac{\partial}{\partial t} N M v_i + \frac{\partial}{\partial x_j} N M v_i v_j = - \frac{\partial p_{ij}}{\partial x_j}. \tag{12}$$

Conservation of kinetic energy requires conservation of the diagonal terms in p_{ij}, requiring conservation of the tensor p_{ij}. So the time rate of change of $\Sigma M u_i u_j$ must be equal to the negative divergence of $\Sigma M u_i u_j u_k$, leading to

$$\frac{\partial}{\partial t} N M v_i v_j + \frac{\partial}{\partial x_k} N M v_i v_j v_k + \frac{\partial p_{ij}}{\partial t} = -\frac{\partial}{\partial x_k}(v_i p_{jk} + v_j p_{ki} + v_k p_{ij}) - \frac{\partial T_{ijk}}{\partial x_k}. \quad (13)$$

Multiply Eq. (11) by v_i and subtract from Eq. (12), providing the familiar Euler equation

$$N M \left(\frac{\partial v_i}{\partial t} + v_j \frac{\partial v_i}{\partial x_j} \right) = -\frac{\partial p_{ij}}{\partial x_j}. \quad (14)$$

Equations (11) and (12) can be used to reduce Eq. (13) to

$$\frac{\partial p_{ij}}{\partial t} + v_k \frac{\partial p_{ij}}{\partial x_k} = -p_{jk} \frac{\partial v_i}{\partial x_k} - p_{ki} \frac{\partial v_j}{\partial x_k} - p_{ij} \frac{\partial v_k}{\partial x_k} - \frac{\partial T_{ijk}}{\partial x_k}. \quad (15)$$

Equations (11), (14), and (15) are the basic equations of HD. The pressure p_{ij} may be anisotropic in the absence of collisions, of course, but the principles are HD.

Several comments are in order here. First of all, the complexity of the right hand side of Eq. (15) is nothing more than a representation of adiabatic heating and cooling in the presence of compression and expansion, respectively, as well as viscosity in the presence of velocity shear. To illustrate the effects consider a uniform plasma density and uniform temperature subject to the one dimension expansion $\partial v_1/\partial x_1$ with all other elements of $\partial v_i/\partial x_j$ equal to zero. The left hand side of Eq. (15) is the Lagrangian derivative dp_{ij}/dt, so that

$$\frac{dp_{11}}{dt} = -3p_{11}\frac{\partial v_1}{\partial x_1}, \quad \frac{dp_{22}}{dt} = -p_{22}\frac{\partial v_1}{\partial x_1}, \quad \frac{dp_{33}}{dt} = -p_{33}\frac{\partial v_1}{\partial x_1}, \quad (16)$$

$$\frac{dp_{12}}{dt} = -2p_{12}\frac{\partial v_1}{\partial x_1}, \quad \frac{dp_{13}}{dt} = -2p_{13}\frac{\partial v_1}{\partial x_1}, \quad \frac{dp_{23}}{dt} = -p_{23}\frac{\partial v_1}{\partial x_1}. \quad (17)$$

Noting that

$$\frac{1}{N}\frac{dN}{dt} = -\frac{\partial v_1}{\partial x_1}, \quad (18)$$

it follows that

$$p_{11} \propto N^3; \; p_{12}, p_{13} \propto N^2; \; p_{22}, p_{33}, p_{23} \propto N. \quad (19)$$

Anisotropic expansion in two and three dimensions provides somewhat more complicated results.

Consider, then, a uniform shear, with $\partial v_1/\partial x_2$ as the only nonvanishing term in $\partial v_i/\partial x_j$. The result is

$$\frac{dp_{11}}{dt} = -2p_{12}\frac{\partial v_1}{\partial x_2}, \quad \frac{dp_{22}}{dt} = \frac{dp_{33}}{dt} = 0, \quad (20)$$

$$\frac{dp_{12}}{dt} = -p_{22}\frac{\partial v_1}{\partial x_2}, \quad \frac{dp_{23}}{dt} = 0, \quad \frac{dp_{31}}{dt} = -p_{23}\frac{\partial v_1}{\partial x_2}. \quad (21)$$

Noting that p_{22}, p_{23}, p_{33} do not vary with time, it follows that

$$p_{12}(t) = p_{12}(0) - p_{22}\frac{\partial v_1}{\partial x_2}t,\ p_{13}(t) = p_{13}(0) - p_{23}\frac{\partial v_1}{\partial x_2}t, \tag{22}$$

so that

$$p_{11}(t) = p_{11}(0) - 2p_{12}(0)\frac{\partial v_1}{\partial x_2}t + p_{22}\left(\frac{\partial v_1}{\partial x_2}\right)^2 t^2. \tag{23}$$

Note that $p_{12}(t)$ represents the transport of the $i = 1$ component of momentum in the $j = 2$ direction across the velocity gradient. It is the viscous stress across the velocity gradient. It increases linearly with time as particles arrive from farther away across the velocity gradient. The increase in the component p_{11} represents the associated viscous heating. Interparticle collisions would limit growth to the time between collisions. (See Parker, 2007, p. 86 for further discussion).

The essential point is that there is nothing mysterious about a collisionless fluid. The interpenetration of differently moving regions of fluid is rapid, as we would expect from their free passage. The effective viscosity increases rapidly with the passage of time as the interpenetration continues. But it all has the nature of HD with initial conditions and boundary conditions determining the evolution of the fluid.

In astrophysical settings a purely collisionless state is a rarity. The ubiquitous magnetic fields limit the particles to cyclotron gyrations in the two dimensions perpendicular to the magnetic field, so the two pressure components are isotropic. Plasma waves scatter the particles between the parallel and perpendicular directions, and any strong anisotropy is unstable to the generation of more plasmas waves. The thermal anisotropies observed in the solar wind at the orbit of Earth are muted by scattering of some sort after leaving the Sun. The principal expansion of the wind is transverse to the radial expansion, and the approximately radial magnetic field, once the wind is up to speed, so a wind without particle scattering would yield a very low transverse temperature, well below 10^4 K. But it is never so low, and sometimes actually exceeds the radial temperature component at time of solar activity. Obviously there is strong transverse wave heating. The scattering and heating is highly variable and cannot be predicted in any useful way, so the theorist usually can do no better than to pursue the HD with a simple scalar pressure p and a plausible estimate of the manner of variation of the temperature, looking for the basic HD behavior of the streaming fluid.

This brings us to our second comment on the HD equations (11), (14), and (15). They do not form a complete set, because of the unknown energy deposition and scattering just mentioned. Computing higher order tensors (computing higher velocity moments of the collisionless Boltzmann equation) does not help, because the physics of the plasma waves in the astrophysical setting is unknown, and no amount of mathematics can conjure it up. As a practical matter we write Eq. (15) as

$$\frac{dp_{ij}}{dt} = -p_{jk}\frac{\partial v_i}{\partial x_k} - p_{ki}\frac{\partial v_j}{\partial x_k} - p_{ij}\frac{\partial v_k}{\partial x_k} + S_{ij}. \tag{24}$$

Where S_{ij} is the total heat input, including thermal conduction, radiative heating, and plasma wave heating. One simply makes the best estimate of S_{ij} and proceeds from there in pursuit of the HD. (For further discussion see Parker, 2007, chap. 8 and 9.)

So far as the momentum equation is concerned, there are generally applied forces F_i, e.g. gravity, magnetic stresses, etc., so that the momentum equation (14) becomes

$$NM\left(\frac{\partial v_i}{\partial t} + v_j \frac{\partial v_i}{\partial x_j}\right) = -\frac{\partial p_{ij}}{\partial x_j} + F_i. \tag{25}$$

In terms of the magnetic stress tensor,

$$M_{ij} = -\delta_{ij}\frac{B^2}{8\pi} + \frac{B_i B_j}{4\pi},$$

representing the isotropic pressure $B^2/8\pi$ and the tension $B^2/4\pi$ along the field, the force per unit volume exerted on the plasma is

$$F_i = \frac{\partial M_{ij}}{\partial x_j}. \tag{26}$$

3 Magnetohydrodynamics

MHD treats the large-scale dynamics of an ionized gas, i.e. an electrically conducting fluid, with a magnetic field **B** throughout. So it is an extension of HD to include the magnetic stresses. We need, then, to write down the appropriate induction equation for the time variation of **B**. The essential feature of an electrically conducting fluid is its inability to sustain a significant electric field in its own moving frame of reference, resulting in the magnetic field being carried along bodily with the moving fluid. The electric current density **j** is determined by the form of **B** and described by Maxwell's equation,

$$4\pi\mathbf{j} + \frac{\partial \mathbf{E}}{\partial t} = c\nabla \times \mathbf{B}$$

in cgs electrostatic units. The development is limited to the non relativistic case ($v \ll c$), from which it is evident that $\partial/\partial t$ is small compared to c curl by $O(v/c)$. It follows from Eq. (32) that **E** is small compared to **B** to the same order, so that is small compared to $c\nabla \times \mathbf{B}$ to second order in v/c. That is the same order of smallness as Lorentz-Fitzgerald contraction and time dilatation, which we also neglect. So Maxwell's equation reduces to Ampere's law,

$$4\pi\mathbf{j} = c\nabla \times \mathbf{B}. \tag{27}$$

Thus a field **B** with a characteristic scale L has a current density of the order of magnitude

$$j = O\left(\frac{c}{4\pi}\frac{B}{L}\right) \tag{28}$$

associated with it. The electrical resistivity of an ionized gas, or plasma, is so small that, over the large scales L appropriate for the astrophysical universe, only a very weak electric field $\mathbf{E}'$ in the moving frame of reference of the plasma is required to drive $\mathbf{j}$. To provide a simple illustrative example, suppose that the plasma is so dense as to be collision dominated, so that the scalar Ohm's law is a useful approximation. Denoting the electrical conductivity by σ it follows that $\mathbf{j} = \sigma \mathbf{E}'$. Then, in order of magnitude,

$$\frac{E'}{B} = O\left(\frac{\eta}{cL}\right), \tag{29}$$

where $\eta = c^2/4\pi\sigma$ is the resistive diffusion coefficient. In ionized hydrogen

$$\eta \approx \frac{4 \times 10^{12}}{T^{\frac{3}{2}}},$$

so that

$$\frac{E'}{B} = O\left[\left(\frac{10^{-4}}{L}\right)\left(\frac{10^4}{T}\right)^{\frac{3}{2}}\right]. \tag{30}$$

As an example, let $T = 10^4$ K and L be as small as 10^7 cm, comparable to the photospheric diameter of a magnetic fibril in the Sun. It follows that $E'/B \approx 10^{-11}$. The electric stresses, $E'^2/8\pi$, are only the fraction 10^{-22} of the magnetic stress, $B^2/8\pi$, and quite negligible. So the electric field in the moving frame of the local plasma plays no role in the large-scale dynamics of the plasma. The dynamics is described by HD in the presence of the magnetic force $F_i = \partial M_{ij}/\partial x_j$ on the right hand side of Eq. (25).

Now in the astrophysical world, where most velocities $\mathbf{v}$ are small compared to the speed of light c, the electric field $\mathbf{E}'$ in the moving frame of the plasma is related to the electric field $\mathbf{E}$ and magnetic field $\mathbf{B}$ in the coordinate system by the nonrelativistic Lorentz transformation

$$\mathbf{E}' = \mathbf{E} + \frac{\mathbf{v} \times \mathbf{B}}{c}, \tag{31}$$

with $\mathbf{B}' = \mathbf{B}$, of course. It is evident from Eq. (30) that $\mathbf{E}'$ can be neglected, with the result that there is an electric field

$$\mathbf{E} = -\frac{\mathbf{v} \times \mathbf{B}}{c} \tag{32}$$

in the frame of the coordinates. This electric field exists *because* there is no electric field in the moving frame of the plasma. The stresses in the electric field $\mathbf{E}$ are much larger than $\mathbf{E}'$ in the moving frame of the plasma, but still quite negligible in the nonrelativistic approximation, being small $O(v^2/c^2)$ compared to the magnetic stresses. So $\mathbf{E}$ plays no dynamical role when $v \ll c$, in spite of an incorrect popular notion that $\mathbf{E}$ drives motions in the terrestrial ionosphere and magnetosphere.

It is unfortunate that there is sometimes confusion as to the appropriate electric field for particle acceleration or for driving an electric current in a moving system.

So we digress briefly to make clear how the problem should be addressed. Consider the question whether there is an electric field pervading the laboratory or office where you, the reader, are working. There is a magnetic field, of course, of 0.3–0.5 Gauss associated with Earth. There may be an atmospheric electric field of the order of 1 Volt/cm as a consequence of the ongoing tropical thunderstorms charging Earth negative by some 3×10^5 Volts relative to the ionosphere. So enclose the laboratory with a fine grounded copper screen to eliminate this natural nuisance. The answer to the question would seem to be that there is no electric field ($\mathbf{E} = 0$) in the laboratory. However, if you rise from your chair and walk across the laboratory with velocity $\mathbf{v}_1$, Eq. (31) indicates that you observe an electric field $\mathbf{E}'(\mathbf{v}_1) = \dfrac{\mathbf{v}_1 \times \mathbf{B}}{c}$. Walking in another direction with velocity $\mathbf{v}_2$ provides an electric field $\mathbf{E}'(\mathbf{v}_2) = \dfrac{\mathbf{v}_2 \times \mathbf{B}}{c}$, etc. It is obvious, then, that a more general answer to the question would be that there are infinitely many distinct electric fields $\mathbf{E}(\mathbf{v}_\mathrm{n})$ in the laboratory, each in its own moving reference frame. This emphasizes that the physically relevant electric field is the field in the reference frame of the particle or of the plasma in which the electric current flows. The electric field is different in different moving frames so the outcome of a theoretical calculation depends on picking the correct electric field out of the infinitely many different fields in the laboratory. This is one of the reasons that attempts to formulate the dynamics in terms of $\mathbf{E}$ and $\mathbf{j}$ become intractable in general time dependent situations.

With these preliminary remarks, consider the dynamics of the large-scale bulk motion of a plasma and magnetic field. With the electric field given by Eq. (32) the Faraday induction equation,

$$\frac{\partial \mathbf{B}}{\partial t} = -c\nabla \times \mathbf{E}, \tag{33}$$

becomes

$$\frac{\partial \mathbf{B}}{\partial t} = \nabla \times (\mathbf{v} \times \mathbf{B}), \tag{34}$$

recognizable as the MHD induction equation. The flux of electromagnetic energy is given by the Poynting vector

$$\begin{aligned} \mathbf{P} &= c\frac{\mathbf{E} \times \mathbf{B}}{4\pi}, \\ &= \mathbf{v}_\perp \frac{B^2}{4\pi}, \end{aligned} \tag{35}$$

where $v_\perp$ is the component of the plasma velocity $\mathbf{v}$ perpendicular to $\mathbf{B}$, and $B^2/4\pi$ is the magnetic enthalpy density. This shows directly that the magnetic field is carried bodily with the plasma, i.e. the magnetic field moves in the frame of reference in which there is no electric field. So MHD is inescapable in the large-scale swirling plasmas that make up the astrophysical universe.

An immediate consequence of the bodily transport of magnetic field in the plasma is the deformation of the magnetic field, producing the nonvanishing Lorentz force

$F_i = \partial M_{ij}/\partial x_j$ on the plasma, as already described by Eqs. (25) and (26). The Lorentz force can also be written

$$\mathbf{F} = -\nabla \frac{B^2}{8\pi} + \frac{(\mathbf{B} \cdot \nabla)\mathbf{B}}{4\pi}, \tag{36}$$

$$= \frac{(\nabla \times \mathbf{B}) \times \mathbf{B}}{4\pi}, \tag{37}$$

$$= \frac{\mathbf{j} \times \mathbf{B}}{c}. \tag{38}$$

Equation (38) is written using Ampere's law, Eq. (27), to replace $\nabla \times \mathbf{B}$ by the current density $\mathbf{j}$. This simple mathematical substitution must not be interpreted as a dynamical role for $\mathbf{j}$, for the current contains no significant energy or stress in itself. The current represents only the slow conduction drift of the electrons relative to the ions, with negligible inertia. It is a passive quantity, driven by $\mathbf{E}'$ so as always to conform to Ampere's law, i.e. to the deformation $\nabla \times \mathbf{B}$ of the magnetic field.

Now the physical universe is not an ideal system and $\mathbf{E}'$ is small but not identically zero. In fact over smaller scales, and particularly in gases with only a slight ionization, $\mathbf{E}'$ may be non-negligible. Shock fronts and the intense current sheets associated with rapid reconnection represent small scales. Then the solar photosphere has but little ionization, so that $\mathbf{E}'$ may be non-negligible over dimensions of no more than 10^7 cm. As an example, consider the case again that the plasma density is sufficiently large that the familiar scalar Ohm's law is a useful approximation. Then with $\mathbf{E}' = \mathbf{j}/\sigma$ the induction equation takes the familiar form

$$\frac{\partial \mathbf{B}}{\partial t} = \nabla \times (\mathbf{v} \times \mathbf{B}) - \nabla \times (\eta \nabla \times \mathbf{B}), \tag{39}$$

reducing to

$$\frac{\partial \mathbf{B}}{\partial t} = \nabla \times (\mathbf{v} \times \mathbf{B}) + \eta \nabla^2 \mathbf{B}$$

when the resistive diffusion coefficient η is uniform. The reciprocal of the magnetic Reynolds number, defined as $R_{\mathrm{m}} = vL/\eta$ in terms of the characteristic velocity v and scale L, is a measure of the smallness of the resistive diffusion term compared to the first term on the right hand side of the induction equation. On astrophysical scales, R_{m} is very large compared to one. So the diffusion term is small and can be neglected so far as the gross dynamics of the system is concerned. It must be appreciated, however, that over time even a very small amount of diffusion may play an important role by reconnecting the field lines, as in the rapid reconnection phenomenon where the field spontaneously develops small scales. Indeed, in all but the simplest field line topologies the Lorentz force pushes the field toward the formation of surfaces of tangential discontinuity, i.e. intense current sheets, where resistive diffusion becomes important (Parker, 1972, 1994). It must be appreciated, too, that the MHD $\alpha\omega$-dynamo requires substantial diffusion if it is to function efficiently in stars and galaxies. It is usually assumed that turbulent diffusion takes care of the problem, but it is not clear how turbulent diffusion can be accomplished in the strong magnetic fields to be found in stars and galaxies (Parker, 1992; Vainshtein *et al.*, 1993). The turbulent motions would appear to be restricted to Alfven waves, whose passage provides little or no diffusion and dissipation of fields. But that is a problem for another time.

A plasma has internal microscopic structure, e.g. the cyclotron gyrations of the ions and electrons, and perhaps only partial ionization. It is sometimes stated that MHD does not apply to a partially ionized gas because MHD ignores the Hall effect and the Pedersen resistivity. As a matter of fact, these effects are the same order in $1/L$ as the resistive diffusion term, and there is no reason why they should not be included in MHD, along with the dominant MHD induction $\nabla \times (\mathbf{v} \times \mathbf{B})$ term, in circumstances where they play a significant role. What is more, the Hall effect exists whether the gas is wholly or only partly ionized and whether there are interparticle collisions or not. To illustrate the variety of effects that arise consider a partially ionized gas, consisting of N neutral atoms per unit volume and n electrons and n singly charged ions per unit volume. The ions, electrons, and neutral atoms all collide with each other, which we represent with linear scattering terms. The time over which an individual ion collides with a neutral atom is denoted by τ_i, with τ_e the time in which an electron does the same. The time over which an electron collides with an ion is τ. The mass of the neutral atom and the ion are both designated by M, and the electron mass is m. Then denote the bulk velocity of the neutral atoms by $\mathbf{v}$, the ions by $\mathbf{w}$, and the electrons by $\mathbf{u}$ (not to be confused with the $\mathbf{u}$, $\mathbf{v}$, $\mathbf{w}$ employed in Section 2 in the discussion of HD). The equation of motion for the neutral atoms becomes

$$NM\frac{d\mathbf{v}}{dt} = -\nabla p + \frac{nM(\mathbf{w}-\mathbf{v})}{\tau_i} + \frac{nm(\mathbf{u}-\mathbf{v})}{\tau_e} + N\mathbf{F}, \tag{40}$$

where $\mathbf{F}$ is the external force, e.g. gravity, exerted on each atom. Treating the ions as a fluid, the equation of motion is

$$nM\frac{d\mathbf{w}}{dt} = -\nabla p_i + ne\left(\mathbf{E} + \frac{\mathbf{w}\times\mathbf{B}}{c}\right) - \frac{nM(\mathbf{w}-\mathbf{v})}{\tau_i} - \frac{nm(\mathbf{w}-\mathbf{u})}{\tau} + n\mathbf{f}_i, \tag{41}$$

where p_i is the ion pressure and $\mathbf{f}_i$ is the external force on each ion. The equation of motion for the electrons in the same fluid approximation is

$$nm\frac{d\mathbf{u}}{dt} = -\nabla p_e - ne\left(\mathbf{E} + \frac{\mathbf{u}\times\mathbf{B}}{\mathbf{c}}\right) - \frac{nm(\mathbf{u}-\mathbf{v})}{\tau_e} + \frac{nm(\mathbf{w}-\mathbf{u})}{\tau} + n\mathbf{f}_e, \tag{42}$$

where p_e is the electron pressure and $\mathbf{f}_e$ is the external force on each electron.

It is convenient to introduce the notation

$$n\mathbf{G} = -\nabla p_i - nM\frac{d\mathbf{w}}{dt} + n\mathbf{f}_i, \tag{43}$$

$$n\mathbf{H} = -\nabla p_e - nm\frac{d\mathbf{u}}{dt} + n\mathbf{f}_e, \tag{44}$$

so that Eqs. (41) and (42) become

$$0 = \mathbf{G} - \frac{M(\mathbf{w}-\mathbf{v})}{\tau_i} - \frac{m(\mathbf{w}-\mathbf{u})}{\tau} + e\left(\mathbf{E} + \frac{\mathbf{w}\times\mathbf{B}}{c}\right), \tag{45}$$

$$0 = \mathbf{H} - \frac{m(\mathbf{u}-\mathbf{v})}{\tau_e} + \frac{m(\mathbf{w}-\mathbf{u})}{\tau} - e\left(\mathbf{E} + \frac{\mathbf{u}\times\mathbf{B}}{c}\right). \tag{46}$$

The electric current density is

$$\mathbf{j} = ne(\mathbf{w}-\mathbf{u}), \tag{47}$$

and Ampere's law can be written

$$\mathbf{u} = \mathbf{w} - \frac{c\nabla\times\mathbf{B}}{4\pi ne}. \tag{48}$$

Add Eqs. (45) and (46), obtaining

$$n\mathbf{G} + n\mathbf{H} = \frac{nM(\mathbf{w}-\mathbf{v})}{\tau_i} + \frac{nm(\mathbf{u}-\mathbf{v})}{\tau_e} - \frac{(\nabla\times\mathbf{B})\times\mathbf{B}}{4\pi}. \tag{49}$$

Then solve Eqs. (48) and (49) for **u** and **w**, obtaining

$$\mathbf{w} = \mathbf{v} + \frac{cm/\tau_e}{4\pi neQ}\nabla\times\mathbf{B} + \frac{(\nabla\times\mathbf{B})\times\mathbf{B}}{4\pi nQ} + \frac{\mathbf{G}+\mathbf{H}}{Q}, \tag{50}$$

$$\mathbf{u} = \mathbf{v} - \frac{cM/\tau_i}{4\pi neQ}\nabla\times\mathbf{B} + \frac{(\nabla\times\mathbf{B})\times\mathbf{B}}{4\pi nQ} + \frac{\mathbf{G}+\mathbf{H}}{Q}, \tag{51}$$

where

$$Q = \frac{M}{\tau_i} + \frac{m}{\tau_e}.$$

Solve Eq. (46) for **E** and use Eqs. (50) and (51) to eliminate **w** and **u**, with the result

$$\mathbf{E} = -\frac{\mathbf{v}\times\mathbf{B}}{c} + \frac{c}{4\pi ne^2}\left[\frac{m}{\tau} + \frac{\frac{m}{\tau_e}\frac{M}{\tau_i}}{Q}\right]\nabla\times\mathbf{B} + \frac{\frac{M}{\tau_i} - \frac{m}{\tau_e}}{4\pi neQ}(\nabla\times\mathbf{B})\times\mathbf{B}$$
$$-\frac{[(\nabla\times\mathbf{B})\times\mathbf{B}]\times\mathbf{B}}{4\pi ncQ} + \frac{\frac{\mathbf{H}M}{\tau_i} - \frac{\mathbf{G}m}{\tau_e}}{eQ} - \frac{(\mathbf{G}+\mathbf{H})\times\mathbf{B}}{cQ}. \tag{52}$$

The next step is to substitute this expression for **E** into the Faraday induction equation. Obviously the first term on the right hand side of Eq. (52) provides the familiar MHD induction term $\nabla\times(\mathbf{v}\times\mathbf{B})$. So MHD is the basic effect. The next three terms involve $\nabla\times\mathbf{B}$, which is smaller by $O(1/L)$ than the first term, where L is the characteristic large scale of variation of **B**. These three terms represent the resistive term, already discussed, the Pedersen effect, and the Hall effect, respectively. The fifth and sixth terms on the right hand side of Eq. (52) are also small $O(1/L)$ in the large scale bulk motion because **G** and **H** are small to that order. So there are these several effects appended to MHD, to be included wherever they may contribute significantly to the large-scale dynamics of the field and fluid.

Consider, then, the relative magnitude of the fifth and sixth terms, involving **G** and **H**, given by Eqs. (43) and (44). In these expressions, the ion and electron pressures may be expressed in terms of the ion and electron temperatures, T_i and T_e, as nkT_i and nkT_e, respectively, so that, for instance,

$$\frac{\nabla p_\mathrm{i}}{n} = \frac{\nabla nkT_\mathrm{i}}{n} = O\left(\frac{kT_\mathrm{i}}{L}\right) = O\left(\frac{Mw_\mathrm{th}^2}{L}\right), \tag{53}$$

where $Mw_\mathrm{th}^2 = kT_\mathrm{i}$ so that w_th represents the ion thermal velocity. A similar order of magnitude expression can be constructed for the electrons. Then we estimate the acceleration terms $d\mathbf{w}/dt$ as

$$\begin{aligned} \frac{d\mathbf{w}}{dt} &= O\left[(\mathbf{w}\cdot\nabla)\mathbf{w}\right], \\ &= O\left(\frac{w^2}{L}\right). \end{aligned} \tag{54}$$

The bulk flow of the electrons is constrained to be comparable to the bulk flow of the ions, so

$$\frac{d\mathbf{u}}{dt} = O\left(\frac{w^2}{L}\right). \tag{55}$$

Assuming, then, that $\mathbf{f}_\mathrm{i}$ and $\mathbf{f}_\mathrm{e}$ are not greater in magnitude than the other terms, they can be ignored in the present estimate, and it follows from Eqs. (43) and (44) that the magnitudes of **G** and **H** are given by

$$G = O\left(M\frac{w_\mathrm{th}^2 + w^2}{L}\right), \tag{56}$$

$$H = O\left(m\frac{u_\mathrm{th}^2}{L}\right). \tag{57}$$

When the ion and electrons temperatures are of the same general order of magnitude, it is apparent that G and H are of comparable magnitude. Noting that the mean free paths for collisions with neutral atoms are comparable for ions and electrons, and that the electron thermal velocity exceeds the ion thermal velocity by a factor of the order of $(M/m)^{1/2}$, it follows that $\tau_\mathrm{i}/\tau_\mathrm{e}$ is of the order of $(M/m)^{1/2}$. Hence

$$Q \cong \frac{M}{\tau_\mathrm{i}}\left[1 + O\left(\frac{m}{M}\right)^{\frac{1}{2}}\right]. \tag{58}$$

It follows that the fifth term on the right hand side of Eq. (52) is of the order of

$$\frac{\dfrac{\mathbf{H}M}{\tau_i} - \dfrac{\mathbf{G}m}{\tau_e}}{eQ} \approx \frac{1}{e}\left(\mathbf{H} - \frac{m}{M}\frac{\tau_i}{\tau_e}\mathbf{G}\right),$$
$$= \frac{1}{e}\left[O\left(\frac{mu_{th}^2}{L}\right) - \left(\frac{m}{M}\right)^{\frac{1}{2}} O\left(\frac{M(w_{th}^2 + w^2)}{L}\right)\right]. \tag{59}$$

With mu_{th}^2 and Mw_{th}^2 of comparable magnitude it is apparent that the second term is small $O(m/M)^{1/2}$ compared to the first term. That is to say,

$$\frac{\dfrac{\mathbf{H}M}{\tau_i} - \dfrac{\mathbf{G}m}{\tau_e}}{eQ} \cong -\frac{\nabla p_e}{ne}\left[1 + O\left(\frac{m}{M}\right)^{\frac{1}{2}}\right]. \tag{60}$$

In a similar manner the sixth term on the right hand side of Eq. (52) can be shown to be of the order of

$$\frac{(\mathbf{G}+\mathbf{H})\times\mathbf{B}}{cQ} = O\left[\frac{B}{cQ}\left(\frac{\nabla p_i + \nabla p_e}{n} + M\frac{d\mathbf{w}}{dt}\right)\right],$$
$$= \frac{\Omega_i \tau_i}{e} O\left(\frac{Mw_{th}^2 + mu_{th}^2 + Mw^2}{L}\right) \tag{61}$$

where Ω_i is the ion cyclotron frequency, eB/Mc, in the magnetic field $\mathbf{B}$. Comparing this with the dominant first term of the right hand side of Eq. (59), it may be seen that the sixth term is of the order of $\Omega_i\tau_i$ times the fifth term. The essential point is that both terms are small $O(1/L)$ compared to the first term $-\mathbf{v}\times\mathbf{B}/c$ on the right hand side of Eq. (52). It follows then that the second through the sixth terms are all small compared to the purely MHD term $-\mathbf{v}\times\mathbf{B}/c$. So the large-scale plasma dynamics is described by the MHD equations with whatever small additional effects are appropriate.

Consider, then, the physics of the second through the sixth terms of the right hand side of Eq. (52). The second term represents the familiar Ohmic dissipation, with the resistive diffusion coefficient

$$\eta = \frac{c^2}{4\pi ne^2}\left(\frac{m}{\tau} + \frac{\dfrac{m}{\tau_e}\dfrac{M}{\tau_i}}{Q}\right)$$
$$\cong \frac{mc^2}{4\pi ne^2}\left(\frac{1}{\tau} + \frac{1}{\tau_e}\right). \tag{62}$$

It contributes the term $-\nabla\times(\eta\nabla\times\mathbf{B})$ on the right hand side of the induction equation (39).

The third and fifth terms on the right hand side of Eq. (52) together provide the *Hall effect*, and the dominant terms are

$$\frac{\frac{M}{\tau_\mathrm{i}}-\frac{m}{\tau_\mathrm{e}}}{neQ}\frac{(\nabla\times\mathbf{B})\times\mathbf{B}}{4\pi}+\frac{\frac{\mathbf{H}M}{\tau_\mathrm{i}}-\frac{\mathbf{G}m}{\tau_\mathrm{e}}}{eQ}$$
$$\cong\frac{1}{ne}\left[-\nabla p_\mathrm{e}+\frac{(\nabla\times\mathbf{B})\times\mathbf{B}}{4\pi}\right]\left[1+O\left(\frac{m}{M}\right)^{\frac{1}{2}}\right]. \tag{63}$$

In this approximation the Hall effect is independent of the collision times τ_i and τ_e between the ions and electrons and the neutral atoms. So the result applies to the collisionless plasma as well, and is readily derived from the fluid equations for a collisionless ion gas and electron gas,

$$nM\frac{d\mathbf{w}}{dt}=-\nabla p_\mathrm{i}+ne\left(\mathbf{E}+\frac{\mathbf{w}\times\mathbf{B}}{c}\right), \tag{64}$$

$$nm\frac{d\mathbf{u}}{dt}=-\nabla p_\mathrm{e}-ne\left(\mathbf{E}+\frac{\mathbf{u}\times\mathbf{B}}{c}\right). \tag{65}$$

In fact we need only Eq. (65) for the electrons, because the Hall effect arises from the electric field required to hold the very fast mobile electrons in the company of the massive sluggish ions. Neglect the slight bulk inertia of the electrons and write Eq. (65) as

$$\mathbf{E}+\frac{\mathbf{w}\times\mathbf{B}}{c}=\frac{1}{ne}\left[-\nabla p_\mathrm{e}+\frac{(\nabla\times\mathbf{B})\times\mathbf{B}}{4\pi}\right], \tag{66}$$

upon using Eq. (47) to express $\mathbf{u}$ in terms of $\mathbf{j}$ and Ampere's law to eliminate $\mathbf{j}$ in favor of $\nabla\times\mathbf{B}$. The left hand side of this equation represents the electric field $\mathbf{E}'$ in the moving frame of reference of the ions that is necessary to tie the electrons to the ions. It contributes to the MHD induction equation

$$\frac{\partial\mathbf{B}}{\partial t}=\nabla\times(\mathbf{w}\times\mathbf{B})-c\nabla\times\left\{\frac{1}{ne}\left[-\nabla p_\mathrm{e}+\frac{(\nabla\times\mathbf{B})\times\mathbf{B}}{4\pi}\right]\right\} \tag{67}$$

only insofar as $\mathbf{E}'$ provides a nonvanishing curl. This occurs when, for instance, $\nabla n\times\nabla T_\mathrm{e}\neq 0$. The Lorentz force,

$$\mathbf{L}=\frac{(\nabla\times\mathbf{B})\times\mathbf{B}}{4\pi}, \tag{68}$$

by itself provides a nondissipative energy circulation or Poynting vector

$$\mathbf{P}=c\frac{\mathbf{L}}{4\pi ne}\times\mathbf{B}. \tag{69}$$

Moving on to the second and sixth terms on the right hand side of Eq. (52), they represent the *Pedersen effect*, also known as *ambipolar diffusion*. Their contribution to the electric field $\mathbf{E}$ is

$$\mathbf{E}_{\mathrm{P}} = -\frac{[(\nabla \times \mathbf{B}) \times \mathbf{B}] \times \mathbf{B}}{4\pi ncQ} - \frac{(\mathbf{G} + \mathbf{H}) \times \mathbf{B}}{cQ}$$
$$= \frac{1}{cQ}\left\{\frac{\nabla p_{\mathrm{i}} + \nabla p_{\mathrm{e}}}{n} + M\frac{d\mathbf{w}}{dt} + m\frac{d\mathbf{u}}{dt} - \frac{1}{n}\frac{(\nabla \times \mathbf{B}) \times \mathbf{B}}{4\pi} - \mathbf{f}_{\mathrm{i}} - \mathbf{f}_{\mathrm{e}}\right\} \times \mathbf{B}. \quad (70)$$

The Pedersen effect arises because the pressure gradients, the Lorentz force, and other forces, exerted on the ions and electrons drive the ions and electrons through the neutral gas. The forced slippage of the ions and electrons relative to the neutral gas is opposed by the friction of the collisions, with characteristic collision times τ_{i} and τ_{e}.

The effect is dissipative, obviously, and allows the magnetic field to relax toward static equilibrium. The effect is sometimes referred to as the *Pedersen resistivity*, although it does not arise from collisions obstructing an electric current. *Ambipolar diffusion*, or just the *Pedersen effect*, is to be preferred.

Equation (70) is easily derived by noting that the slow drift of the ions and electrons, as they are pushed through the neutral gas, is balanced by the frictional forces. The balance of forces is described by adding Eqs. (41) and (42) and assuming that $\mathbf{u}$ and $\mathbf{w}$ are equal in the inertial terms, while their slight difference provides the current $\mathbf{j}$, to be replaced by $\nabla \times \mathbf{B}$ through Ampere's law. The result is

$$\mathbf{w} - \mathbf{v} = \frac{1}{nQ}\left[-\nabla p_{\mathrm{i}} - \nabla p_{\mathrm{e}} + \frac{(\nabla \times \mathbf{B}) \times \mathbf{B}}{4\pi} - n(M+m)\frac{d\mathbf{w}}{dt} + n\mathbf{f}_{\mathrm{i}} + n\mathbf{f}_{\mathrm{e}}\right]. \quad (71)$$

This extra velocity $\mathbf{w} - \mathbf{v}$ represents the drift of the ions relative to the neutral atoms, in addition to the bulk transport velocity $\mathbf{v}$ of the neutral atoms. So we would write

$$\mathbf{E} = -\frac{\mathbf{v} \times \mathbf{B}}{c} - \frac{(\mathbf{w} - \mathbf{v}) \times \mathbf{B}}{c} \quad (72)$$

in Eq. (52) to account for the Pedersen effect. It is obvious by inspection of Eqs. (70), (71), and (72) that the term $(\mathbf{w} - \mathbf{v}) \times \mathbf{B}/c$ represents the right hand side of Eq. (70) when the ions and electrons move together, $\mathbf{w} = \mathbf{u}$.

4 Popular Misconseptions

With the foregoing brief exposition of the dynamics of a swirling magnetized plasma, we review some of the more popular misconceptions encountered in the literature. For instance, it is claimed that the electric field $\mathbf{E} = -\mathbf{v} \times \mathbf{B}/c$ in the solar wind, as seen by an observer in the reference frame of Earth, actively penetrates into the geomagnetic field $\mathbf{B}_{\mathrm{E}}$ where it drives the plasma toward the electric drift velocity

$$\mathbf{v}_{\mathrm{D}} = c\frac{\mathbf{E} \times \mathbf{B}_{\mathrm{E}}}{B_{\mathrm{E}}^2}.$$

It is claimed that this effect is responsible for driving the convection of the magnetosphere. We have already pointed out that the electric stress density $E^2/8\pi$ is small

$O(v^2/c^2)$ compared to the magnetic stress density $B^2/8\pi$, and therefore negligible. So there cannot be any active dynamical effect to be attributed to the **E** in the Earth frame of reference. It should be appreciated, too, that there is no significant electric field $\mathbf{E}'$ in the moving frame of the solar wind, so if the calculation of the alleged dynamical effect were carried out in that frame of reference, the result would be quite different. But, of course, there is no effect at all in either reference frame.

The concern is sometimes expressed that the use of MHD overlooks the boundary conditions on the current (Melrose, 1995). In fact, the initial conditions and the boundary conditions together provide a *unique* solution for the magnetic field, which automatically provides the correct boundary conditions for the current. It has to be that way because there is no further freedom to adjust the solution if the current boundary conditions were not properly satisfied (Parker, 1996). It has to be remembered that Newton's equations and Maxwell's equations, i.e. Ampere's law, are fundamental laws of Nature. So they are mutually consistent. In particular, Newton's equations automatically provide the currents required by Ampere and Maxwell.

A particularly popular notion concerns the behavior of the electric current associated through Ampere's law with the magnetic field in a swirling plasma. The current flows through a slightly resistive medium and has magnetic fields with large energy associated with them. So it is proposed that the flow of the current can be described by the familiar electric circuit equations for an electric circuit containing resistors, inductances, and perhaps capacitors (Alfven and Carlquist, 1967). The inductance L is deduced by equating the magnetic energy to $LI^2/2$ where I is the total current flowing around the circuit. The current I in the circuit then controls the associated **B** through Ampere's law. This idea is not derived from Newton and Maxwell, being considered so self evident as to need no derivation. The idea is appealing because it reduces the solution of the partial differential equations of MHD to the solution of the ordinary differential equations of laboratory circuit theory.

In fact the electric current in a swirling magnetized plasma flows under quite different conditions from the current in the laboratory circuit. Thus, the laboratory circuit is fixed in the laboratory so the topology (connectivity) of the current loops is fixed, and the currents experience the electric field in the frame of reference of the laboratory. In contrast, the electric current in the swirling plasma is carried with the plasma velocity **v**, and in that moving frame there is only the insignificant electric field $\mathbf{E}'$, given by Eqs. (30) and (31). So there can be no inductive effects on the current **I**. What is more, the topology of the electric current in the swirling plasma is determined from **B** through Ampere's law and may change as the magnetic field is carried along in the plasma and deformed with the passage of time. Thus, for instance, the abrupt current interruption that might be caused by the onset of plasma turbulence and anomalous resistivity, or an electric double layer, does not provide the large *emf* sometimes imagined to accelerate ions and electrons to high velocities. The sudden appearance of an obstacle in the current path causes the current to deviate around the obstacle, so there is no current interruption, only an immediate rerouting.

It is interesting to explore a simple example of a current interruption to see how the plasma side steps the electric field associated with the change in current flow pattern. Imagine an infinite space filled with an incompressible, infinitely conducting, inviscid fluid and a uniform magnetic field **B**. Orient the z-axis of the coordinate system along the direction of the field, and denote radial distance from the z-axis by $\omega = (x^2 + y^2)^{1/2}$. There is an azimuthal magnetic field $B_\phi(\omega)$ circling the z-axis from $z = 0$ out to $z = a$, and vanishing beyond $\omega \geq a$. That is to say, consider a twisted flux bundle of radius a lying along the z-axis. The field outside the bundle is uniform and parallel to the bundle.

Ampere's law dictates that there is an electric current $j_z(\omega)$ in $\omega < a$, given by

$$j_z = \frac{c}{4\pi}\frac{1}{\omega}\frac{d}{d\omega}(\omega B_\phi).$$

The total current carried in the bundle is identically zero, of course, for if it were not, then $B_\phi(\omega)$ would not vanish beyond $\omega = a$. In the simplest case, there is a net current **I** flowing one way along the flux bundle in the neighborhood of the axis and in the opposite direction at larger ω. We can imagine that the flux bundle extends along the z-axis between distant infinitely conducting end plates $z = \pm h$, where $h >>> a$. So we have an electric circuit with current I, and, according to the electric circuit analog, the inductance per unit length in the z-direction is related to $B_\phi(\omega)$ by the energy relation

$$\frac{1}{2}LI^2 = 2\pi \int_0^a d\omega\omega \frac{B_\phi^2(\omega)}{8\pi}.$$

The total inductance $2hL$ in the circuit is large for large h.

At time $t = 0$ slice across the flux bundle with a sheet of nonconducting material of thickness 2ϵ, so that $j_z(\omega) = 0$ throughout $-\epsilon < z < +\epsilon$, $0 \leq \omega < a$, thereby blocking the flow of I at $z = 0$. The electric circuit theory predicts that the sudden blocking of the current I flowing through an inductance $2hL$ provides a potential difference V given by

$$V = 2hL\frac{dI}{dt},$$

which can be very large if the interruption of the current is sudden.

In fact what happens is quite different. Within the insulating sheet, $-\epsilon < z < +\epsilon$, the field $B_\phi(\omega)$ is decoupled from the plasma, becoming a ring of magnetic field in a dielectric, which propagates away as an electromagnetic wave at the speed of light in the dielectric. This leaves the longitudinal field $z = \pm\epsilon$, which connects at $z = \pm\epsilon$ into the severed ends of the twisted flux bundle. According to Ampere's law the kink in the field lines at $z = \pm\epsilon$ causes a radial current to flow that connects between the I in the central region of the bundle and the opposite I in the outer region, thereby reconnecting the flow of the current I at $z = \pm\epsilon$. So I is not interrupted by the insertion of the insulating sheet. It is rerouted.

Then it must be realized that the torque $T(\omega)d\omega$ about the z-axis in the annulus $(\omega, \omega + d\omega)$ in the twisted flux bundle is

$$T(\omega)d\omega = \frac{B_\phi(\omega)B_z(\omega)}{4\pi}2\pi\omega^2 d\omega.$$

This torque is interrupted at $z = \pm\epsilon$ and is zero throughout $-\epsilon < z < +\epsilon$ where $B_\phi(\omega)$ vanishes. So the torque at the severed ends of the twisted bundle at $z = \pm\epsilon$ sets the plasma in motion, providing torsional Alfven waves which propagate away in both directions along the twisted flux bundle. The torsional waves unwind the twisted bundle, so that

$$\frac{1}{2}\rho v_\phi^2(\omega) = \frac{B_\phi^2(\omega)}{8\pi},$$

and

$$v_\phi(\omega) = \frac{B_\phi(\omega)}{(4\pi\rho)^{\frac{1}{2}}}.$$

The magnetic energy associated with the current I, then, is converted directly into the kinetic energy of rotation of the plasma about the axis of the bundle as the torsional wave fronts sweep away in both directions. Note then that no electric potential differences are created in the moving plasma. The plasma in the wave motion avoids even the modest electric field $\mathbf{E} = -\mathbf{v} \times \mathbf{B}/c$ in the laboratory frame. Electric circuit theory simply does not apply. It is MHD alone that describes the large-scale dynamical behavior of the magnetized plasma.

The reader is referred to Parker (2007), Vasyliunas (1999, 2001, 2005a, b), and Vasyliunas and Song (2005) for further and more detailed exploration of the basic physics.

References

Alfven, H. and P. Carlquist, Currents in the solar atmosphere and a theory of solar flares, *Sol. Phys.*, **1**, 220–228, 1967.

Melrose, D. B., Current paths in the corona and energy release in solar flares, *Astrophys. J.*, **451**, 391–401, 1995.

Parker, E. N., Topological dissipation and the small-scale fields in turbulent gases, *Astrophys. J.*, **174**, 499–510, 1972.

Parker, E. N., Fast dynamos, cosmic rays, and the galactic magnetic field, *Astrophys. J.*, **401**, 137–145, 1992.

Parker, E. N., *Spontaneous current sheets in magnetic fields*, Oxford University Press, New York, 1994.

Parker, E. N., Comment on "Current paths in the corona and energy release in solar flares", *Astrophys. J.*, **471**, 489–496, 1996.

Parker, E. N., *Conversations of electric and magnetic fields in the cosmos*, Princeton University Press, Princeton, 2007.

Vainshtein, S. I., R. Rosner, and E. N. Parker, On the generation of "strong" magnetic fields, *Astrophys. J.*, **404**, 773–780, 1993.

Vasyliunas, V. M., A note on current closure, *J. Geophys. Res.*, **104**, 25143–25144, 1999.

Vasyliunas, V. M., Electric field and plasma flow. What drives what?, *Geophys. Res. Lett.*, **28**, 2177–2180, 2001.

Vasyliunas, V. M., Time evolution of electric fields and currents and generalized Ohm's law, *Ann. Geophys.*, **28**, 1347–1354, 2005a.

Vasyliunas, V. M., Relation between magnetic fields and currents in plasmas, *Ann. Geophys.*, **25**, 2589–2507, 2005b.

Vasyliunas, V. M. and P. Song, Meaning of Joule heating, *J. Geophys. Res.*, **110**, A02301, doi:10.1029/2004JA010615, 2005.

Climate and Weather of the Sun-Earth System (CAWSES): Selected Papers from the 2007 Kyoto Symposium,
Edited by T. Tsuda, R. Fujii, K. Shibata, and M. A. Geller, pp. 41–62.

The 1960s—A decade of remarkable advances in middle atmosphere research

Marvin A. Geller

School of Marine and Atmospheric Sciences,
Stony Brook, New York 11794-5000, USA
E-mail: marvin.geller@sunysb.edu

The 1960s was a decade of great advances in middle atmosphere research. In this paper, I briefly discuss some of the seminal papers and events that greatly influenced middle atmosphere research and continue to do so to this day. This paper also gives a primer on many of the basics of middle atmosphere wave-mean flow interaction.

1 Introduction

The 1960s was a decade of great social change and scientific advances. The ends of World War II and the Great Depression were more than a decade in the past. The Cold War was ongoing, and the conflicts in Southeast Asia intensified during that decade. Technology was progressing rapidly, with the applications of the World War II technologies of rockets and radar as well as newly developed satellite observations of our environment. Computers were becoming common research tools. Governments appreciated the important role of science and technology in World War II and were increasing their expenditures in these areas, particularly given Cold War competition. It was also a period where young people were enticed by science and technology, and were entering these fields in great numbers, often with government support for their education.

2 1960s Middle Atmosphere Advances

The decade of the 1960s saw many important advances in observations, theory, and numerical modeling of the middle atmosphere. Different people will come up with different lists of the seminal papers that appeared during this period. Table 1 shows my selection. Of course, there is some arbitrariness here. One is that papers published 1959, or before, and 1970, or after, are not discussed here. Also, of course, I have made the selection, and therefore, some important papers are not mentioned that might have been chosen by others, so no doubt some very important papers have not made my list. I would argue, however, that all of the papers in Table 1 are seminal!

2.1 Hines (1960)

Hines' (1960) paper was a tour de force. He presented observational evidence for gravity waves in the atmosphere. He developed the linear theory for these waves.

Table 1.

Paper(s) or Event	Significance
Hines (1960)	Developed the theory for acoustic-gravity waves.
Construction of Jicamarca Radio Observatory (1960–61)	The infrastructure for the demonstration of MST radar by Woodman and Guillen (1974).
Reed *et al.* (1961) and Veryard and Ebdon (1961)	Discovery of the QBO.
Murgatroyd and Singleton (1961)	Derivation of the middle atmosphere diabatic circulation.
Charney and Drazin (1961)	First theory for planetary wave propagation.
Eliassen and Palm (1961) and Charney and Drazin (1961)	Derivation of the non-interaction theorem.
Hampson (1964) and Hunt (1966)	First suggestion of catalytic loss for stratospheric ozone.
Leovy (1964)	First successful model for mesospheric structure.
Reed (1965)	Discovery of the SAO.
Kato (1966) and Lindzen (1966)	Discovery of negative equivalent depth Hough functions.
Matsuno (1966)	Derivation of the structure of equatorially trapped Kelvin and mixed Rossby-gravity waves.
Yanai and Maruyama (1966) and Wallace and Kousky (1968)	Observational identification of equatorially trapped Kelvin and mixed Rossby-gravity waves.
Booker and Bretherton (1967)	Theory for gravity wave critical levels.
Hodges (1967)	Beginnings of the theory for gravity wave breaking.
Manabe and Hunt (1968) and Hunt and Manabe (1968)	First troposphere-stratosphere general circulation model.
Wallace and Holton (1968), Lindzen and Holton (1968)	First successful theory for the QBO.
Dickinson (1969)	Framework for planetary wave-mean flow interactions.

He discussed some of their effects, and he predicted which waves could be observed in ionospheric regions. Hines (1960) showed that there were two distinct classes of waves in a compressible, gravitationally stratified atmosphere: internal gravity waves with frequencies less than the Brunt-Vaisälä frequency and acoustic-gravity waves with frequencies higher than the acoustic cut-off frequency. Further, he showed that the internal gravity waves had the asymptotic behavior of internal gravity waves in incompressible fluids for low frequencies and the acoustic-gravity waves had the asymptotic behavior of sound waves for frequencies much higher than the acoustic cut-off frequency.

One of the most fundamental results of the Hines (1960) paper was that the vertical component of an internal gravity wave's phase velocity is opposite to the vertical component of the internal gravity wave's group velocity, the speed at which the gravity wave energy propagates. This is shown in Fig. 1 here, which is figure 2 in Hines (1960).

2.2 Construction of the Jicamarca radar

The Jicamarca Radio Observatory was originally constructed in Peru by the Central Radio Propagation Laboratory (CRPL), which was then part of the United States

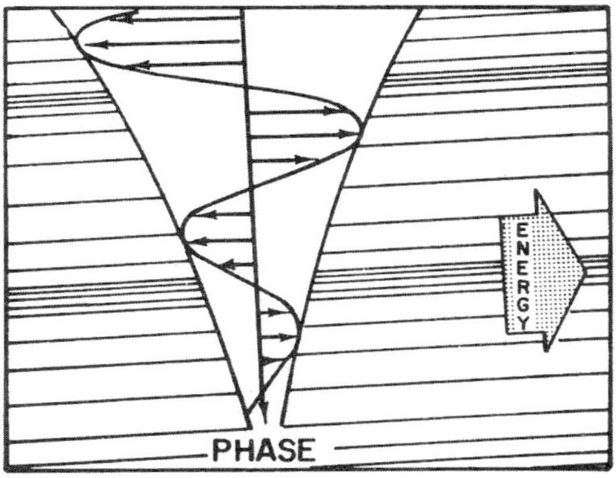

Fig. 1. Pictorial representation of internal gravity waves. Instantaneous velocity vectors are shown, as are their instantaneous and overall envelopes. Density variations are depicted by a background lying in surfaces of constant phase. The vertical component of the phase velocity is downward while energy is being propagated upward. Note that gravity is directed vertically downward. (From Hines (1960).)

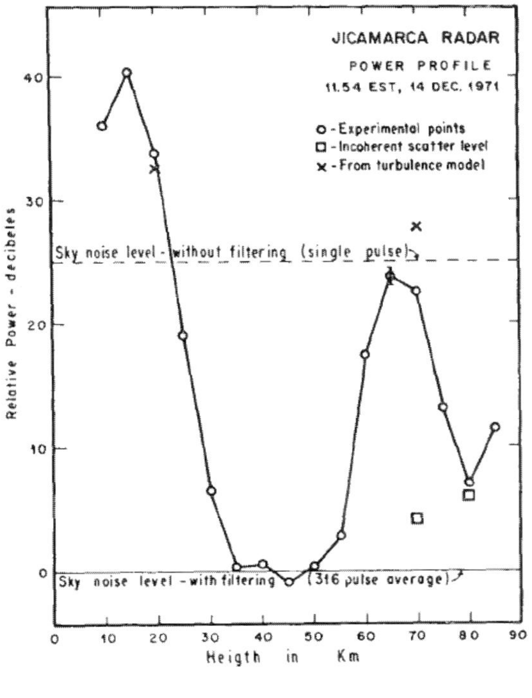

Fig. 2. Backscatter power profile from the Jicamarca radar as reported by Woodman and Guillen (1974).

Bureau of Standards. The CRPL later was absorbed into ESSA, the Environmental Science Service Administration, which ultimately became NOAA, the National Oceanic and Atmospheric Administration. Since 1979, the Jicamarca Radio Observatory has received financial support from a US National Science Foundation grant through Cornell University. Jicamarca is a unique research facility in many ways. It is essentially located on the magnetic equator (1° dip angle). It has the largest antenna of any atmospheric radar in the world (18,432 half-wave dipoles arranged over an area of a little less than 85,000 square meters). Its principal use was envisaged to be, and still is, incoherent scatter measurement of the equatorial ionosphere, but Woodman and Guillen (1974) showed that the Jicamarca radar could be used to obtain neutral winds in the stratosphere and mesosphere. Figure 2, from Woodman and Guillen (1974), shows the backscatter power profile as a function of altitude from a single pulse and from a 316 pulse average. Thus, with suitable averaging, the Jicamarca radar can measure winds throughout the middle atmosphere.

The construction of the Jicamarca radar in 1960–61 paved the way for the construction of MST radars worldwide, so at the present time, there are one or more MST radars operating on every continent except for Africa and Antarctica, and the Japanese have plans for constructing an MST radar in Antarctica. Measurements from MST radars have yielded a great deal of information on middle atmosphere dynamics, especially on gravity waves where their ability to sample frequently and with excellent altitude resolution is unique.

2.3 Discovery of the quasi-biennial oscillation (QBO)

The QBO was discovered independently by Reed *et al.* (1961) and by Veryard and Ebdon (1961). Figure 3 shows its structure over the Equator. Note a number of features seen in Fig. 3. Most evident is a quasi-periodic pattern of descending easterlies (unshaded) followed by descending westerlies (shaded). Looking more closely, one sees that the average period of a complete cycle is about 28 months, but that the period varies considerably, being about 21 months in 1972–1974 and about 35 months in 1983–1986. Moreover, the westerlies descend more quickly than the easterlies. The maximum amplitude of the QBO is about 20 m s^{-1}, occurring in the middle stratosphere.

Following the discovery of the QBO, there were many attempts to explain why this phenomenon occurred, but it was not until the late 1960s that a successful explanation was given. This will be discussed later.

2.4 Murgatroyd and Singleton (1961)

In a pioneering publication, Murgatroyd and Singleton (1961) attempted to calculate the middle atmosphere mean meridional circulation by using observations of temperature, calculating the net diabatic heating rates, and assuming that the dynamical heating resulting from the mean meridional circulation balances the radiative terms. This approach ignores the eddy heat and momentum transport terms, so gives an incorrect picture for the Eulerian mean meridional circulation. Dunkerton (1978) realized, however, that the Murgatroyd and Singleton (1961) circulation gave what we now call the "diabatic circulation," which is a good first approximation to the Lagrangian mean meridional circulation that correctly portrays mass transports. It

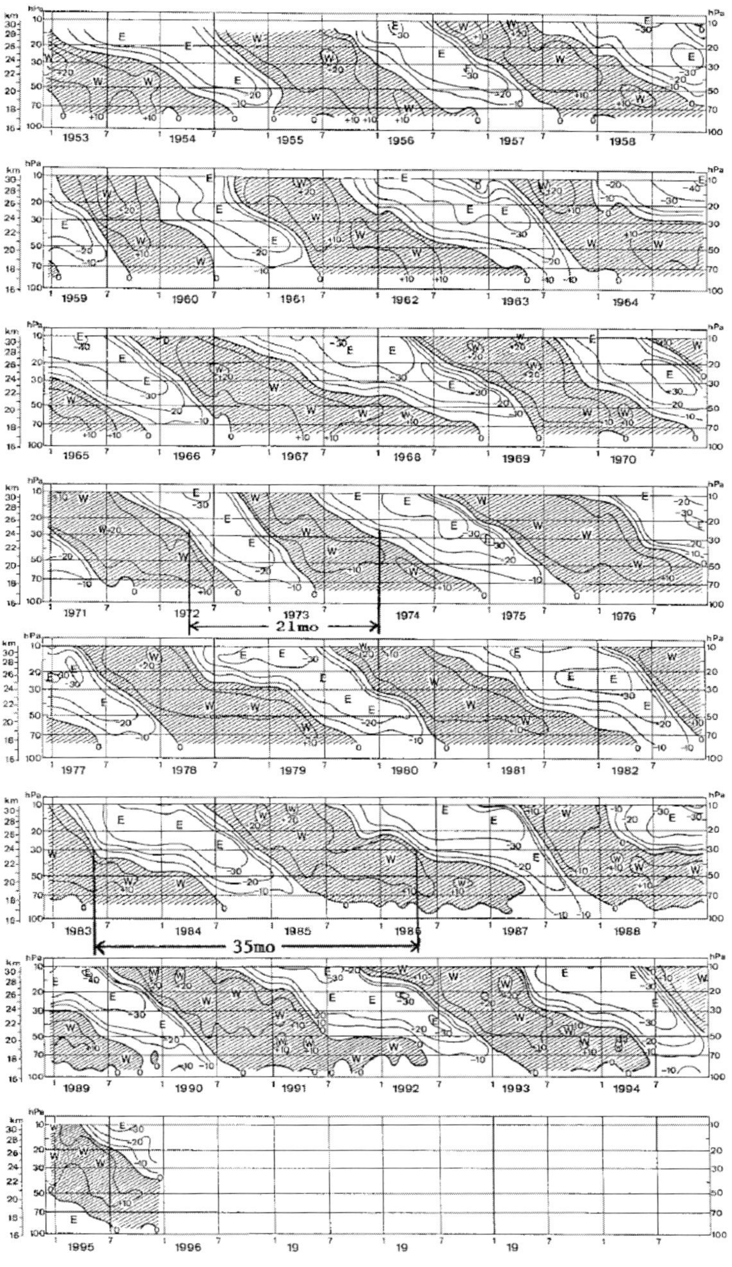

Fig. 3. Time-height section of the monthly mean zonal winds (in m s^{-1}) over equatorial stations. (From Geller *et al.* (1997), which is an update from Naujokat (1986).)

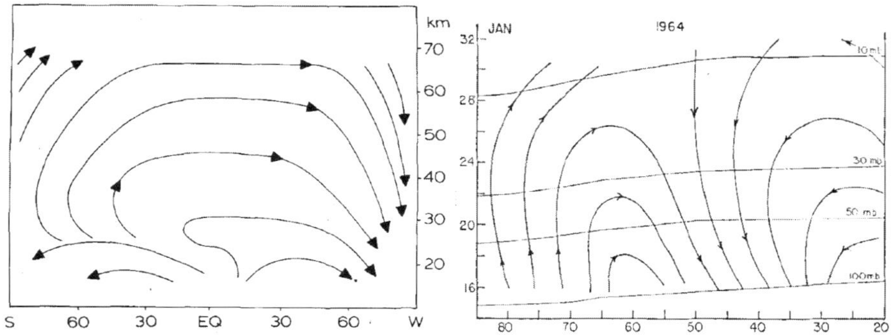

Fig. 4. (Left)—The middle atmosphere diabatic circulation from Dunkerton (1978). This is essentially the circulation computed by Murgatroyd and Singleton (1961) for Northern Hemisphere winter solstice. (Right)—The January 1964 Northern Hemisphere, lower stratosphere, Eulerian circulation from Vincent (1968). Note that very different height ranges are shown in the two panels, and the winter pole is on the right in the left panel but on the left in the right panel.

is interesting to compare the discussions on pages 12 and 14 in Holton (1975) with that on pages 303–305 in Andrews *et al.* (1987) to see the evolution in time of our understanding on this subject.

Figure 4 shows a comparison between the middle atmosphere diabatic circulation from Dunkerton (1978), essentially the Murgatroyd and Singleton (1961) calculation, and the lower stratosphere Eulerian circulation, as computed by Vincent (1968). Both represent January conditions. Comparing the circulations in the lower stratosphere, one sees ascending air through the tropical tropopause and descending air at high latitudes in the left panel, but the right panel shows ascending air both in the tropics and at polar latitudes with descending air at middle latitudes. Thus, the left panel is consistent with the Brewer-Dobson circulation that is required to explain observed stratospheric water vapor and ozone distributions, while the right panel is not. This is reasonable since we now know that the left panel represents the Lagrangian circulation that transports constituents while the right panel does not represent this "transport circulation." Why this is so will become clearer in the following discussion of the "non-interaction" and "non-transport" theorems.

2.5 Charney and Drazin (1961)

During the 1960s, observations indicated that the summer hemisphere polar vortex showed little longitudinal variation while the winter hemisphere polar vortex was very asymmetric (see Hare, 1968, for example). This was explained in the classic paper of Charney and Drazin (1961).

They used the linear, quasi-geostrophic equations on a β-plane to derive the following equation for the vertical structure of planetary wave disturbances.

$$\frac{d^2V}{dz^2} - \frac{1}{H}\frac{dV}{dz} - \frac{\beta N^2(u_0 - c - u_c)}{f_0^2 u_c(u_0 - c)}V = 0 \tag{1}$$

where $u_c \equiv \dfrac{\beta}{k^2+l^2}$, z being the altitude, H the pressure scale height, β the latitudinal derivative of the Coriolis parameter, u_0 the mean zonal flow, c the planetary wave zonal phase velocity, f_0 the Coriolis parameter, V the vertical variation of the planetary wave northward velocity, k and l the zonal and meridional planetary wave wavenumbers, respectively, and N the Brunt Vaisälä frequency. The solution to Eq. (1) is

$$V = \left(Ae^{inz} + Be^{-inz}\right) e^{\frac{z}{2H}}, \text{ where } n^2 = -\frac{1}{4H^2} - \frac{N^2}{f_0^2}\left[\left(k^2+l^2\right) - \frac{\beta}{u_0 - c}\right]. \quad (2)$$

Now, Eq. (2) implies that for vertical planetary wave propagation, n must be real, which implies n^2 must be greater than zero, which in turn implies that

$$0 < u_0 - c < U_c, \text{ where } U_c = \frac{\beta}{k^2+l^2} + \frac{f_0^2}{4H^2N}. \quad (3)$$

Condition (3) implies that for stationary planetary waves ($c = 0$), the mean zonal flow must be westerly (or eastward) and less than a critical wind speed U_c. This immediately accounts for the undisturbed summer polar vortex since the mean zonal winds are easterly (westward) during summer. It also accounts for the fact that even in winter, the asymmetries are large scale relative to the situation in the troposphere since k and l are larger for the smaller scale waves, implying U_c is smaller. One difficulty though is that even when the mean zonal winds are quite large, stationary planetary waves are seen to be present, but improvements in the theory to address this were done by Dickinson (1968) and Matsuno (1970).

Finally, Charney and Drazin (1961) show that for steady waves, in the absence of dissipation, the planetary wave heat and momentum fluxes are so configured that there is no interaction between the vertically propagating planetary waves and the mean zonal flow. This, so-called non-interaction theorem is discussed next.

2.6 Eliassen and Palm (1961)

While Charney and Drazin (1961) treated the non-interaction theorem for planetary waves, Eliassen and Palm (1961) focus on the non-interaction theorem more generally. A more general form of their equation (2.13), for the case of a steady gravity wave propagating in a shear flow in the absence of diabatic effects, is

$$\overline{p'w'} = -\rho_0(u_0 - c)\overline{u'w'}, \quad (4)$$

where p, u and w are pressure and horizontal and vertical velocities, the overlines denote averaging over wave phase, and the primes indicate the wave perturbations. Equation (4) is sometimes referred to as Eliassen and Palm's 1st Theorem. It implies that for upward wave energy flux ($\overline{p'w'} > 0$), the wave momentum flux ($\rho_0\overline{u'w'}$) is negative when the mean flow, u_0, is greater than the wave phase velocity c, and is positive when $u_0 < c$. Thus, any physical process that leads to a decrease of the wave amplitude as it propagates (e.g., dissipation) will force the mean flow toward the wave phase velocity.

For gravity waves with phase velocity c, Eliassen and Palm's 2nd Theorem can be written as

$$\rho_0(u_0 - c)\overline{u'w'} = \text{constant} \tag{5}$$

in the case of no wave transience, no diabatic effects, and $u_0 - c \neq 0$. Thus, in this case, there is no gravity wave interaction with the mean flow. This is consistent with the non-interaction planetary wave result of Charney and Drazin (1961), which was also obtained by Eliassen and Palm (1961).

The implications of Eliassen and Palm's 1st and 2nd theorems are far-reaching. These indicate that unless there are dissipation, other diabatic effects, wave transience, or $u_0 = c$, atmospheric waves do not interact with the main flow. Conversely, if any of these are present, the waves do interact with the mean flow, and this interaction gives rise to a deceleration or acceleration of the mean flow toward the wave's phase velocity. These interactions are fundamental in understanding why the quasi-biennial and semi-annual oscillations occur in the equatorial middle atmosphere and what gives rise to stratospheric sudden warmings.

2.7 Leovy (1964)

By the 1960s, it was known that summer mesopause temperatures were much colder than winter mesopause temperatures, which was opposite to the influence of solar heating—the summer mesopause being constantly sunlit while the winter mesopause is subject to continuous darkness. Leovy (1964) constructed the first sim-

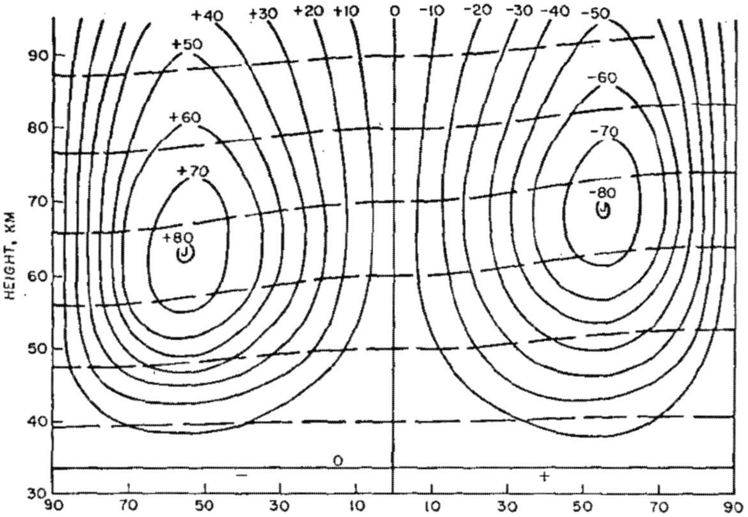

Fig. 5. Mean zonal wind computed by Leovy (1964) with Newtonian cooling representation for radiation and Rayleigh drag for momentum dissipation. Sloping dashed lines represent constant pressure contours.

ple model of the zonally averaged mesospheric circulation. He carried out numerical experiments where the middle atmosphere circulation was forced by realistic solar differential heating. He used two different formulations of mechanical damping, one being Rayleigh friction and the other being momentum diffusion by viscosity. Figure 5 shows the zonal wind simulation for the Rayleigh friction case. Leovy (1964) also showed results for viscosity and thermal conduction. He was able to get reasonable results, but only with viscosities and thermal conduction coefficients that were excessively large.

Leovy's (1964) calculations were followed up much later by Schoeberl and Strobel (1978) and by Holton and Wehrbein (1980), at which time there was a greater appreciation that Rayleigh drag was likely a crude representation of gravity wave effects. Research into proper representation of gravity wave effects on mean flows then became a popular subject for research following the pioneering works of Matsuno (1982) and Lindzen (1981).

2.8 Early research on stratospheric ozone loss by catalytic reactions

Chapman (1930) suggested the following set of chemical reactions might explain the stratospheric ozone layer.

$$O_2 + h\nu(\lambda < 240 \text{ nm}) \rightarrow O + O \tag{6}$$

$$O + O_2 + M \rightarrow O_3 + M \tag{7}$$

$$O_3 + h\nu(\lambda < 320 \text{ nm}) \rightarrow O + O_2 \tag{8}$$

$$O + O_3 \rightarrow 2O_2 \tag{9}$$

Thus, ultraviolet radiation dissociates molecular ozone in (6); ozone is formed in the presence of a third body, M, in (7); ozone is photochemically dissociated in (8), but the atomic oxygen quickly transforms back into ozone by (7); and the ozone is lost by (9).

Benson and Axworthy's (1957) measurements of reaction (9), however, indicated that reaction proceeded much more slowly than had been assumed, implying that reactions (6)–(9) give excessive stratospheric ozone. Hampson (1964) suggested that the following catalytic chemical reactions might resolve the issue.

$$OH + O_3 \rightarrow HO_2 + O_2 \tag{10}$$

$$HO_2 + O_3 \rightarrow OH + 2O_2 \tag{11}$$

Reactions (10) and (11) constitute a set of reactions for a catalytic cycle for stratospheric ozone loss, since OH initiates the cycle and remains after the cycle is completed. Hunt (1966) suggested a set of reaction rates for (10) and (11) that was consistent with observed values for stratospheric ozone. In fact, these reactions were not found to be very important in the stratosphere, but are important in the mesosphere.

Crutzen (1970) and Johnston (1971) showed that the nitrogen catalytic cycles are very important in the natural stratosphere, and Molina and Rowland (1974) pointed out that the chlorine resulting from the photodissociation of the industrially produced chlorofluoromethanes would likely destroy significant amounts of stratospheric ozone. Of course, Paul Crutzen, Sherwood Rowland, and Mario Molina went on to receive the 1995 Nobel Prize in Chemistry for their contributions, but I believe it is fair to say that this line of investigation into catalytic destruction of stratospheric ozone had its genesis during the 1960s.

2.9 Discovery of the semi-annual oscillation by Reed (1965)

Reed (1965) used a new two-year time series of wind measurements over Ascension Island (70°55′S, 14°25′W) to examine how the quasi-biennial oscillation evolved with increasing altitude between 30 and 50 km. Figure 6, from his paper, shows his results.

Note that the altitude scale for Fig. 6 essentially starts at the top of Fig. 3. Looking at Fig. 6, we see a gradual transition from a long period oscillation at 28 km, to shorter-term variability at 40 km, to a clear 6-month oscillation at 52 km. Thus, while the equatorial lower stratosphere is characterized by the quasi-biennial oscillation, the upper stratosphere—lower mesosphere is characterized by the semi-annual oscillation.

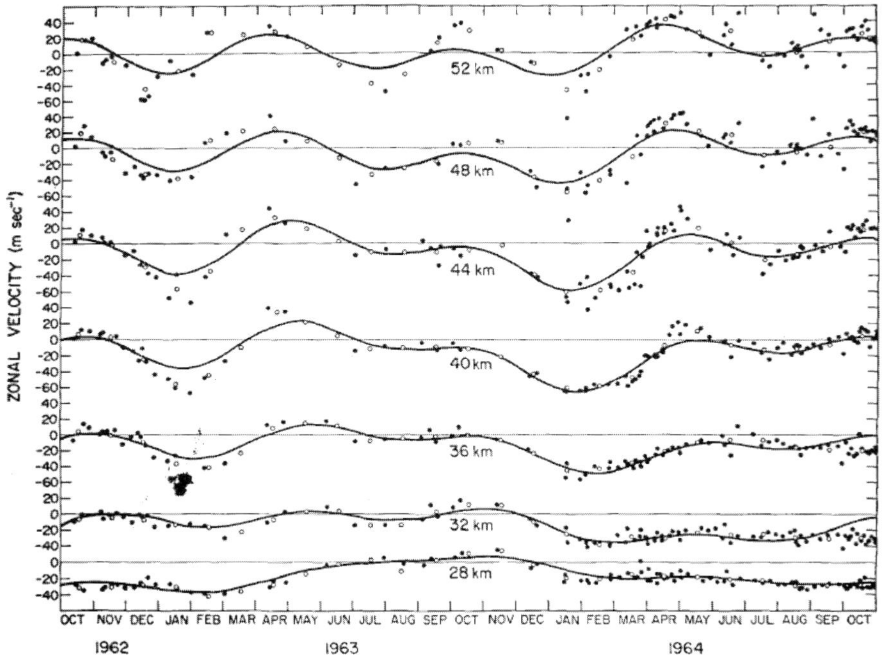

Fig. 6. Two-year time series for measured zonal wind over Ascension Island at 28, 32, 36, 40, 44, 48, and 52 km. Circles indicate measurements, and curves are harmonic curve fits. (From Reed (1965).)

2.10 Discovery of missing tidal modes by Kato (1966) and Lindzen (1966)

The theory for atmospheric tides goes back to the work of Laplace (1799), who derived the idealized equations for the free and forced oscillations for a thin atmosphere on a spherical planet. While the Moon's gravitation forces the oceanic tide, which is semidiurnal, it is the Sun's heating that forces the Earth's atmospheric tides. Under simplifying assumptions, Laplace used the traditional method of separation of variables to obtain an ordinary differential equation for the latitudinal structure of the tide as well as one for its vertical structure. The equation for the latitude structure is called Laplace's tidal equation. Its eigenvalues are often referred to in terms of equivalent depths (an analogy to the ocean), and its eigenfunctions are called Hough functions. The equivalent depth then occurs as a parameter in the vertical structure equation. One can then expand the latitudinal structure of the solar heating in terms of these Hough functions.

Early observations contradicted simple intuition, in that the surface pressure showed tidal variations that were predominantly semidiurnal, while the solar forcing is predominantly diurnal. Figure 7, from Chapman and Lindzen (1970), illustrates two things. One is that surface pressure variations in connection with extratropical systems are much greater than surface pressure variations in the tropics, except when tropical cyclones occur. The other is that tidal variations in the tropics are larger in the tropics and are predominantly semidiurnal rather than diurnal, as expected.

Initial suggestions to explain the dominance of the semidiurnal tide in surface pressure over the expected diurnal tide were that the semidiurnal solar forcing excited a resonance in the atmosphere while the diurnal forcing was off resonance. As more and more was learned about atmospheric tides and their forcing, this did not prove to be the case. It was not until Kato (1966) and Lindzen (1966) independently discovered the existence of negative equivalent depth eigenmodes to Laplace's tidal equation that this problem was solved. The solution is best explained with the aid

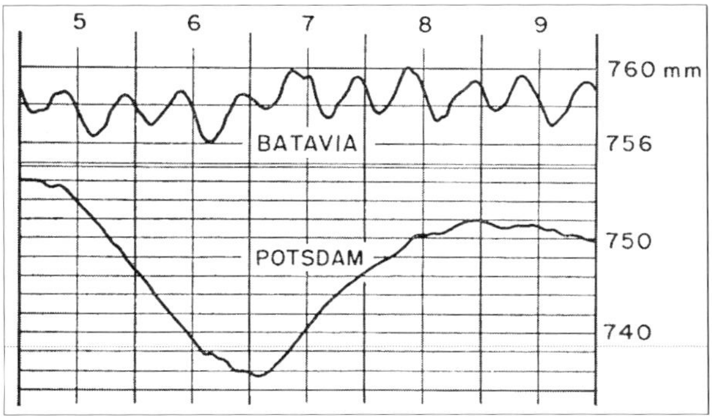

Fig. 7. Barometric variations (on two different scales) at Batavia (the present Jakarta at 6°S) and Potsdam (52°N) in November 1919. This figure is from Chapman and Lindzen (1970), who in turn took it from Bartels (1928).

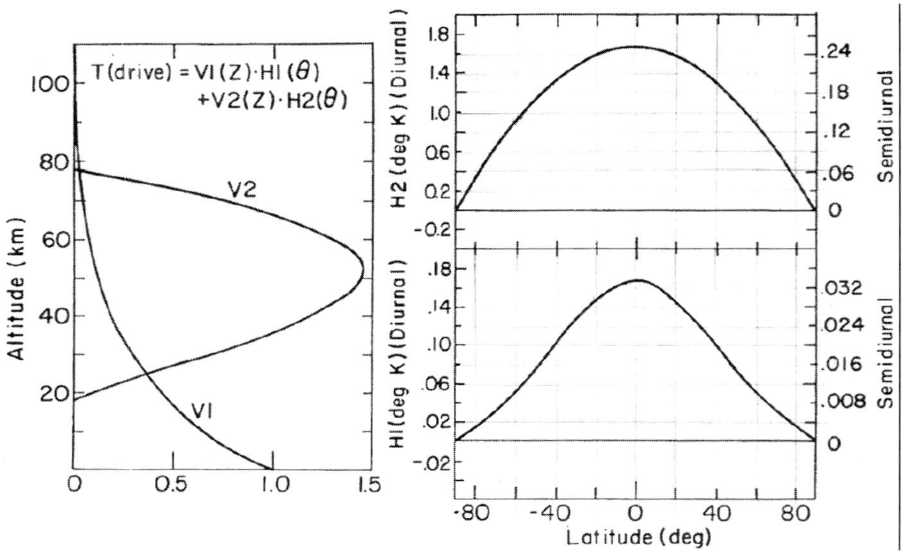

Fig. 8. Vertical and horizontal variations of solar heating of atmospheric water vapor and ozone. From Chapman and Lindzen (1970), who in turn took this from Lindzen (1968).

of Fig. 8, from Chapman and Lindzen (1970). In this figure, the two profiles represent the vertical variation of the solar heating of atmospheric water vapor (V1) and ozone (V2), along with the latitudinal variations of these heating functions. Note the different scales for the diurnal and semidiurnal heating functions. When these heating functions are expanded in the diurnal and semidiurnal Hough function solutions of Laplace's tidal equation, it turns out that the semidiurnal heating only gives rise to positive equivalent depth modes, the principal one of which has a very long vertical wavelength of about 100 km, while the diurnal heating gives rise to negative equivalent depth modes, which cannot propagate vertically and principally a positive equivalent depth mode of rather short vertical wavelength (about 25 km). The ozone heating principally forces a negative equivalent depth mode that cannot propagate down to the surface plus a short vertical wavelength mode that does reach the surface. In contrast, the semidiurnal ozone heating principally forces a long wavelength mode that does reach the surface. For the semidiurnal solar tide, the vertical wavelength is longer than the depth of the ozone heating and it propagates down to the surface, so the tidal pressure variations forced by ozone and water vapor add to produce a sizable response in surface pressure. For the diurnal ozone heating, not only does the negative equivalent depth mode not propagate down to the surface, but there is also destructive interference over the depth of the ozone heating for the short vertical wavelength solution at low latitudes. Therefore, the sizable semidiurnal variation in surface pressure is a consequence of the additive solutions from water vapor and ozone heating while for the diurnal tide, there is very little surface response to the ozone heating.

Not only did Lindzen's (1966) and Kato's (1966) papers lead to a resolution of the long-standing questions about the solar atmospheric tides, it also led to an appreciation of the completeness of the Hough functions. This, in turn, led to Longuet-Higgins (1968) paper that examined the family of oscillations on a fluid envelope on a rotating sphere, which gave a theoretical framework for understanding global atmospheric wave motions.

2.11 Matsuno's (1966) discovery of equatorial wave modes

Matsuno (1966) examined the quasi-geostrophic wave modes that exist on an equatorial β-plane. His lowest-order wave modes correspond to what we now refer to as the equatorially trapped Kelvin and mixed Rossby-gravity waves. The structures of these waves are shown in Fig. 9. Note that both the Kelvin and mixed Rossby-gravity waves become geostrophic in that the winds are parallel to the geopotential contours off the Equator. Also, note that the Kelvin wave is symmetric around the Equator while the mixed Rossby-gravity wave is asymmetric about the Equator. Not so obvious from this figure is the fact that the Kelvin wave propagates to the east ($c > 0$) and the mixed Rossby-gravity wave propagates to the west ($c < 0$).

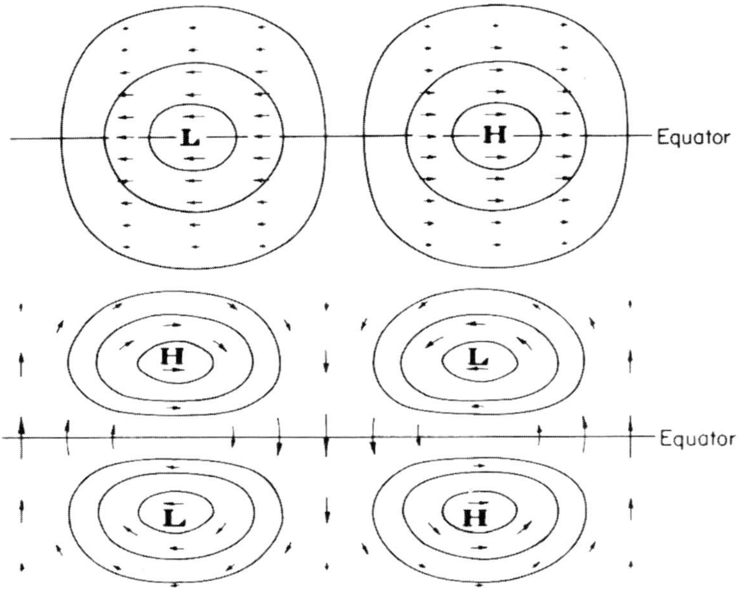

Fig. 9. Schematic illustration of the geopotential and wind fields for the equatorial trapped Kelvin (top) and mixed Rossby-gravity (bottom) waves. (Adapted from *Middle Atmosphere Dynamics*, Andrews *et al.*, Copyright 1987 Academic Press, with permission from Elsevier. Andrews *et al.* (1987) in turn adapted theirs from Matsuno (1966).)

Table 2. Characteristics of the dominant observed planetary-scale waves in equatorial lower stratosphere. (Reprinted from *Middle Atmosphere Dynamics*, Andrews *et al.*, Copyright 1987 Academic Press, with permission from Elsevier.)

Theoretical description	Kelvin wave	Rossby-gravity wave
Discovered by	Wallace and Kousky (1968)	Yanai and Maruyama (1966)
Period (ground-based) $2\pi\omega^{-1}$	15 days	4–5 days
Zonal wave number $s = ka \cos \phi$	1–2	4
Vertical wavelength $2\pi m^{-1}$	6–10 km	4–8 km
Average phase speed relative to ground	$+25$ m s^{-1}	-23 m s^{-1}
Observed when mean zonal flow is	Easterly (max. ≈ -25 m s^{-1})	Westerly (max. $\approx +7$ m s^{-1})
Average phase speed relative to maximum zonal flow	$+50$ m s^{-1}	-30 m s^{-1}
Approximate observed amplitudes		
u'	8 m s^{-1}	2–3 m s^{-1}
v'	0	2–3 m s^{-1}
T'	2–3 K	1 K
Approximate inferred amplitudes		
Φ'/g	30 m	4 m
w'	1.5×10^{-3} m s^{-1}	1.5×10^{-3} m s^{-1}
Approximate meridional scales		
$\left(\frac{2N}{\beta\lvert m\rvert}\right)^{1/2}$	1300–1700 km	1000–1500 km

Not so long after Matsuno's discovery of these wave modes, they were identified observationally in the stratosphere. Yanai and Maruyama (1966) identified the mixed Rossby-gravity wave and Wallace and Kousky (1968) identified the Kelvin wave. Table 2, from Andrews *et al.* (1987) shows the observed characteristics of these waves in the equatorial lower stratosphere. Note that these waves are of planetary scale—wave number 1–2 for the Kelvin wave and 4 for the mixed Rossby-gravity wave. Note also that they are trapped within a couple of thousand km of the Equator.

2.12 Gravity wave critical level theory by Bretherton (1966) and Booker and Bretherton (1967)

We saw in the earlier discussion of the Eliassen and Palm (1961) paper that one of the conditions for non-interaction between gravity waves and the mean flow is that the wave phase velocity is not equal to the mean zonal flow, or $u_0 \neq c$. Bretherton (1966) examined the case of a gravity wave in a shear flow where $u_0 = c$ (the critical level) and the Richardson number is very large. He found that in this case the gravity wave vertical group velocity $\to 0$ as $u_0 \to c$. Thus, the gravity wave energy flux $\overline{p'w'}$ vanished on the far side of the critical level, in which case the momentum flux $\rho_0\overline{u'w'}$ was also zero. Since there was no wave interaction below the critical level, this implies a convergence (or divergence) of the wave momentum flux at the critical level.

Booker and Bretherton (1967) generalized this result to the case of finite Richardson number, in which case they derived the results that in passing through the critical level, the wave momentum flux is attenuated by a factor of $e^{-2\pi\sqrt{Ri-\frac{1}{4}}}$. Thus, at a gravity wave critical level, the absorption of the wave will tend to bring the mean flow toward the wave phase velocity.

There followed a period of very active research into the nature of gravity wave critical levels. Hazel (1967) showed that the Booker and Bretherton (1967) result was essentially correct in the case of a fluid with viscosity and heat conduction. Breeding (1971) suggested that nonlinear effects might lead to some wave reflection in addition to absorption, and Geller *et al.* (1975) suggested that as the wave approached a critical level, it produces turbulence that would likely lead to wave absorption before nonlinear effects would lead to wave reflection.

2.13 Hodges (1967) and gravity wave breaking

In an isothermal atmosphere, the density decreases exponentially with increasing altitude z as $e^{-\frac{z}{H}}$, H being the pressure scale height. Without dissipation or critical levels, the gravity wave kinetic energy per unit volume $\rho_0\overline{\vec{v}^2}$ should remain constant, in which case the amplitude of the wave's horizontal velocity (and as it turns out temperature) fluctuations should grow as $e^{+\frac{z}{2H}}$. This being the case, the wave eventually becomes unstable. Hodges (1967) was the first to point out that this will be a source of turbulence in the middle and upper atmosphere.

This provided the starting point for Lindzen's (1981) seminal paper that suggested a self-consistent way of parameterizing the effects of unresolved gravity waves in climate models. The principle for this parameterization is illustrated in Fig. 10. On the right is illustrated a gravity wave whose wind and temperature amplitudes are exponentially increasing with height, and as pictured, the wave momentum flux $\rho_0\overline{u'w'}$ is

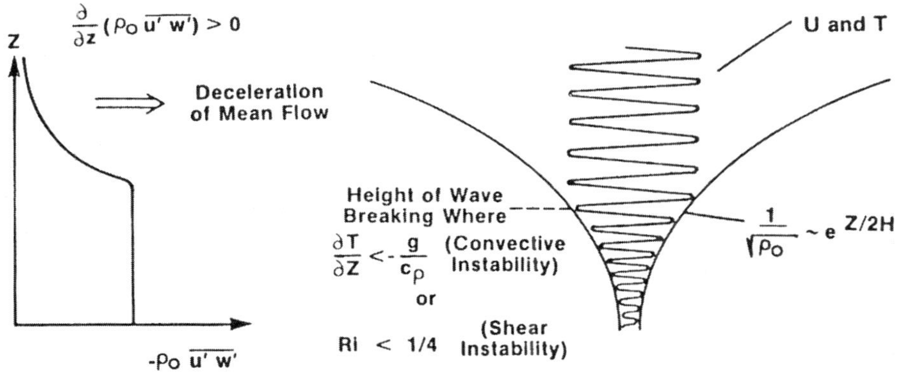

Fig. 10. Schematic of wave breaking, with the resultant convergence of gravity wave momentum flux. (With kind permission from Springer Science+Business Media: *Space Sci. Rev.*, Dynamics of the middle atmosphere (tutorial lecture), **34**, 1983, 359–375, Geller, M. A., figure 12.)

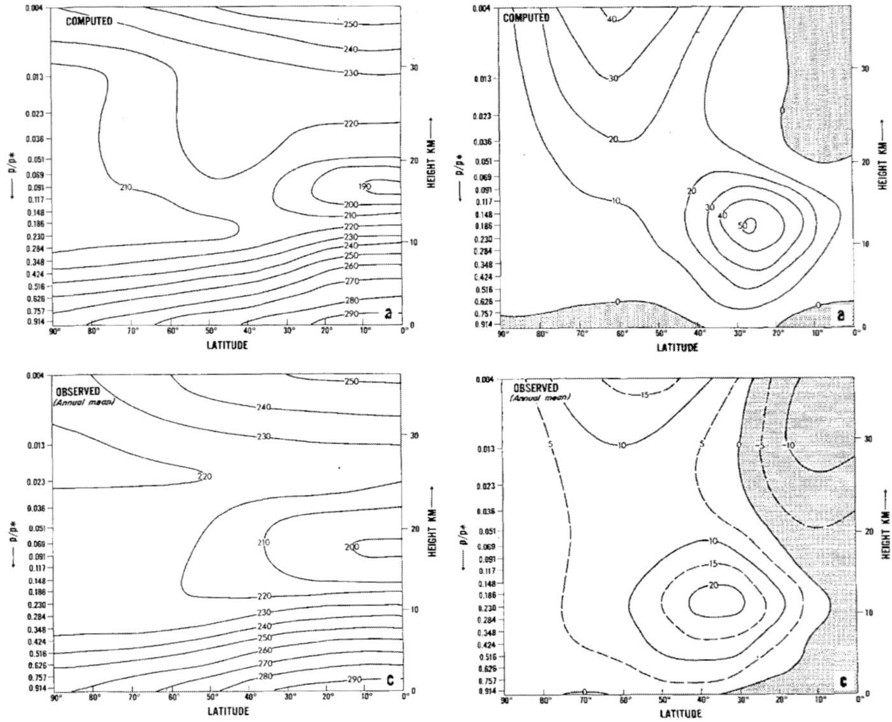

Fig. 11. Observed (lower) and computed (upper) zonally averaged Northern Hemisphere temperatures (left) and eastward winds (right) from Manabe and Hunt's (1968) troposphere-stratosphere general circulation model. (Adapted from Manabe and Hunt (1968).)

constant with height. Since the vertical wavelength of this wave is fixed $\frac{\partial \vec{v}}{\partial z}$ and $\frac{\partial T}{\partial z}$ also increase with height exponentially, as illustrated by the outer envelope, eventually, the wave becomes either convectively unstable or shear unstable and breaks down. Lindzen (1981) made the assumption that above the level where the wave breaks down, it loses just enough energy to turbulence to keep the wave amplitude constant above that level, as illustrated. This means that $\rho_0 \overline{u'w'}$ decreases with height above the breaking level so that there is a divergence of wave momentum flux above the level where the gravity wave begins to break, also as pictured in Fig. 10. Of course, one could make different assumptions of what occurs above the breaking level. For instance, Alexander and Dunkerton (1999) assume that gravity waves deposit all of their momentum at the breaking level.

Developing and implementing ways of parameterizing the effects of unresolved gravity waves in climate models is a research topic of great current interest, but one might say that this had its intellectual roots in the papers of Eliassen and Palm (1961),

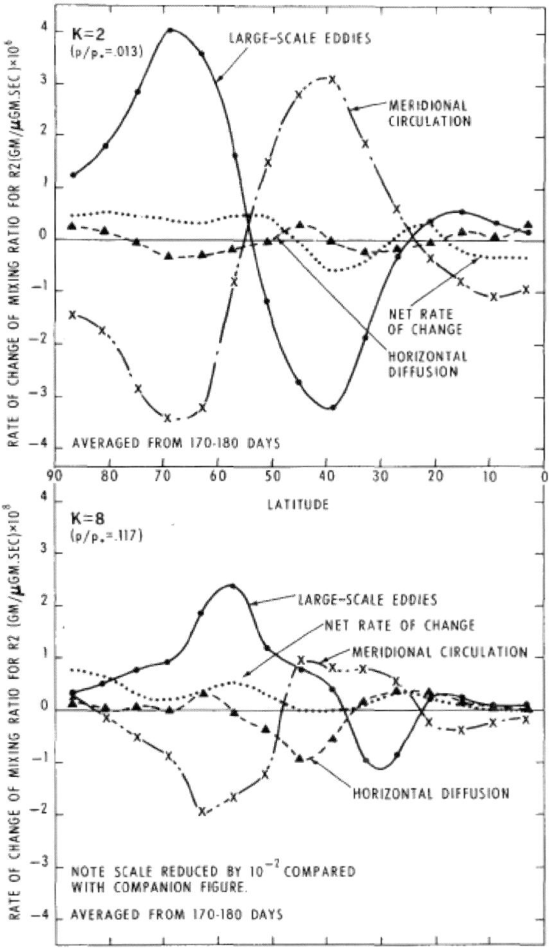

Fig. 12. The contributions from the mean meridional circulation, the large-scale eddies, and horizontal diffusion to the continuity equation for an ozone-like tracer at two levels. The net rate of change of this ozone-like tracer is also shown. (From Hunt and Manabe (1968).)

Hodges (1967), and Booker and Bretherton (1967).

2.14 The beginnings of troposphere-stratosphere general circulation models

Manabe and Hunt (1968) and Hunt and Manabe (1968) published the results of the first troposphere-stratosphere general circulation models. These papers used a model that contained a full treatment (for that time) of tropospheric physics and also had transport tracers. Figure 11 shows the results of their simulations of the zonally averaged temperature and zonal wind for January for their 18-level model that had its top at 37.5 km. The agreement between the model temperatures and winds and observations are pretty good. Of course, there are noticeable differences between the model results and observations, some of which are the modeled tropopause jet is too strong and too far poleward; the modeled tropopause temperatures are too cold; and the polar stratosphere is too cold, which leads to the lower portion of the polar night jet being too strong in comparison with observations.

Hunt and Manabe (1968) conducted some tracer experiments with this model. One of their results is shown in Fig. 12. These results are for a passive ozone-like tracer that has its source in the tropical upper stratosphere. Note the near cancellation of the flux divergence terms from the mean meridional circulation and the large-scale eddies. This suggests that, in an analogous fashion to the non-interaction theorem, there is also a non-transport theorem; that is to say, that for steady state conditions, no dissipation, no critical levels, and for a conserved tracer, there is no net transport by the planetary-scale eddies since these eddies give rise to canceling transport effects by the mean meridional circulation.

2.15 Theory for the QBO

After the discovery of the quasi-biennial oscillation, there followed a number of suggestions for its cause. The first successful explanation for the QBO was given by Lindzen and Holton (1968). This followed a very important precursor paper by Wallace and Holton (1968), who showed that one could not explain the QBO by periodically varying heating terms. Rather, they showed that the forcings for the QBO must be periodically varying terms in the momentum equation. Lindzen and Holton (1968) first advanced the theory that, with few modifications, is accepted as the explanation of the QBO today. Basically, they proposed that waves carrying eastward and westward momentum upward were forced in the tropics and their interaction with the mean flow gave rise to alternating momentum source terms for zonal mean momentum, leading to the QBO. This theory was altered a bit in Holton and Lindzen (1972) when dissipation replaced critical level interaction as the mechanism for the wave mean flow momentum interactions.

This theory is best explained by the schematic shown in Fig. 13, which is taken from Plumb (1984). If there exists two vertically propagating gravity waves, one with phase velocity $+c$ and the other with phase velocity $-c$, and if the flow is slightly biased to the eastward direction, the group velocity for the $+c$-wave is diminished relative to that of the $-c$-wave. Therefore, in the presence of any dissipation, there will be preferential momentum absorption of the $+c$-wave. Eliassen and Palm's 1st theorem implies acceleration of the mean flow toward $+c$. Profile (2) shows that the absorption of the $+c$-wave momentum flux leads to descending eastward flow. Since the $-c$-wave is not being absorbed at lower altitudes, it propagates freely to higher levels where it is also subject to dissipation, thus leading to higher level westward flow in profile (2). Profile (3) shows a later stage where the eastward flow has descended further, and is close to $+c$, while the upper level westward flow is tending toward $-c$ and is also descending. In this idealized situation, the eastward flow descends toward the forcing level, and the shears become so sharp that diffusion brings the flow toward zero, but at this point, there is a descending westward flow, which is

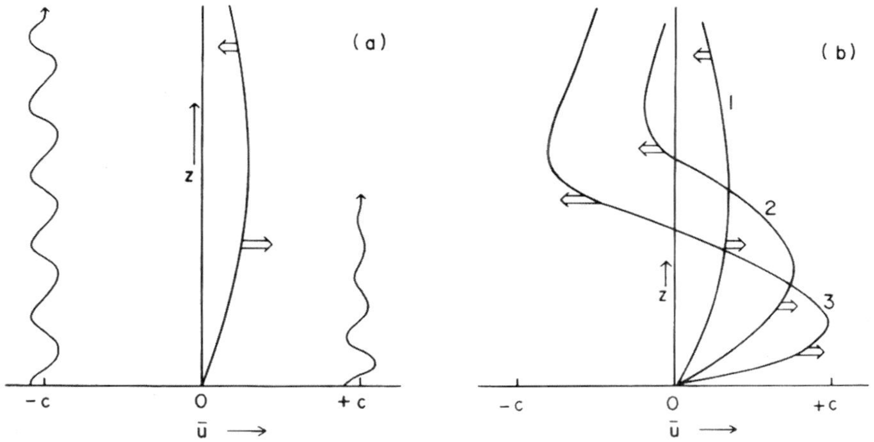

Fig. 13. Schematic representation of the Lindzen and Holton (1968)/Holton and Linden (1972) theory for the QBO. Panel (a) shows the initial state, which corresponds to profile (1) in panel (b). Profiles (2) and (3) show successive stages of evolution, as explained in the text. (After Plumb (1984).)

similar to the mirror image of panel (a), i.e., a westward bias to the flow, and the mirror image of the cycle proceeds, producing one complete cycle of alternating descent of eastward flow followed by a descent of westward flow, and so on.

Interestingly enough, the waves that carry eastward and westward momentum need not be equatorial waves, rather it is the very small Coriolis force near the Equator that makes the wave momentum fluxes influence $\frac{\partial \overline{u}}{\partial t}$, rather than forcing a merdional flow. This has been elegantly shown by Haynes (1998).

2.16 Dickinson's (1969) paper on planetary wave-mean flow interactions

There have been several generalizations of the Eliassen-Palm (1961) and Charney-Drazin (1961) noninteraction theorems. Most of these have occurred outside of our window of the 1960s (e.g., Holton, 1974; Andrews and McIntyre, 1976; Boyd, 1978), but one such generalization that did occur during the 1960s was by Dickinson (1969). He used a quasi-geostrophic formulation for small amplitude planetary waves on a β-plane geometry to demonstrate that there must exist eddy transport of quasi-geostrophic potential vorticity for the waves to cause an acceleration of the mean zonal flow, and that such potential vorticity transports will occur only in the presence of critical lines (where the wave phase speed equals the mean flow) or in the presence of dissipation.

It turns out that this condition for non-zero quasi-geostrophic potential vorticity transports is quite equivalent to the modern-day Eliassen-Palm flux divergence. For example, Andrews *et al.* (1987) show on pages 129–130 that

$$\overline{v'q'} = \frac{1}{\rho_0} \vec{\nabla} \cdot \vec{F}, \tag{12}$$

where v' is the eddy northward velocity, q' is the eddy quasi-geostrophic potential vorticity, the overbar is a zonal average, ρ_0 is the basic state density, and $\vec{F}$ is the quasi-geostrophic Eliassen-Palm flux vector. Thus, a framework for today's Eliassen-Palm flux diagnostics had been established during the 1960s.

3 Some Closing Comments

The 1960s were a very special time for me since that period spanned the time I finished high school, did my undergraduate education, earned my PhD, and became an Assistant Professor. I was very fortunate to have known, and still know, many of the individuals whose work I have cited here. It is no coincidence that I have worked on many of the topics discussed here since such an excellent foundation for my work had already been established by others by the time I entered the field.

While preparing this paper, I realized that what began as a historical review, also might serve as a good introduction to middle atmosphere dynamics. It was my pleasure to prepare this paper.

Acknowledgments. Support for M. Geller, while preparing this paper, came from the National Science Foundation's Large-scale Dynamics program and from the National Aeronautics and Space Administration's Atmospheric Chemistry and Modeling Program.

References

Alexander, M. J. and T. J. Dunkerton, A spectral parameterization of mean-flow forcing due to breaking gravity waves, *J. Atmos. Sci.*, **56**, 4167–4182, 1999.

Andrews, D. G. and M. E. McIntyre, Planetary waves in horizontal and vertical shear: the generalized Eliassen-Palm relation and the mean zonal acceleration, *J. Atmos. Sci.*, **33**, 2031–2048, 1976.

Andrews, D. G., J. R. Holton, and C. B. Leovy, *Middle Atmosphere Dynamics*, 489 pp., Academic Press, San Diego, California, 1987.

Bartels, J., Gezeitenschwingungen der Atmosphäre, *Handbuch der Experimentalphysik*, **25**, 163–210, 1928.

Benson, S. W. and A. E. Axworthy, Mechanism of the gas phase. Thermal decomposition of ozone, *J. Chem. Phys.*, **26**, 1718–1726, 1957.

Booker, J. R. and F. P. Bretherton, The critical layer for internal gravity waves in shear flow, *J. Fluid Mech.*, **27**, 513–539, 1967.

Boyd, J. P., The noninteraction of waves with the zonally averaged flow on a spherical earth and the interrelationships of eddy fluxes of energy, heat, and momentum, *J. Atmos. Sci.*, **33**, 2285–2291, 1978.

Breeding, R. J., A non-linear investigation of critical levels for internal gravity waves, *J. Fluid Mech.*, **50**, 545–563, 1971.

Bretherton, F. P., The propagation of groups of internal gravity waves in a shear flow, *Quart. J. R. Meteorol. Soc.*, **92**, 466–480, 1966.

Chapman, S., A theory of upper atmospheric ozone, *R. Meteorol. Soc. Mem.*, **3**, 103–125, 1930.

Chapman, S. and R. S. Lindzen, *Atmospheric Tides*, 200 pp., Reidel, Dordrecht, Netherlands, 1970.

Charney, J. G. and P. G. Drazin, Propagation of planetary-scale disturbances from the lower into the upper atmosphere, *J. Geophys. Res.*, **66**, 83–109, 1961.

Crutzen, P. J., The influence of nitrogen oxides on the atmospheric ozone content, *Quart. J. R. Meteor. Soc.*, **96**, 320–325, 1970.

Dickinson, R. E., Planetary Rossby waves propagating vertically through weak westerly wind wave-guides, *J. Atmos. Sci.*, **25**, 984–1002, 1968.

Dickinson, R. E., Theory of planetary wave-mean flow interaction, *J. Atmos. Sci.*, **26**, 73–81, 1969.

Dunkerton, T. J., On the mean meridional mass motions of the stratosphere and mesosphere, *J. Atmos. Sci.*, **35**, 2325–2333, 1978.

Eliassen, A. and E. Palm, On the transfer of energy in stationary mountain waves, *Geofys. Publ.*, **22**, 1–23, 1961.

Geller, M. A., Dynamics of the middle atmosphere (tutorial lecture), *Space Sci. Rev.*, **34**, 359–375, 1983.

Geller, M. A., H. Tanaka, and D. C. Fritts, Production of turbulence in the vicinity of critical levels for internal gravity waves, *J. Atmos. Sci.*, **32**, 2125–2135, 1975.

Geller, M. A., W. Shen, and W. Wu, Calculations of the stratospheric QBO for time-varying wave forcing, *J. Atmos. Sci.*, **54**, 883–894, 1997.

Hampson, J., *C. A. R. D. E. (Canadian Armament Research and Development Establishment) TN, 1627/64*, 280 pp., Valvartier, Quebec, Canada, 1964.

Hare, F. K., The arctic, *Quart. J. R. Meteor. Soc.*, **94**, 439–459, 1968.

Haynes, P. H., The latitudinal structure of the quasi-biennial oscillation, *Quart. J. Roy. Meteor. Soc.*, **124**, 2645–2670, 1998.

Hazel, P., The effect of viscosity and heat conduction on internal gravity waves at a critical level, *J. Fluid Mech.*, **30**, 775–784, 1967.

Hines, C. O., Internal atmospheric gravity waves at ionospheric heights, *Can. J. Phys.*, **38**, 1441–1481, 1960.

Hodges, R. R. Jr., Generation of turbulence in the upper atmosphere by internal gravity waves, *J. Geophys. Res.*, **72**, 3455–3458, 1967.

Holton, J. R., Forcing of mean flows by stationary waves, *J. Atmos. Sci.*, **31**, 942–945, 1974.

Holton, J. R., The dynamic meteorology of the stratosphere and mesosphere, *Meteorol. Monogr. No. 37*, 218 pp., Am. Meteorol. Soc., Boston, Massachusetts, 1975.

Holton, J. R. and R. S. Lindzen, An updated theory for the quasi-biennial cycle of the tropical stratosphere, *J. Atmos. Sci.*, **29**, 1076–1080, 1972.

Holton, J. R. and W. M. Wehrbein, A numerical model of the zonal mean circulation of the middle atmosphere, *Pageoph*, **118**, 284–306, 1980.

Hunt, B. G., The need for a modified photochemical theory of the ozonosphere, *J. Atmos. Sci.*, **23**, 88–95, 1966.

Hunt, B. G. and S. Manabe, Experiments with a stratospheric general circulation model, II. Large-scale diffusion of tracers in the stratosphere, *Mon. Wea. Rev.*, **96**, 503–539, 1968.

Johnston, H. S., Reduction of stratospheric ozone by nitrogen oxide catalysts from SST exhaust, *Science*, **173**, 517–522, 1971.

Kato, S., Diurnal atmospheric oscillation, 1, eigenvalues and Hough functions, *J. Geophys. Res.*, **71**, 3201–3209, 1966.

Laplace, P. S., *Mécanique céleste*, Paris, 1799.

Leovy, C. B., Simple models of thermally driven mesospheric circulation, *J. Atmos. Sci.*, **21**, 327–341, 1964.

Lindzen, R. S., On the theory of the diurnal tide, *Mon. Wea. Rev.*, **94**, 295–301, 1966.

Lindzen, R. S., The application of classical atmospheric tidal theory, *Proc. Roy. Soc.*, **A303**, 299–316, 1968.

Lindzen, R. S., Turbulence and stress owing to gravity wave and tidal breakdown, *J. Geophys. Res.*, **86**, 9707–9714, 1981.

Lindzen, R. S. and J. R. Holton, A theory of the quasi-biennial oscillation, *J. Atmos. Sci.*, **25**, 1095–1107, 1968.

Longuet-Higgins, M. S., The eigenfunctions of Laplace's tidal equations over a sphere, *Philos. Trans. R. Soc. London. Ser. A*, **262**, 511–607, 1968.

Manabe, S. and B. G. Hunt, Experiments with a stratospheric general circulation model. I. radiative and dynamical aspects, *Mon. Wea. Rev.*, **96**, 477–502, 1968.

Matsuno, T., Quasi-geostrophic motions in the equatorial area, *J. Meteorol. Soc. Jpn.*, **44**, 25–43, 1966.

Matsuno, T., Vertical propagation of stationary planetary waves in the winter northern hemisphere, *J. Atmos. Sci.*, **27**, 871–883, 1970.

Matsuno, T., A quasi one-dimensional model of the middle atmosphere circulation interacting with internal gravity waves, *J. Meteorol. Soc. Jpn.*, **60**, 215–226, 1982.

Molina, M. J. and F. S. Rowland, Stratospheric sink for chlorofluoromethanes: chlorine atom-catalyzed destruction of ozone, *Nature*, **249**, 810–812, 1974.

Murgatroyd, R. J. and F. Singleton, Possible meridional circulations in the stratosphere and mesosphere, *Q. J. R. Meteorol. Soc.*, **87**, 125–135, 1961.

Naujokat, B., An update of the observed quasi-biennial oscillation of the stratospheric winds over the tropics, *J. Atmos. Sci.*, **43**, 1873–1877, 1986.

Plumb, R. A., The quasi-biennial oscillation, in *Dynamics of the Middle Atmosphere*, edited by J. R. Holton and T. Matsuno, 217–251, TERRAPUB/D. Reidel Pub. Co., Tokyo/Dordrecht, 1984.

Reed, R. J., The quasi-biennial oscillation of the atmosphere between 30 and 50 km over Ascension Island, *J. Atmos. Sci.*, **22**, 331–333, 1965.

Reed, R. J., W. J. Campbell, L. A. Rasmussen, and D. G. Rogers, Evidence of downward-propagating annual wind reversal in the equatorial stratosphere, *J. Geophys. Res.*, **66**, 813–818, 1961.

Schoeberl, M. R. and D. F. Strobel, The response of the zonally averaged circulation to stratospheric ozone reductions, *J. Atmos. Sci.*, **35**, 1751–1757, 1978.

Veryard, R. G. and R. A. Ebdon, Fluctuations in tropical stratospheric winds, *Meteorol. Mag.*, **90**, 125–143, 1961.

Vincent, D. G., Mean meridional circulation in the Northern Hemisphere lower stratosphere during 1964 and 1965, *Quart. J. Roy. Meteor. Soc.*, **94**, 333–349, 1968.

Wallace, J. M. and J. R. Holton, A diagnostic numerical model of the quasi-biennial oscillation, *J. Atmos. Sci.*, **25**, 280–292, 1968.

Wallace, J. M. and V. E. Kousky, Observational evidence of Kelvin waves in the tropical stratosphere, *J. Atmos. Sci.*, **25**, 900–907, 1968.

Woodman, R. F. and A. Guillen, Radar observations of winds and turbulence in the stratosphere and mesosphere, *J. Atmos. Sci.*, **31**, 493–505, 1974.

Yanai, M. and T. Maruyama, Stratospheric wave disturbances propagating over the equatorial Pacific, *J. Meteorol. Soc. Jpn.*, **44**, 291–294, 1966.

Climate and Weather of the Sun-Earth System (CAWSES): Selected Papers from the 2007 Kyoto Symposium,
Edited by T. Tsuda, R. Fujii, K. Shibata, and M. A. Geller, pp. 63–75.

Hinode "a new solar observatory in space"

Saku Tsuneta[1], Louise K. Harra[2], and Satoshi Masuda[3]

[1]*National Astronomical Observatory, Japan*
E-mail: saku.tsuneta@nao.ac.jp
[2]*Mullard Space Science Laboratory, University College London, UK*
[3]*Solar-Terrestrial Environment Laboratory, Nagoya University, Japan*

Since its launch in September 2006, the Japan-US-UK solar physics satellite, Hinode, has continued its observation of the sun, sending back solar images of unprecedented clarity every day. Hinode is equipped with three telescopes, a visible light telescope, an X-ray telescope, and an extreme ultraviolet imaging spectrometer. The Hinode optical telescope has a large primary mirror measuring 50 centimeters in diameter and is the world's largest space telescope for observing the sun and its vector magnetic fields. The impact of Hinode as an optical telescope on solar physics is comparable to that of the Hubble Space Telescope on optical astronomy. While the optical telescope observes the sun's surface, the Hinode X-ray telescope captures images of the corona and the high-temperature flares that range between several million and several tens of millions of degrees. The telescope has captured coronal structures that are clearer than ever. The Hinode EUV imaging spectrometer possesses approximately ten times the sensitivity and four times the resolution of a similar instrument on the SOHO satellite. The source of energy for the sun is in the nuclear fusion reaction that takes place at its core. Here temperature drops closer to the surface, where the temperature measures about 6,000 degrees. Mysteriously, the temperature starts rising again above the surface, and the temperature of the corona is exceptionally high, several millions of degrees. It is as if water were boiling fiercely in a kettle placed on a stove with no fire, inconceivable as it may sound. The phenomenon is referred to as the coronal heating problem, and it is one of the major astronomical mysteries. The Hinode observatory was designed to solve this mystery. It is expected that Hinode would also provide clues to unraveling why strong magnetic fields are formed and how solar flares are triggered. An overview on the initial results from Hinode is presented. Dynamic video pictures captured by Hinode can be viewed on the website of the National Astronomical Observatory of Japan (NAOJ) at http://hinode.nao.ac.jp/index_e.shtml.

1 Hinode Spacecraft

The Hinode spacecraft (Kosugi *et al.*, 2007; see Fig. 1), previously known as Solar-B, was successfully launched in September 2006 from the Uchinoura Space Center in Japan using an M-V launch vehicle. On 25 October 2006, it started its scientific operation. Its orbit is a sun-synchronous orbit that provides 9-month continu-

Fig. 1. Image of the Hinode spacecraft in orbit. (Courtesy of NAOJ, JAXA, and NASA.)

ous observations. It comprises an observatory style set of instruments that function together with the aim of answering the following mission science goals:

- To understand the process of magnetic field generation and transport.
- To investigate the processes responsible for energy transfer from the photosphere to the corona.
- To determine the mechanisms responsible for eruptive phenomena such as flares and coronal mass ejections.

Hinode is designed to answer the fundamental question of how magnetic fields are formed and how they interact with the atmosphere to create the activity that we observe on the sun. This subsequently addresses all phenomena that have an impact on the Sun-Earth system, such as the formation of the solar winds (both slow and fast), triggering of flares and coronal mass ejections, formation and maintenance of filaments and prominences. This review summarizes how the new results from Hinode are addressing these critical questions as well as probing fundamental physical processes that will have applications in many other scenarios across the universe.

There are three instruments onboard Hinode, the Solar Optical Telescope (SOT), the X-ray Telescope (XRT), and the EUV Imaging Spectrometer (EIS), each measuring critical parts of the sun's atmosphere from the surface (photosphere) to the chromosphere, the transition region and finally the outer and hottest part of the atmosphere, the corona.

1.1 The Solar Optical Telescope (SOT)

SOT is the first large optical telescope flown in space to observe the sun (Tsuneta *et al.*, 2008a). It has an aperture of 50 cm and achieves an angular resolution of $0.25''$ (175 km on the sun) covering a wavelength range from 480–650 nm. It consists of

2 components, the Optical Telescope Assembly (OTA: Suematsu *et al.*, 2008) and the Focal Plane Package (FPP). The FPP includes the narrow-band (NFI) and the wide-band (BFI) filtergraphs and the Stokes spectropolarimeter (SP). This complex instrument allows measurements of the magnetic field both in the longitudinal and transverse directions, Doppler shifts, and imaging in the range from the low photosphere through to the chromosphere very accurately under precise calibration (Ichimoto *et al.*, 2008).

1.2 The X-ray Telescope (XRT)

XRT is an advanced solar X-ray telescope (Golub *et al.*, 2007; Kano *et al.*, 2008) with the highest angular resolution of 2″. It is a Wolter Type III grazing incidence telescope that uses 2 reflections to focus soft X-rays onto a CCD array. It can provide both full sun and partial disk images. Filters ranging in thickness by a factor of 10,000 provide a huge dynamic range able to measure very weak features in coronal holes and very large flares. The temperature range extends from 1 million K to 30 million K. This temperature range is much wider than that of the soft X-ray Telescope on board Yohkoh.

1.3 The EUV Imaging Spectrometer (EIS)

EIS is an imaging spectrometer (Culhane *et al.*, 2007) built by a consortium consisting of UCL-MSSL, RAL, NRL, GSFC, UiO, and NAOJ. EIS was designed to probe the dynamics and composition of the solar atmosphere with a spatial resolution of 1″. For observations, it uses two EUV wave-bands which were chosen to measure plasma with temperatures ranging from 50,000 K to 20 million K. EIS is a flexible instrument with the ability to measure high-resolution spectra along with imaging achieving velocity resolutions of several km/s, and it can also observe fast cadence (seconds) monochromatic images. The two wave-bands cover a total of 90 Å containing at least 500 spectral lines of which 55% are identified to come from previously known atomic transitions (Brown *et al.*, 2007).

In the following sections, highlights of the new results from Hinode during its initial operational phase are introduced.

2 New Views of the Solar Atmosphere (Photosphere/Chromosphere/Corona)

Figure 2 shows images of the sunspot and the sun's surface captured by the optical telescope. The sizes of the sunspot range from between several thousand kilometers to several tens of thousands of kilometers. By enlarging the sun's image further, structures called granules can be seen, as shown in Fig. 2, in which bubbling gases flow in swirls of convection over an area measuring about 1,000 km. Small white spots seen between the granules correspond to strong magnetic fields. It is still a mystery how these structures were formed, but thanks to Hinode, it has become possible to observe these changing features for the first time. Figure 3 shows a side view of the sun with a sunspot positioned along the edge. Bright emissions appear frequently around the sunspot, and material jets up like a fountain to an altitude of 20,000 km.

Hinode shows magnetic field structures in a new light. A spectacular example is solar prominences. These are large-scale, cool structures that lie surrounded by the hot corona. It has been suggested that these are caused by plasma trapped in hor-

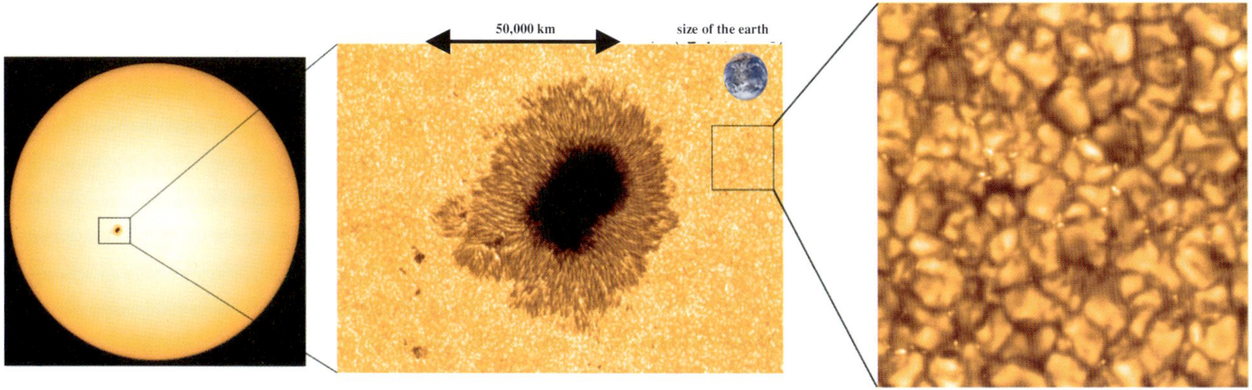

Fig. 2. Sunspot on the sun's surface captured by the optical telescope onboard Hinode. The observations were made at a wavelength of 460 nm. (The colors in the image have been artificially enhanced and differ from the actual colors.) Convection called granules can be seen in the photosphere (right). Small spots between the convection cells are magnetic elements (the minimum unit of a magnetic field) that have an exceptionally strong magnetic field. (Courtesy of NAOJ, JAXA, and NASA.)

Fig. 3. Dynamic solar activities around the sunspot as seen through the optical telescope; substances of temperatures of several tens of thousands of degrees are being sprayed into the solar atmosphere. (Courtesy of NAOJ, JAXA, and NASA.)

izontal magnetic field-line configurations. Observations show horizontal threads of plasma which have oscillatory behavior with periods around 170s which is consistent with Alfvén wave propagation which may possibly heat the surrounding corona (Okamoto *et al.*, 2007). What comes as a surprise is that these threads show dark upward flows that are turbulent and move at speeds of 20 km/s (Berger *et al.*, 2008). The existence of these flows is a real challenge to the current MHD understanding of prominences as they are inconsistent with the idea of a low-Beta (where Beta is the ratio of gas to magnetic pressure) plasma lying in horizontal magnetic fields.

While the optical telescope observes the photosphere/chromosphere, the X-ray telescope captures images of the corona and the high-temperature flares that range from between several million and several tens of millions of degrees. The telescope has achieved the greatest spatial resolution in history, and it has captured coronal structures that are clearer than ever. The sun viewed through the X-ray (Fig. 4) is completely different from that seen in the visible light range. The area around the sunspot, where the strong magnetic fields exist, is referred to as an active region, from which X-rays further radiate. Many stripes can be seen around the active region, which represent the magnetic lines of force that spread upward from the photosphere.

The source of energy for the sun is in the nuclear fusion reaction that takes place at its core. The temperature drops closer to the surface where the temperature measures about 6,000 degrees. Mysteriously, the temperature starts rising again above the surface and the temperature of the corona is exceptionally high at several millions of degrees. This is the so-called coronal heating problem, and it is one of the major astronomical mysteries involving something of a lower temperature being able to heat something of a higher temperature.

There is now the possibility that the results of the observations of Hinode may be

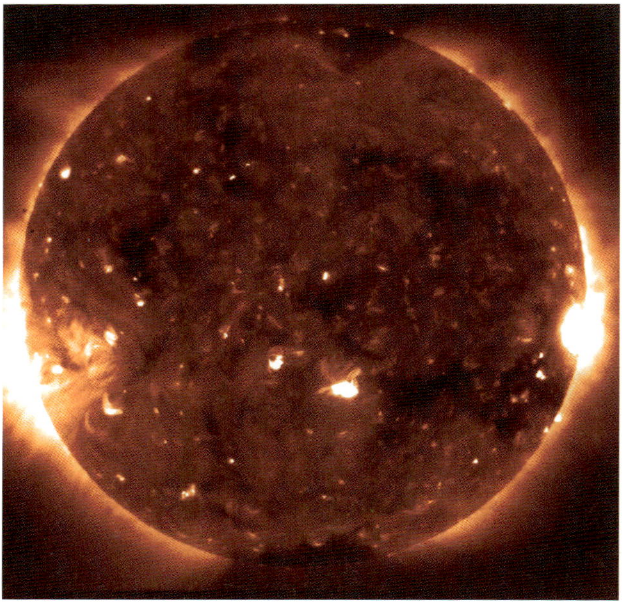

Fig. 4. X-ray images of the entire solar surface photographed by the X-ray telescope onboard Hinode. A punctuate structure called an X-ray bright point was identified for the first time, which revealed that the corona is made up of magnetic loops. (Courtesy of NAOJ, JAXA, and NASA.)

used to solve the problem. Hinode has revealed there are strong magnetic fields and strong corona activity in areas without the sunspot. Sources of X-rays are scattered here and there on the photosphere, and it still remains a mystery as to how they were formed. Scientists expect to understand the solar corona heating mechanism by comparing such information with information on the magnetic fields obtained by the optical telescope. The current hypothesis is that there are areas on the photosphere where there is a strong magnetic field, and the corona is being vigorously heated by the magnetic energy. An explanation of the specific mechanism is now anticipated. In the near future, there may also be clues to unraveling why such strong magnetic fields are formed and how solar flares are generated.

EIS has the capability of obtaining physical information of the coronal plasma, such as temperature, density, and velocity. Broad EUV emission lines might correspond to some turbulent or non-thermal processes critical to understand energy transfer, for example, coronal heating. The fact that the most turbulent regions which we have been studying for decades can actually be seen outside the bright loops has huge implications on how the atmosphere of the sun is heated and sustained. The processes that make loops bright may not actually be the key to understanding the atmosphere as a whole. This is also seen in the quiet corona where weak emission regions have measured outflows of 100 km/s (Dere *et al.*, 2007). Another discovery is that the density of very weak emission regions can be high, which suggests that the observed active region is surrounded by weakly emitting high-density loops. The weak regions often show large line widths in the EUV spectral lines (Doschek *et al.*, 2007).

3 Alfvén Waves and Solar Wind Source

The activity on the sun is known to be driven by the magnetic fields that are prevalent everywhere. Hinode has higher temporal, spatial and velocity resolution than any satellite previously and is probing wavelength regimes that have never had such continuous time coverage available. This has allowed us to measure waves in the atmosphere in a way we have been unable to do before. In 1947 Alfvén predicted the existence of magnetic waves caused by the constant movement due to convection on the surface of the sun. The convection disturbs the magnetic fields causing waves and may then be damped in the corona providing an energy source that may create enough heating for the atmosphere and energy to accelerate the solar wind. Attempts to measure Alfvén waves have been ambiguous in previous missions, but Hinode now appears to be opening the door to these waves being observed in many different circumstances.

The chromosphere is highly structured as can be seen in Fig. 2. At the solar limb, spiky features known as spicules are seen in abundance. Now with Hinode, observations can be made continuously without being concerned about seeing conditions as is the case with ground-based telescopes. De Pontieu *et al.* (2007a) have analyzed these spicules and found that there are two types. The first type is formed when global oscillations and convective flows leak into the atmosphere causing shocks. These have timescales of minutes and show persistent upward and downward motions. The type II spicules (named straws) have lifetimes of seconds, are much thinner and move plasma at speeds of over 100 km/s. These seem to be related directly to the magnetic reconnection process. However both types of spicules show a swaying behavior strongly indicative of Alfvén waves. They show amplitudes and periods that are consistent with simulations of the generation, propagation and dissipation of Alfvén waves from the photosphere to Earth orbit (De Pontieu *et al.*, 2007b), and they have enough energy to power the solar wind.

Jets of collimated, hot plasma have been observed for many years on the sun. These can occur in the active region loops that lie above sunspot groups or within the coronal holes. One explanation of the formation of jets involves a scenario in which emerging flux from below the surface interacts with coronal holes. Hinode now has the ability to observe jets in significantly more detail to study basic physical processes (Shimojo *et al.*, 2007). Jets in coronal holes have been observed to have speeds of 200 km/s. However, a new component that occurs at the beginning of the jet formation has been discovered, high speed, reaching 800 km/s, close to the Alfvén speed in the corona (Cirtain *et al.*, 2007). The characteristics of these jets are consistent with plasma being ejected at the Alfvén speed during a relaxation period following magnetic reconnection. These small jets carry as much as a tenth of the mass necessary for the solar wind.

The sun supplies a huge amount of plasma into interplanetary space as solar wind.

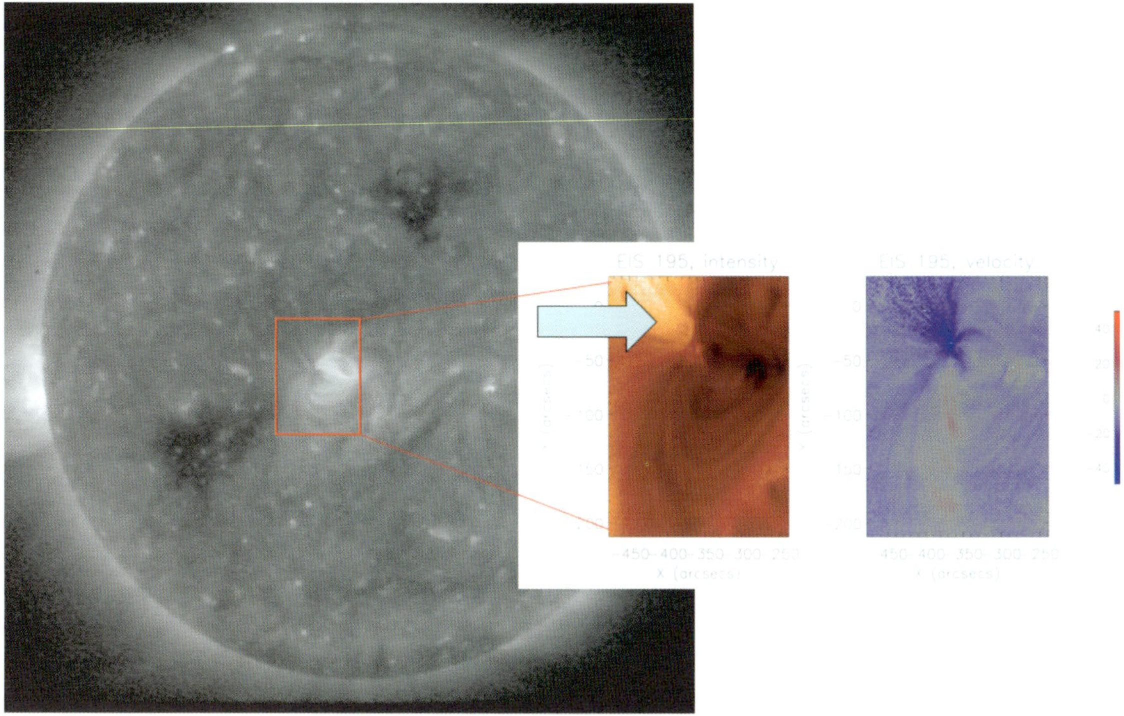

Fig. 5. The right hand side shows a full sun XRT image with the box showing the EIS FOV. The left EIS image is the intensity and the right image is the Doppler velocity. The blue arrow shows the source of the strongest outflow (i.e., from Sakao *et al.*, Continuous Plasma Outflows from the Edge of a Solar Active Region as a Possible Source of Solar Wind, *Science*, **318**, 1585–1588, 2007, reprinted with permission from AAAS; Harra *et al.*, 2008, reproduced by permission of the AAS).

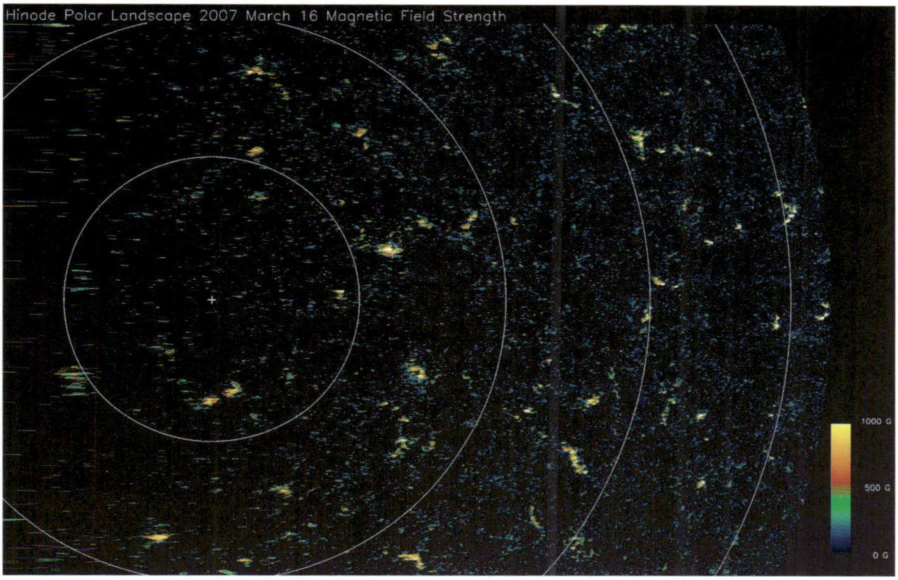

Fig. 6. South polar view of the magnetic field strength taken at 12:02:19–14:55:48 UT on 16 March 2007. The original observing field of view is 327.52″ (east-west) by 163.84″ (north-south) and was converted to a map seen from above the south pole. East is to the left, west is to the right, and the observation was carried out from the top down. Spatial resolution is lost near the extreme limb (i.e., near the bottom of the figure). The field of view for the line-of-sight direction (163.84″) expands to 472.96″ as a result of correction for foreshortening. The pixel size is 0.16″. Latitudinal lines for 85°, 80°, 75°, and 70° are shown as large circles, while the plus sign marks the south pole. The magnetic field strength is obtained for pixels meeting a given threshold. (From Tsuneta *et al.* (2008b). Reproduced by permission of the AAS.)

It is well known that the solar wind consists of two components, a fast solar wind that comes from predominantly coronal holes that have open magnetic fields, and a slow solar wind that comes from other open magnetic fields, such as the boundary of large-scale coronal holes and small-scale coronal holes even though its source has been debated for many years. Alfvén waves as mentioned above are also critical for solar wind formation. However, in this paper we focus on the source of solar wind.

The slow solar wind has fluctuating speeds between 200–500 km/s. XRT has observed persistent and steady flows directly at the edges of active regions (Sakao *et al.*, 2007) with speeds of over 100 km/s that persist for days. The mass loss rate was estimated to be 25% of the slow solar wind. Using EIS alongside this dataset, the real Doppler flows can be then observed. Figure 5 shows the XRT image for context alongside the EIS data. The EIS data show the intensity of the corona on the left-hand side. Now for the first time, the Doppler velocities can also be observed (on the right hand side of Fig. 5). The region of weakest emission at the top of the active region shows the strongest velocities (Harra *et al.*, 2008). Probing the magnetic field shows that this region is highly extended, most likely open fields caused by reconnection with a much smaller active region to the west of the one studied. These large-scale reconnections are of significant importance when aiming to understand the slow solar wind.

We have also new observations related to a fast solar wind. Tsuneta *et al.* (2008b) reported SOT observations of the magnetic landscape of the polar region of the sun with an extremely high spatial resolution and high polarimetric precision. Using a Milne-Eddington inversion, it was found that many vertically oriented magnetic flux tubes with field strengths as strong as 1 kG were scattered in latitude between 70° and 90° (Fig. 6). They all have the same polarity, consistent with the global polarity of the polar region. The field vectors have been observed to diverge from the centers of the flux elements, consistent with a view of magnetic fields that are expanding and fanning out with height. The polar region has also been found to have ubiquitous horizontal fields. The polar regions are the source of the fast solar wind, which is channeled along unipolar coronal magnetic fields whose photospheric source is evidently rooted in the strong-field, vertical patches of flux. We conjecture that vertical flux tubes with large expansion around the photospheric-coronal boundary serve as efficient chimneys for Alfvén waves that accelerate the solar wind.

4 Solar Flare: 13 November 2006 Event

Also observed on the surface of the sun are explosive phenomena called solar flares that last for several minutes to several hours and the eruption of substances referred to as high-speed jets caused by the explosions. A magnetic field controls the activities in the corona, and large explosions take place frequently near the sunspot where magnetic field lines are in close proximity.

Although the sun is now in at the minimum of its activity level, we are still occasionally treated to an active region which produces solar flares. Such a region appeared in December 2006 just after the Hinode scientific operation started. Days before the large X-flare on 13 December 2006, the active region had been monitored. There was a large negative polarity sunspot, and to the south of it emerged a smaller positive polarity sunspot. Over the days of this evolution, the smaller sunspot rotated dramatically, colliding into the pre-existing sunspot (Kubo *et al.*, 2007). The impact on the corona above of this shearing motion could be seen (Su *et al.*, 2007). Just as the smaller sunspot was emerging, the coronal loops were lying perpendicular to the inversion line (marking out the separation between positive and negative polarity). At this stage the loops were essentially potential with little stored energy. By 12 December 2006, the story was completely different. At this stage the new sunspot had been emerging and rotating at the surface, dragging the field lines around, and causing the magnetic loops to become so sheared that they were now lying parallel to the inversion line. A few hours after this, a huge X-class flare took place there. Figure 7 shows an image of a gigantic flare captured by Hinode, which stretches to several tens of thousands of kilometers. Even more dynamic video pictures captured by Hinode can be viewed on the website of the National Astronomical Observatory of Japan (NAOJ) at http://hinode.nao.ac.jp/index_e.shtml.

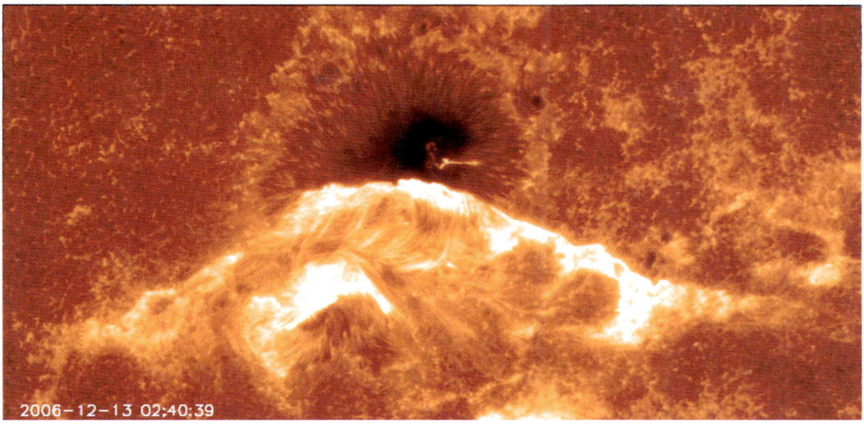

Fig. 7. Large solar flare captured by SOT on 13 December 2006. Two-Ribbon structure and some of post-flare loops between the two ribbons are clearly seen. (Courtesy of NAOJ, JAXA, and NASA.)

During this X-class flare, we measured directly for the first time what part of the atmosphere erupted away from the sun. Imada *et al.* (2007) analyzed the flare on 13 December 2006 and found that the strongest outflows were away from the main flare site. The force of the flare had ripped off some of the plasma from the sun. Interestingly, there seems to be a strong relationship between this outflow and the temperature of the plasma. The strongest outflows are from the hottest plasma which provides a constraint to the mechanism that forms this fast component to the solar wind. Speeds of 1000 km/s were measured at ACE from this event.

The sun is currently in a period of minimum activity. Solar flares will start to occur more frequently as the sun heads toward its period of maximum activity around 2012, when it will then undergo a transformation that will make it appear as a different star. Hinode is sure to change fundamentally the academic perspectives on solar observation through the analysis of clear and high-resolution images and spectroscopic data during the period of maximum activity that received for the first time.

Acknowledgments. The author expresses sincere thanks to the ISAS/JAXA Solar-B launch team for their exceptional achievement. Hinode is a Japanese mission developed and launched by ISAS/JAXA, collaborating with NAOJ as a domestic partner and NASA and STFC (UK) as international partners.

References

Berger, T. E., R. A. Shine, G. L. Slater, T. D. Tarbell, A. M. Title, T. J. Okamoto, K. Ichimoto, Y. Katsukawa, Y. Suematsu, S. Tsuneta, B. W. Lites, and T. Shimizu, Hinode SOT Observations of Solar Quiescent Prominence Dynamics, *Astrophys. J.*, **676**, L89–L92, 2008.

Brown, C. M., H. Hara, S. Kamio, U. Feldman, J. F. Seely, G. A. Doschek, J. T. Mariska, C. M. Korendyke, J. Lang, K. P. Dere, L. Culhane, R. J. Thomas, and J. M. Davila, Wavelength Determination for Solar Features Observed by the EUV Imaging Spectrometer on Hinode, *PASJ*, **59**, S865–S869, 2007.

Cirtain, J. W., L. Golub, L. Lundquist, A. van Ballegooijen, A. Savcheva, M. Shimojo, E. DeLuca, S. Tsuneta, T. Sakao, K. Reeves, M. Weber, R. Kano, N. Narukage, and K. Shibasaki, Evidence for Alfvén Waves in Solar X-ray Jets, *Science*, **318**, 1580–1582, 2007.

Culhane, J. L., L. K. Harra, A. M. James, K. Al-Janabi, L. J. Bradley, R. A. Chaudry, K. Rees, J. A. Tandy, P. Thomas, M. C. R. Whillock, B. Winter, G. A. Doschek, C. M. Korendyke, C. M. Brown, S. Myers, J. Mariska, J. Seely, J. Lang, B. J. Kent, B. M. Shaughnessy, P. R. Young, G. M. Simnett, C. M. Castelli, S. Mahmoud, H. Mapson-Menard, B. J. Probyn, R. J. Thomas, J. Davila, K. Dere, D. Windt, J. Shea, R. Hagood, R. Moye, H. Hara, T. Watanabe, K. Matsuzaki, T. Kosugi, V. Hansteen, and Ø. Wikstol, The EUV Imaging Spectrometer for Hinode, *Sol. Phys.*, **243**, 19–61, 2007.

De Pontieu, B., S. McIntosh, V. H. Hansteen, M. Carlsson, C. J. Schrijver, T. D. Tarbell, A. M. Title, R. A. Shine, Y. Suematsu, S. Tsuneta, Y. Katsukawa, K. Ichimoto, T. Shimizu, and S. Nagata, A Tale of Two Spicules: The Impact of Spicules on the Magnetic Chromosphere, *PASJ*, **59**, S655–S662, 2007a.

De Pontieu, B., S. W. McIntosh, M. Carlsson, V. H. Hansteen, T. D. Tarbell, C. J. Schrijver, A. M. Title, R. A. Shine, S. Tsuneta, Y. Katsukawa, K. Ichimoto, Y. Suematsu, T. Shimizu, and S. Nagata, Chromospheric Alfvénic Waves Strong Enough to Power the Solar Wind, *Science*, **318**, 1574–1577, 2007b.

Dere, K. P., G. A. Doschek, J. T. Mariska, V. H. Hansteen, L. K. Harra, K. Matsuzaki, and R. J. Thomas, The Structure and Dynamics of the Quiet Corona from Observations with the Extreme Ultraviolet Imaging Spectrometer on Hinode, *PASJ*, **59**, S721–S726, 2007.

Doschek, G. A., J. T. Mariska, H. P. Warren, C. M. Brown, J. L. Culhane, H. Hara, T. Watanabe, P. R. Young, and H. E. Mason, Nonthermal Velocities in Solar Active Regions Observed with the Extreme-Ultraviolet Imaging Spectrometer on Hinode, *Astrophys. J.*, **667**, L109–L112, 2007.

Golub, L., E. DeLuca, G. Austin, J. Bookbinder, D. Caldwell, P. Cheimets, J. Cirtain, M. Cosmo, P. Reid, A. Sette, M. Weber, T. Sakao, R. Kano, K. Shibasaki, H. Hara, S. Tsuneta, K. Kumagai, T. Tamura, M. Shimojo, J. McCracken, J. Carpenter, H. Haight, R. Siler, E. Wright, J. Tucker, H. Rutledge, M. Barbera, G. Peres, and S. Varisco, The X-Ray Telescope (XRT) for the Hinode Mission, *Sol. Phys.*, **243**, 63–86, 2007.

Harra, L. K., T. Sakao, C. H. Mandrini, H. Hara, S. Imada, P. R. Young, L. van Driel-Gesztelyi, and D. Baker, Outflows at the Edges of Active Regions: Contribution to Solar Wind Formation?, *Astrophys. J.*, **676**, L147–L150, 2008.

Ichimoto, K., B. Lites, D. Elmore, Y. Suematsu, S. Tsuneta, Y. Katsukawa, T. Shimizu, R. Shine, T. Tarbell, A. Title, J. Kiyohara, K. Shinoda, G. Card, A. Lecinski, K. Streander, M. Nakagiri, M. Miyashita, M. Noguchi, C. Hoffmann, and T. Cruz, Polarization Calibration of the Solar Optical Telescope onboard Hinode, *Sol. Phys.*, **249**, 233–261, 2008.

Imada, S., H. Hara, T. Watanabe, S. Kamio, A. Asai, K. Matsuzaki, L. K. Harra, and J. T. Mariska, Discovery of a Temperature-Dependent Upflow in the Plage Region During a Gradual Phase of the X-Class Flare, *PASJ*, **59**, S793–S799, 2007.

Kano, R., T. Sakao, H. Hara, S. Tsuneta, K. Matsuzaki, K. Kumagai, M. Shimojo, K. Minesugi, K. Shibasaki, E. E. DeLuca, L. Golub, J. Bookbinder, D. Caldwell, P. Cheimets, J. Cirtain, E. Dennis, T. Kent, and M. Weber, The Hinode X-Ray Telescope (XRT): Camera Design, Performance and Operations, *Sol. Phys.*, **249**, 263–279, 2008.

Kosugi, T., K. Matsuzaki, T. Sakao, T. Shimizu, Y. Sone, S. Tachikawa, T. Hashimoto, K. Minesugi, A. Ohnishi, T. Yamada, S. Tsuneta, H. Hara, K. Ichimoto, Y. Suematsu, M. Shimojo, T. Watanabe, S. Shimada, J. M. Davis, L. D. Hill, J. K. Owens, A. M. Title, J. L. Culhane, L. K. Harra, G. A. Doschek, and L. Golub, The Hinode (Solar-B) Mission: An Overview, *Sol. Phys.*, **243**, 3–17, 2007.

Kubo, M., T. Yokoyama, Y. Katsukawa, B. Lites, S. Tsuneta, Y. Suematsu, K. Ichimoto, T. Shimizu, S. Nagata, T. D. Tarbell, R. A. Shine, A. M. Title, and D. Elmore, Hinode Observations of a Vector Magnetic Field Change Associated with a Flare on 2006 December 13, *PASJ*, **59**, S779–S784, 2007.

Okamoto, T. J., S. Tsuneta, T. E. Berger, K. Ichimoto, Y. Katsukawa, B. W. Lites, S. Nagata, K. Shibata, T. Shimizu, R. A. Shine, Y. Suematsu, T. D. Tarbell, and A. M. Title, Coronal Transverse Magnetohydrodynamic Waves in a Solar Prominence, *Science*, **318**, 1577–1580, 2007.

Sakao, T., R. Kano, N. Narukage, J. Kotoku, T. Bando, E. E. DeLuca, L. L. Lundquist, S. Tsuneta, L. K. Harra, Y. Katsukawa, M. Kubo, H. Hara, K. Matsuzaki, M. Shimojo, J. A. Bookbinder, L. Golub, K. E.

Korreck, Y. Su, K. Shibasaki, T. Shimizu, and I. Nakatani, Continuous Plasma Outflows from the Edge of a Solar Active Region as a Possible Source of Solar Wind, *Science*, **318**, 1585–1588, 2007.

Shimojo, M., N. Narukage, R. Kano, T. Sakao, S. Tsuneta, K. Shibasaki, J. W. Cirtain, L. L. Lundquist, K. Reeves, and A. Savcheva, Fine Structures of Solar X-Ray Jets Observed with the X-Ray Telescope aboard Hinode, *PASJ*, **59**, S625–S630, 2007.

Su, Y., L. Golub, A. Van Ballegooijen, E. E. DeLuca, K. K. Reeves, T. Sakao, R. Kano, N. Narukage, and K. Shibasaki, Evolution of the Sheared Magnetic Fields of Two X-Class Flares Observed by Hinode/XRT, *PASJ*, **59**, S785–S791, 2007.

Suematsu, Y., S. Tsuneta, K. Ichimoto, T. Shimizu, M. Otsubo, Y. Katsukawa, M. Nakagiri, M. Noguchi, T. Tamura, Y. Kato, H. Hara, M. Kubo, I. Mikami, H. Saito, T. Matsushita, N. Kawaguchi, T. Nakaoji, K. Nagae, S. Shimada, N. Takeyama, and T. Yamamuro, The Solar Optical Telescope of Solar-B (Hinode): The Optical Telescope Assembly, *Sol. Phys.*, **249**, 197–220, 2008.

Tsuneta, S., K. Ichimoto, Y. Katsukawa, S. Nagata, M. Otsubo, T. Shimizu, Y. Suematsu, M. Nakagiri, M. Noguchi, T. Tarbell, A. Title, R. Shine, W. Rosenberg, C. Hoffman, B. Jurcevich, G. Kushner, M. Levay, B. Lites, D. Elmore, T. Matsushita, N. Kawaguchi, H. Saito, I. Mikami, L. D. Hill, and J. K. Owens, The Solar Optical Telescope for the Hinode Mission: An Overview, *Sol. Phys.*, **249**, 167–196, 2008a.

Tsuneta, S., K. Ichimoto, Y. Katsukawa, B. W. Lites, K. Matsuzaki, S. Nagata, D. Orozco Suárez, T. Shimizu, M. Shimojo, R. A. Shine, Y. Suematsu, T. K. Suzuki, T. D. Tarbell, and A. M. Title, The Magnetic Landscape of the Sun's Polar Region, *Astrophys. J.*, **688**, 1374–1381, 2008b.

Climate and Weather of the Sun-Earth System (CAWSES): Selected Papers from the 2007 Kyoto Symposium,
Edited by T. Tsuda, R. Fujii, K. Shibata, and M. A. Geller, pp. 77–120.

Coronal mass ejections and space weather

Nat Gopalswamy

NASA Goddard Space Flight Center, Greenbelt, Maryland, USA
E-mail: Nat.Gopalswamy@nasa.gov

Solar energetic particles (SEPs) and geomagnetic storms are the two primary space weather consequences of coronal mass ejections (CMEs) and their interplanetary counterparts (ICMEs). I summarize the observed properties of CMEs and ICMEs, paying particular attention to those properties that determine the ability of CMEs in causing space weather. Then I provide observational details of two the central issues: (i) for producing geomagnetic storms, the solar source location and kinematics along with the magnetic field structure and intensity are important, and (ii) for SEPs, the shock-driving ability of CMEs, the Alfven speed in the ambient medium, and the connectivity to Earth are crucial parameters.

1 Introduction

Space weather can be described as the physical conditions in space that affects human technology in space and on the ground as well as life on Earth in a number of ways including satellite drag, satellite sensor degradation, effects of geomagnetically induced currents on the power grid and pipelines, radiation threat to crews of high-flying aircraft and astronauts, high-frequency communication outages in the polar regions (see articles in Song *et al.*, 2001). Space weather is mainly caused by the Sun because of the variability in its mass and photon emissions on various time scales. The electromagnetic emission consists of the quasi-steady irradiance and flares. While the long-term variability in photon emission may be related to climate effects, the short-term variation (solar flares) affects Earth's ionosphere resulting in a number of communication problems. The ionization is increased by the flash of enhanced X-ray and EUV emission during flares. The mass emission from the Sun has three components: the ubiquitous solar wind, coronal mass ejections (CMEs), and solar energetic particles (SEPs). CMEs and the solar wind carry the coronal magnetic field into the heliosphere. CMEs propagate into the solar wind and drive shocks; the shocks accelerate SEPs. Flares also accelerate particles, but generally over shorter duration and to lower intensity levels. When CMEs arrive at Earth, they interact with Earth's magnetosphere causing geomagnetic storms, which have multitude of effects from the magnetosphere to the ground. This paper discusses two main aspects of CMEs related to space weather: SEPs and geomagnetic storms (Gosling, 1993; Reames, 1999). Geomagnetic storms were known for a long time before the discovery of CMEs. Even Carrington (1860) suspected that the flare he discovered on 1859 September 1,

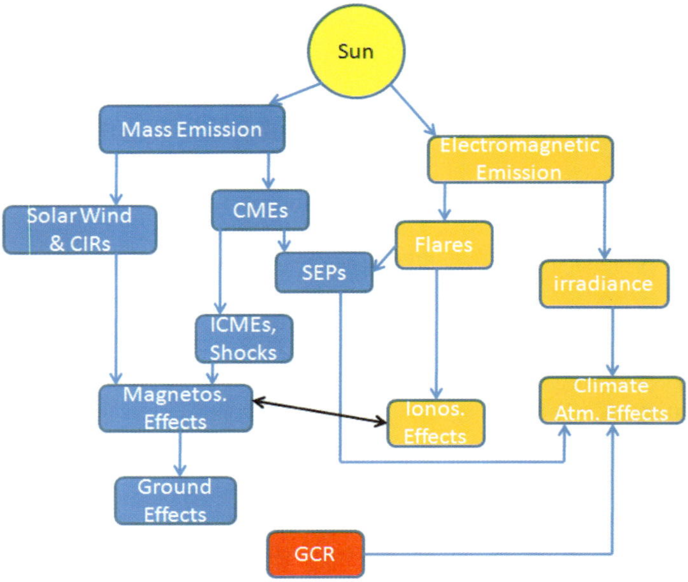

Fig. 1. The two primary outputs (mass and electromagnetic emissions) from the Sun, whose variability affects the heliosphere in general and the geospace in particular. The mass emission manifests as the solar wind, coronal mass ejections (CMEs), and solar energetic particles. CMEs and the solar wind also carry magnetic fields with them. The galactic cosmic rays (GCRs) enter the heliosphere from outside and interact with the solar wind and CMEs. Neutral matter also enters the heliosphere, gets ionized and accelerated as anomalous cosmic rays (not shown).

may have something to do with the huge geomagnetic storm that occurred the next day (but decided against the association). SEPs were detected in 1942 and reported in 1946 (Forbush, 1946). An interesting discussion on the development of the term Space Weather can be found in Kane (2006).

Figure 1 illustrates how the two primary outputs (mass and electromagnetic emissions) from the Sun, whose variability affects the heliosphere in general and the geospace in particular. Both the solar wind and CMEs cause geomagnetic storms, but the CMEs are the primary cause of SEPs. In addition to being the source of SEPs, CMEs also deflect the galactic cosmic rays (GCRs) that enter the heliosphere from outside, thus producing the well known Forbush effect. Neutral matter also enters the heliosphere, gets ionized and accelerated as anomalous cosmic rays (not shown). This article focuses on issues related to CMEs. Interaction between fast and slow solar wind streams form corotating interaction regions (CIRs) that also can cause geomagnetic storms of lower intensity and accelerate particles at several AU from the Sun; these will not be discussed here.

Section 2 summarizes the basic properties of CMEs derived from remote-sensing observations by coronagraphs, in particular the Large Angle and Spectrometric Coronagraph (LASCO) on board the Solar and Heliospheric Observatory (SOHO). The SOHO mission has observed more than 13,000 CMEs during cycle 23, which form the basis of the description provided here. Section 3 describes CMEs in the interplanetary (IP) medium (ICMEs for short), mostly observed in the near-Earth space by spacecraft such as Wind and the Advanced Composition Explorer (ACE), emphasizing on the connection between CMEs and ICMEs. Section 4 presents CMEs as a source of geomagnetic storms and how the ICME properties such as speed and magnetic content influence the intensity of the resulting geomagnetic storms. Section 5 presents the association between CMEs and SEP events, especially those relevant to space weather. Finally, Section 6 summarizes the article.

2 Properties of Coronal Mass Ejections

CMEs are magnetized plasma structures that erupt from closed field regions on the Sun (active regions and quiescent filament regions) and propagate into the heliosphere. Recent reviews on CMEs can be found in Gopalswamy (2004, 2006a, b), Kahler (2006), Kunow *et al.* (2006). In white-light observations, CMEs appear as large-scale features moving away from the Sun. Coronagraphs image the CMEs in Thomson-scattered photospheric light, by blocking the direct sunlight using an occulting disk. Therefore, details of the source region are usually obtained using non-coronagraphic observations such as in X-rays, EUV, microwaves, and H-alpha (Gopalswamy, 1999; Hudson and Cliver, 2001). Photospheric magnetograms show that the CME source regions have generally enhanced magnetic fields (consisting of opposite polarity patches) compared to the quiet Sun regions. Magnetic field lines extending into the corona are inferred to be connecting the two polarities. It is believed that free energy can be stored in the coronal field lines, which are then released in the form of CMEs.

2.1 Morphology

Figure 2 shows two CMEs imaged by SOHO/LASCO. The first CME (2001 December 20) shows the three-part structure of CMEs: the core, void, and frontal structure (Illing and Hundhausen, 1986). The void is supposed to be a region of high magnetic field. In some CMEs the void region appears to have a flux rope structure (Chen *et al.*, 1997). In the difference images one can see perturbations around the CME and the deflection of a nearby streamer. The second event (2003 November 18) CME, on the other hand, shows just a single feature. However, the difference image shows a faint feature surrounding the CME, which is likely to be the sheath of the shock driven by the CME. While shocks are readily inferred from radio-burst observations, it is generally difficult to identify them from white-light observations (Sheeley *et al.*, 2000; Vourlidas *et al.*, 2003). The second CME in Fig. 2 originated from close to the disk center contrary to the first event, which erupted from close to the southeast limb. The viewing angle can significantly affect the appearance of CMEs. The bright core of the CME is the eruptive prominence (also known as eruptive filament when happening on the disk), that lies along and above the photospheric neutral line in the pre-eruption phase.

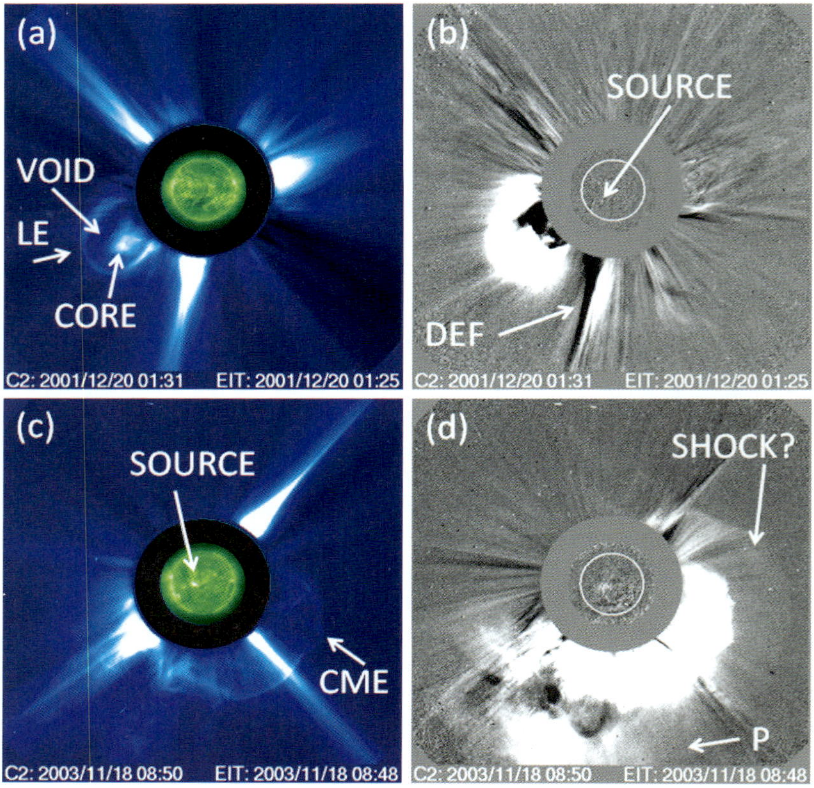

Fig. 2. CME morphology as illustrated using two CMEs, one originating from close to the limb (top panels) and the other from close to the disk center (bottom panels). The left images are direct images with SOHO's Extreme-ultraviolet imaging telescope (EIT) images superposed on the LASCO images. The source locations of the eruptions are seen as compact bright features in the EIT images (SOURCE). The right images are difference images with the previous frames subtracted to see the changes taking place in the corona. The three-part structure is evident in the first CME (2001 December 20—(a, b)), while only a single structure is seen in the second (2003 November 18—(c, d)). The CME leading edge (LE) also known as the frontal structure is the outermost feature followed by a void region (also known as the cavity) and the prominence core. The difference images show perturbations around the CMEs. The dark regions in the difference images correspond to material depletion, which normally indicates the displacement of a structure between the two frames used. In the first CME, one can see streamer deflections (DEF) and in the second case, there seems to be a shock and the associated sheath surrounding the CME. A previous CME (P) is in progress when the second CME was ejected.

2.2 Physical properties

CMEs are multithermal plasmas containing coronal material at a temperature of $\sim$ a few $\times 10^6$ K in the frontal structure and cavity. The prominence core is at a temperature $\sim$8000 K in the early phases. The prominence gets heated as time progresses and may contain plasma of multiple temperatures. Heated plasma ($\sim 10^7$ K) is also thought to be injected into CME structures, observed as material with high charge states at 1 AU (Henke *et al.*, 2001; Reinard, 2005; Aguilar-Rodriguez *et al.*, 2006). The density in the frontal structure of CMEs close to the Sun is estimated to be in the range 10^{8-9} cm^{-3} and is expected to decrease as the CME expands into the interplanetary space. The prominence is denser by three orders of magnitude, which is the reason it appears as the brightest in white light images. The void region is inferred to have a lower density compared to the frontal structure. Direct measurement of the magnetic field strength in CMEs has not been possible yet. However, magnetic field in prominences (before eruption) has been measured, which ranges from a few G to ~100 G. Moving type IV radio bursts produced by energetic electrons gyrating in CME structures or producing plasma emission imply that the magnetic field strength should be in the range 2–10 G in CMEs (Dulk and McLean, 1978; Gopalswamy and Kundu, 1989; Bastian *et al.*, 2001), which overlaps with prominence field strengths. The electrons must be trapped in the void region of the CMEs, which is thought to be a flux rope. The frontal structure is of higher density and lower magnetic field strength compared to the void region.

2.3 Kinematic properties

Speed, angular width, and acceleration are considered to be the basic attributes of a CME, measured from a series of images taken during an eruption. CME properties summarized here were obtained from the SOHO data. The speed measured in the sky plane varies from ~20 km/s to >3000 km/s. The speed distribution of more than 10,000 CMEs observed during solar cycle 23 (see Fig. 3) has an average value of ~470 km/s. The highest speed recorded is ~3387 km/s for a CME on 2004 November 10. The CME width (W) ranges from $<5°$ to $360°$. Widths down to ~2° are measured, but many such narrow CMEs might have been missed especially during the solar maximum. CMEs with $W = 360°$ (apparent) are the halo CMEs (Howard *et al.*, 1982). The true width of halos ($W = 360°$) and partial halos ($120° < W < 360°$) are unknown, so the average width (~54°) in Fig. 3(b) includes only CMEs narrower than 120°. Halo CMEs constitute only ~3% of all CMEs, while CMEs with $W \geq 120°$ account for ~11%. Although discovered in the Solwind era, the importance of halo CMEs for space weather was recognized only in the SOHO era (Webb *et al.*, 2000; Zhao and Webb, 2003; Gopalswamy *et al.*, 2007). Halo CMEs are more energetic than average CMEs (average speed ~1000 km/s) and when they originate on the frontside of the Sun, they can directly impact Earth (see Section 3.3).

The above values of speed and width are consistent with those obtained from all the previous coronagraphs (see e.g., Howard *et al.*, 1985; Hundhausen, 1997; Gopalswamy, 2004; Kahler, 2006). In many cases the speed changes within the coronagraphic field of view because CMEs are subject to propelling and retarding forces. The distribution of measured accelerations in Fig. 3(c) shows that on the average, the acceleration is almost zero. However, CME acceleration within the coronagraphic field of view is speed-dependent (Gopalswamy *et al.*, 2001a; Vršnak *et al.*, 2004; Yashiro *et al.*, 2004; Gopalswamy, 2006a). CMEs faster (slower) than the solar wind accelerate (decelerate), while those moving at the solar wind speed show little acceleration. The sky-plane measurements shown in Fig. 3(a–c) are subject to projec-

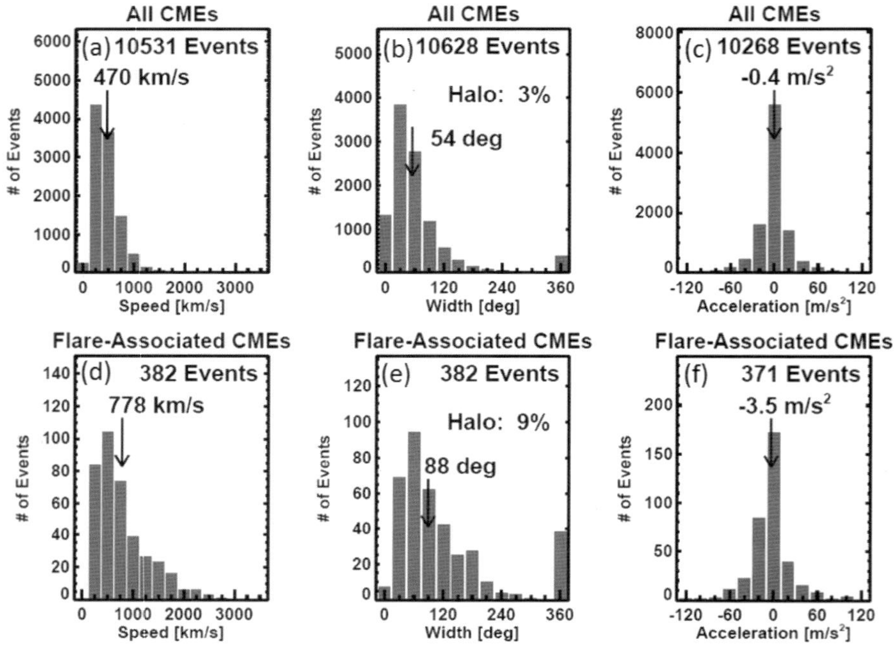

Fig. 3. Distributions of speed, width, and acceleration of all CMEs during 1996–2005 (top) and a subset of CMEs that originated within 30° from the limb (bottom). The CMEs in the bottom panel were chosen based on their association with flares above C3.0 class.

tion effects, which can be avoided when CMEs occurring at large central meridian distance (CMD) ($60^\circ \leq \mathrm{CMD} < 90^\circ$) are considered. The speed, width, and acceleration distributions of the subset of CMEs are shown in Fig. 3(d–f). The average speed and width of the limb CMEs are higher and the acceleration is more negative. For definitiveness, we selected only CMEs associated with flares of X-ray importance C3.0 and higher, which might have contributed to the higher average speed of CMEs, in addition to the fact that the limb CMEs are not subject to projection effects.

2.4 Mass and kinetic energy

The mass content of CMEs ranges from $\sim 10^{12}$ g to $> 10^{16}$ g, wider CMEs being more massive. The mass is estimated from coronagraphic images as the excess mass above the pre-event corona. The mass below the occulting disk is not included. Each CME is assigned a single mass value, which is the mass attained in the outer corona after an initial increase with time, similar to the CME width (Vourlidas *et al.*, 2002). The average value of CME mass is $\sim 4\times 10^{14}$ g. The CME kinetic energy ranges from $\sim 10^{27}$ erg to $> 10^{33}$ erg (average $\sim 5.4\times 10^{29}$ erg), making CMEs the most energetic phenomenon in the heliosphere (Gopalswamy, 2004).

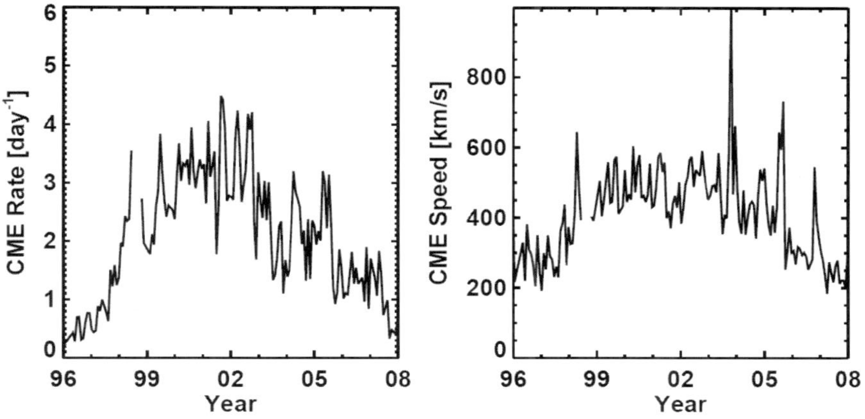

Fig. 4. CME rate (per day) and speed averaged over Carrington rotation periods. CMEs with width $\geq 30°$ are considered. The large spikes in the speed plot are due to CMEs from super active regions that produced energetic CMEs in quick succession. Note that the CME rate varies over a factor of 10 from solar minimum to maximum, while the mean speed increases by a factor $\sim$2. (Updated from Gopalswamy *et al.* (2006).)

2.5 CME rate

CMEs are rather frequent: the daily CME rate averaged over Carrington rotation periods ranges from $<$0.5 per day (solar minimum) to $>$6 CMEs per day (solar maximum). It is not uncommon that more than a dozen CMEs occur on individual days (Gopalswamy *et al.*, 2003a). The CME daily rate also shows solar-cycle variation with notable deviations from that of the sunspot number. CMEs originate from closed magnetic field regions, which may or may not have sunspots. For example, quiescent filament regions have no sunspots, but can produce CMEs. In fact, the high-latitude ($>60°$) CMEs associated with polar-crown filaments can contribute significantly to the total CME rate during solar maxima (Hundhausen, 1993; Gopalswamy *et al.*, 2003b). Figure 4 shows the daily CME rate and speed averaged over Carrington rotation periods including only those CMEs with width $\geq 30°$. This restriction eliminates the non-uniformity in identifying weak and narrow CMEs (Yashiro *et al.*, 2008a). The rate jumps by an order of magnitude from the minimum to the maximum phase of the solar cycle. CMEs also seem to be more energetic during solar maxima: the average speed of CMEs increases by a factor of $>$2 from $\sim$250 km/s during solar minimum to $\sim$550 km/s during solar maximum. The large spikes in the speed plot correspond to energetic CMEs from some super active regions that were prolific producers of CMEs and flares (Gopalswamy *et al.*, 2006).

2.6 Associated phenomena

CMEs are associated with a number of phenomena occurring in the chromosphere, corona, and interplanetary space. Flares, shocks, radio bursts, SEP events, ICMEs, and geomagnetic storms are all associated with CMEs. Most of these phe-

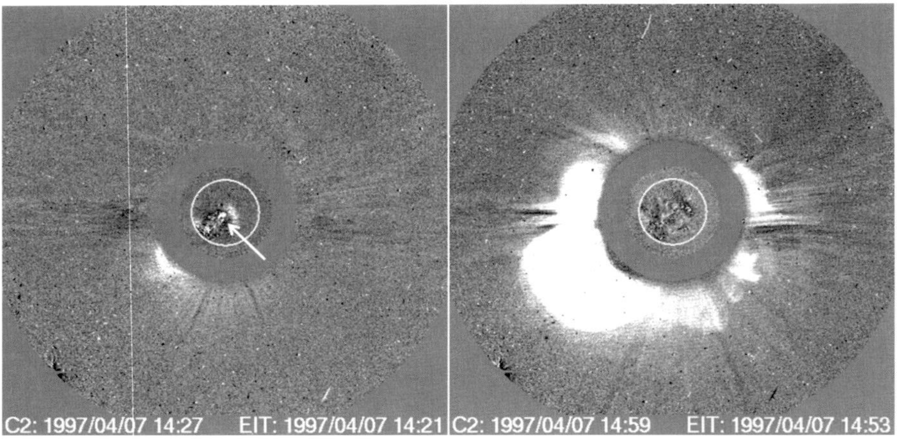

Fig. 5. Example showing how CMEs are related to the disk activity using EUV and white light difference images from SOHO. The arrow points to the location of eruption. The CME occurred on 1997 April 07.

nomena are observed in H-alpha, soft X-rays, EUV, radio, hard X-rays, and gamma rays in images and/or spectra. Radio bursts of type II (produced by nonthermal electrons accelerated in shocks) and type IV (produced by nonthermal electrons trapped in moving magnetic structures), are closely related to CMEs. H-alpha flare ribbons (Munro *et al.*, 1979), coronal dimming (Rust, 1983; Hudson and Webb, 1997; Zarro *et al.*, 1999; Thompson *et al.*, 2000), EUV wave transients (Biesecker *et al.*, 2002; Cliver *et al.*, 2005; Chen, 2006; Veronig *et al.*, 2006) and post-eruption arcades (Kahler, 1977; Tripathi *et al.*, 2004) are the disk signatures routinely used in identifying the source regions of CMEs on the Sun. Figure 5 illustrates how the disk activity in EUV relates to the white-light CME. Changes in the corona are usually best observed in difference images obtained by subtracting a previous frame from a given frame. In Fig. 5, the EUV difference image at 14:21 UT shows a large-scale disturbance originating from a location pointed by the arrow. The same disturbance can be seen in the LASCO difference image taken 6 minutes later. The EUV disturbance can be seen over the entire solar disk in the next image at 14:53 UT and the CME has already expanded to cover most of the coronagraph field of view. The EUV disturbance is thought to be a wave in the corona surrounding the CME (Neupert, 1989; Thompson *et al.*, 1999); the wave, on occasions becomes a shock resulting in a type II radio burst (Cliver *et al.*, 1999; Gopalswamy, 2006d). The EUV waves are also observed in the chromosphere as Moreton waves (Moreton, 1960; Narukage *et al.*, 2002; Warmuth *et al.*, 2005).

2.6.1 CMEs eruptive prominences

Prominence eruptions are known from the late 1800s from the scientific observations of Secchi and de la Rue (see, e.g., Tandberg-Hanssen, 1995) and recognized as mass eruptions from the Sun. After the discovery of white light CMEs (see e.g.,

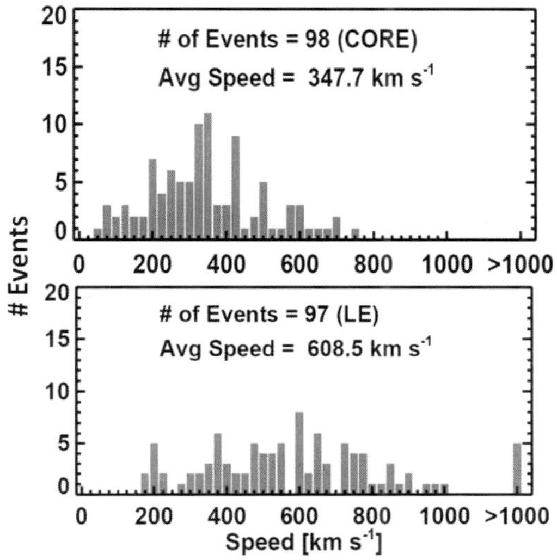

Fig. 6. Speed distributions of the leading edge and prominence core of CMEs observed in the SOHO/LASCO field of view. The average leading edge speed is much higher (~609 km/s) than that of the core (348 km/s). (Adapted from Gopalswamy *et al.* (2003c). Reproduced by permission of the AAS.)

Koomen *et al.*, 1974), the eruptive prominence has been found to be the inner core of CMEs (see, e.g., House *et al.*, 1981; Gopalswamy *et al.*, 1998; Moore *et al.*, 2001) as illustrated in Fig. 2. When the eruption happens on the disk, dark filaments in H-alpha erupt and disappear. Filaments are the same as prominences but observed on the disk. The eruptive prominence is an integral part of an eruption, with a close correspondence between the onset times (Gopalswamy *et al.*, 2003c). CMEs selected on the basis of their association with eruptive prominences provide the true distribution of CME speeds because these CMEs are not subject to projection effects. Figure 6 shows the speed distributions of nearly 100 CMEs and their prominence cores as obtained from LASCO observations (Gopalswamy *et al.*, 2003c). The speed of the CME leading edge is nearly twice that of the associated prominence cores (609 km/s vs. 348 km/s). The lower speed of the prominences compared to the CME leading edge is considered as evidence that prominences may not be driving the CMEs (Hundhausen, 1999). However, some authors think that CMEs can be caused by prominence eruptions (Filippov, 1998). Prominence eruption is linked to the reconnection process between magnetic field that restrains the prominences and the newly emerging field in the CME source regions (Feynman and Martin, 1995; Shibata *et al.*, 1995). The average CME leading edge speed in Fig. 6 is similar to that of the limb population in Fig. 3(d), rather than that of the general population (Fig. 3(a)). This is because prominences indicate eruptions close to the limb, so the associated CMEs are subject to minimal projection effects (see also Burkepile *et al.*, 2004).

2.6.2 CMEs and flares

Like prominence eruptions, flares are also an integral part of CMEs. Flares associated with CMEs are known as eruptive flares as opposed to flares that are not associated with mass motion. Figure 7 illustrates a typical eruptive flare using H-alpha observations and the associated structures. One of the typical characteristics of an eruptive flare is the H-alpha ribbons (R1, R2 in Fig. 7) connected by post-eruption loops (PEL) straddling the polarity inversion line. When observed in EUV and X-rays, the ribbons + PEL are observed as post-eruption arcades (PEA). The ribbons correspond to the locations where energetic particles accelerated in the corona are supposed to deposit their energy thus exciting the H-alpha line. In a subset of the flare loops, precipitating energetic electrons produce hard X-ray emission, while trapped electrons produce microwave emission. The reconnection process that forms the PEA also forms a flux rope, which is ejected into the IP space as a CME. The axis of the flux rope (FR) is sketched in Fig. 7. The legs of the flux rope are thought to be anchored in a pair of dimming regions (D1, D2) usually observed in EUV images (see Fig. 7). The area of the flare ribbons defines the size of a flare in H-alpha referred to as optical importance of a flare. The soft X-ray flux from the PEA defines the X-ray importance of a flare. The X-ray importance is denoted by the letters X, M, C, B, and A from the largest (peak flux $>10^{-4}$ W m^{-2}) to smallest (peak flux $<10^{-7}$ W m^{-2}),

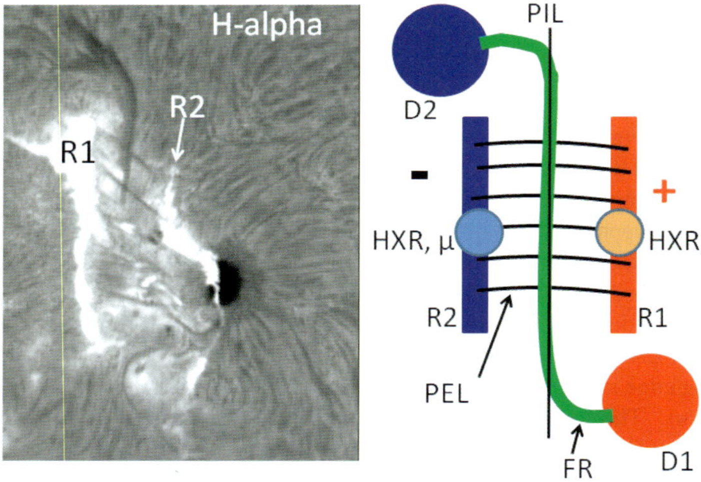

Fig. 7. (Left) H-alpha ribbons (R1, R2) with post-eruption loops (PEL) for a large-eruption that occurred on 2005 May 13. (Right) The eruption geometry sketched based on a number of observations from the photosphere to the interplanetary medium. In addition to the flare ribbons and post-eruption loops, the polarity inversion line (PIL) separating the positive and negative polarities in the active region, the dimming regions (D1, D2), and the sources of hard X-ray (HXR) and microwave (μ) emissions are also marked. The axis of a flux rope (FR) with its legs rooted in the dimming regions is also sketched. The flux rope and the PELs are formed by the reconnection process.

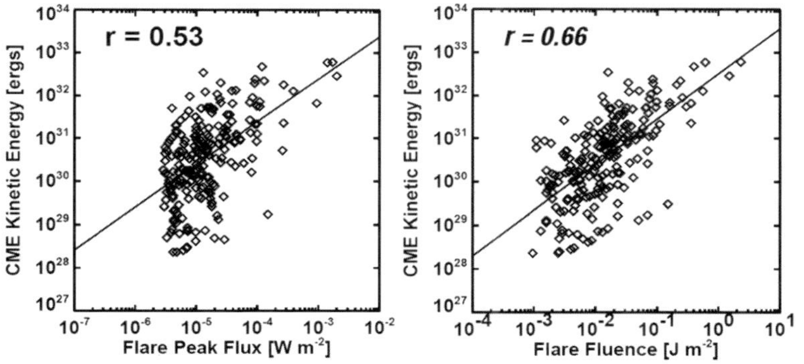

Fig. 8. Scatter plot between CME kinetic energy and peak soft X-ray flux (left) and the fluence of the associated GOES soft X-ray flux. The correlation coefficients and the regression lines are also shown on the plots.

consecutive levels differing by an order of magnitude.

The sketch showed in Fig. 7 represents the top view of the standard eruption model known as the CSHKP (Carmichael, Sturrock, Hirayama, Kopp and Pneuman) model (see e.g., Švestka and Cliver, 1992; Hanaoka *et al.*, 1994; Gopalswamy, 2006c). The eruption geometry shown has some symmetry with the hard X-ray and microwave sources occupying the central location. The post-eruption arcade is the largest flare structure. The separation between the dimming regions is the largest structure, which represents the CME scale. The prominence core is supposed to be entrained at the bottom of the flux rope, which appears as the bright core in white light. The white light CME is actually the compressed coronal material that surrounds the flux rope. The symmetry suggested by the CSHKP model was not found in the statistical studies that investigated the spatial relationship between CME position angle and flare location in the pre-SOHO era: flare was thought to be located anywhere under the span of the CME (e.g., Harrison, 2006). These studies involved only a small number of CME-flare pairs. A recent investigation involving nearly 500 flare-CME pairs in the SOHO era found that the flare is typically located radially below the CME leading edge for limb CMEs (Yashiro *et al.*, 2008b). This finding calls for a closer flare-CME relationship as implied by the CSHKP eruption model and other observations connecting CME kinematics and soft X-ray flares (Zhang *et al.*, 2001).

How does the close spatial relationship between flares and CMEs figure in the energetics? Several authors have found varying degrees of correlation between CME speed and peak soft X-ray flux (Hundhausen, 1997; Moon *et al.*, 2002; Burkepile *et al.*, 2004; Vršnak *et al.*, 2005; Yeh *et al.*, 2005). Some used speed, others used CME kinetic energy. Not all took care of projection effects. Hundhausen (1997) reported a weak correlation ($r = 0.53$) between CME kinetic energy and peak soft X-ray flux of the associated flares for all the events observed by the Solar Maximum

Mission (SMM). Burkepile *et al.* (2004) reanalyzed the SMM data including only those CMEs originating close to the limb (to avoid projection effects) and found a better correlation ($r = 0.74$). Other attempts to correct for projection effects in fact worsened the correlation (Yeh *et al.*, 2005). Figure 8 shows another attempt to find the relation between flare and CME energies for a set of 379 SOHO CMEs that originated in the CMD interval 60°–90°. The data set is similar to that of Burkepile *et al.* (2004) with minimal projection effects. It is clear that the resulting correlation coefficient is identical to that of Hundhausen (1997). However, the correlation between CME kinetic energy and flare fluence is much better ($r = 0.66$). It is possible that the higher correlation coefficient obtained by Burkepile *et al.* (2004) is due to a smaller sample size. The weaker correlation may not be entirely due to projection effects and may be related to the basic process by which the energy released in an eruption is partitioned between heating (soft X-ray flux) and mass motion (CME kinetic energy). For example, in confined flares the entire free energy goes into heating, with no mass motion in the radial direction. The overlying magnetic structure in the eruptive region may also influence the amount of mass ejected and hence the resulting CME kinetic energy.

It must be pointed out that only less than half of all flares are associated with CMEs. Is there any fundamental difference between flares with and without CMEs? Yashiro *et al.* (2006) studied two sets of flares one with and the other without CMEs. The number of flares as a function of peak X-ray flux, fluence, and duration in both sets followed power laws. Surprisingly, the power law index was >2 for flares without CMEs, while <2 for flares with CMEs. In flares without CMEs, the released energy seems to go entirely into heating, which suggests that tiny flares may contribute significantly to coronal heating as suggested by Hudson (1991). Another study found that flares without CMEs indicate a higher temperature (Kay *et al.*, 2003), which is consistent with the view that the entire released energy goes into heating.

2.6.3 CMEs and shocks

CMEs moving with super-Alfvenic speeds drive fast mode MHD shocks. The earliest indication of shocks in the corona is the metric type II radio bursts (Nelson and Melrose, 1985). Type II radio bursts are thought to be produced by nonthermal electrons accelerated in the shock front by the plasma emission mechanism (Langmuir waves generated by the nonthermal electrons get converted into electromagnetic radiation at the fundamental and harmonic of the local plasma frequency). Metric type II bursts (observed by ground based radio telescopes) start typically at 150 MHz and drift down to lower frequencies, corresponding to the heliocentric distances <2–$3R_o$ (solar radii) in the corona. Spaceborne radio instruments such as the Wind/WAVES experiment are needed to observe at frequencies below the ionospheric cutoff (~15 MHz). Type II radio bursts have been observed down to ~20 kHz, which corresponds to the plasma frequency in the vicinity of the observing spacecraft.

Figure 9 shows a type II radio burst and the associated CME, flare, and SEP activities. The continuous track of type II radio burst implies that the shock continuously accelerates electrons throughout the IP medium until the shock reaches the Wind spacecraft. Early in the event, type III bursts are produced, which are due to

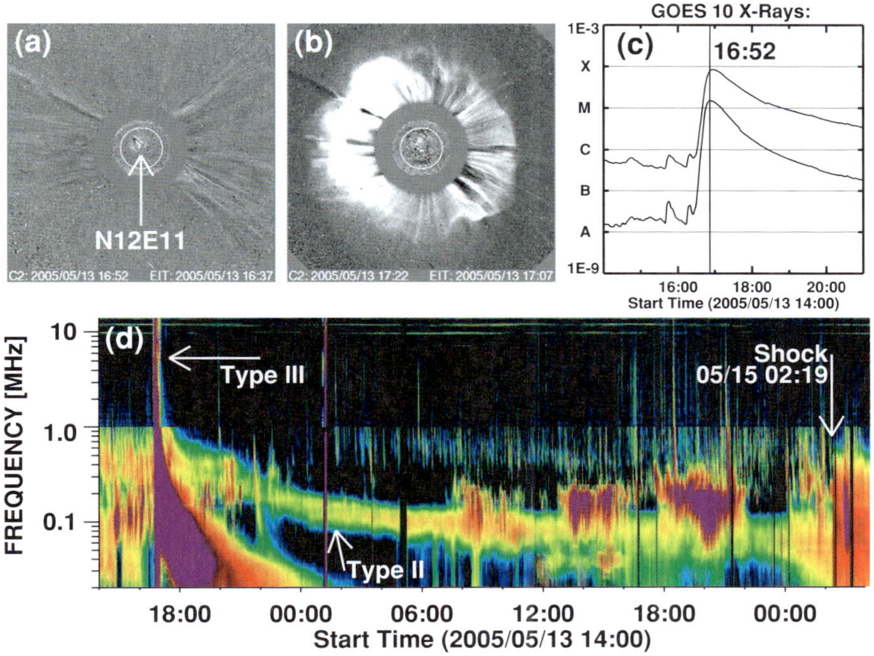

Fig. 9. A complete Sun-to-Earth event with (a) solar source near the disk center (N12E11), (b) a halo CME, (c) a M-class GOES soft X-ray flare, (d) shock arrival as indicated by the SEP time profile, and (d) an IP type II extending from the outer corona through the IP space to the observing spacecraft.

electrons accelerated in the flare site (Fig. 9(a, c)) located below the CME (Fig. 9(b)). The SOHO coronagraphs track CMEs only up to $32R_o$ (the outer edge of the field of view), whereas the Type II bursts can track the CME-driven shocks all the way to Earth. Note that the type II radio emission occurs at $\sim$40 KHz when the shock arrives at the spacecraft. When the Type II burst is at decameter-hectometric (DH) wavelengths (14–1 MHz), the shock is typically in the near-Sun interplanetary medium (heliocentric distance in the range 3–$10R_o$). When type II bursts are observed in the DH and longer wavelengths, it is highly likely that the underlying shocks and CMEs reach Earth (provided the CME source is close to the disk center). The lower the frequency reached by a type II, the higher is the energy of the associated CME, implying that such CMEs drive shocks throughout the inner heliosphere (Gopalswamy *et al.*, 2005a; Gopalswamy, 2006d). The GOES SEP intensity profiles (not shown) reveal that SEPs are produced soon after the CME lifts off at the Sun with a sharp rise when the shock arrives at the observing satellite. Such a rise is known as the energetic storm particle (ESP) event because it coincides with the sudden commencement of geomagnetic storms (see e.g., Cohen, 2006 for a review). SEPs will be discussed in more detail in Section 5. Shocks are also observed in in-situ data marked by a sudden jump in density, temperature, magnetic field, and flow speed. Shocks are important for space weather because they mark the first arrival of a CME-associated disturbance and the ESP event. Shocks also compress the magnetosphere making the magnetosphere shrink thereby exposing satellites in geostationary orbits to the solar wind.

3 Coronal Mass Ejections in the IP Medium

In the decade before the first detection of white-light CMEs, people contemplated about IP structures that are not too different from what are observed today. In particular, Parker (1957) concluded that "magnetic clouds may be ejected from a magnetic field with velocities as high as the Alfven wave velocity...". Gold (1962) had sketched (see Fig. 10) the "Idealized configuration in space, showing solar plasma cloud, the drawn-out field and the shock wave ahead". Koomen *et al.* (1974) found similarity between the Gold magnetic bottle and CMEs observed by OSO-7 coronagraph. While most CMEs fade to the background level within the coronagraphic field of view (below $\sim 32R_{\rm o}$), $\sim$10% of them seem to go past the edge of the field of view and propagate as disturbances in the IP medium as envisaged by Gold (1962). The only difference between Gold's picture and modern picture is that the magnetic field of the plasma cloud should be in the form of a flux rope. The modern version of the Gold plasma cloud is known as ICMEs. Driver gas, ejecta, magnetic clouds (MCs) and flux ropes are other names for ICMEs. Although the Solar Terrestrial Relations Observatory (STEREO) mission is capable of observing CMEs over the entire Sun-

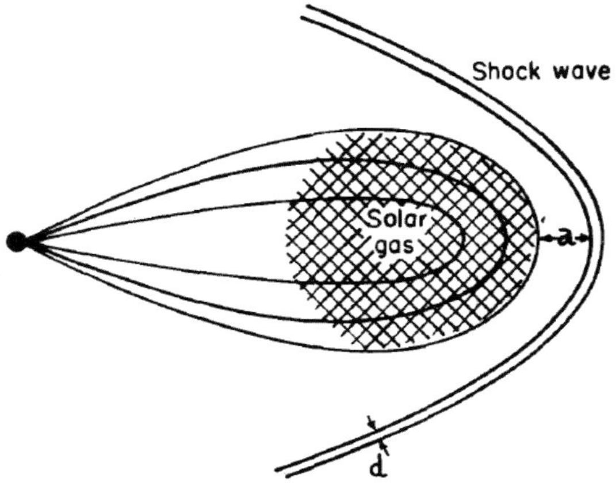

Fig. 10. The spatial structure of an ICME as sketched by Gold (1962). Note the enormous size of the magnetized "plasma cloud" connected to the Sun and driving a shock that has a thickness *d* and stands ahead at a distance *a* from the cloud. This is pretty much the structure we know today, except for the flux-rope nature of the magnetized gas cloud. (With kind permission from Springer Science+Business Media: *Space Sci. Rev.*, Magnetic storms, **1**, 1962, 100–114, Gold, T.)

Earth distance (Harrison *et al.*, 2008), the current knowledge on ICMEs comes mainly from in situ observations.

3.1 ICME signatures

ICMEs are identified in a variety of ways: solar wind plasma signatures (temperature, plasma beta, flow speed), magnetic signatures (field strength, field rotation), compositional and charge state signatures, and particle (thermal and nonthermal) flux signatures (see Gosling, 1990; Neugebauer and Goldstein, 1997 for a review). Arrival of shocks at the spacecraft is also a good indicator of the impending ICME arrival except when the associated CME propagates at a large angle to the Sun-spacecraft line. In situ observations by the Helios and Pioneer Venus Orbiter spacecraft (at distances <1 AU), by Voyager spacecraft (at 10s of AU) and Ulysses spacecraft at high latitudes have shown that ICMEs are observed throughout the heliosphere (see e.g., Wang *et al.*, 2005 and references therein). Techniques such as IP scintillation observations and IP type II burst observations also provide information on parts of CMEs in the heliosphere (see e.g., Tokumaru *et al.*, 2007 and references therein).

Magnetic-field signatures are most useful to identify flux ropes: enhancement of the magnetic field magnitude, smooth rotation of the field along the trajectory of the observing spacecraft, and low proton temperature (or beta) (Burlaga *et al.*, 1981). MCs are also referred to as flux ropes, but the latter need not have low plasma beta. Bidirectional flux of thermal electrons, ions and energetic particles are signatures indicating that the legs of the ICME are connected to the solar corona. When the flux rope expands from near the Sun, the density and temperature decrease. Depressed proton temperature (with respect to an expected value—see Lopez and Freeman, 1986) is often used as a signature to identify intervals of ICMEs in solar wind data (Cane and Richardson, 2003; Neugebauer *et al.*, 2003; Elliott *et al.*, 2005). Charge state signatures include the enhancement of the average Fe charge state within the ICME (Lepri *et al.*, 2001), the density ratio of O^{7+} and O^{6+} ions (O^{7+}/O^{6+}—see Henke *et al.*, 2001; Reinard 2005; Aguilar-Rodriguez *et al.*, 2006), and the density ratio of helium ions to protons (N_α/N_p).

Figure 11 shows an ICME illustrated using plots of the magnetic field magnitude (B_t) and its Y and Z components (B_y, B_z), the solar wind bulk flow speed (V), proton temperature (T_p), and the alpha to proton density ratio (N_α/N_p) in GSE coordinates (from the ACE data). Also plotted are the expected solar wind proton temperature (T_{exp}) from Lopez and Freeman (1986). Ejecta intervals are those with T_p depressed significantly with respect to T_{exp}. In Fig. 11, T_p is depressed between 14:15 on 2005 January 16 to 06:30 on the next day. The ratio of helium density to proton density (N_α/N_p) is also enhanced during the ejecta interval. In addition, the magnetic field is enhanced, the B_z component shows rotation, and the B_y component points to the east throughout the ejecta interval. The white light CME responsible for the ICME was very fast ($>$2050 km/s) and originated close to the disk center (N16E04) on 2005 January 15 around 06:30 UT. The type II radio burst indicative of interplanetary shock is also shown in the figure. The shock was observed ahead of the ICME at 11 UT on 2005 January 16 in the in situ data. Nearly 200 such CME-ICME pairs have been identified during solar cycle 23, which helped assess the importance of

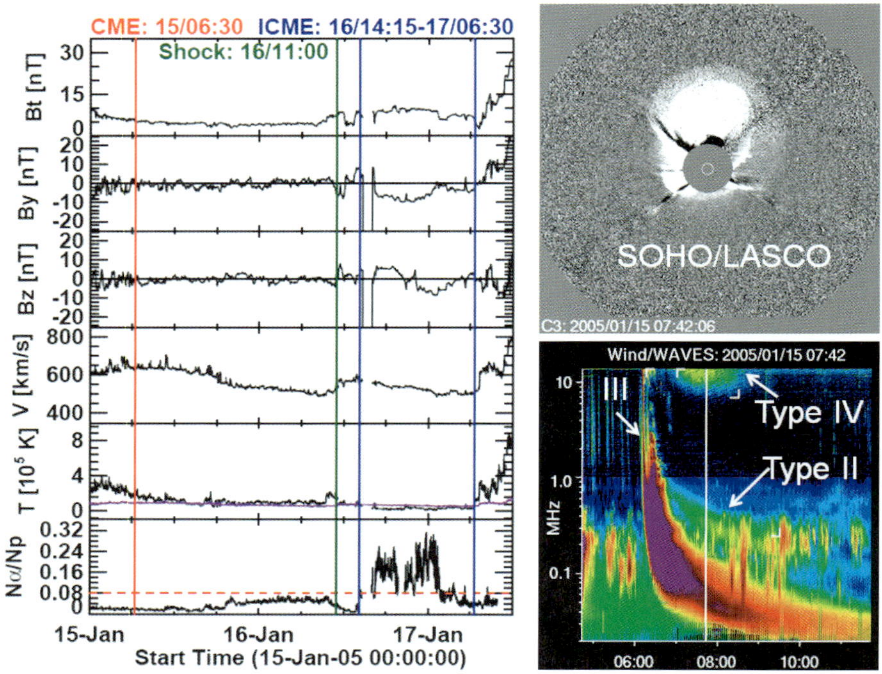

Fig. 11. Plots of the magnetic field magnitude (B_t) with its Y and Z components (B_y, B_z), the solar wind bulk flow speed (V), proton temperature (T), and the alpha to proton density ratio (N_α/N_p) in GSE coordinates (from the ACE data). The expected solar wind proton temperature (T_{exp}) from Lopez and Freeman (1986) is overlaid on the T plot. Ejecta intervals are those with T_p depressed significantly with respect to T_{exp}. The associated CME and the radio bursts are also shown.

CMEs in causing space weather effects.

3.2 Charge states in ICMEs

The solar wind ions maintain their chemical and charge state identity as they propagate into the IP medium because the expansion time scale is much larger than the ionization and recombination time scales. Thus the charge state of ions in the solar wind is a good indicator of the conditions in the corona where this time-scale inequality happens (also known as the freezing-in altitude). Since the source region of the CMEs is hotter than that of the normal solar wind, one expects higher charge states in ICMEs (Henke *et al.*, 2001; Reinard, 2005; Aguilar-Rodriguez *et al.*, 2006). In a study involving more than 150 ICMEs, Aguilar-Rodriguez *et al.* (2006) found that elements inside ICMEs have higher ionization states compared to those in the ambient solar wind. The required higher temperature in the CME source region can be traced to the flare plasma, which can exceed 10 MK: according to the standard model of CMEs associated with eruptive flares, the CME flux rope forms out of sheared active region field lines and the heated plasma from the reconnection region is injected into the flux rope. As the flux rope expands, the ionization states corresponding to the flare plasma remain frozen. They also found that MCs possess the enhanced charge states more often than the non-cloud ejecta. Such a scenario is supported by the observed correlation between the flare size and charge state enhancement for eruptions close to the disk center (Reinard *et al.*, 2005). Off the disk center CMEs are likely to be observed by an observer along the Sun-Earth line as non-cloud ejecta, consistent with the observations of Aguilar-Rodriguez *et al.* (2006).

3.3 Solar sources of ICMEs

CMEs propagate radially outward from the Sun, so those directly impacting Earth must originate close to the disk center. It is well known that CMEs associated with MCs have their sources close to the disk center (see e.g., Gopalswamy *et al.*, 2000). The other extreme is the case of shocks arriving at Earth without discernible ejecta. These "driverless shocks" are associated with CMEs that travel approximately orthogonal to the Sun-Earth line, so their solar sources are expected to be close to the limb. Non-cloud ejecta must have their solar sources at intermediate longitudes. The source locations of the MCs, non-cloud ICMEs, and driverless shocks shown in Fig. 12 confirm the expected pattern: there is a heavy concentration of MC sources near the disk center, non-cloud ejecta sources are mostly at larger central meridian distances, and the driverless shock sources are mostly near the limb. There are significant exceptions to this pattern, especially for non-cloud ejecta and driverless shocks. Driverless shocks with their sources near the disk center are quite unexpected. A recent investigation revealed that the associated CMEs are deflected away from the Sun-

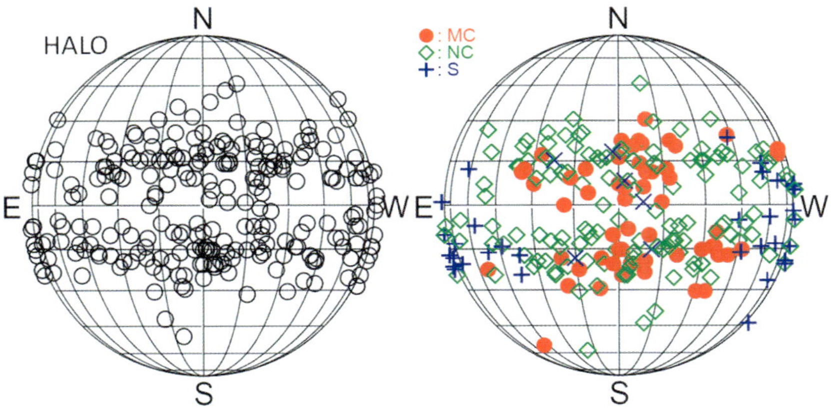

Fig. 12. (Left) Solar source locations of halo CMEs and (right) those of CMEs associated with magnetic clouds (filled circles), non-cloud ICMEs (NC, diamonds) and shocks without discernible drivers (plus and cross symbols). Although there are a few exceptions, the overall distribution of sources is similar in both cases. There is a heavy concentration of halo CME sources near the disk center. Similarly there is a heavy concentration of MC sources near the disk center. While there are many non-cloud ejecta from near the disk center, most of them are at large central meridian distances. The shocks are mostly near the limb, except for a handful of events near the disk center (cross symbols).

Earth line by coronal holes located in the vicinity of the eruption region (Gopalswamy *et al.*, 2008b), thus forcing the CMEs behave like limb events. Such deflections happened only during the declining phase of the solar cycle, when large low-latitude coronal holes occur frequently. A similar deflection is suspected in the case of non-cloud ICMEs with disk-center sources, but further investigation is needed to confirm this. This will also explain why some CMEs originating close to the disk center do not become MCs. According to the above geometrical argument, one would expect the central one third of ICMEs become MCs, assuming that most energetic CMEs have a width of $\sim 60^\circ$. In reality, the fraction of MCs changes with the phase of the solar cycle influenced heavily by the global field during solar minima (Riley *et al.*, 2006; Gopalswamy *et al.*, 2008a).

Figure 12 also compares the solar source locations of MCs, non-cloud ICMEs, and driverless shocks with those of halo CMEs. There is a heavy concentration of halo CME sources near the disk center, consistent with the result that ~86% MCs are associated with halos or partial halos (Gopalswamy *et al.*, 2008a). Halos at larger CMD are expected to produce non-cloud ejecta at Earth. Finally, most of the limb halos produce driverless shocks. In the case of limb halos, the CME originates close to the limb, but the disturbance (most likely CME-driven shock) extends above the opposite limb because these CMEs are very energetic and expand rapidly to produce a signal above the opposite limb. As we noted before, CMEs associated with ICMEs are generally more energetic, similar to the halo CMEs.

3.4 Statistical properties of ICMEs

Figure 13 summarizes the properties of ICMEs separated into MCs and non-cloud events. For definitiveness we have included only those non-cloud ICMEs that drive shocks. The maximum magnetic field strength in the two types of ICMEs ranges from a few nT to several tens of nT. The average values are about three times that of the solar wind (~5 nT), which confirms that enhanced magnetic field is an important ICME signature. The mean speeds of MCs (478 km/s) and non-cloud ICMEs (466 km/s) are also similar and only slightly larger than the typical slow solar wind speed (437 km/s—see Gopalswamy, 2006c). The average values of duration and proton thermal speed are smaller for MCs. The larger duration of non-cloud ICMEs may be due to the passage of the measuring spacecraft away from the nose of the flux rope. The MC duration implies a spatial scale of ~0.3 AU for the thickness of the flux rope (see Lepping and Berdichevski, 2000; Gopalswamy, 2006c). The higher proton thermal speed may also indicate a spacecraft trajectory along the flanks of the ICME.

A compilation of ICME observations at various radial distances (R) made from many spacecraft (Helios-1 and 2, Pioneer Venus Orbiter, ACE, and Ulysses) reveal a rapid decline of density ($\sim R^{-2.4}$) and magnetic field ($\sim R^{-1.5}$) inside ICMEs compared to the solar wind (Wang *et al.*, 2005). The temperature, while smaller than in the solar wind, declines less rapidly ($\sim R^{-0.7}$). Extrapolating the 1-AU values of magnetic field, density and temperature in CMEs seem to agree with the expected values at the Sun.

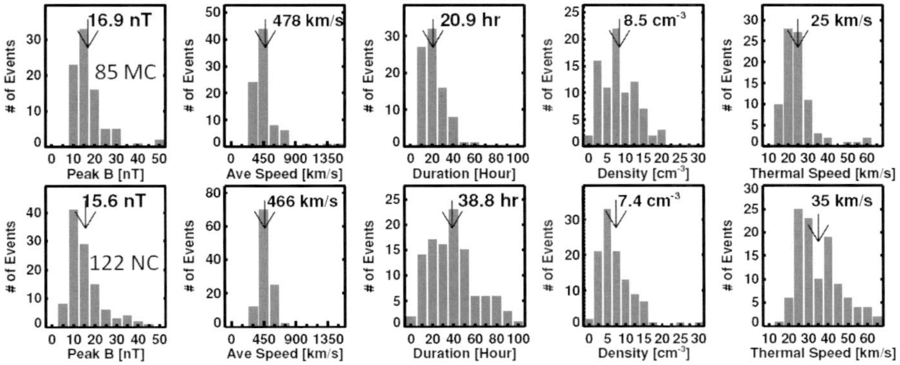

Fig. 13. Distributions of properties of MCs and non-cloud (NC) ICMEs observed during cycle 23: magnetic field strength, flow speed, duration, proton density and proton thermal speed. The average values of the distributions are marked on the plots. (With kind permission from Springer Science+Business Media: *Space Sci. Rev.*, Properties of interplanetary coronal mass ejections, **24**, 2006, 145–168, Gopalswamy, N., figure 4.)

3.4.1 CME and ICME speeds

Overall, the ICME speeds are not too different from the speed of the slow solar wind and the average speed of the general population of CMEs observed near the Sun. On the other hand the average speeds of CMEs associated with MCs and non-cloud ICMEs is close to 1000 km/s (Gopalswamy *et al.*, 2007). This is a direct consequence of the interaction between CMEs and the solar wind in the IP medium. Once a CME lifts off from the Sun, its subsequent evolution is determined by its interaction with its surroundings. After the propelling force ceases at a few solar radii, the interaction with the solar wind becomes the dominant force acting on the CME. This is normally described by the drag force, which depends on the flow properties of the ambient medium and the CME. The scatter plot in Fig. 14 shows how the speeds measured at the Sun and at 1 AU are related. While the CME speeds at the Sun varies from $\sim$100 km/s to $>$3000 km/s, the corresponding speeds at Earth range from $\sim$300 km/s to $\sim$1500 km/s. This is clearly an indication that slow CMEs at the Sun eventually accelerate, while the fast CMEs decelerate to produce the observed narrower range of speeds at Earth. In fact, there are only three ICMEs with speeds exceeding 1000 km/s. CMEs near the bisector do not change in speed because they are simply carried by the solar wind speed. Such a scatter plot was first obtained by Lindsay *et al.* (1999) using Helios and Pioneer Venus Orbiter data in the solar wind and coronagraphic observations near the Sun (Solar Maximum Mission and P-78 satellite data). They found a linear correlation. A reanalysis of the same data indicated that the relationship is not linear (Gopalswamy *et al.*, 2001b; Gopalswamy, 2002).

The interplanetary acceleration can be quantified using pairs of CMEs and ICMEs (Gopalswamy *et al.*, 2000, 2001b, 2005e): observational data devoid of projection

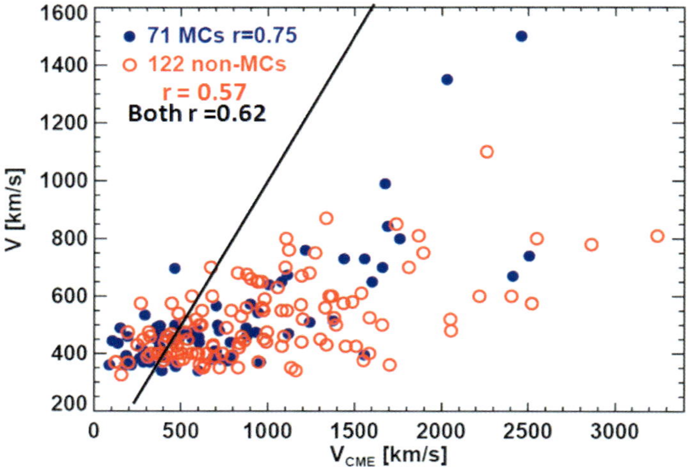

Fig. 14. Scatter plot between CME speeds measured near the Sun (V_{CME}) and ICME speeds (V) measured at 1 AU. Magnetic cloud (MC) and non-cloud ICMEs are denoted by closed and open circles, respectively. The solid line is the bisector. The correlation coefficient is significantly higher for MCs ($r = 0.75$) than for non-cloud ICMEs ($r = 0.57$); the combined set has an intermediate value ($r = 0.62$). (Reprinted from *J. Atmos. Sol.-Terr. Phys.*, **70**, Gopalswamy, N., Solar connections of geoeffective magnetic structures, 2078–2100, Copyright 2008, with permission from Elsevier.)

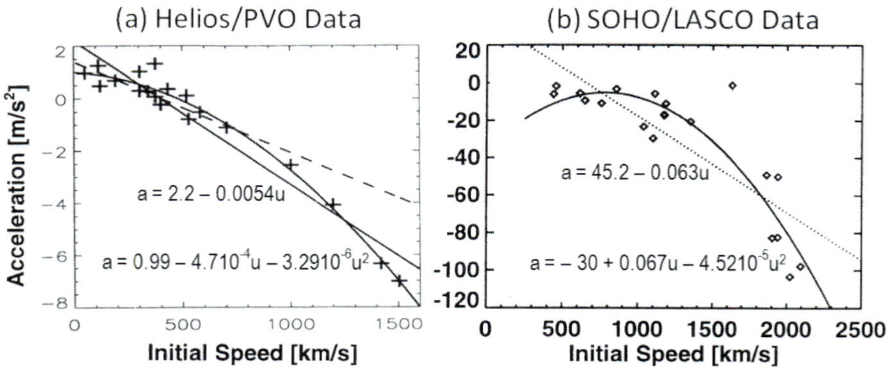

Fig. 15. Dependence of acceleration as a function of initial speed of CMEs (a) over the Sun-Earth distance (b) within the first 32 solar radii (from SOHO/LASCO).

effects indicate a linear relationship between acceleration (a) and the CME initial speed (u): $a = 2.2 - 0.0054u$ or $a = -0.0054(u - 406)$, with a in m s^{-2} and u in km/s (Gopalswamy *et al.*, 2001b). In this relation, the speed of 406 km/s corresponds to the solar wind speed. From Fig. 15(a) we see that this linear relationship is quite reasonable for speeds below ~1000 km/s. At higher speeds, the quadratic relation seems to work better. In fact, within the coronagraphic field of view, the deceleration seems to be quadratic for limb CMEs that produce IP type II radio emission (see Fig. 15(b)).

While the acceleration profile is devoid of projection effects, its utility depends on obtaining the actual speed of CMEs in the Earth direction. Unfortunately, CMEs originating close to the disk center are the ones important for space weather, but they are subject to severe projection effects. At present there are two ways to eliminate projection effects. One is to fit the observed CME in the coronagraphic field of view to a cone and get the radial speed and width of the CME (Hundhausen *et al.*, 1994; Xie *et al.*, 2004; Michalek, 2006; Zhao, 2008). The other method is to obtain an empirical relation between the radial speed ($V_{\rm rad}$) and the expansion speed ($V_{\rm exp}$) of a set of CMEs occurring near the limb and then use the relation for CMEs for which the expansion speed can be measured. Dal Lago *et al.* (2003) obtained a relationship $V_{\rm rad} = 0.88 V_{\rm exp}$. One difficulty is that we do not know the actual width of halo CMEs, so the empirical relation leads to large uncertainties (Gopalswamy *et al.*, 2008e). Other possibilities include using a set of well observed CMEs when the separation between the STEREO spacecraft is ~90° and/or when they are at 90° from the Sun-Earth line to validate the acceleration profiles and test their validity over the entire Sun-Earth distance.

3.5 Magnetic field structure

When a ICME is shock driving, one observes the following temporal ordering of structures: the shock, the sheath, and the ejecta in a typical time-series observation at 1 AU. In the case of MCs, the ejecta portion has a well defined structure with B_z showing a smooth rotation from north to south (NS) or south to north (SN) as illustrated in Fig. 16. This happens when the MC axis is close to the ecliptic ($Z = 0$ in GSE coordinates). B_z is the azimuthal component of the flux rope that defines the

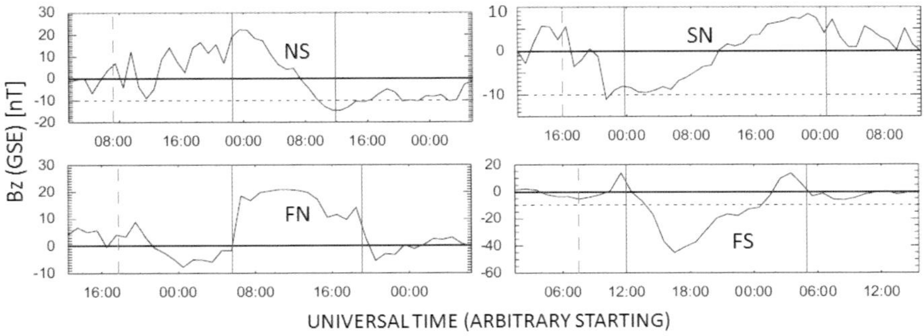

Fig. 16. The four basic types of MCs defined based on the direction of the leading field from the B_z component in GSE coordinates: North-south (NS), south-north (SN), fully north (FN), and fully south (FS). The vertical dashed line denotes the shock. The MC interval is between the two vertical lines. The region between the shock and MC is the sheath region.

MCs. When the MC axis makes an angle >45° to ecliptic, B_z is due to the axial field. For such high-inclination MCs, the azimuthal field (B_y) shows smooth rotation in the east-west direction. High-inclination MCs are known as unipolar MCs (Mulligan *et al.*, 1998) because the axis points fully to the north (FN) or south (FS) throughout the MC interval (see Fig. 16). Thus the SN, NS, FN, and FS MCs constitute the four basic types of flux ropes. Note that the each MC type can have two subtypes: the bipolar MCs can have their axes pointing to the east (SEN, NES) or west (SWN, NWS); the unipolar MCs can have their rotation from east to west (ESW, ENW) or west to east (WSE, WNE).

The flux rope types and their connection to the solar source regions have been extensively discussed in the literature (Bothmer and Schwenn, 1994; Mulligan *et al.*, 1998; Li and Luhmann, 2004; Yurchyshyn, 2008). In general, MCs have their axis parallel to the neutral line in the source active region on the Sun. However, significant deviations are known, which may be due to the influence of the overlying heliospheric magnetic field structure (Yurchyshyn, 2008 and references therein). The internal structure of MCs is helpful in understanding their solar origin. They also have important implications for their geoeffectiveness, as will be discussed in the next section.

4 CMEs and Geomagnetic Storms

Geomagnetic storms can be defined as disturbances of Earth's magnetosphere caused by the impact of IP magnetic field structures. The disturbance storm time ($D_{\rm st}$) index, which is the average change in the horizontal component of Earth's magnetic field brought about by the geomagnetic storm as measured (in nT) at four low-latitude stations (see http://swdcwww.kugi.kyoto-u.ac.jp/dstdir/dst2/onDstindex.html for details), quantifies the intensity of the storms. In a typical magnetic storm, the $D_{\rm st}$ index suddenly increases from the zero level (the storm sudden commencement or SSC), drops to the zero level and then decreases to a minimum (negative) value before recovering back to the zero level. The minimum $D_{\rm st}$ value reached is considered as the intensity of the storm. Geomagnetic storms can be classified into five groups based on the minimum value of $D_{\rm st}$: weak (−30 to −50 nT), moderate (−50 to −100 nT), strong (−100 to −200 nT), severe (−200 to −350 nT), and great (<−350 nT). It is also common to designate $D_{\rm st} \le -50$ nT as storm level (Loewe and Prolss, 1997). NOAA space weather scale uses the $K_{\rm p}$ index (planetary geomagnetic index) defined on a scale of 0 to 9 and the intensity levels are termed minor ($K_{\rm p} = 5$), moderate ($K_{\rm p} = 6$), strong ($K_{\rm p} = 7$), severe ($K_{\rm p} = 8$), and extreme ($K_{\rm p} = 9$). Details can be found in http://www.swpc.noaa.gov/NOAAscales/. In our discussions we adopt a simpler scheme: $D_{\rm st} \le -50$ nT denotes storm level; -100 nT $< D_{\rm st} \le -50$ nT denotes moderate storms while $D_{\rm st} \le -100$ nT denotes intense storms. The relation between $K_{\rm p}$ and $D_{\rm st}$ is discussed in Vennerstroem (2001).

In order to get an idea on the frequency and intensity range of geomagnetic storms, we have shown the distribution of $D_{\rm st}$ values at or below −100 nT in Fig. 17. The number of such storms is typically ∼100 in each solar cycle, based on the data available at the World Data Center in Kyoto (http://swdcwww.kugi.kyoto-u.ac.jp/

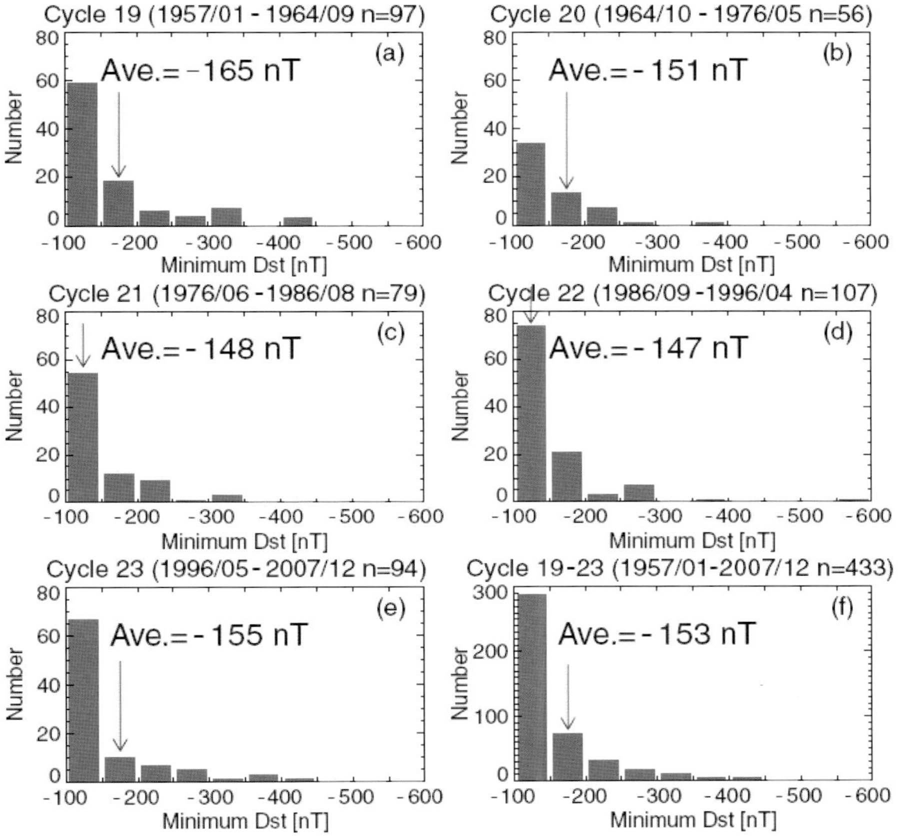

Fig. 17. D_{st} distribution for large geomagnetic storms ($D_{st} \leq -100$ nT) during the past five solar cycles (19–23). The duration and the number (n) of storms in cycle are given at the top of each plot.

dstdir/). Cycle 20 had the smallest number of storms (56), while cycle 22 had the largest (107). The combined set in Fig. 17(f) shows that the number of storms with $D_{st} \leq -300$ nT is rather rare. The strongest storms had $D_{st} < -600$ nT. It must be pointed out that not all the ≤ -100 nT storms are caused by ICMEs. It is known that ~10–15% of the large storms may be due to CIRs (Zhang *et al.*, 2007), but the D_{st} values are generally at the lower end (~−100 nT), well below the average value of all the large storms.

The southward component of the IP magnetic field (B_s) reconnects with Earth's magnetic field in the dayside followed by another reconnection in the night-side enabling the coupling of the solar wind to the magnetosphere ultimately resulting in the buildup of the ring current. The D_{st} index essentially reflects the build up and decay of the ring current. The mechanism was first elucidated by Dungey (1961).

The CME connection to geomagnetic storms stems from the fact that CMEs are a source of B_S. While the heliospheric magnetic field under quiescent conditions does not have an out-of-the-ecliptic component ($B_z \sim 0$ except for some Alfvenic fluctuations), a CME introduces B_z by virtue of the flux rope structure and the sheath field that drapes the CME. The three distinct features of an ICME event, viz., the shock, the sheath, and the ejecta have corresponding signatures in D_{st} plots as SSC, the sheath storm and the ejecta (or cloud) storm (Tsurutani *et al.*, 1988). The ability of a CME or any IP structure to produce a geomagnetic storm is known as geoeffectiveness and such a structure is said to be geoeffective. Once a CME leaves the Sun, its ability to cause space weather effects depends on several factors: its arrival at Earth, its speed at Earth, and its magnetic field content (location and strength of B_S). These are discussed below in more detail.

4.1 CME arrival at Earth

CMEs have to reach Earth before causing geomagnetic storms. This happens when CMEs originate close to the disk center as discussed in Section 3.3 (see also Fig. 12). Occasionally, CMEs originating close to the disk center may not arrive at Earth because of deflection by nearby coronal holes (Gopalswamy *et al.*, 2008b) or faster CMEs merging with preceding slower ones. CMEs originating farther from the

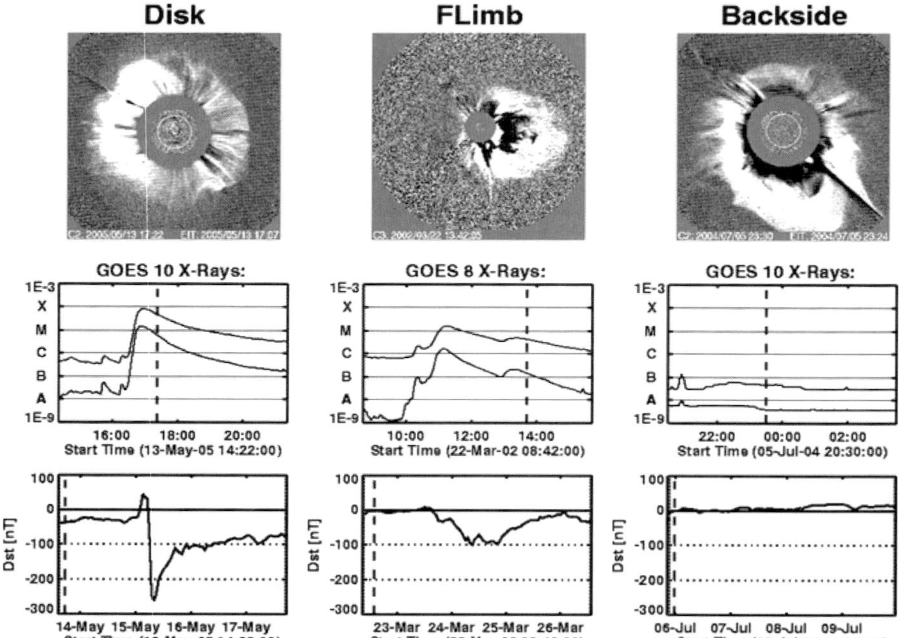

Fig. 18. Effect of solar source location on geoeffectiveness: (left) Halo CME from the disk center, (b) Halo CME originating from close to the limb but frontsided (FLimb), and (right) backside halo.

disk center also can cause geomagnetic storms, but what causes the storm is the B_s in the ICME sheath (Gosling *et al.*, 1990; Vennerstroem, 2001; Gopalswamy *et al.*, 2007; Gopalswamy *et al.*, 2008a and references therein). Since there is less chance for the sheath to have B_s than the whole ICME event, the geoeffectiveness of CMEs declines as a function of their solar source longitude (Gopalswamy *et al.*, 2007). This is illustrated in Fig. 18 using three halo CMEs originating from (i) close to the disk center, (ii) close to the limb but frontsided (Flimb), and (iii) backside of the Sun. The GOES soft X-ray light curve shows no changes for the backside CME because the flare is completely occulted. The disk-center halo results in a super storm (D_{st} = -263 nT), the Flimb halo results in a moderately intense storm ($D_{st} \sim -100$ nT), while the backside halo has no geomagnetic consequence at all ($D_{st} \sim 0$ nT). This result has also been confirmed statistically: ~75% of disk halos and ~60% of limb halos are geoeffective (Gopalswamy *et al.*, 2007) confirming the importance of the solar source location of CMEs. According to Fig. 12, we see that MC associated CMEs also originate close to the disk center, so the majority of MCs (~60%) are geoeffective.

One might wonder why all the disk halos and all the MCs are not geoeffective even though they originate close to the disk center. There are several possibilities: (i) CME cannibalism—fast halos catch up with slow ones, so only one ejecta arrives at Earth, (ii) CMEs are deflected away from the Sun-Earth line by coronal holes or by other CMEs, (iii) the ICMEs have no B_s (fully north pointing MCs) that they do not produce any geomagnetic storms, (iv) the magnitude of B_s is not high enough to cause a significant ($D_{st} \leq -50$ nT) storm. These exceptions need to be properly understood before assessing the geoeffectiveness of CMEs. It must be noted that cannibalism and coronal-hole deflection can also increase the geoeffectiveness of CMEs because interacting CMEs can enhance the southward component of the IP magnetic field, while coronal holes can deflect CMEs toward the Sun-Earth line.

4.2 Importance of CME speed

The geoeffectiveness also depends on the speed with which CMEs arrive at Earth. The D_{st} index is correlated with the ICME speed (provided the magnetic field has a B_s). Figure 19 shows that the D_{st} in the sheath and cloud portions correlate reasonably with the corresponding speeds ($r = 0.67$ for sheath D_{st} vs. sheath speed and $r = 0.65$ for MC D_{st} vs. MC speed). Interestingly, when we replace the sheath and MC speeds by the near-Sun speed of the associated CMEs, the correlation remains positive albeit weak ($r = 0.49$ for sheath D_{st} vs. CME speed and $r = 0.46$ for MC D_{st} vs. CME speed). The last set of correlations occurs because the near-Sun and near-Earth CME speeds are related via the IP acceleration discussed in Section 3.

4.3 Importance of CME magnetic field

Like the speed, the magnetic field content of the ICMEs also plays an important role. By this we mean the magnitude of B_s. The ICME magnetic field is also related to its speed: faster ICMEs tend to have higher magnetic field strength (Gonzalez *et al.*, 1998; Echer *et al.*, 2005; Gopalswamy *et al.*, 2008a). This is also true when the sheath and cloud portions of MCs are considered separately. Furthermore, faster CMEs near the Sun seem to result in MCs and sheaths with higher magnetic field

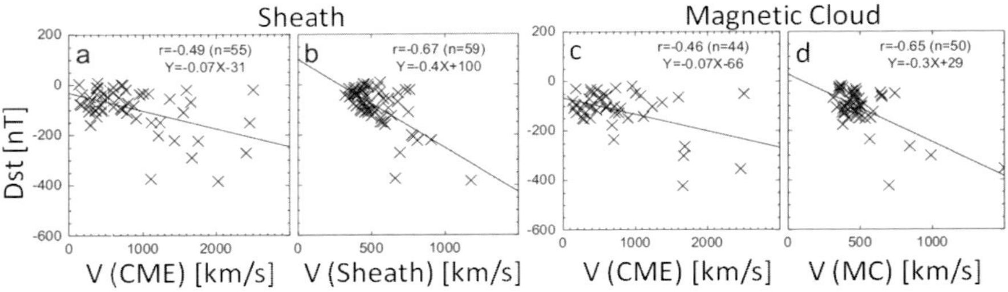

Fig. 19. D_{st} vs. speed plots for Sheaths and MCs. In (a) and (b), the D_{st} values correspond to the sheaths. In (c) and (d), the D_{st} values correspond to the cloud portion. The regression lines and the correlation coefficients (r) are shown on the plots.

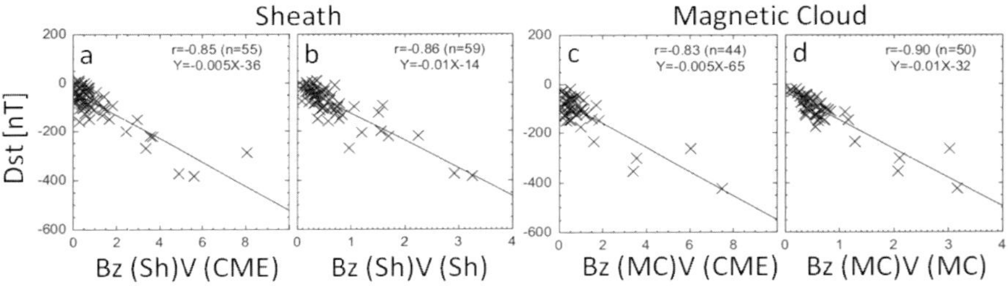

Fig. 20. D_{st} vs. speed plots for Sheaths and MCs. In (a) and (b), the D_{st} values correspond to the sheaths. In (c) and (d), the D_{st} values correspond to the cloud portion. The regression lines and the correlation coefficients (r) are shown on the plots.

(Gopalswamy *et al.*, 2008a). The minimum D_{st} values are better correlated with the product of the field strength (or B_s) and speed in ICMEs, as shown in Fig. 20. Such a product is related to the well-known Epsilon function (Akasofu, 2002). Note that D_{st} has a high correlation with the speed-magnetic field (B_z) product in both the cloud and sheath portions with $r = 0.90$ and 0.86, respectively. As before, when we the replace the sheath and MC speeds by the near-Sun CME speed, the correlations remain very high: $r = 0.83$ and 0.85, respectively (see Fig. 20). This is an interesting result for space weather applications, especially because we do not have direct measurement of magnetic field inside CMEs near the Sun. If one can estimate the strength and orientation of the magnetic field in CMEs near the Sun, it will prove to be useful in predicting the occurrence of geomagnetic storms. It must be pointed out that we have considered only MCs and their sheaths in the above discussion. The result should be applicable to non-cloud ICMEs also.

4.4 Magnetic structure of MCs and geomagnetic storms

The four types of MCs discussed in Section 3.5 have important implications to the D_{st} profile of the resulting storms. Figure 21 shows the D_{st} profiles corresponding to

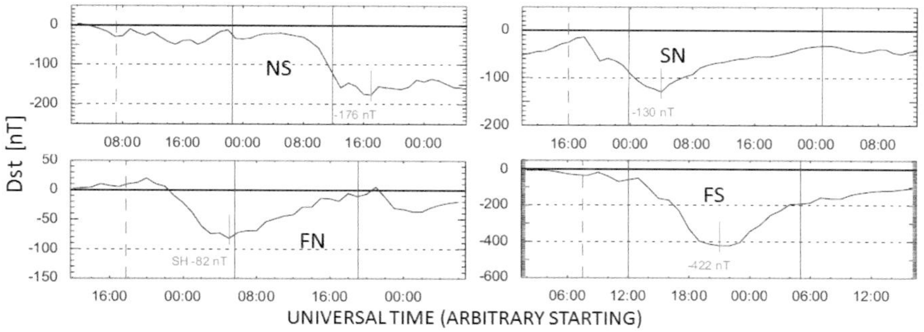

Fig. 21. Typical D_{st} profiles associated with the four types of MCs (NS: north-south, SN: south-north, FN: fully north, and FS: fully south) shown in Fig. 16. The cloud portions correspond to the interval between the vertical solid lines. The vertical dashed line indicates the shock. The D_{st} minimum values are notes on the plots, which represent the intensity of the geomagnetic storms associated with the MCs. For the NS MC, the storm peaks beyond the MC interval because the southward field is present only in the tail end of the MC. In the case of FN MC, the storm peak is before the MC interval because the southward field is present only in the sheath portion.

the four types of MCs described in Fig. 16. For NS MCs, the B_s occurs only in the trailing portion of the MC, so the storm occurs only during or after the tail portion. For SN MCs, the B_s occurs in the leading portion of the MC, so the storm peaks in the leading portion. For FS MCs, B_s is present throughout the MC, so the storm starts promptly upon arrival of the MC and the recovery starts only at the end of the MC interval (Gopalswamy *et al.*, 2005b). For FN MCs, there is no B_s during the MC interval, so they are not geoeffective at all (Yurchyshyn *et al.*, 2001). Thus one expects a large variability in the time at which the D_{st} minimum occurs with respect to the MC arrival at Earth. Additional time structure is caused by the presence of B_s in the sheaths preceding the MCs. If the sheath field has B_s and turns north before the MC interval, then one gets a double-dip storm (see e.g., Kamide *et al.*, 1998) for all MC types except for the FN MCs; in the case of FN MCs, all we get is a sheath storm.

Figure 22 shows the distribution of minimum D_{st} values in the cloud and sheath portions of MCs identified during solar cycle 23. The mean values of D_{st} in the cloud (-109 nT) and sheath (-90 nT) portions are at the intense storm level and comparable to each other. As noted above, the FN MCs are not geoeffective (average $D_{st} = -26$ nT, probably due to noise), while the FS MCs are most geoeffective (average $D_{st} = -125$ nT). The NS and SN MCs have similar average D_{st} values (-104 and -107 nT, respectively). The average D_{st} values in the sheath portions are at the storm level of all MC types with the FN MC sheaths being the most geoeffective (average $D_{st} = -132$ nT) and the FS MC sheaths the least geoeffective (average $D_{st} = -66$ nT). The NS and SN MC sheath storms are at the intermediate level: -75 nT and -85 nT, respectively.

The delay times of the D_{st} minima with respect to the arrival times of MCs are

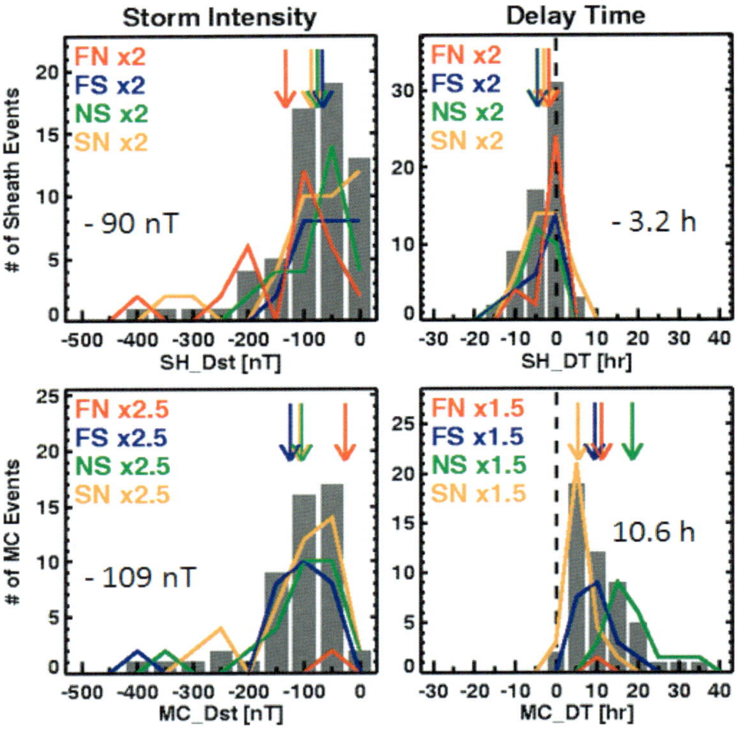

Fig. 22. Distributions of storm intensity (left) and delay time (right) for various MC types (colored lines) and all MCs (gray histogram). The values corresponding to the individual cloud types are multiplied by a suitable factor as indicated. The arrows point to the average values of the distributions. The average values corresponding to all events are also given on the plots. The vertical line on the delay time plots indicate the arrival time of MCs at the observing spacecraft.

also shown in Fig. 22. Sheaths arrive before the clouds, so the sheath storms peak before the MC arrival (average delay $= -3.2$ h). The cloud storms peak after MC arrival, as expected (average delay $= +10.6$ h). Among MC types, the delay time of cloud storms varies by a factor of $\sim$3, the smallest being for SN MCs (average delay $\sim +5.5$ h) and the largest for NS MCs (average delay $\sim +18.6$ h). The FS MCs have an intermediate delay (average delay $\sim +9.3$ h). In a few cases, the sheath storms have their D_{st} minima after the MC arrival. This is because the sheath B_s occurs too close to the MC arrival and the D_{st} minimum typically occurs $\sim$2 h after $|B_s|$ attains its maximum value (Gonzalez and Echer, 2005). It must be noted that the magnetospheric tail current also contributes to the D_{st} minimum (Ohtani *et al.*, 2001, 2005, 2007).

5 CMEs and Solar Energetic Particle Events

SEPs are so-called because their kinetic energy is much higher than that of the solar wind particles and they are of solar origin (in flares and CMEs) to distinguish them from cosmic rays, which are of galactic origin. SEP ions and electrons may suddenly have access to the inner magnetosphere during geomagnetic storms, where they can get trapped in a new radiation belt (Lorentzen *et al.*, 2002). During the largest SEP events, the energetic protons can penetrate all the way to Earth's atmosphere leading to significant ozone depletion (Jackman *et al.*, 2005). SEPs can damage the electronics of spacecraft, can pose radiation hazard to passengers and crew in high-flying aircraft, can harm humans in space, and may render satellites useless (see e.g., Turner, 2006) depending on the intensity of the SEPs. The SEP intensity is defined in terms of particle flux units (pfu, 1 pfu = 1 particle cm^{-2} s^{-1} sr^{-1}). SEPs of intensity exceeding 10 pfu in the >10 MeV energy channel are considered to be important for space weather. Occasionally the 10 MeV SEP intensity can exceed 10^4 pfu (see Gopalswamy *et al.*, 2005c).

5.1 Illustrative example

Figure 23 shows a large SEP event, plotted as the intensity of protons in three energy channels (>10 MeV, >50 MeV, and >100 MeV) as detected by the GOES satellite. Around the time of the sudden increase in SEP intensity (early on 2006 December 13), one can see a fast (1774 km/s) halo CME and a major solar flare (X-ray importance X3.4) in progress. The solar source of the flare was in the southwest quadrant (S06W23) on the Sun. The SEP intensity declines slowly in the three energy channels, until the ESP event that marks the arrival of the shock (which has been producing SEPs ever since it left the Sun). The shock arrival is also recorded by the ground based magnetometers and recognized as a positive excursion of the D_{st} index (SSC). The geomagnetic storm follows if the shock sheath or the driving ICME has a B_{s} component. The rapid decline of the SEP intensity after the shock arrival is due to the exclusion of the particles by the ICME analogous to the Forbush decrease.

5.2 Origin of SEPs: shock vs. flare

The connection between solar flares and SEPs was recognized early on (Forbush, 1946; Meyer *et al.*, 1956), and shocks of solar origin have also been thought to be another source of energetic particles (Uchida, 1960; Wild and Smerd, 1972). Kahler *et al.* (1978) established the link between CMEs and SEPs by identifying the CMEs as the drivers of the shocks that accelerate SEPs. Flares occur without CMEs, and produce energetic electrons (responsible for hard X-ray, microwave, and type III bursts) as well as protons. ESP events provide direct evidence for ion acceleration by shocks. Thus both flares and shocks can produce energetic particles. The current paradigm for SEPs is that impulsive, short-lived SEP events are due to flares and the large, gradual, long-lived events are accelerated in CME-driven shocks (see, e.g. Lin, 1987; Reames, 1999). Since all CMEs are associated with flares and higher-energy CMEs are associated with larger flares (see Fig. 8) one expects both flare and shock mechanisms operate simultaneously in every large eruptive event. However, since the spatial scale of flares is generally small (~15° heliographic) one expects to observe both flare and shock components only during the well-connected events. Shocks are

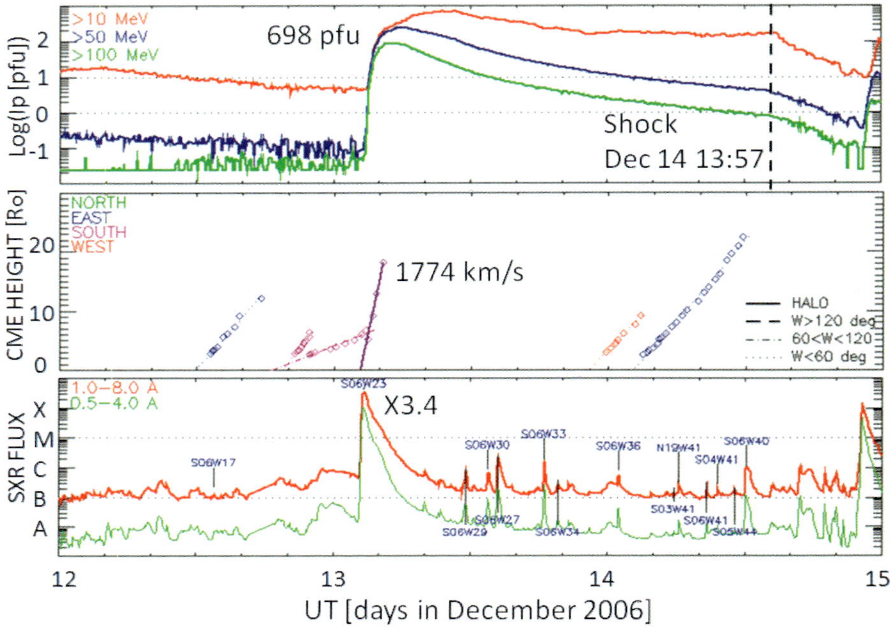

Fig. 23. The gradual SEP event on 2006 Dec 13. (Top) Proton intensity in three energy channels (>10 MeV, >50 MeV, and >100 MeV). The proton intensity reached a maximum value of 698 particle flux units (pfu) in the >10 MeV energy channel (1 pfu = 1 particle per (cm^2.s.sr)). The SEP intensity dropped rapidly after the shock arrived at 1 AU on December 14 at 13:57 UT (marked by the vertical dashed line). (Middle) Height-time plots of all CMEs that occurred over a three-day period. The fast (1774 km/s) halo CME on December 13 at 02:54 was responsible for the SEP event in question. (Bottom) GOES soft X-ray light curves in two energy channels (1.0–8.0 Å and 0.5–4.0 Å). Several flares occurred during the three-day period most of them coming from the active region NOAA 0930. The X3.4 flare associated with the SEP event occurred on December 13 at 02:40 UT.

more extended so the solar source can occur in wide range of longitudes, still favoring the western location because of the Parker-spiral structure of the interplanetary magnetic field that carries the particles to a detector near Earth. Figure 24(a) shows the solar sources of shock-producing CMEs inferred from DH type II radio bursts, with the SEP-producing subset distinguished. The shock-driving CMEs need to originate from the central meridian to western longitudes to produce an SEP event detectable at Earth. Figure 24(b) shows that the SEP association rate of radio-emitting CMEs peaks in the western hemisphere at $\sim 69^\circ$ longitude with a broad longitudinal distribution. Note that more than half of the CMEs originating from behind the west limb can result in SEP events, which is possible only when the shocks accelerate the particles. SEP events that result in high energy particles penetrating Earth's atmosphere are known as ground level enhancement (GLE) events (Cliver, 2006; Lopate, 2006). The solar sources of the GLE events are generally well-connected and hence a flare

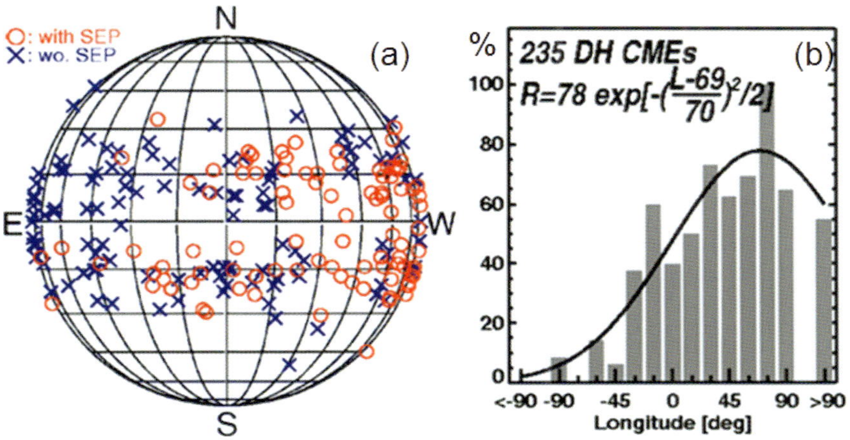

Fig. 24. (Left) Heliographic coordinates of CMEs that produce type II radio bursts in the decameter-hectometric wavelengths (all symbols). The subset associated with SEP events (with intensities ≥ 1 pfu) are denoted by the open circles. (Right) The SEP association rate (R) of CMEs that produce DH type II bursts, plotted as a function of CME source longitude (L). The curve fitted to the distribution is given on the plot. The distribution peaks at $L = 69°$ and has a width of $70°$. Note that there are no SEP events associated with CMEs from behind the east limb, but many from CMEs originating from behind the west limb.

origin is often assumed. However, several GLE events with poor connectivity have been reported in the literature (Cliver, 1981, 2006; Kudela *et al.*, 1993). In cycle 23, a single GLE event (2001 April 18) originated from $\sim 30°$ behind the west limb, yet the GLE event was well observed and ranked the fifth largest in cycle 23 (Gopalswamy, 2008a). It is highly unlikely that the flare component was observed in this GLE event.

Extensive comparison between SEPs and CMEs became possible only during cycle 23, which led to the result that large SEP events are invariably associated with fast and wide CMEs (Fig. 25). The average sky-plane speed of SEP-associated CMEs is $\sim$1623 km/s, much larger than the average speed of all CMEs (see Fig. 3). Most of the SEP-associated CMEs are also halo or partial halo CMEs confirming their higher energy. The speeds of GLE-associated CMEs are the highest, suggesting that they constitute the high-energy subset of the SEP-associated CMEs. Among the 16 GLE events detected in cycle 23, only two CMEs had speed less than 1500 km/s (938 km/s on May 02 and 1099 km/s on May 06 in 1998). Two such CMEs have also been reported for cycle 21 GLEs. Cliver (2006) noted that the two eruptions occurred when SEP background was elevated, which might imply an efficient shock acceleration as suggested by Gopalswamy *et al.* (2004). Another factor to be considered is the Alfven speed distribution in the corona (Gopalswamy *et al.*, 2001c; Mann *et al.*, 2003), which can vary anywhere from a few hundred km/s to more than 1600 km/s (Gopalswamy *et al.*, 2008c, d) in the outer corona where the GLEs and SEPs are released. Therefore, one cannot rule out the possibility that the lower speed

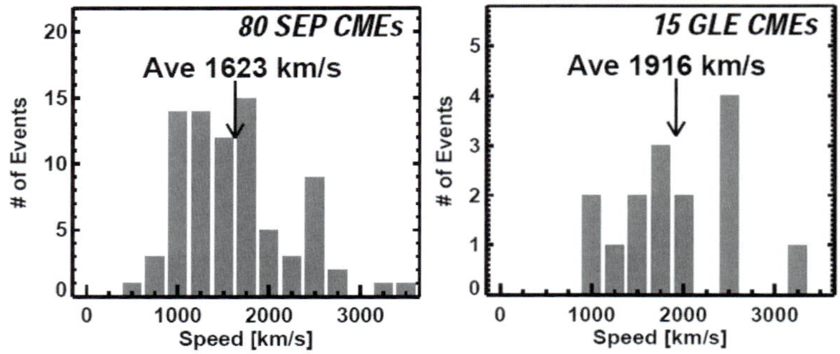

Fig. 25. Speed distributions for CMEs associated with SEP events (left) and GLE events (right) from solar cycle 23. The average values of the distributions are noted on the plots. There were actually 16 GLE events in cycle 23, but one did not have CME observations.

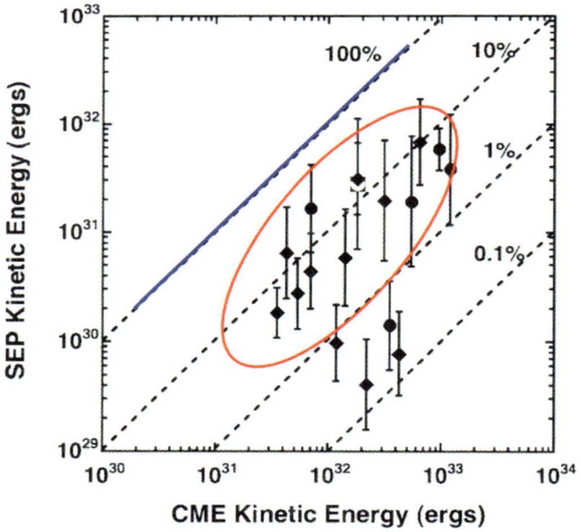

Fig. 26. Scatter plot between the CME kinetic energy and the SEP kinetic energy for a set of events during solar cycle 23. The blue line is the bisector. The dashed lines represent the fraction of CME kinetic energy (100%, 10%, 1%, and 0.1%) that goes into SEP kinetic energy. The red ellipse encircles the events that lie near the 10% level. (With kind permission from Springer Science+Business Media: *Space Sci. Rev.*, Solar energetic particle composition, energy spectra, and space weather, **124**, 2006, 303–316, Mewaldt, R. A., figure 5.)

CMEs associated with GLEs propagate into a medium with low Alfven speed. This also applies to some SEP events with low-speed CMEs (Gopalswamy *et al.*, 2008d). Another observational evidence for the close connection between CMEs and SEPs is the fraction of CME kinetic energy that goes into SEPs: For a set of 16 well observed

SEP events, the SEP kinetic energy constitutes ~10% CME kinetic energy (Fig. 26), which represents a high efficiency of particle acceleration by shocks (Mewaldt, 2006). Since the typical kinetic energy of SEP-producing CMEs is $>10^{32}$ erg, we see that large SEP events have an energy content of at least 10^{31} erg.

5.3 SEP release sites from shocks and flares

The presence of a type II burst in all SEP events implies the presence of a shock early in the event (Gopalswamy *et al.*, 2002; Cliver *et al.*, 2004). When the type II burst starts in the corona, the CME leading edge is typically in the heliocentric distance range of 1.7–$2.0R_o$, which means the CME-driven shock is also located at similar heights (Gopalswamy *et al.*, 2005a). On the other hand, the SEP release seems to happen when the CME is farther out in the corona, in the range 4–$8R_o$ (Kahler, 1994; Gopalswamy *et al.*, 2005d; Gopalswamy, 2008b). However, flare energy release takes place in the inner corona, typically at a heliocentric distance of $\sim 1.03R_o$. Since the shock is more extended than the CME located radially above the flare site, the flare particles (if at all they escape) have to pass through the shock front and may be further accelerated. Under the standard CSHKP eruption scenario, the CME flux rope forms due to reconnection, so particles accelerated in the reconnection region flow downwards to produce gamma ray lines and pion continuum (when the particle energy is very high—hundreds of MeV). As we noted before, the particles precipitate over a narrow range of angles, and may be injected over the same range in the direction away from the Sun (Cliver and Ling, 2007). It appears that the SEPs will be trapped in the CME flux rope and may not freely propagate into the IP medium, unless the accelerated ions have access to open field lines. Evidence for high energy (100s of keV to MeV) electrons trapped in moving CME structures is known for a long time from moving type IV radio bursts (see e.g. Gopalswamy and Kundu, 1989; Bastian *et al.*, 2001), and moving hard X-rays sources (Hudson *et al.*, 2001). But ions do not produce electromagnetic signatures, so we do not know if they are trapped in the flux rope. On the other hand, every large SEP event is associated with a type III bursts in the beginning (Cane *et al.*, 2002; Gopalswamy *et al.*, 2005d), which means energetic electrons from the reconnection site do escape from the flare site along open field lines (unless these electrons are also from the shock). There is no reason why the ions should not propagate along these field lines, thus indicating the possibility of a flare component to the large SEP events.

5.4 Challenges to shock acceleration paradigm

Despite the close correspondence between CMEs and SEP events, there are some discrepancies. For example, the CME speed and SEP intensity are reasonably correlated, yet the scatter is very large (Kahler, 2001; Gopalswamy *et al.*, 2003d). CME-driven shocks take in what lies ahead of them to accelerate and inject back into the heliosphere. This so-called source material is not just the ordinary solar wind, but can be ions from impulsive solar flares and previous gradual events, CIR events, pickup ions, CME ejecta, and the suprathermal tail of the solar wind (Gopalswamy *et al.*, 2004; Tylka *et al.*, 2005; Mewaldt, 2006). In fact, CMEs associated with most of the large SEP events seem to be propagating through a medium disturbed and distorted by preceding CMEs, a process described as preconditioning (Gopalswamy *et*

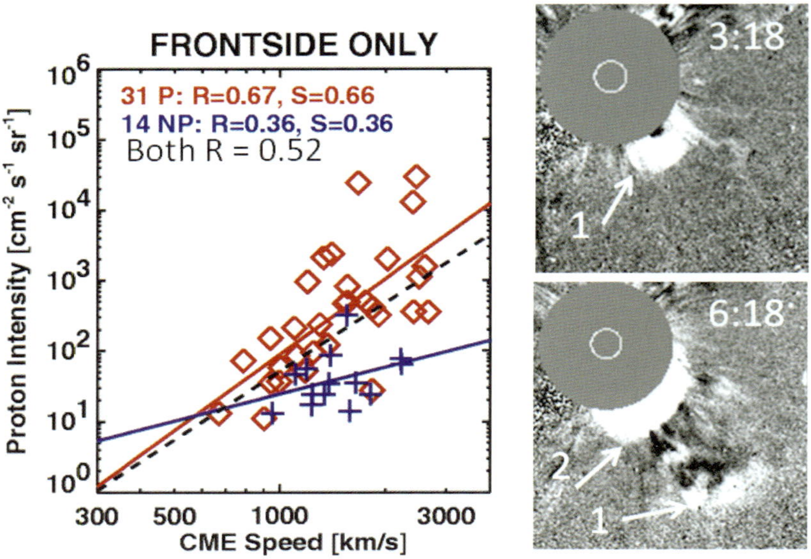

Fig. 27. (Left) Scatter plot between SEP intensity and CME speed. CMEs with (P) and without (NP) preceding CMEs are distinguished by different symbols. The correlation coefficients are shown for the two populations and for the combined set ($R = 0.52$). (Right) CMEs corresponding to one of the events used in the scatter plot, which occurred on 2001 October 1. The first CME (marked 1) appeared at 03:18 UT from an active region close to the southwest limb. While CME 1 was in progress the second one (marked) appeared at 06:18 UT from the same active region. Note that CME 2 propagates into the preceding CME and the associated disturbance rather than through the normal interplanetary medium.

al., 2003d, 2004). In order to investigate how the preceding CMEs affect the SEP intensity, Fig. 27 shows a scatter plot between CME speed and the SEP intensity; events with preceding CMEs (P) and no preceding CMEs (NP) are distinguished. The combined set of all SEP events has a correlation coefficient (r) of only 0.52. On the other hand, the P events have a much better correlation ($r = 0.67$) and the NP events have a poorer correlation ($r = 0.36$). In addition, P events consistently have a higher intensity than the NP events, so that the P and NP population occupy distinct regions in the scatter plot, thus explaining the large scatter in the CME speed vs. SEP intensity plot. Figure 27 also shows the CMEs involved in one of the SEP events used in the scatter plot: CME 2 is ejected directly behind a preceding CME (marked 1) that left from the same active region a few hours earlier. In the 06:18 UT SOHO/LASCO frame, one can see that what lies ahead of the SEP producing CME is the preceding CME, not the ordinary solar wind as in the 03:18 UT frame. Kahler and Vourlidas (2005) confirmed the above result, but attributed the difference between high and low intensity SEP events to the high and low brightness, respectively of the associated CMEs. Low brightness can also result if a CME propagates through a tenuous medium so it cannot sweep up more material.

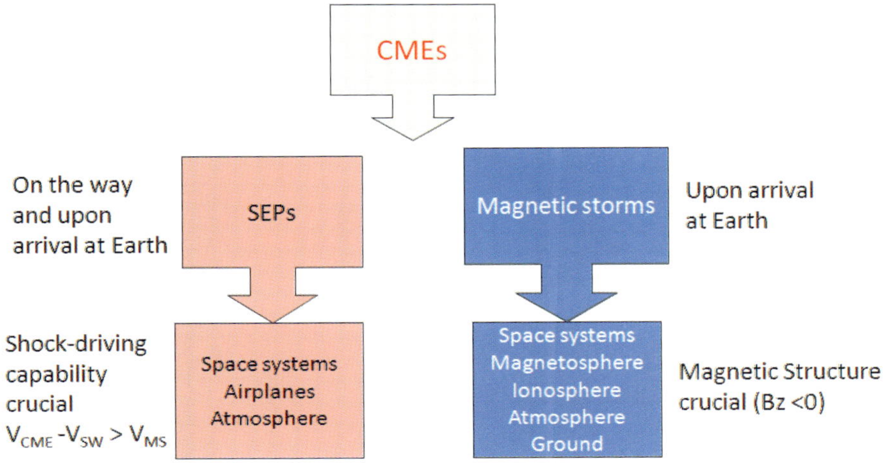

Fig. 28. The two primary interplanetary consequences of CMEs: SEPs and magnetic storms. SEPs are produced by CME-driven shocks on the way and upon arrival at Earth. CME speed (V_{CME}) in excess of the solar wind speed (V_{SW}) must exceed the magnetosonic speed (V_{MS}) to drive shocks. The SEPs affect space systems (with or without humans), passengers in high-altitude airplanes, and Earth's atmosphere. Geomagnetic storms are produced only when CMEs arrive at Earth with the additional condition that they possess south-pointing magnetic fields ($B_z < 0$). Geomagnetic storms have consequences to space systems, the entire geospace and even to the ground.

6 Summary

The two key space weather aspects of CMEs (SEPs and geomagnetic storms) and their geospace consequences are summarized in Fig. 28. CMEs responsible for SEPs and magnetic storms have different paths in the heliosphere, with some overlap. The spatial domains where the space weather effects are generated are also different. SEP production typically starts when the CME-driven shock is within a few solar radii from the Sun. The shock continues to produce SEPs in the IP medium so long as it is strong enough. When the shock arrives at the observing spacecraft, a sudden increase in SEP intensity is observed (the ESP event). Production of geomagnetic storms, on the other hand, happens only upon the arrival of B_s in the CME and/or the sheath at Earth's magnetosphere.

The shock-driving ability of a CME is crucial for the production of SEPs and the shocks need to be strong enough to efficiently accelerate SEPs. The coronal and interplanetary environment also plays a role in determining the strength of the shocks because the Alfven speed can vary by a factor of 4 in this region. The shock is also important for geoeffective CMEs in compressing the magnetosphere, but the intensity and duration of the southward component of the magnetic field in the shock sheath and ICME determine the strength of the resulting magnetic storm. Thus the magnetic structure of the ICME is important for geomagnetic storms, but not for SEPs.

Table 1 shows a compilation of the numbers of solar and interplanetary events

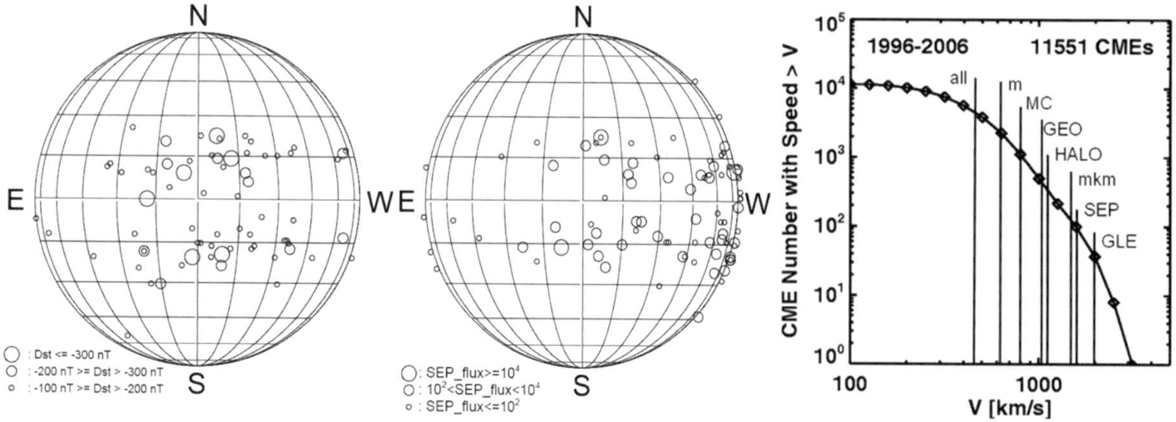

Fig. 29. Comparing the source location and speed of CMEs producing SEPs and geomagnetic storms. (Left) Source locations of geoeffective CMEs with large, medium and small circles representing storms with $D_{st} \leq -300$ nT, -200 nT $\geq D_{st} > -300$ nT, and -100 nT $\geq D_{st} > -200$ nT, respectively. (Middle) Source locations of SEP-associated CMEs with large, medium and small circles representing SEP events with peak flux $\geq 10^4$ pfu, in the range 10^2 to 10^4 pfu, and $\leq 10^2$ pfu, respectively. (Right) Cumulative distribution of CMEs, with the average speeds of various subsets: m (metric type II bursts), MC (magnetic clouds), GEO (geomagnetic storms), HALO (halo CMEs), mkm (type II bursts extending from metric to kilometric wavelengths), SEP (solar energetic particles), and GLE (ground level enhancement). The average speed of all CMEs is also shown for comparison (all).

Table 1. List of solar and interplanetary events important for space weather.

Event type	Rise	Max	Decl	Total	V_{CME} (km/s)
All CMEs ($W > 30°$)	962	3516	3261	7739	450[a]
Magnetic clouds	20	29	20	69	774[b]
Non-cloud ICMEs	17	72	35	124	960[c]
Frontside Halos	30	116	91	237	1052[b]
Major storms	12	32	25	69	1042[b]
Shocks (in situ)	34	110	72	226	987[d]
IP Type II bursts	28	125	104	257	1532[e]
Large SEP events	5	38	33	76	1623[f]

[a]For CMEs in Fig. 4; [b]from Gopalswamy *et al.* (2008a); [c]events in Fig. 14; [d]from Gopalswamy *et al.* (2009) (under preparation); [e]from Gopalswamy *et al.* (2008d); [f]from Fig. 25.

relevant for space weather. For each event type, the average sky-plane speed of the associated LASCO CMEs is given in the last column. Clearly only a tiny fraction of all CMEs (~3% or less) are important for space weather. The numbers are also binned into the three phases of the solar cycle 23. There is a general solar-cycle dependence of the numbers, the most coming from the maximum phase. The numbers of in situ shocks, IP type II bursts, and halo CMEs are similar (and much higher than the others); these phenomena happen when the associated CMEs meet special conditions in terms of source location on the Sun and magnetic connectivity to Earth. The last column shows that the speeds of the associated CMEs range from ~770 km/s to 1600 km/s, suggesting that these CMEs are very energetic. The CMEs causing major geomagnetic storms (geoeffective CMEs) have an average speed of ~1000 km/s, intermediate between those of halo CMEs and MC-associated CMEs. On the other hand, the speed of SEP associated CMEs is about 50% higher. This may not be a major difference if we take into account of the projection effects. Only a subset of SEP associated CMEs are close to the disk center, while most of the geoeffective CMEs originate there. Therefore, the geoeffective CMEs are subject to projection effects and their true speed is likely to be higher. Thus, we can conclude that both SEP producing and geoeffective CMEs are very energetic.

Figure 29 summarizes the source and kinematic properties of CMEs that produce SEPs and geomagnetic storms in comparison with other energetic CME populations. Storm-producing CMEs originate close to the disk center of the Sun, while the SEP-associated CMEs are mostly confined to the western hemisphere. Exceptions are the smaller events in both categories. The source locations of the two populations are distinct because the CME and SEP arrival at Earth take different paths: SEPs follow the magnetic field lines, while CMEs propagate radially away from the Sun. The cumulative number of CMEs in Fig. 29 drops precipitously at speeds >3000 km/s. In fact, there were only 9 CMEs with speed >2500 km/s among the 13000 + CMEs detected until the end of 2007. There were only two CMEs with speed exceeding 3000 km/s. This tells us that there is a limit to the CME kinetic energy. This limit can be traced to the amount of free energy that can be stored in the source active region. Since the amount of free energy is of the order of potential energy of the source magnetic fields, this translates into the maximum field strength in large active regions, known to be within ~5000 G. An important consequence of the speed limit is that the associated shocks take a finite time to reach Earth. Historical data on fast-transit shocks (Cliver *et al.*, 1990) and empirical modeling (Gopalswamy *et al.*, 2005c) suggest such a limit to the shock travel time, thus allowing a warning of at least about half a day.

Acknowledgments. Work supported by NASA's LWS program. The author thanks: S. Yashiro, H. Xie, S. Akiyama, and P. Mäkelä for help with some of the figures; E. W. Cliver and an anonymous referee for helpful comments on the manuscript; the organizers of the CAWSES meeting for travel support.

References

Aguilar-Rodriguez, E., X. Blanco-Cano, and N. Gopalswamy, Composition and magnetic structure of interplanetary coronal mass ejections at 1 AU, *Adv. Space Res.*, **36**, 522, 2006.

Akasofu, S.-I., Predicting geomagnetic storms as a space weather project, in *Space Weather Study using Multi-point Techniques*, edited by L.-H. Lyu, p. 3, COSPAR Colloquia series 12, 2002.

Bastian, T. S., M. Pick, A. Kerdraon, D. Maia, and A. Vourlidas, The coronal mass ejection of 1998 April 20: Direct imaging at radio wavelengths, *Astrophys. J.*, **558**, L65, 2001.

Biesecker, D. A. *et al.*, Solar phenomena associated with "EIT waves", *Astrophys. J.*, **569**, 1009, 2002.

Bothmer, V. and R. Schwenn, Eruptive prominences as sources of magnetic clouds in the solar wind, *Space Sci. Rev.*, **70**, 215, 1994.

Burlaga, L., E. Sittler, F. Mariani, and R. Schwenn, Magnetic loop behind an interplanetary shock—Voyager, Helios, and IMP 8 observations, *J. Geophys. Res.*, **86**, 6673, 1981.

Burkepile, J. T., A. J. Hundhausen, A. L. Stanger *et al.*, Role of projection effects on solar coronal mass ejection properties: 1. A study of CMEs associated with limb activity, *J. Geophys. Res.*, **109**, A03103, 2004.

Cane, H. V. and I. Richardson, Interplanetary coronal mass ejections in the near-Earth solar wind during 1996-2002, *J. Geophys. Res.*, **108**, SSH 6-1, 2003.

Cane, H. V., W. C. Erickson, and N. P. Prestage, Solar flares, type III radio bursts, coronal mass ejections, and energetic particles, *J. Geophys. Res.*, **107**, 1315, 2002.

Carrington, R. C., Description of a singular appearance seen in the Sun on September 1, 1859, *MNRAS*, **20**, 13, 1860.

Chen, J., R. A. Howard, G. E. Brueckner *et al.*, Evidence of an erupting magnetic flux rope: LASCO coronal mass ejection of 1997 April 13, *Astrophys. J.*, **490**, L191, 1997.

Chen, P. F., The relation between EIT waves and solar flares, *Astrophys. J.*, **641**, L153, 2006.

Cliver, E. W., Prompt injection of relativistic protons from the September 1, 1971 solar flare, *Sol. Phys.*, **75**, 341, 1981.

Cliver, E. W., The unusual relativistic solar proton events of 1979 August 21 and 1981 May 10, *Astrophys. J.*, **639**, 1206, 2006.

Cliver, E. W. and A. G. Ling, Electrons and protons in solar energetic particle events, *Astrophys. J.*, **658**, 1349, 2007.

Cliver, E. W., J. Feynman, and H. B. Garrett, An estimate of the maximum speed of the solar wind, 1938-1989, *J. Geophys. Res.*, **95**, 17103, 1990.

Cliver, E. W., D. F. Webb, and R. A. Howard, On the origin of solar metric type II bursts, *Sol. Phys.*, **187**, 89, 1999.

Cliver, E. W., S. W. Kahler, and D. V. Reames, Coronal shocks and solar energetic proton events, *Astrophys. J.*, **605**, 902, 2004.

Cliver, E. W., M. Laurenza, M. Storini, and B. J. Thompson, On the origins of solar EIT waves, *Astrophys. J.*, **631**, 604, 2005.

Cohen, C. M. S., Observations of energetic storm particles: An overview, *AGU Monograph 165*, 275, 2006.

Dal Lago, A., R. Schwenn, and W. D. Gonzalez, Relation between the radial speed and theexpansion speed of coronal mass ejections, *Adv. Space Res.*, **32**, 2637, 2003.

Dulk, G. and D. J. McLean, Coronal magnetic fields, *Sol. Phys.*, **57**, 279, 1978.

Dungey, J. W., Interplanetary magnetic field and the auroral zones, *Phys. Rev. Lett.*, **6**, 47, 1961.

Echer, E., M. V. Alves, and W. D. Gonzalez, A statistical study of magnetic cloud parameters and geoeffectiveness, *J. Atmos. Sol.-Terr. Phys.*, **67**, 839, 2005.

Elliott, H. A., D. J. McComas, N. A. Schwadron *et al.*, An improved expected temperature formula for identifying interplanetary coronal mass ejections, *J. Geophys. Res.*, **110**, A04103, 2005.

Feynman, J. and S. F. Martin, The initiation of coronal mass ejections by newly emerging magnetic flux, *J. Geophys. Res.*, **100**, 3355, 1995.

Filippov, B. P., Causes and effects of filament eruptions (review), in *New Perspectives on Solar Prominences (IAU Colloquium 167)*, edited by D. Rust, D. F. Webb, and B. Schmieder, Vol. 150, p. 342, 1998.

Forbush, S., Three unusual cosmic-ray increases possibly due to charged particles from the Sun, *Phys. Rev.*, **70**, 771, 1946.

Gold, T., Magnetic storms, *Space Sci. Rev.*, **1**, 100, 1962.

Gonzalez, W. D., A. L. Clua de Gonzalez, A. Dal Lago *et al.*, Magnetic cloud field intensities and solar wind velocities, *Geophys. Res. Lett.*, **25**(7), 963, 1998.

Gonzalez, W. D. and E. Echer, A study on the peak Dst and peak negative Bz relationship during intense geomagnetic storms, *Geophys. Res. Lett.*, **32**, L18103, 2005.

Gopalswamy, N., X-ray and microwave signatures of coronal mass ejections, in *Solar Physics with Radio Observations*, edited by T. Bastian, N. Gopalswamy, and K. Shibasaki, NRO Report No. 479, p. 141, 1999.

Gopalswamy, N., Relation between coronal mass ejections and their interplantary counterparts, *COSPAR Colloquia Series 14*, 157, 2002.

Gopalswamy, N., A global picture of CMEs in the inner heliosphere, in *The Sun and the Heliosphere as an Integrated System*, edited by G. Poletto and S. T. Suess, p. 201, Kluwer, Boston, 2004.

Gopalswamy, N., Coronal mass ejections of solar cycle 23, *J. Astrophys. Astron.*, **27**, 243, 2006a.

Gopalswamy, N., Consequences of coronal mass ejections in the heliosphere, *Sun and Geosphere*, **1**(2), 5, 2006b.

Gopalswamy, N., Properties of interplanetary coronal mass ejections, *Space Sci. Rev.*, **124**, 145, 2006c.

Gopalswamy, N., Coronal mass ejections and type II radio bursts, *AGU Monograph 165*, 207, 2006d.

Gopalswamy, N., Solar connections of geoeffective magnetic structures, *J. Atmos. Sol.-Terr. Phys.*, **70**, 2078, doi:10.1016/j.jastp.2008.06.010, 2008a.

Gopalswamy, N., Type II radio emission and solar energetic particle events, *AIP Conference Proceedings*, **1039**, p. 196, 2008b.

Gopalswamy, N. and M. R. Kundu, Radioheliograph and white-light coronagraph studies of a coronal mass ejection event, *Sol. Phys.*, **122**, 145, 1989.

Gopalswamy, N., Y. Hanaoka, T. Kosugi *et al.*, On the relationship between coronal mass ejections and magnetic clouds, *Geophys. Res. Lett.*, **25**, 2485, 1998.

Gopalswamy, N., A. Lara, R. P. Lepping, M. L. Kaiser, D. Berdichevsky, and O. C. St. Cyr, Interplanetary acceleration of coronal mass ejections, *Geophys. Res. Lett.*, **27**, 145, 2000.

Gopalswamy, N., S. Yashiro, M. L. Kaiser, R. A. Howard, and J.-L. Bougeret, Characteristics of coronal mass ejections associated with long-wavelength type II radio bursts, *J. Geophys. Res.*, **106**, 29219, 2001a.

Gopalswamy, N., A. Lara, S. Yashiro, M. L. Kaiser, and R. A. Howard, Predicting the 1-AU arrival times of coronal mass ejections, *J. Geophys. Res.*, **106**, 29207, 2001b.

Gopalswamy, N., A. Lara, M. L. Kaiser, and J.-L. Bougeret, Near-Sun and near-Earth manifestations of solar eruptions, *J. Geophys. Res.*, **106**, 25,261-25,278, 2001c.

Gopalswamy, N., S. Yashiro, G. Michalek *et al.*, Interacting coronal mass ejections and solar energetic particles, *Astrophys. J.*, **572**, L103, 2002.

Gopalswamy, N., A. Lara, S. Yashiro, S. Nunes, and R. A. Howard, Coronal mass ejection activity during solar cycle 23, *ESA SP-535*, 403, 2003a.

Gopalswamy, N., A. Lara, S. Yashiro, and R. A. Howard, Coronal mass ejections and solar polarity reversal, *Astrophys. J.*, **598**, L63, 2003b.

Gopalswamy, N., M. Shimojo, W. Lu, S. Yashiro, K. Shibasaki, and R. A. Howard, Prominence eruptions and coronal mass ejection: A statistical study using microwave observations, *Astrophys. J.*, **586**, 562, 2003c.

Gopalswamy, N., S. Yashiro, A. Lara *et al.*, Large solar energetic particle events of cycle 23: A global view, *Geophys. Res. Lett.*, **30**(12), SEP 3-1, 2003d.

Gopalswamy, N., S. Yashiro, S. Krucker, G. Stenborg, and R. A. Howard, Intensity variation of large solar energetic particle events associated with coronal mass ejections, *J. Geophys. Res.*, **109**, A12105, 2004.

Gopalswamy, N., E. Aguilar-Rodriguez, S. Yashiro *et al.*, Type II radio bursts and energetic solar eruptions, *J. Geophys. Res.*, **110**, A12S07, 2005a.

Gopalswamy, N., S. Yashiro, G. Michalek, H. Xie, R. P. Lepping, and R. A. Howard, Solar source of the largest geomagnetic storm of cycle 23, *Geophys. Res. Lett.*, **32**, L12S09, 2005b.

Gopalswamy, N., S. Yashiro, Y. Liu *et al.*, Coronal mass ejections and other extreme characteristics of the 2003 October-November solar eruptions, *J. Geophys. Res.*, **110**, A09S15, 2005c.

Gopalswamy, N., H. Xie, S. Yashiro, and I. Usoskin, Coronal mass ejections and ground level enhancements, *Proc. 29th International Cosmic Ray Conference*, Pune, p. 169, 2005d.

Gopalswamy, N., A. Lara, P. K. Manoharan, and R. A. Howard, An empirical model to predict the 1-AU arrival of interplanetary shocks, *Adv. Space Res.*, **36**(12), 2289, 2005e.

Gopalswamy, N., S. Yashiro, and S. Akiyama, Coronal mass ejections and space weather due to extreme events, in *Proceedings of the ILWS Workshop*, edited by N. Gopalswamy and A. Bhattacharyya, Quest, Mumbai, p.79, 2006.

Gopalswamy, N., S. Yashiro, and S. Akiyama, Geoeffectiveness of halo coronal mass ejections, *J. Geophys. Res.*, **112**, A06112, 2007.

Gopalswamy, N., S. Akiyama, S. Yashiro, G. Michalek, and R. P. Lepping, Solar sources and geospace consequences of interplanetary magnetic clouds observed during solar cycle 23, *J. Atmos. Sol.-Terr. Phys.*, **70**, 245, 2008a.

Gopalswamy, N., P. Mäkelä, H. Xie, S. Akiyama, and S. Yashiro, CME interaction with coronal holes and their interplanetary consequences, *J. Geophys. Res.*, doi:10.1029/2008JA013686, 2008b (in press).

Gopalswamy, N., S. Yashiro, H. Xie *et al.*, Radio-quiet fast and wide coronal mass ejections, *Astrophys. J.*, **674**, 560, 2008c.

Gopalswamy, N., S. Yashiro, S. Akiyama, P. Mäkelä, H. Xie, M. L. Kaiser, R. A. Howard, and J.-L. Bougeret, Coronal mass ejections, type II radio bursts, and solar energetic particle events in the SOHO era, *Ann. Geophys.*, **26**, 1, 2008d.

Gopalswamy, N., A. Dal Lago, S. Yashiro, and S. Akiyama, The expansion and radial speeds of coronal mass ejections, *Cent. Eur. Astrophys. Bull. 1*, 2008e (in press).

Gosling, J. T., Coronal mass ejections and magnetic flux ropes in interplanetary space, in *Physics of Magnetic Flux Ropes, AGU Monograph 58*, p. 343, 1990.

Gosling, J. T., The solar flare myth, *J. Geophys. Res.*, **98**, 18937, 1993.

Gosling, J. T., D. J. McComas, J. L. Phillips, and S. J. Bame, Coronal mass ejections and large geomagnetic storms, *Geophys. Res. Lett.*, **17**, 901, 1990.

Hanaoka, Y., H. Kurokawa, S. Enome *et al.*, Simultaneous observations of a prominence eruption followed by a coronal arcade formation in radio, soft X-rays, and H-alpha, *PASJ*, **46**, 205, 1994.

Harrison, R. A., Bursting the solar bubble: the flare - coronal mass ejection relationship, in *Solar Eruptions and Energetic Particles, AGU Monograph Series 165*, 73, 2006.

Harrison, R. A., C. J. Davis, C. J. Eyles *et al.*, First imaging of coronal mass ejections in the heliosphere viewed from outside the Sun Earth line, *Sol. Phys.*, **247**, 171, 2008.

Henke, T., J. Woch, R. Schwenn *et al.*, Ionization state and magnetic topology of coronal mass ejections, *J. Geophys. Res.*, **106**, 10,597, 2001.

House, L. L., W. J. Wagner, E. Hildner, C. Sawyer, and H. U. Schmidt, Studies of the corona with the Solar Maximum Mission coronagraph/polarimeter, *Astrophys. J.*, **244**, L117, 1981.

Howard, R. A., D. J. Michels, N. R. Sheeley Jr., and M. J. Koomen, The observation of a coronal transient directed at earth, *Astrophys. J.*, **263**, L101, 1982.

Howard, R. A., D. Michels, N. R. Sheeley, and M. J. Koomen, Coronal mass ejections—1979-1981, *J. Geophys. Res.*, **90**, 8173, 1985.

Hudson, H. S., Solar flares, microflares, nanoflares, and coronal heating, *Sol. Phys.*, **133**, 357, 1991.

Hudson, H. S. and D. F. Webb, Soft X-ray signatures of coronal ejections, in *Coronal Mass Ejections, AGU Monograph 99*, p. 27, 1997.

Hudson, H. S. and E. W. Cliver, Observing coronal mass ejections without coronagraphs, *J. Geophys. Res.*, **106**, 25199, 2001.

Hudson, H. S., T. Kosugi, N. V. Nitta, and M. Shimojo, Hard X-radiation from a fast coronal ejection, *Astrophys. J.*, **561**, L211, 2001.

Hundhausen, A. J., Sizes and locations of coronal mass ejections—SMM observations from 1980 and 1984-1989, *J. Geophys. Res.*, **98**, 13177, 1993.

Hundhausen, A. J., An introduction, in *Coronal Mass Ejections, AGU Monograph 99*, p.1, 1997.

Hundhausen, A. J., Coronal mass ejections, in *The Many Faces of the Sun*, edited by K. T. Strong, J. L. R. Saba, and B. M. Haisch, p. 143, Springer-Verlag, New York, 1999.

Hundhausen, A. J., J. T. Burkepile, and O. C. St. Cyr, Speeds of coronal mass ejections: SMM observations from 1980 and 1984-1989, *J. Geophys. Res.*, **99**(A4), 6543, 1994.

Illing, R. M. E. and A. J. Hundhausen, Disruption of a coronal streamer by an eruptive prominence and coronal mass ejection, *J. Geophys. Res.*, **91**, 10951, 1986.

Jackman, C. H., N. Achilleos, E. J. Bunce *et al.*, Neutral atmospheric influences of the solar proton events in October-November 2003, *J. Geophys. Res.*, **110**, A09S27, 2005.

Kahler, S., The morphological and statistical properties of solar X-ray events with long decay times, *Astrophys. J.*, **214**, 891, 1977.

Kahler, S. W., Injection profiles of solar energetic particles as functions of coronal mass ejection heights, *Astrophys. J.*, **428**, 837, 1994.

Kahler, S. W., The correlation between solar energetic particle peak intensities and speeds of coronal mass ejections: Effects of ambient particle intensities and energy spectra, *J. Geophys. Res.*, **106**, 20947, 2001.

Kahler, S. W., Observational properties of coronal mass ejections, in *Solar Eruptions and Energetic Particles, AGU Monograph 165*, 21, 2006.

Kahler, S. W. and A. Vourlidas, Fast coronal mass ejection environments and the production of solar energetic particle events, *J. Geophys. Res.*, **110**, A12S01, 2005.

Kahler, S. W., E. Hildner, and M. A. I. van Hollebeke, Prompt solar proton events and coronal mass ejections, *Sol. Phys.*, **57**, 429, 1978.

Kamide, Y., N. Yokoyama, W. Gonzalez *et al.*, Two-step development of geomagnetic storms, *J. Geophys. Res.*, **103**, 6917, 1998.

Kane, R. P., The idea of Space Weather A historical perspective, *Adv. Space Res.*, **37**, 1261, 2006.

Kay, H. R. M., L. K. Harra, S. A. Matthews, J. L. Culhane, and L. M. Green, The soft X-ray characteristics of solar flares, both with and without associated CMEs, *Astron. Astrophys.*, **400**, 779, 2003.

Koomen, M., R. Howard, R. Hansen, and S. Hansen, The coronal transient of 16 June 1972, *Sol. Phys.*, **34**, 447, 1974.

Kudela, K., M. A. Shea, D. F. Smart, and L. C. Gentile, Relativistic solar particle events recorded by the Lomnicky Stit neutron monitor, *Proc. 1993 ICRC 3*, 71, 1993.

Kunow, H., N. U. Crooker, J. A. Linker, R. Schwenn, and von R. Steiger, *Coronal mass ejections*, Springer, 2006.

Lepping, R. P. and D. Berdichevski, Interplanetary magnetic clouds: Sources, properties, modeling, and geomagnetic relationship, *Recent. Res. Geophys.*, **3**, 77, 2000.

Lepri, S. T., T. Zurbuchen, L. Fisk, I. Richardson, H. V. Cane, and G. Gloeckler, Iron charge distribution as an identifier of interplanetary coronal mass ejections, *J. Geophys. Res.*, **106**, 29231, 2001.

Li, Y. and J. G. Luhmann, Solar cycle control of the magnetic cloud polarity and the geoeffectiveness, *J. Atmos. Sol.-Terr. Phys.*, **66**, 323, 2004.

Lin, R. P., Solar particle acceleration and propagation, *Rev. Geophys.*, **25**, 676, 1987.

Lindsay, G. M., J. G. Luhmann, C. T. Russell, and J. T. Gosling, Relationships between coronal mass ejection speeds from coronagraph images and interplanetary characteristics of associated interplanetary coronal mass ejections, *J. Geophys. Res.*, **104**, 12515, 1999.

Loewe, C. A. and G. W. Prolss, Classification and mean behavior of magnetic storms, *J. Geophys. Res.*, **102**, 14209, 1997.

Lopate, C., Fifty years of ground level solar particle event observations, in *Solar Eruptions and Energetic Particles, AGU Monograph 165*, 283, 2006.

Lopez, R. E. and J. W. Freeman, Solar wind proton temperature-velocity relationship, *J. Geophys. Res.*, **91**, 1701, 1986.

Lorentzen, K. R., J. E. Mazur, M. D. Looper, J. F. Fennell, and J. B. Blake, Multisatellite observations of MeV ion injections during storms, *J. Geophys. Res.*, **107**(A09), 1231, 2002.

Mann, G., A. Klassen, H. Aurass, and H.-T. Classen, Formation and development of shock waves in the solar corona and the near-Sun interplanetary space, *Astron. Astrophys.*, **400**, 329, 2003.

Mewaldt, R. A., Solar energetic particle composition, energy spectra, and space weather, *Space Sci. Rev.*, **124**, 303, 2006.

Meyer, P., E. N. Parker, and J. A. Simpson, Solar cosmic rays of February, 1956 and their propagation through interplanetary space, *Phys. Rev*, **104**, 768, 1956.

Michalek, G., An asymmetric cone model for halo coronal mass ejections, *Sol. Phys.*, **237**, 101, 2006.

Moon, Y.-J., G. Choe, H. Wang, Y. Park, N. Gopalswamy, G. Yang, and S. Yashiro, A statistical study of two classes of coronal mass ejections, *Astrophys. J.*, **581**, 694, 2002.

Moore, R. L., A. C. Sterling, H. S. Hudson, and J. R. Lemen, Onset of the magnetic explosion in solar flares and coronal mass ejections, *Astrophys. J.*, **552**, 833, 2001.

Moreton, G. E., Hα observations of flare-initiated disturbances with velocities ~1000 km/sec, *Astron. J.*, **65**, 494, 1960.

Mulligan, T., C. T. Russell, and J. G. Luhmann, Solar cycle evolution of the structure of magnetic clouds in the inner heliosphere, *Geophys. Res. Lett.*, **25**, 2959, 1998.

Munro, R. H., J. T. Gosling, E. Hildner, R. M. MacQueen, A. I. Poland, and C. L. Ross, The association of coronal mass ejection transients with other forms of solar activity, *Sol. Phys.*, **61**, 201, 1979.

Narukage, N., H. S. Hudson, T. Morimoto *et al.*, Simultaneous observation of a Moreton wave on 1997 November 3 in Hα and soft X-rays, *Astrophys. J.*, **572**, L109, 2002.

Nelson, G. J. and D. B. Melrose, Type II bursts, in *Solar radiophysics*, p. 333, Cambridge University Press, New York, 1985.

Neugebauer, M. and B. E. Goldstein, Particle and field signatures of coronal mass ejections in the solar wind, in *Coronal Mass Ejections, AGU Monograph 99*, 245, 1997.

Neugebauer, M., J. T. Steinberg, R. L. Tokar *et al.*, Genesis on-board determination of the solar wind flow regime, *Space Sci. Rev.*, **105**, 661, 2003.

Neupert, W. M., Transient coronal extreme ultraviolet emission before and during the impulsive phase of a solar flare, *Astrophys. J.*, **344**, 504, 1989.

Ohtani, S., M. Nosé, G. Rostoker, H. Singer, A. T. Y Lui, and M. Nakamura, Storm-substorm relationship: Contribution of the tail current to Dst, *J. Geophys. Res.*, **106**, 21199, 2001.

Ohtani, S., P. C. Brandt, D. G. Mitchell *et al.*, Storm-substorm relationship: Variations of the hydrogen and oxygen energetic neutral atom intensities during storm-time substorms, *J. Geophys. Res.*, **110**, A07219, 10.1029/2004JA010954, 2005.

Ohtani, S., H. Korth, P. C. Brandt *et al.*, Cluster observations in the inner magnetosphere during the 18 April 2002 sawtooth event: Dipolarization and injection at r = 4.6 RE, *J. Geophys. Res.*, **112**, A08213, doi:10.1029/2007JA012357, 2007.

Parker, E. N., The gross dynamics of a hydromagnetic gas cloud, *Astrophys. J.*, **S 3**, 51, 1957.

Reames, D., Particle acceleration at the Sun and in the heliosphere, *Space Sci. Rev.*, **90**, 413, 1999.

Reinard, A. A., Comparison of interplanetary CME charge state composition with CME-associated flare magnitude, *Astrophys. J.*, **620**, 501, 2005.

Riley, P., C. Schatzman, H. V. Cane, I. G. Richardson, and N. Gopalswamy, On the rates of coronal mass ejections: Remote solar and in situ observations, *Astrophys. J.*, **647**, 648, 2006.

Rust, D. M., Coronal disturbances and their terrestrial effects /Tutorial Lecture/, *Space Sci. Rev.*, **34**, 21, 1983.

Sheeley, N. R., W. N. Hakala, and Y.-M. Wang, Detection of coronal mass ejection associated shock waves in the outer corona, *J. Geophys. Res.*, **105**, 5081, 2000.

Shibata, K., S. Masuda, M. Shimojo *et al.*, Hot-plasma ejections associated with compact-loop solar flares, *Astrophys. J.*, **451**, L83, 1995.

Song, P., H. J. Singer, and G. L. Siscoe, Space weather, *AGU Monograph 125*, 2001.

Švestka, Z. and E. W. Cliver, History and basic characteristics of eruptive flares, in *Eruptive Solar Flares*, edited by Z. Švestka, B. V. Jackson and M. E. Machado, p. 1, Springer, New York, 1992.

Tandberg-Hanssen, E., *The Nature of Solar Prominences*, Kluwer, Dordrecht, 1995.

Thompson, B. J., J. Gurman, W. Neupert *et al.*, SOHO/EIT observations of the 1997 April 7 coronal transient: Possible evidence of coronal Moreton waves, *Astrophys. J.*, **517**, L151, 1999.

Thompson, B. J., E. W. Cliver, N. Nitta, C. Delannée, and J.-P. Delaboudiniére, Coronal dimmings and energetic CMEs in April-May 1998, *Geophys. Res. Lett.*, **27**, 1431, 2000.

Tokumaru, M., M. Kojima, K. Fujiki, M. Yamashita, and B. V. Jackson, The source and propagation of the interplanetary disturbance associated with the full-halo coronal mass ejection on 28 October 2003, *J. Geophys. Res.*, **112**, A05106, 2007.

Tripathi, D., V. Bothmer, and H. Cremades, The basic characteristics of EUV post-eruptive arcades and their role as tracers of coronal mass ejection source regions, *Astron. Astrophys.*, **422**, 337, 2004.

Tsurutani, B. T., E. J. Smith, W. D. Gonzalez, F. Tang, and S.-I. Akasofu, Origin of interplanetary southward magnetic fields responsible for major magnetic storms near solar maximum (1978-1979), *J. Geophys. Res.*, **93**, 8519, 1988.

Turner, R. E., Space weather challenges intrinsic to the human exploration of space, in *Solar Eruptions and Energetic Particles, AGU Monograph 165*, 367, 2006.

Tylka, A. J., C. M. S Cohen, W. F. Dietrich *et al.*, Shock geometry, seed populations, and the origin of variable elemental composition at high energies in large gradual solar particle events, *Astrophys. J.*, **625**, 474, 2005.

Uchida, Y., On the exciters of type II and type III solar radio bursts, *PASJ*, **12**, 376, 1960.

Vennerstroem, S., Interplanetary sources of magnetic storms: A statistical study, *J. Geophys. Res.*, **106**, 29175, 2001.

Veronig, A., M. Temmer, B. Vršnak, and J. Thalmann, Interaction of a Moreton/EIT wave and a coronal hole, *Astrophys. J.*, **647**, 1466, 2006.

Vourlidas, A., D. Buzasi, R. A. Howard, and E. Esfandiari, Mass and energy properties of LASCO CMEs, in *Solar variability: from core to outer frontiers*, edited by A. Wilson, ESA SP-506, Vol. 1, Noordwijk: ESA Publications Division, p. 91, 2002.

Vourlidas, A., S. T. Wu, A. H. Wang, P. Subramanian, and R. A. Howard, Direct detection of a coronal mass ejection-associated shock in large angle and spectrometric coronagraph experiment white-light images, *Astrophys. J.*, **598**, 1392, 2003.

Vršnak, B., D. Ruzdjak, D. Sudar, and N. Gopalswamy, Kinematics of coronal mass ejections between 2 and 30 solar radii. What can be learned about forces governing the eruption?, *Astron. Astrophys.*, **423**, 717, 2004.

Vršnak, B., D. Sudar, and D. Ruzdjak, The CME-flare relationship: Are there really two types of CMEs?, *Astron. Astrophys.*, **435**, 1149, 2005.

Wang, C., D. Du, and J. D. Richardson, Characteristics of the interplanetary coronal mass ejections in the heliosphere between 0.3 and 5.4 AU, *J. Geophys. Res.*, **110**, A10107, 2005.

Warmuth, A., B. Vršnak, H. Aurass, and A. Hanslmeier, Moreton waves and their relation with EIT waves, in *Proceedings of the Second Solar Cycle and Space Weather Euroconference, ESA SP 477*, 195, 2002.

Webb, D. F., E. W. Cliver, N. U. Crooker, O. C. St. Cyr, and B. J. Thompson, Relationship of halo coronal mass ejections, magnetic clouds, and magnetic storms, *J. Geophys. Res.*, **105**, 7491, 2000.

Wild, J. P. and S. F. Smerd, Radio bursts from the solar corona, *Ann. Rev. Astron. Astrophys.*, **10**, 159, 1972.

Xie, H., L. Ofman, and G. Lawrence, Cone model for halo CMEs: Application to space weather forecasting, *J. Geophys. Res.*, **109**, 3109, 2004.

Yashiro, S., N. Gopalswamy, G. Michalek *et al.*, A catalog of white light coronal mass ejections observed by the SOHO spacecraft, *J. Geophys. Res.*, **109**, A07105, 2004.

Yashiro, S., S. Akiyama, N. Gopalswamy, and R. A. Howard, Different power-law indices in the frequency distributions of flares with and without coronal mass ejections, *Astrophys. J.*, **650**, L143, 2006.

Yashiro, S., G. Michalek, S. Akiyama, N. Gopalswamy, and R. A. Howard, Spatial relationship between solar flares and coronal mass ejections, *Astrophys. J.*, **673**, 1174, 2008a.

Yashiro, S., G. Michalek, and N. Gopalswamy, A comparison of coronal mass ejections identified by manual and automatic methods, *Ann. Geophys.*, **26**, 3103, 2008b.

Yeh, C.-T., M. D. Ding, and P. F. Chen, Kinetic properties of CMEs corrected for the projection effect, *Sol. Phys.*, **229**, 313, 2005.

Yurchyshyn, V., Relationship between EIT posteruption arcades, coronal mass ejections, the coronal neutral line, and magnetic clouds, *Astrophys. J.*, **675**, L49, 2008.

Yurchyshyn, V., H. Wang, P. R. Goode, and Y. Deng, Orientation of the magnetic fields in interplanetary flux ropes and solar filaments, *Astrophys. J.*, **563**, 381, 2001.

Zarro, D., A. C. Sterling, B. J. Thompson, H. S. Hudson, and N. Nitta, SOHO EIT observations of extreme-ultraviolet "dimming" associated with a halo coronal mass ejection, *Astrophys. J.*, **520**, L193, 1999.

Zhang, J., K. P. Dere, R. A. Howard, M. R. Kundu, and S. M. White, On the temporal relationship between coronal mass ejections and flares, *Astrophys. J.*, **559**, 452, 2001.

Zhang, J., I. Richardson, D. F. Webb *et al.*, Solar and interplanetary sources of major geomagnetic storms (Dst ≤ -100 nT) During 1996–2005, *J. Geophys. Res.*, **112**, A10102, 2007.

Zhao, X. P., Inversion solutions of the elliptic cone model for disk frontside full halo coronal mass ejections, *J. Geophys. Res.*, **113**, A02101, 2008.

Zhao, X. P. and D. F. Webb, Source regions and storm effectiveness of frontside full halo coronal mass ejections, *J. Geophys. Res.*, **108**(A6), 1234, 2003.

Climate and Weather of the Sun-Earth System (CAWSES): Selected Papers from the 2007 Kyoto Symposium,
Edited by T. Tsuda, R. Fujii, K. Shibata, and M. A. Geller, pp. 121–170.

Magnetotail after Geotail, Interball and Cluster: Thin current sheets, fine structure, force balance and stability

Lev Zelenyi[1], Helmi Malova[1,2], Anton Artemyev[1], Victor Popov[1,3], Anatoly Petrukovich[1], Dominique Delcourt[4], and Alexey Bykov[3]

[1]*Space Research Institute of Russian Academy of Sciences, Russia*
E-mail: lzelenyi@iki.rssi.ru
[2]*Scobeltsyn Institute of Nuclear Physics of Moscow State University, Russia*
[3]*Physics Department of Moscow State University, Russia*
[4]*CETP-CNRS-IPSL, Saint-Maur des Fossés, France*

We present a short review devoted to some scientific achievements in the physics of Earth's magnetotail current sheets based on results of Geotail, Interball and Cluster missions. The structure and dynamics of very thin current sheets (TCSs) that are often observed at the night side of the Earth's magnetosphere are discussed in details. The development of theoretical ideas and models, stimulated by numerous experimental findings in the terrestrial (and other) magnetospheres, outlined to formulate the state of the art in this field of physics in the "pre-THEMIS" era.

1 Early Space Investigations: the Discovery of the Earth's Magnetotail

The launch of the first soviet satellite Sputnik in 1957 started a new era of space exploration. The earliest magnetospheric FINDING was the discovery of the Earth's inner and outer radiation belts (Vernov, 1958; Van Allen and Frank, 1959). Later further measurements revealed that the Earth's giant magnetic dipole actively interacts with the flow of solar plasma. This flow, lately romantically named "solar wind", was predicted by Parker (1958) and Biermann *et al.* (1951), but for the first time was measured *in situ* by Gringauz and his colleagues (Gringauz *et al.*, 1960) at the soviet spacecraft Luna launched towards the Moon. In 1959 Thomas Gold of Cornell University proposed the name "magnetosphere" for the magnetic cavity formed around the Earth (Gold, 1959). Very soon Normann Ness (1965) on the basis of the existing experimental observations described the Earth's magnetosphere to be similar to very compressed dipole at the dayside, but at the nightside having very different non-dipole shape. Nightside magnetosphere appeared to have very elongated magnetic structure resembling the one of the visible cometary tail due to its strong stretching by the solar wind. This structure was named "magnetotail" by Normann Ness. In the central part of magnetotail the electric currents flow from the dawn to dusk direction. The thick current sheet (with thickness about one to several Earth radii) supports the oppositely directed magnetic fields existing in a vast region named "magnetotail

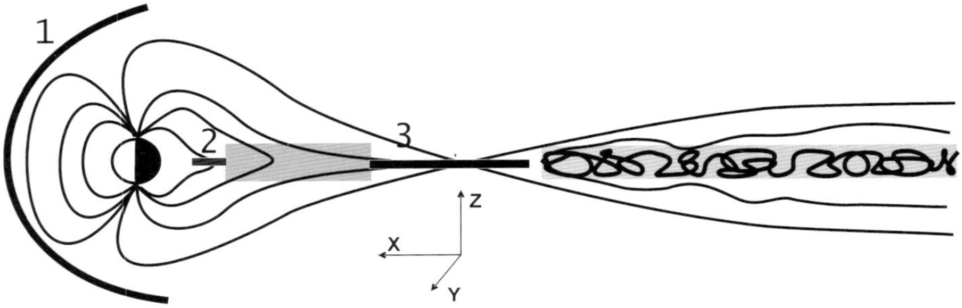

Fig. 1. The schematic view of the Earth's magnetosphere in GSM coordinate system. Bold black line 1 signifies the dayside magnetopause, regions 2 (near-Earth edge of current sheet) and 3 (distant magnetotail region) indicate the areas where thin current sheets have been observed most frequently. The neutral "X"-line separates two different magnetotail regions: in the near-Earth's part of the tail where the influence of the dipole B_z-component of the magnetic field is still appreciably strong, the magnetic field lines have closed topology and current sheet has relatively regular structure. Tailward from "X"-line the dipole influence becomes negligibly weak and downstream magnetic field strongly fluctuates.

lobes". The maximum length of the tail is not exactly known. The trace of the tail-like magnetic structure was observed far downstream of the lunar orbit ($\sim 60 R_e$), and, in some measurements, at the distances about 1.5 millions of kilometers. The radius of magnetotail cross-section has a tendency to flare downstream from the Earth and reach about $(25\text{–}30) R_e$ at the distances about $200 R_e$.

Hereafter we will use the standard Geocentric Solar Magnetospheric System (GSM), where the X-axis is directed from the Earth center to the center of the Sun, the Z-axis coincides with the direction of the Earth dipole and Y-axis is directed correspondingly from the dawn to dusk side. The schematic illustration of magnetotail structure is shown in Fig. 1.

During last 15 years few international space missions including Geotail, Interball and Cluster have been launched and performed the important investigations of the structure and dynamics of the Earth's magnetosphere and solar-terrestrial interactions. Nowadays the new generation of space missions based on new technologies and concepts is prepared to be launched (or were recently launched) for further investigations of the Earth's environment. As an example one can mention the project THEMIS (launched in 2007), which is studying the physical processes controlling large-scale phenomenology of magnetosphere, such as magnetotail dynamics, various energy transformations and substorm initiations. Next multipoint magnetospheric mission to study physical processes responsible for magnetic reconnection will be MMS (magnetospheric multiscale mission) which NASA plans to launch in 2014. Because very important site of energy conversion is the magnetotail, the significant attention in the past studies have been devoted to investigations of the fine structure, intrinsic properties and response to external perturbations of the "magnetotail heart"—its current sheet.

Three abovementioned core spacecraft missions have contributed a lot to the physics of space plasma and stimulated theoretical modeling of a numerous magnetospheric processes. One should mention here the fruitful investigations of the processes in the vicinity of X-line made by Geotail and Cluster teams (Nagai *et al.*, 2001; Baumjohann, 2002) and "GEM-Reconnection challenge" devoted to theoretical modeling of the X-line formation (Birn *et al.*, 2001) and subsequent plasma dynamics timing done by Interball (Petrukovich *et al.*, 1998, 1999, 2003) which confirmed the role of magnetotail as the "seed" region for the beginning of substorms. Studies of plasmoid formation by Interball allowed to estimate the properties of plasmoids and their propagation speed in the midtail (Yermolaev, 1998) and even provided the evidence that plasmoids could be strongly asymmetric in the North-South direction (Borodkova *et al.*, 1999). Another outstanding breakthrough was the discovery of ESW (electrostatic solitary waves) in the plasma sheet boundary layer, which after Fourier transform looks like the BEN (broadband electrostatic noise) (Matsumoto *et al.*, 1994; Kojima *et al.*, 1997). In a frame of this single review we will not attempt to address all these interesting area of scientific investigations, instead we will try to focus both on the experimental findings as well as their theoretical interpretations related with the only one, but key region of magnetosphere—its current sheet (CS) which was thoroughly investigated during Geotail, Interball and Cluster missions.

The coupling of the solar wind mass, momentum and energy to the Earth's magnetosphere takes place at the magnetopause (bold black line 1 in Fig. 1) separating the solar wind and the compressed magnetic field of the Earth and controls the entire magnetotail. The solar wind impacts the magnetosphere by hot solar plasma with a "frozen" magnetic field. At the magnetopause magnetic reconnection often leads to the transfer of magnetic flux into the magnetosphere. The large-scale structure of this interplanetary magnetic field (IMF) is usually sector-like with arbitrary changing directions. When the IMF orientation is southward the active reconnection begins between the magnetic field of the solar wind and the Earth's magnetosphere at the dayside magnetopause. According to the well-known classical scenario by Dungey the reconnected field lines flow downstream towards the tail with the solar wind flow (Dungey, 1961). This leads to an increase of magnetic flux in the magnetotail and to the growth of the convective electric field E_y, which produces current sheet thinning and stretching the tail in downstream direction. During the growth phase of substorms a TCS could be formed in the near Earth edge of the overall current sheet (region 2 in Fig. 1). The thickness of this current sheet could be between 250 to 2500 km (Sergeev *et al.*, 1993; Pulkkinen *et al.*, 1994; Runov *et al.*, 2006), which is significantly thinner than the one of current sheet under quiet conditions (up to 10 times). This process of TCS formation may have a duration about 30 min–2 hours (growth phase of substorm), and finally TCS becomes the reservoir of stored magnetic energy. The peculiar property of this TCS is metastability (Galeev and Zelenyi, 1976), i.e., current sheet can be stable for quite a long time, but abrupt release of the stored energy destroys it. Magnetic energy is converted to the kinetic energy of plasma flows and electromagnetic waves. Among the processes contributing to the evolution of the current sheet structure, the generally accepted scenario is magnetic reconnection and the formation of a neutral "X"- line (Lui, 1993, 1996; Baker, 1996; Lui *et al.*, 1998).

It explains the destruction of the TCS by the beginning of fast reconnection while the "X"-line might be considered as an element of a long-wave tearing mode. However there is also substantial evidence for another non-reconnection triggering due to disruption of magnetotail current sheet in the near-Earth part of the magnetotail. This disruption also is often associated with the development of various instabilities driven by transverse magnetotail currents (Lui, 1993, 1996, 2004). The choice between these two scenarios (or development of a unified (synthetic) one) is till now the subject of very intense debates hopefully the issue will be resolved by global THEMIS observations.

This aspect of current sheet evolution through development of various instabilities has been actively pursued during the last decade. While many instabilities have been identified and studied, the picture is still unclear. The development of a substorm beyond the onset stage however is better known: the stored magnetic energy is released in a form of plasma flows, electromagnetic waves, and plasmoid formation. At the earthward side of a neutral line, a part of the plasma flows directed to the ionosphere, redistributes a system of ionospheric currents and fields, and dissipates in the ionosphere producing as a byproduct bright auroral emissions. At the tailward region near the neutral line, a giant plasmoid (about 10 to several tens of R_E) could be formed and expelled downstream, returning to the solar wind some part of substorm energy. TCS processes obviously play the principal role in all these phenomena. The experimental investigations of TCSs began later than observations of other magnetospheric structures. This is related with the difficulty to separate spatial and temporal variations and the necessity to restore pattern of currents in the plasma sheet. For the first time these problems were solved (to a certain extent) by four-point Cluster measurements. However, spatial separation of Cluster spacecraft limits the possible accuracy to restore fine structure of cross-tail currents. Therefore, there exist many questions concerning the structure, stability, and evolution of the TCS which do not have yet clear answers. Below we will outline the main concepts, experimental facts and models concerning the magnetotail current sheet, its fine structure and some dynamical properties.

2 Geotail, Interball, Cluster and Other ISTP

In this section we will describe shortly three key space missions, which provided the very significant advance in our knowledge of current sheet during last 1.5 decades: Geotail, Interball and Cluster (see Fig. 2). In order to accomplish the detailed investigations of the tail region of the magnetosphere, the Institute of Space and Astronautical Science of Japan (ISAS) and the NASA undertook a joint project to develop, launch, and operate a scientific **Geotail** satellite. The Geotail mission measured global energy flow and transformations in the magnetotail to increase our understanding of fundamental magnetospheric processes, including the physics of the magnetopause, plasma sheet, reconnection and neutral line formation. These objectives required spacecraft measurements in two orbits: a nightside double lunar swingby Geotail orbit to distances of $220R_E$ and a low inclination orbit at geocentric distances of about 8 to $30R_E$. The Geotail satellite was placed into the initial distant

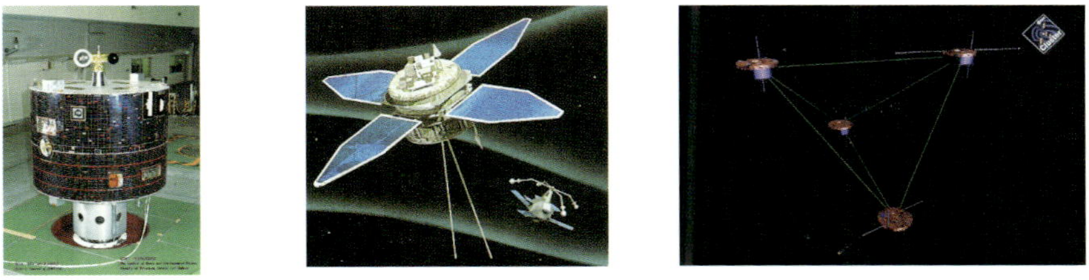

Fig. 2. Geotail, Interball and Cluster devices.

tail orbit for approximately two and a half years; thereafter, it had been maneuvered to a second orbit where it will operate for the remainder of its lifetime.

Since its launch Geotail has been operating satisfactorily and has given a variety of new information that has substantially advanced our knowledge of the Earth's magnetotail.

The Russian Space Agency (Lavochkin Association) and Space Research Institute (IKI) of Russian Academy of Sciences launched **Interball 1-2** and Magion 4-5 satellites in the same time frame. The aim of Interball project was to study various plasma processes in the near-Earth space, especially solar-terrestrial physical coupling. This project included two pairs of spacecrafts (satellite-subsatellite) at high altitude orbits: to 200 000 km for the Tail Probe pair and to 20 000 km for the Auroral Probe pair. Tail Probe and Auroral Probe were successfully launched, correspondingly, on 3 August 1995 and 29 August 1996. The Tail Probe orbit was launched to elliptical orbit to reach high altitude cusp and subsolar magnetopause regions at the dayside and then, the magnetotail lobes and the neutral sheet in the nightside near-Earth tail. The Auroral Probe orbit was optimized for the magnetic conjunctions between the two pairs of satellites around the midnight meridian.

The Interball project involved the efforts of a large international community of Russia together with 15 countries including France, Bulgaria, Cuba, Poland, Chech republics and Slovak. Spacecrafts were operated from DSN station in Eupatoria (Ukraine). GEOTAIL and Interball/Tail satellites represent a complementary set of spacecraft for the magnetotail study since Interball/Tail was often in the tail lobes while GEOTAIL was designed to be mostly in the plasma sheet most of the time. The comparison with the POLAR observations was also actively conducted. Upstream measurements were provided by Wind spacecraft (NASA). The joint measurements in a form of a few coordinated observation campaigns were coordinated by IACG (Inter-Agency Consulting Group). IACG consisted of the heads of ESA, NASA, ISAS and Russian Space Agency and a few groups of scientific experts from these countries. IACG was originally formed in 1980-ies to coordinate Halley comet missions. IACG played an outstanding role in promotion the ISTP (International Solar-Terrestrial Physics Program) besides numerous excellent examples of effective international cooperation provided a set of unique scientific results concerning the global characteristics of solar-terrestrial relations.

Besides the close cooperation between these projects ground-based geophysical community provided timely information on the current solar and geophysical conditions which are indispensable to put the satellite data in the global solar-terrestrial perspective. Although in 2000 year the Interball-Tail spacecraft reentered the atmosphere, the operation was successful, especially its results on the cause-and-effect relationships of various plasma processes in the magnetotail and auroral regions.

The **Cluster** mission launched by ESA (European Space Agency) includes four identical spacecraft flying in a tetrahedral configuration. Cluster spacecrafts were successfully launched in 2000 by Russian Soyuz-Fregat rockets (after the tragic Ariane-5 failure in 1996). The Cluster mission is aimed to investigate the Earth's magnetic environment and its interaction with the solar wind in few critical physical regions, including the current sheet of the Earth's magnetotail. Four-point Cluster measurements enable the separation of spatial and temporal effects and provide a detailed three-dimensional map of the current systems in the magnetosphere. The distance between the four satellites varies along the orbit, from a few hundreds of km near perigee up to a few thousands of km near apogee.

Cluster is the first scientific mission flying in a constellations enabling to map the Earth's environment in 4 dimensions, both space and time. This experimental breakthrough provided for scientists around the world great opportunity to study the topology of current systems and their fine scale structure. Of course, "fine" for Cluster means at best the resolution of only ion scales (c/ω_{pi}, ρ_{i}). The electron scale processes ($c/\omega_{\mathrm{pe}} \sim$ tens of km) are averaged out in Cluster measurements due to their relatively large separation. New generation of missions expected to operate during the next decade: MMS, Cross-scale, ROY will hopefully resolve this issue and provide next qualitative jump in our understanding of space plasma.

3 General Properties of Thin Current Sheets in the Magnetotail

The spacecraft data obtained onboard of Geotail, Interball and Cluster spacecraft provided an unprecedented opportunity to receive a new knowledge and develop a new generation of models to investigate current sheets in the magnetosphere. In recent years the attention of many researchers has been focused on extremely thin current sheets with thickness about ion Larmor radius or less, which are observed in the magnetotail at different distances from the Earth (Mitchell *et al.*, 1990; Pulkkinen *et al.*, 1993; Sergeev *et al.*, 1993; Pulkkinen *et al.*, 1994; Hoshino *et al.*, 1996; Sergeev *et al.*, 1998). The region at the near-Earth edge of the magnetotail current sheet is of particular interest, because, first of all, it is more easily accessible by spacecraft, secondly, it is the expected site of substorm non-stationary evolution and energy transformations. TCSs in this region may appear as a result of the enhanced plasma convection in the course of the substorm growth phase and, as we have mentioned above, might play a role of the reservoir of magnetic energy, which could be released after current sheet disruption at the beginning of the substorm onset phase.

The structure and properties of thin current sheets in the Earth's magnetosphere are now investigated by many spacecraft-borne in situ experiments, including recent multi-point Cluster measurements (Runov *et al.*, 2003a, b; Sergeev *et al.*, 2003). The

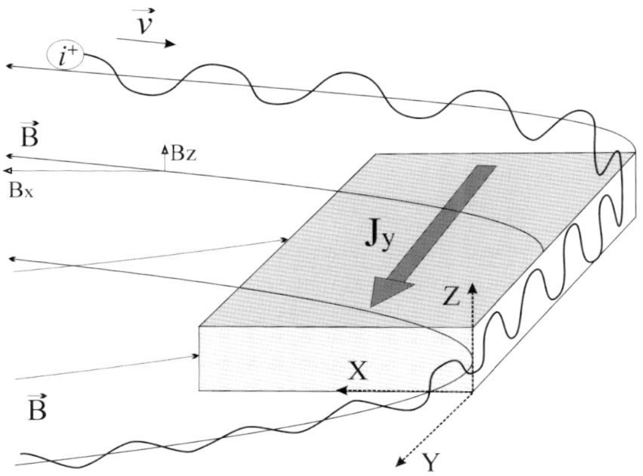

Fig. 3. A fragment of TCS configuration with a single Speiser ion at open trajectories from/to the infinity. Near the neutral plane the magnetized ion motion becomes to be demagnetized and particle moves along serpentine-like orbit around current sheet plane. GSM system of coordinates, the direction of magnetic field lines and of the current J_y are shown in the figure.

essential properties of TCS are very different from those of current sheets which are observed under the quiet conditions in the magnetosphere. Satellites investigations revealed the following particular properties of TCS:

1. very small thicknesses $L \sim \rho_{\rm i} \sim 250$–1000 km, where $\rho_{\rm i}$ is the ion Larmor gyroradius;

2. large value of current density, which could be 10–20 times larger than the one under usual conditions in the tail ($\sim 10\ \mu\mathrm{A/M^2}$);

3. very stretched shape of magnetic field lines, with $B_z/B_x \sim 0.1$;

4. the main current carriers across the sheet are as a rule the ions at the open so called Speiser orbits, shown in Fig. 3 (named in honor of Theodor W. Speiser (1965), who was the first who described these particles). Mitchell *et al.* (1990) proposed that at the late substorm phase the main fraction of current is carried by electrons with energies about 1 keV. Later it was shown that the dominated current carriers in TCS are nevertheless ions due to substantial difference of temperatures $T_i/T_e \geq 5$ (Sergeev *et al.*, 1998);

5. The current density profile does not coincide with plasma density one, and thin current sheet is embedded within more broad plasma sheet (e.g., papers by Sergeev *et al.* (1993) and Sanny *et al.* (1994) based on ISEE1 and ISEE2 data). Thin current sheets ($\delta \sim 0.2R_{\rm E}$) were found to be immersed in these experiments into much broad plasma sheet ($\Delta \sim 3$–$4R_{\rm E}$);

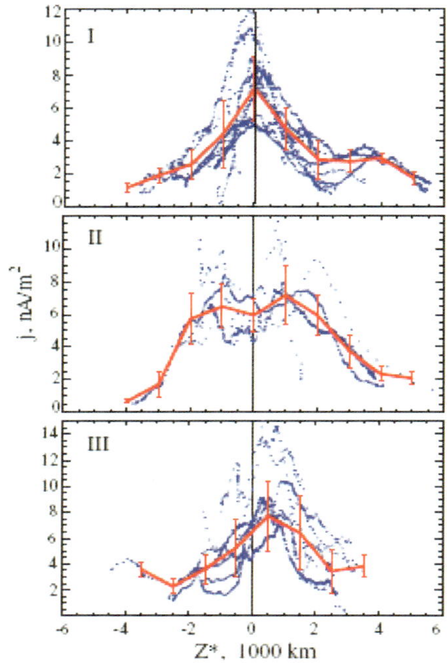

Fig. 4. Average vertical profiles for center peaked (I), bifurcated (II) and asymmetrical current sheets (III). (From Runov *et al.* (2006).)

6. The newest Cluster findings in a near-Earth edge of the current sheet (Runov, 2003a, b) showed that TCS can have not only classical bell-shape current profiles, but these profiles could be bifurcated, with two maxima of current density at the edges and minima in a center (see Fig. 4 from Runov *et al.*, 2006);

7. In some observations one can see unusual current sheets with peaks shifted from zero of the magnetic field and having asymmetrical shape of current density. The reasons of this asymmetry are not known yet (Asano *et al.*, 2005; Runov *et al.*, 2005). Below we develop the models where we will try to explain all these observational features.

4 Equilibrium Current Sheets Models

First analytical and mathematically very elegant model of the 1D current sheet was suggested by E. G. Harris in 1959. Model assumes shifted Maxwellian distributions both for ions and electrons

$$f_\alpha(v) = n_\alpha(z)\left(\frac{m_\alpha}{2\pi T_\alpha}\right)^{\frac{3}{2}} \exp\left\{-m_\alpha \frac{v_x^2 + \left(v_y - V_{\alpha y}\right)^2 + v_z^2}{2T_\alpha}\right\}$$

$$
\begin{aligned}
&= n_0 \left(\frac{m_\alpha}{2\pi T_\alpha}\right)^{\frac{3}{2}} \exp\left\{-\frac{W_0}{T_\alpha} + \frac{V_{\alpha y} P_y}{T_\alpha} - \frac{m_\alpha V_{\alpha y}^2}{2T_\alpha}\right\} \qquad (1)\\
n_\alpha(z) &= n_0 \exp\left\{\frac{e_\alpha V_{\alpha y} A_y(z)}{cT_\alpha}\right\}\\
\frac{dB_x}{dz} &= \frac{4\pi}{c} \sum_{\alpha=i,e} \int \vec{v} f_\alpha(\vec{v}, z) d\vec{v}\\
\sum_{\alpha=i,e} q_\alpha n_\alpha &= 0,\ \varphi(z) = 0
\end{aligned}
$$

(T_α is the temperature of the $\alpha = i, e$ kind of particles, m_α is the mass, $V_{\alpha y}$ is the flow velocity, $W_0 = m_\alpha v^2/2$ is the total particle energy, $P_y = m_\alpha v_y + (e/c)A_y(z)$ is the generalized momentum) which resulted in a self-consistent profiles of current, magnetic field and plasma density:

$$
\begin{aligned}
B_x(z) &= B_0 \tanh\left(\frac{z}{L}\right)\\
n(z) &= \frac{n_0}{\cosh^2\left(\frac{z}{L}\right)} \qquad (2)\\
j_y(z) &= \frac{J_0}{\cosh^2\left(\frac{z}{L}\right)}
\end{aligned}
$$

where half-thickness of current sheet is equal $L = 2(T_i + T_e)/\{cB_0(V_{iy} - V_{ey})\}$.

The principal property of this equilibrium is the absence of embedding, i.e., scales of the current and plasma density profiles are exactly equal to each other.

First 2D models of magnetotail current sheet were developed for thick current sheets (with thicknesses larger than ion gyroradius) at the beginning of 1970-ies. Birn *et al.* (1975) were the first to formulate the basics of such 2D current sheet equilibria ($B_n \neq 0$) in a frame of MHD approximation where the corresponding tensor of plasma pressure is assumed to be isotropic. In such isotropic 2D models the tension of the magnetic field lines should be balanced by the gradient of plasma pressure along the Earth-Sun direction (Soop and Schindler, 1973). The semifluid magnetotail models allow anisotropic ($T_\parallel \neq T_\perp$) but diagonal pressure tensor (Tsyganenko and Suslikov, 1982; Cheng, 1992). This approach cannot be applied inside the central current sheet and within TCSs where the radius of magnetic field line curvature becomes comparable to ion gyroradius and guiding center approximation is violated. In such cases full pressure tensor with both effects of anisotropy and non-gyrotropy is needed to obtain TCS balance.

The pioneering ideas about particle dynamics in thin current sheet were formulated by Speiser (1965) who qualitatively described TCS equilibrium configurations supported by ions moving along open quasi-adiabatic trajectories. The models of extremely thin current sheet supported by Speiser's ions were elaborated later by Eastwood (1972), Francfort and Pellat (1976), Pritchett *et al.* (1992), Kropotkin and

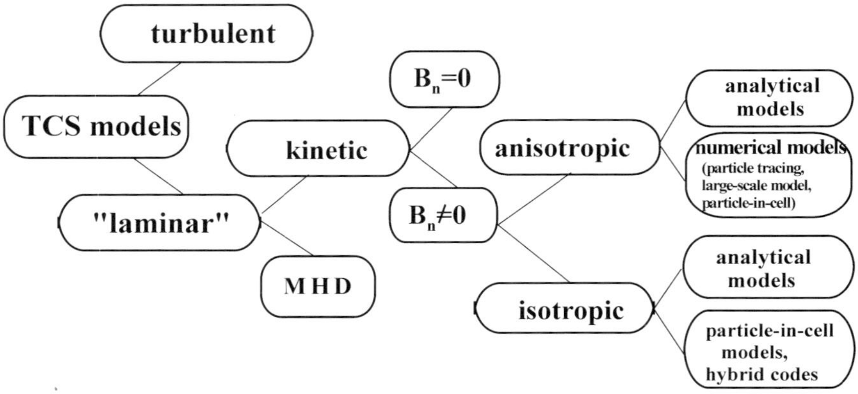

Fig. 5. Different classes of current sheet models and methods of their investigations: turbulent and laminar models. Laminar models can be kinetic and MHD. Class of kinetic models are divided on the models with nonzero and zero normal components of the magnetic field. The last kind of models is divided to anisotropic and isotropic. At the right one can see the methods to find equilibrium TCS solutions.

Domrin (1995), Sitnov *et al.* (2000a), Zelenyi *et al.* (2000). They studied a new type of plasma equilibria with CS having very small thicknesses ($L \sim \rho_L$, where ρ_L is the ion Larmor gyroradius) and 1D magnetic configurations almost homogeneous along Earth-Sun direction (as was shown later in 2D simulation by Ashour-Abdalla *et al.* (1991), Pritchett and Coroniti (1992) the limitation of the one dimensionality is not very restrictive for such quasi-adiabatic sheets). In such configurations, equilibrium is supported not by pressure gradient but by the finite inertia of ion beams moving along sharply bended field lines and at the same time maintaining the cross-tail current. Contrary to isotropic models of thick current sheets (both MHD and kinetic), the models of thin current sheets are principally anisotropic. The structure of the pressure tensor is determined by the structure of a phase space of particle motion, which is shaped by non-adiabatic effects. Ashour-Abdalla *et al.* (1994) have shown that even in their test particle model of TCS pressure tensor has significant off-diagonal $\mathbf{P}_{xz}$ terms at the edges of the current sheet, which are necessary for the maintenance of the general force balance. The corresponding velocity distribution functions of plasma sheet ions at different locations in the magnetotail demonstrate that as a rule such distributions have non-isotropic (as well as non-gyrotropic) non-Maxwellian shape. In a recent paper by Mingalev *et al.* (2007) the characteristics of anisotropic pressure tensor in already self-consistent kinetic model of TCS were investigated and result appeared to be qualitatively very similar to those obtained much earlier by Ashour-Abdalla *et al.* (1994).

The "tree" of a various current sheet models is shown in Fig. 5. On one hand, there exist a branch of "laminar" current sheet models, where particle motion is controlled by regular time-independent (or slowly changing) magnetic field. On another hand the researchers developed numerous models of turbulent current sheets (Milovanov *et al.*, 1996; Zimbardo *et al.*, 2003), which are more applicable for the distant regions of magnetotail (Fig. 1). The "laminar" models are based on two different methods of plasma description: MHD (Syrovatsky, 1971; Soop and Schindler, 1973; Birn *et al.*, 1975) and kinetic ones.

Historically, kinetic models were developed earlier, but in this paper we will concentrate mostly at the models using kinetic description of plasma particles dynamics which are most adequate to TCS description, because their characteristics (size of magnetic inhomogeneity; ion gyroradius etc) brings them beyond the range of validity of MHD description. For kinetic models one should distinguish the models with zero value of the normal component of the magnetic field (Harris, 1962) and another models with nonzero normal component (Voronina and Kan, 2001). Harris model gives a single equilibrium solution, which could not be reduced to another ($B_n \neq 0$) equilibria even in the limit $B_n \rightarrow 0$. This is very principal point often overlooked in an earlier studies. Even infinitely small B_n qualitatively changes the topology of magnetic configuration and equilibria with $B_z = 0$ and the one with even infinitely small B_z are principally different. At the right of Fig. 5 (from the top to the bottom) one can see list of various analytical and numerical methods which are usually applied to study both isotropic and anisotropic current sheets: particle-in-cell method (Pritchett and Coroniti, 1992), particle tracing one (Eastwood, 1974), hybrid codes (Kuznetsova *et al.*, 1996), large-scale kinetic modeling (Ashour-Abdalla *et al.*, 1994).

5 TCS Fine Structure: Possible Mechanisms of Its Formation

From the MHD point of view TCS solution could be considered as singularities appeared in the course of evolution of MHD system. The formation of such singular structures was described in an early pioneering works by S. I. Syrovatsky (1971). In this sense TCSs could be considered as collapsing MHD solutions forming a planar discontinuity which at the kinetic level has a thickness of the order of the ion Larmor radius (Mitchell *et al.*, 1990; Pulkkinen *et al.*, 1993; Sergeev *et al.*, 1993; Pulkkinen *et al.*, 1994; Hoshino *et al.*, 1996; Sergeev *et al.*, 1998; Zelenyi *et al.*, 2000; Runov *et al.*, 2003a, b). TCSs may exhibit a multitude of profiles of current density that is principally different from the well known bell-shaped profile of the Harris-type plasma configuration (Harris, 1962). For example, satellite measurements sometimes revealed abovementioned bifurcated ("double-humped") profile of current density in TCS (Baumjohann and Nakamura, 2002; Runov *et al.*, 2003a, b) or even a triple-humped profile with a current density that has a sharp peak in the center and two peaks on the sides, as well as rectangular and asymmetric shapes (Nakamura *et al.*, 2004).

The physical mechanisms underlying the different self-consistent TCS structures are not well understood. It should be noted that trajectories of ions and electrons within TCS are very different. While the behavior of the thermal (usually a few keVs) ions is principally nonadiabatic near the equatorial plane (Büchner and Zelenyi, 1989), the electrons (having energies 5 or 6 times lower than that of ions and at least two orders of magnitude smaller gyroradii) are mostly magnetized in the current layer (except for a very narrow regions of about an electron gyroradius wide, in close

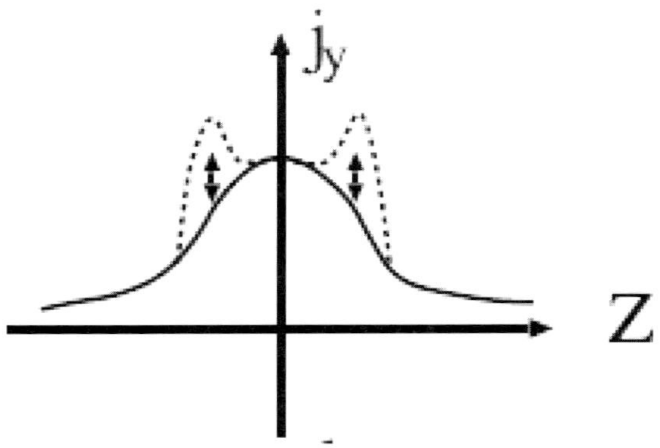

Fig. 6. Scheme of the current density distribution along the Z axis. Bifurcated current sheet is ion-dominated in the center and electron dominated at the edges. (Adapted from Asano *et al.*, Evolution of the thin current sheet in a substorm observed by Geotail, *J. Geophys. Res.*, **108**, 1189, 2003. Copyright 2003 American Geophysical Union. Reproduced by permission of American Geophysical Union.)

vicinity of X- and O-lines). It was shown in experimental work by Pulkkinen *et al.* (1994) that during the growth phase of substorm the electron pressure in magnetotail might become anisotropic ($p_{\parallel} - p_{\perp} > 0$).

Numerical simulations using full kinetic, hybrid, Hall-MHD codes have shown that electrons may be responsible for the characteristic bifurcated structures of TCS and could carry substantial currents in the narrow vicinity of the weak field region, e.g., near X- and O-lines (Pritchett and Coroniti, 1995; Hesse *et al.*, 1996; Birn *et al.*, 1998; Asano, 2001; Pritchett, 2001; Yin *et al.*, 2002; Asano *et al.*, 2003). Hoshino *et al.* (1996) and Asano *et al.* (2003) developed a numerical current sheet model where electron currents dominate at the edges of the TCS due to the $\vec{E} \times \vec{B}$ drift near the X-line, thus leading to a characteristic double-humped structure of the current density profile. This is illustrated in Fig. 6 (from the paper by Asano *et al.*, 2003), where the fine structure of current sheet near X-line is shown. Here the central part of the sheet is ion-dominated and supported by demagnetized ions, but the current sheet periphery is dominated by electron drift currents, supporting characteristic two-peaked, or bifurcated current sheet structure.

Greco *et al.* (2002) suggested that in turbulent current sheet current bifurcation may result from chaotic particle scattering at magnetic fluctuations. Zelenyi *et al.* (2002, 2003) developed a self-consistent model where thin current sheets are gradually "deteriorated" by quasi-trapped particles due to nonadiabatic scattering of Speiser ions. This process, referred to as current sheet "aging", leads to the splitting (i.e., bifurcation) to a double-humped shape, at the timescale comparable to one of substorm growth phase. This raises questions about the relative contributions of electrons and ions to the TCS formation. In a recent study Zelenyi *et al.* (2004)

investigated the role of electron current bifurcation using a 1D self-consistent analytical model in the limit of isotropic pressure. A detailed analysis of the results favors the ion-dependent "aging" process rather than electron drift currents as an efficient mechanism of current sheet bifurcation.

Although only one example of a "triple" peaked structure of current density has been reported up to now (Nakamura *et al.*, 2004), it provides clues of the variety of possible current sheet configurations. A very thin region of electron flow at the center of the current sheet embedded in a thick ion layer may be the origin of the triple-peaked current sheet. In this paper, we use a self-consistent 1D kinetic model of thin current sheets (Sitnov *et al.*, 2000a; Zelenyi *et al.*, 2000) to examine electrostatic effects in the case of anisotropic electron pressure. The problem of the electron current in 1D TCS has not been addressed so far. Sitnov *et al.* (2000b) adopted a simple approach where the influence of electrons is taken into account only to redistribute ions in the electrostatic potential arising from quasi-neutrality and the electron contributions to the net current were not considered. A procedure to include the electron drift in a similar self-consistent numerical TCS models has been described by Peroomian *et al.* (2002) in a study dedicated to large-scale kinetic modeling of the magnetotail. In this reviews we discuss results of the analysis of various self-consistent TCS structures as well as important effects obtained in a frame of (non-selfconsistent) large scale kinetics. Theoretical results could explain many TCS properties concerning their structure, embedded, double-triple-peaked current densities and, correspondingly, CS marginal stability in real conditions of the Earth's magnetosphere.

6 Particle Trajectories and Quasi-adiabatic Approximation

In general, two exact integrals of motion are conserved when charged particle moves within two-dimensional TCS configuration: the total particle energy $W = mv^2/2 + e\tilde{\varphi}(z)$ ($\tilde{\varphi}$ is the electrostatic potential) and the canonical momentum $P_y = mv_y - (e/c)A_y(x, z)$ which is conserved due to the absence of explicit Y-dependence in our 2D model. Under the condition $\kappa = \sqrt{R_c/\rho_{\max}} < 1$ where $\rho_{\max}$ is the maximum gyroradius of the ion, and R_c is the minimum curvature radius (see Büchner and Zelenyi, 1989) the quasiadiabatic approach could be applied, i.e., the approximate integral of motion $I_z = (1/2\pi)\oint p_z dz \approx \text{const}$. This condition holds in the magnetotail as was shown by numerous TCS observations (Pulkkinen *et al.*, 1994). As a matter of fact, Lui (1993) and Sergeev *et al.* (1993) estimated the parameter κ in the midtail to be of the order of 0.1 for ions and above 3 for electrons. The κ parameter of the ions is thus small enough to ensure the possibility of quasi-adiabatic description, whereas that of electrons is large enough to consider these particles as being a magnetized component of the plasma. Therefore in the quasi-adiabatic case, the approximate integral of "fast" ion motion along Z-coordinate (I_z) is roughly conserved (Sonnerup, 1971; Francfort and Pellat, 1976; Büchner and Zelenyi, 1989; Whipple *et al.*, 1990). At the same time we could use the standard guiding center description for magnetized electrons where electron magnetic moment $\mu = mv_\perp^2/2B$ could be used as corresponding invariant of motion. This principal difference of ion and electron trajectories in thin current sheets explains many of their characteristic

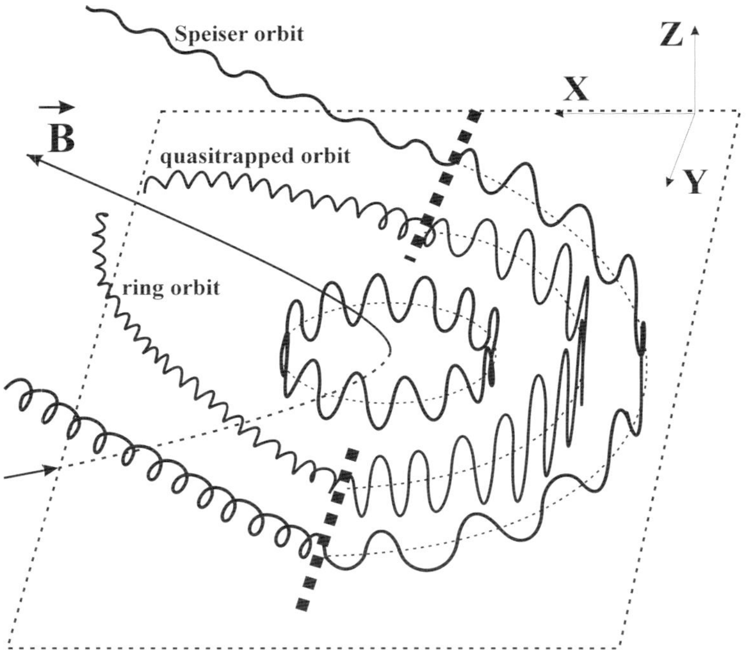

Fig. 7. Three types of ion trajectories in TCS. The thick dotted lines indicate schematically the position of separatrix of motion.

features, in particular, the generation of a significant ambipolar electrostatic fields necessary to maintain the quasineutrality of plasma.

All classes of particle populations in the magnetotail may actually be classified according to their I_z values (Büchner and Zelenyi, 1989), as it is shown in Fig. 7. The smallest I_z values are characteristic for transient ions named Speiser's ones, the larger I_z values belong to quasi-trapped ions at "cucumber"-like orbits and the largest adiabatic invariant I_z define the trapped motion, i.e., ions trapped at "ring"-like orbits in the vicinity of current sheet midplane. Contrary to exactly integrable ring orbits, the transient and quasi-trapped particles change the type of their motion interacting with the current sheet when they cross separatrix, schematically shown in Fig. 5. This happens twice at each motion of cycle. The position of separatrix of motion is shown in Fig. 7. During separatrix crossing the values of adiabatic invariants changes by small jumps, but the values of jumps are smaller than the values of invariants I_z itself.

7 Basic Approach for the Development of Self-consistent Anisotropic TCS Models

We present here an analytical 1D self-consistent model of a thin current sheet, taking into account both ion and electron plasma populations. The current sheet is so thin in comparison with its width (along Y in GSM system) and length (along X)

that one can consider it to be homogeneous along the X and Y directions (Sitnov *et al.*, 2000a; Zelenyi *et al.*, 2000). Therefore we consider a simplified magnetic field $\mathbf{B} = \{B_x(Z), 0, B_n\}$ in which only $B_x(Z)$ is dependent on Z and reverses sign across the $Z = 0$ plane, as it is shown in Fig. 3. The Z-component B_n, which is the normal component of magnetic field, is assumed to have constant value. Within such sheet, magnetic field line tension could be balanced by the finite inertia of ions rather than by plasma pressure gradients as in the case of Harris-like equilibria. The particles, e.g., coming from the mantle, are streaming from the edges of TCS toward the neutral sheet plane ($Z = 0$). The ion population consists of transient particles moving along Speiser orbits

$$f(\vec{v}) = \frac{n_0}{\left(\sqrt{\pi}\, v_{\mathrm{T}}\right)^3 \left\{1 + \mathrm{erf}\left(\frac{v_D}{v_{\mathrm{T}}}\right)\right\}} \exp\left\{-\frac{\left(v_\parallel - v_D\right)^2 + v_\perp^2}{v_{\mathrm{T}}^2}\right\} d\vec{v} \tag{3}$$

(the initial Maxwellian distribution, being characterized by the thermal velocity v_{T} and the bulk flow speed v_D). We assume that ion population could be described by the quasiadiabatic approach (see previous paragraph) so that the action integral of ion motion $I_z = (m/2\pi) \oint v_z dz$ will be approximately conserved.

TCS is already not “thin” for electrons having gyroradius smaller than the minimum curvature radius of magnetic field lines in a neutral sheet. It makes then possible to describe transverse motion of electrons by hydrodynamic equations with anisotropic (or sometimes isotropic) pressure tensor. The electron motion along magnetic field lines is assumed to be fast enough to support a quasi-equilibrium Boltzmann distribution in the presence of an ambipolar electrostatic potential and mirror forces. Plasma quasi-neutrality $n_i \approx n_e$ supported by these electrostatic fields. Such an approach represents a generalization of the 1D self-consistent TCS model developed earlier (Sitnov *et al.*, 2000a; Malova *et al.*, 2000; Zelenyi *et al.*, 2000). Here we will consider the model taking into account both ions with Speiser's orbits (which are the main current carriers) and almost trapped “cucumber” ions which do not provide any net contributions to the total cross-current current due to closeness of their trajectories. The effect of the self-consistent interaction of these ion populations, investigated earlier in studies by Zelenyi *et al.* (2002, 2003) will be also discussed below.

8 Vlasov-Maxwell System of Equations

For the 1D TCS model that we consider here, the self-consistent system of Vlasov-Maxwell equations has the following simple form:

$$\frac{dB_x}{dz} = \frac{4\pi}{c} \left\{ j_{yi}(z) + j_{ye}(z) \right\} \tag{4}$$

$$j_{yi} = \frac{4\pi}{c} \sum_i e \int v_y f_i(z, \vec{v}) d\vec{v} \tag{5}$$

Here f_i is the ion distribution function (3) and j_{yi} is the total ion current in the y-direction. Note that the ion distribution function at any location inside the current sheet could be obtained by mapping the source distribution from the edges of TCS to the neutral plane using Liouville's theorem.

One could introduce then the dimensionless variables $\vec{r} = \vec{R}/(\omega_0 v_D \epsilon^{4/3})$, $\zeta = z\omega_0/\epsilon^{4/3} V_D$, $\vec{w} = \overleftarrow{v}/(v_D \epsilon^{2/3})$, $I = I_z \epsilon^{2/3}\omega_0/(m_i v_T)$, $\varphi = m v_D^2 \epsilon^{4/3}\tilde{\varphi}/2e$, $\vec{E} = c\vec{E}/B_0 v_D \epsilon^{2/3}$, $\tilde{p}_{\perp,\parallel} = p_{\perp,\parallel}/N_0 T_{\perp,\parallel 0}\tilde{j} = J/en_0 v_D \epsilon^{2/3}$, $n = \tilde{n}/N_0$, $\tilde{b}_x = B_x/B_0$, $\tilde{b} = B/B_0$, where $\vec{r}$ is a normalized distance (here, ζ is the dimensionless z coordinate), $\vec{w}$ is the dimensionless particle velocity vector, I is the dimensionless quasi-adiabatic invariant, $\epsilon = v_T/v_D$ is the ion source anisotropy parameter, $\omega_0 = eB_0/mc$ is the ion gyrofrequency in the magnetic field B_0 at the edges of the sheet, b is the normalized B_x component of the magnetic field, φ is the dimensionless electrostatic potential and j_e is the normalized electron current in the y direction: $j_e = eN_e v_D \epsilon^{2/3}\tilde{j}_e$. Taking into account the conservation of particle magnetic moments outside current sheet $\mu \equiv m v_\perp^2/B \cong \text{const}$ and the relation between the usual magnetic moment and quasi-adiabatic invariant at the infinity $I_z = (2mc/e)\mu$ one can rewrite the normalized ion distribution function (3) in the form:

$$f_i(\vec{w}, I(\zeta)) = \frac{n_0 \exp\left(-\epsilon^{-\frac{2}{3}} I\right) \times \exp\left(\epsilon^{-\frac{2}{3}}\left[\left\{\sqrt{w_0^2 + \varphi - I(\zeta)} - \epsilon^{-\frac{2}{3}}\right\}^2 + I(\zeta)\right]\right)}{\pi^{\frac{3}{2}} v_T^3 \left\{1 + \mathrm{erf}(\epsilon^{-1})\right\}} \tag{6}$$

In dimensionless form, the Maxwell equation (6) then becomes

$$\frac{db_x}{d\zeta} = \frac{4\epsilon}{\pi^{\frac{3}{2}}}\left(\frac{v_D}{v_A}\right)^2 \left(\int \frac{w_y}{1 + \mathrm{erf}\left(\epsilon^{-1}\right)} \cdot \exp\left[\epsilon^{-\frac{2}{3}}\left\{\left(\sqrt{w_0^2 + \varphi - I} - \epsilon^{-\frac{2}{3}}\right)^2 + I\right\}\right] d^3 w + \frac{\epsilon \pi^{\frac{3}{2}}}{4}\tilde{j}_e\right) \tag{7}$$

We simplify this equation by replacing the ζ-coordinate with the dimensionless vector potential $\eta = \epsilon^{2/3}\int_0^\zeta b(\zeta')d\zeta'$ and by integrating over the η-coordinate:

$$b_x(\eta) = \frac{2\sqrt{2}\epsilon^{\frac{1}{6}}}{\pi^{\frac{3}{4}}}\left(\frac{v_D}{v_A}\right)\left\{\int_0^\eta \left(\int \frac{w_y}{1 + \mathrm{erf}\left(\epsilon^{-1}\right)} \cdot \exp\left[\epsilon^{-\frac{2}{3}}\left\{\left(\sqrt{w_0^2 + \varphi - I} - \epsilon^{-\frac{2}{3}}\right)^2 + I\right\}\right] d^3 w + \frac{\epsilon \pi^{\frac{3}{2}}}{4}\tilde{j}_e\right) d\eta'\right\}^{\frac{1}{2}} \tag{8}$$

The usual boundary normalization condition $b(\infty) = 1$ is used additionally to Eq. (8). Taking into account the equations of particle motion and the conservation of the integral P_y, quasi-adiabatic invariant I_z may be expressed as

$$I_z(\vec{v}, z) = 2\frac{m}{\pi}\int_{z_0}^{z_1}\left[v_y^2 + v_z^2 + 2e\left\{\tilde{\varphi}(z) - \tilde{\varphi}(z')\right\} - \left\{v_y + \frac{e}{mc}\int_{z}^{z'} B_x(z'')dz''\right\}^2\right]^{\frac{1}{2}} dz' \quad (9)$$

with the following limits of integration

$$v_y \pm \sqrt{v_y^2 + v_z^2 + 2e\frac{\tilde{\varphi}(z) - \tilde{\varphi}(z')}{m}} = -\frac{e}{mc}\int_{z_0}^{z_{0,1}} B_x(z'')dz''. \quad (10)$$

In the case when the solution of Eq. (10) is negative (i.e., particle crosses the tail midplane $z = 0$, the value z_0 should be set equal to 0.

9 Electrostatic Potential and Electric Field

Let us now describe the calculation of the electron current $\tilde{j}_e$, ambipolar electrostatic field E and potential φ in our kinetic TCS model. For simplicity, we assume that the electron distribution depends upon the magnetic moment μ as $f_e(\mu) \sim \delta(\mu - \mu^*)$, i.e., we assume that the electron component could be represented by the distribution of particles having the same value of magnetic moment μ^* (μ^* could be considered as some average value over an ensemble) with the parallel pressure $P_\parallel$ and perpendicular pressure $P_\perp$ as the diagonal components of the total electron gyrotropic pressure tensor. Based on such a semi-hydrodynamic description, the electron motion within TCS could be described as:

$$m_e\frac{d\vec{v}_{e\perp}}{dt} = -e\left(\vec{E}_\perp + \frac{[\vec{v}_e \times \vec{B}]}{c}\right) - \frac{\nabla_\perp p_e}{n_e} \quad (11)$$

$$m_e\frac{d\vec{v}_{e\parallel}}{dt} = -e\vec{E}_\parallel - \frac{\vec{\nabla}_\parallel p_e}{n_e} - \mu\vec{\nabla}_\parallel B \quad (12)$$

($-e$) being the electron charge. We use a fluid approach to describe the electron motion across the magnetic field lines and a guiding center approximation for the parallel motion. The last term in Eq. (12) is the repulsive diamagnetic force acting on the electrons with magnetic moment μ, $p_{ij} = p_\perp\delta_{ij} + (p_\parallel - p_\perp)h_i h_j$, $\vec{h} = \vec{B}/B$ is the electron pressure tensor, and $B = \left(B_x^2 + B_z^2\right)^{1/2}$ is the total magnetic field. Neglecting the electron inertia for the fast parallel motion and substituting $E_\parallel = -\nabla\varphi$, we obtain from (12)

$$\vec{\nabla}\varphi(s) = \frac{1}{en_e}\vec{\nabla}_\parallel p_e + \frac{\mu\vec{\nabla}_\parallel B(s)}{e}, \quad \varphi_0 \equiv \varphi(\infty) = 0 \quad (13)$$

where s is the curvilinear coordinate along the field line, and $n_e(\vec{r}, \varphi(\zeta)) = n_i(\vec{r}, \varphi(\zeta))$ $= n$ is a result of quasi-neutrality assumption. In the 1D case, this equation can be easily integrated if we assume for example an isothermal (T_e = const) equation of state for electron component:

$$p_{\parallel e} = n_e T_{\parallel e} \tag{14}$$

Note that the quantity $\tau_{\parallel} = T_i / T_{e\parallel}$ here becomes an important characteristic parameter. For isothermal electrons Eq. (12) can be easily rewritten in a form corresponding to the Boltzmann distribution

$$\frac{n_e(s)}{n_0} = \exp\left[\frac{e\{\varphi(s) - \varphi_0\} - \mu\{B(s) - B_0\}}{T_e}\right] \tag{15}$$

10 Electron Currents

For the cross-field particle motion, if we will neglect the higher order drifts, it is easy to get from Eq. (11)

$$\vec{v}_{e\perp} = c\frac{\vec{E}'_{\perp} \times \vec{B}}{B^2}, \quad \vec{E}'_{\perp} = \vec{E}_{\perp} + \frac{\nabla_{\perp} p_e}{e n_e} \tag{16}$$

In the case of isotropic pressure the cross-field electron current has the following simple form:

$$j_{e\perp} = -e n_e v_{e\perp} \tag{17}$$

In the case of anisotropic pressure, we may use a more general form of this expression (Krall and Trivelpiece, 1973). Taking into account the influence of curvature drifts that are of particular importance here, we get

$$j_{e\perp} = -e n_e c\frac{\left[\vec{E} \times \vec{h}\right]}{B} + \frac{c}{B}\left[\vec{h} \times \vec{\nabla}_{\perp} p_{\perp e}\right] + \frac{c}{B}\left(p_{\parallel e} - p_{\perp e}\right)\left[\vec{h} \times \left(\vec{h}\vec{\nabla}\right)\vec{h}\right], \; \vec{h} = \frac{\vec{B}}{B} \tag{18}$$

To calculate the electron current, one must determine the distributions of electric fields and plasma pressures (both perpendicular and parallel). Since we examine the equilibrium of a conservative system, we will use energy conservation condition to calculate the spatial variation of a parallel pressure:

$$p_{\parallel} = nmv_0^2 - nmv_{\perp}^2 = nmv_0^2\left\{1 - q^2\frac{B(z)}{B_0}\right\} \tag{19}$$

Perpendicular pressure could be taken from the first CGL (Chew-Golberger-Low) equation of state (Chew *et al.*, 1956)

$$\frac{p_{\perp}}{nB} = \text{const} \tag{20}$$

11 Pressure Balance in TCS

Important check of consistency of any equilibrium plasma is the calculation of a stress balance. For isotropic 2D model of CS equilibrium (Kan, 1973; Lembege

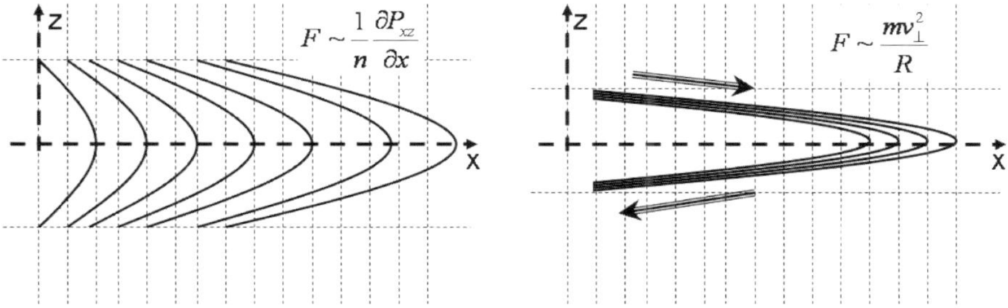

Fig. 8. Pressure balance mechanisms: thick isotropic CS at left picture and thin anisotropic CS at right picture.

and Pellat, 1982) this balance could be easily provided by pressure gradients in x-direction $j_y B_z = \partial p/\partial x$. For 1D configuration $\partial/\partial x \equiv 0$, stress balance could be supported only by non-gyrotropic terms in the pressure tensor. As was shown by Rich *et al.* (1972) off-diagonal pressure tensor components are required to maintain the stress balance. For our 1D equilibrium stresses should be balanced along z and x-directions:

$$p_{zz} + \frac{B_x^2}{8\pi} = \text{const} \tag{21}$$

$$-p_{xz} + \frac{B_z B_x}{4\pi} = \text{const} \tag{22}$$

More trivial Eq. (21) was checked earlier in a paper by Zelenyi *et al.* (2000). Therefore here we present previously unpublished calculation of horizontal stresses. The appearance of off-diagonal terms in local pressure tensor is a key factor (Fig. 8).

Equation (21) has the first order over the parameter of smallness B_z/B_0 of our model and therefore we should calculate additionally next order corrections to equilibrium parameters. If $v_\parallel$ is the parallel velocity component of ion motion in the magnetic field $\mathbf{B} = \{B_x(z), 0, B_z\}$, the ion pressure balance can be written as (Burkhart *et al.*, 1992):

$$p_{xz} = \int_{v_\parallel<0} v_x v_z f_{\text{in}}\, d^3v + \int_{v_\parallel>0} v_x v_z f_{\text{out}}\, d^3v \tag{23}$$

Where input and output distribution functions have the form $f_{\text{in}}(v_\perp, v_\parallel) \sim \exp[-v_{Ti}^{-2} \{v_\perp^2 + (v_\parallel - v_D)^2\}]$ and $f_{\text{out}}(v_\perp, v_\parallel) = f_{\text{in}}(v_\perp, -v_\parallel)$ (Zelenyi *et al.*, 2000).

To calculate non-gyrotropic terms we should perform earlier calculations (Zelenyi *et al.*, 2000) keeping higher order corrections. For ion moving in a field $\mathbf{B} = \{B_x(z), 0, B_z\}$ exact equation of motion has the form:

$$\begin{cases} \dot{\tilde{v}}_x = \omega_z \tilde{v}_y \\ \dot{\tilde{v}}_y = \omega_x \tilde{v}_z - \omega_z \tilde{v}_x \\ \dot{\tilde{v}}_z = -\omega_x \tilde{v}_y \end{cases} \tag{24}$$

where $\omega_z = (eB_z)/(mc)$ and $\omega_x = \{eB_x(z)\}/(mc)$. Therefore we get for the x component of ion velocity $\tilde{v}_x = v_x + \omega_z \Delta y$ and for the y component $\tilde{v}_y = v_y + (e/mc)\int_z^{\tilde{z}} B_x(z')dz' - \omega_z \Delta x$. Displacement of coordinates Δy and Δx in terms of integrals along trajectory can be written as:

$$\Delta x(z_0, z_1) = v_x \int_{z_0}^{z_1} \left[h - v_x^2 - \left\{ v_y + \frac{e}{mc} \int_z^{\tilde{z}} B_x(z')dz' \right\}^2 \right]^{-\frac{1}{2}} d\tilde{z}$$

$$\Delta y(z_0, z_1) = \int_{z_0}^{z_1} \left\{ v_y + \frac{e}{mc} \int_z^{\tilde{z}} B_x(z')dz' \right\} \cdot \left[h - v_x^2 - \left\{ v_y + \frac{e}{mc} \int_z^{\tilde{z}} B_x(z')dz' \right\}^2 \right]^{-\frac{1}{2}} d\tilde{z} \qquad (25)$$

Where h is total energy: $h = v_x^2 + v_y^2 + v_z^2 + (2e/m)\{\varphi(z) - \varphi(\tilde{z})\}$. After using these expressions invariant I_z can be written as:

$$I_z = (2\pi)^{-1} m \int_{z_-}^{z_+} \left[h - (v_x + \omega_z \Delta y)^2 - \left\{ v_y + \frac{e}{mc} \int_z^{\tilde{z}} B_x(z')dz' - \omega_z \Delta x \right\}^2 \right]^{\frac{1}{2}} d\tilde{z} \qquad (26)$$

Taking into account that parameter ω_z is small in comparison with ω_x one can rewrite Eq. (26) in a new improved form $I_{z1} = I_{z0} - (2\pi)^{-1} m\omega_z (t_y v_x - t_x v_y - \Delta)$. Here I_{z0} is quasi-adiabatically invariant as it was introduced by Sonnerup (1971) and was used in an earlier model of thin CS (Zelenyi *et al.*, 2000). Additional terms are described by formulas:

$$t_x = \int_{z_-}^{z_+} \Delta x \left[h - v_x^2 - \left\{ v_y + \frac{e}{mc} \int_z^{\tilde{z}} B_x(z')dz' \right\}^2 \right]^{-\frac{1}{2}} d\tilde{z}$$

$$t_y = \int_{z_-}^{z_+} \Delta y \left[h - v_x^2 - \left\{ v_y + \frac{e}{mc} \int_z^{\tilde{z}} B_x(z')dz' \right\}^2 \right]^{-\frac{1}{2}} d\tilde{z} \qquad (27)$$

$$\Delta = \frac{e}{mc} \int_{z_-}^{z_+} \Delta y \int_z^{\tilde{z}} B_x(z')dz' \left[h - v_x^2 - \left\{ v_y + \frac{e}{mc} \int_z^{\tilde{z}} B_x(z')dz' \right\}^2 \right]^{-\frac{1}{2}} d\tilde{z}$$

With these next order corrections taken into account one could easily calculate the additional off-diagonal components p_{xz} of the ion pressure tensor.

For magnetized electrons with gyrotropic pressure tensor $p_{ij} = p_\perp \delta_{ij} + (p_\parallel - p_\perp)(B_i B_j / B^2)$ the calculation of contributions to stress balance in x-direction is

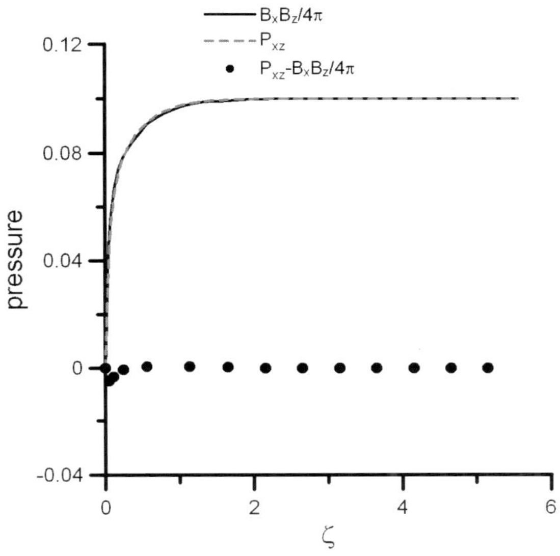

Fig. 9. Pressure balance along xz in CS without electrons.

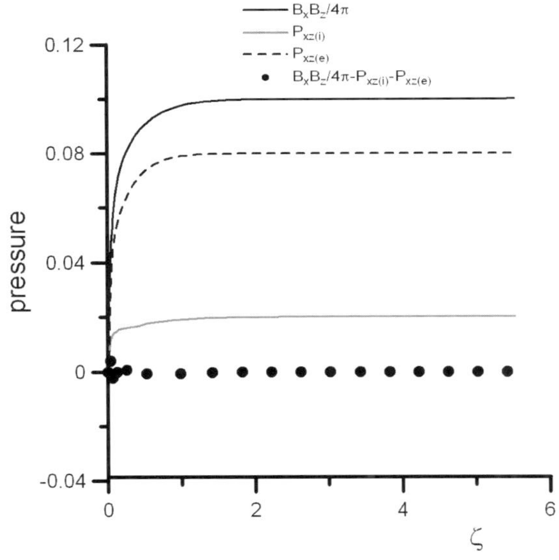

Fig. 10. Pressure balance along xz in CS with electrons ($T_i/T_e = 3$).

straightforward and will give the result $p_{xz(e)} = (p_{\|} - p_{\perp})(B_x B_z/B^2)$. Combining this equation with Eq. (27) one could now check numerically that Eq. (22) is very well satisfied in two cases: isotropic electrons (Fig. 9), anisotropic electrons (Fig. 10).

This analytical result was confirmed by recent PIC simulations (Mingalev *et al.*, 2007). Thus we see that inertia force could replace pressure gradients and support even infinitely long (theoretically) homogeneous current sheet. This is very important for understanding extremely stretched (almost one-dimensional) magnetotail plasma configurations, even having finite transverse (B_z) components of equilibrium field.

This new CS model has new structural properties like embedding, different profiles of electron and ion currents. Therefore it is interesting and challenging problem to investigate stability of such "new" anisotropic plasma configuration to the same extent as it was done last decades for Harris configuration (Schindler, 1974; Galeev and Zelenyi, 1976; Daughton, 1999). Results of such investigation are shown in the next section.

12 Formation of Embedded Current Density Profile: Role of Electron Currents

The solution of self-consistent system of Eqs. (4)–(20) with corresponding boundary conditions allows to obtain self-consistent profiles of partial and total current densities, plasma density, electric and magnetic fields as well as distribution of electrostatic potential are obtained. We select anisotropy parameter ϵ to be equal to which is realistic condition for the Earth's magnetotail. Figure 11 shows the self-consistent TCS current density for different initial values of b_n and for the case of isotropic and anisotropic electron pressure. Here a remarkably strong b_n dependence of electron

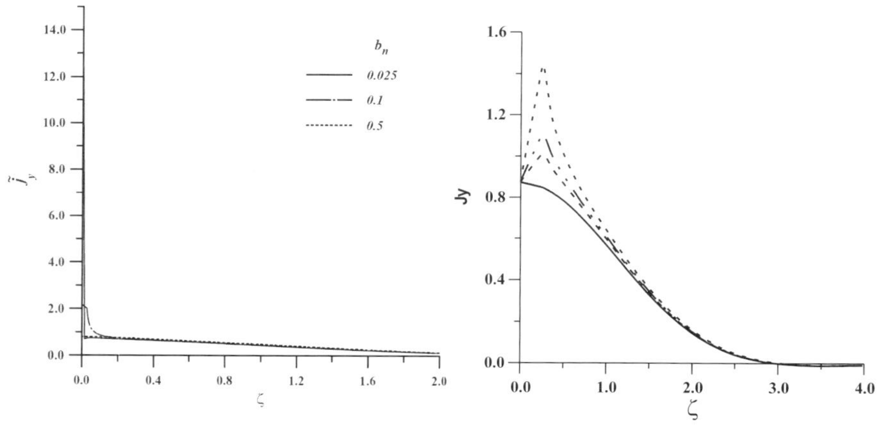

Fig. 11. The profiles of TCS current density in TCS with anisotropic distribution of electrons ($\tilde{j}_y(\zeta)$ at the left) and isotropic one ($J_y(\zeta)$ at the right) as a function of the ratios $b_n = 0.025, 0.1, 0.5$. At the right the current profile is presented at $b_n = 0.1$ (dotted line), 0.25 (dotted-dashed line), 0.5 (dashed line); $\tau = 5$, $q^2 = 0.5$, $\epsilon = 1$; ion current is shown by solid line. One can clearly see a multi-layer structure, viz., electron currents are embedded within a much broader region where the ion contribution is dominant.

currents (both in central and off-central regions) can be seen. Electron currents in case of isotropic electron pressure lead to a current sheet splitting and the appearance of a characteristic bifurcated (double-humped) structure, because in this (isotropic) case $\mathbf{E} \times \mathbf{B}$ ($\mathbf{E} = -\nabla\varphi$) and grad $\mathbf{B}$ electron drifts (important at CS edges) dominate.

In contrast to this, in the case of anisotropic electron pressure electrons support a very narrow current maximum in the vicinity of the neutral plane (Fig. 11), a feature which is embedded inside a much thicker ion current. This peak is entirely due to the curvature drift (last term on the right hand side of Eq. (18)) of electrons in the central region of very large magnetic field line curvature. The thickness of this nearly singular embedded electron sheet is about $L_e \sim \rho_{L_i} b_n^{4/3}$ (Zelenyi *et al.*, 2000). One can see that the maximum of electron current in the very center of the sheet varies roughly in inverse proportion to the value of normal magnetic field component b_n, since this component controls the curvature radius of the field lines. This leads to a substantial steepening of the magnetic field profile near the neutral plane. It is important to note that although locally electron current might strongly dominate over the ion one in a global sense total electron contribution ($\int j_e dz$) to the cross tail current for realistic parameters ($b_n \geq 0.05$, $T_i/T_e \approx 5$–6) could at best be of the order of 20–25% of total cross-tail current, mostly supported by ions. These local electron peaks sometimes observed by "in situ" particle measurements (Hoshino *et al.*, 1996) should not be misinterpreted as an evidence of electron current domination for supporting magnetotail configuration.

13 The Formation of Multiple Embedded Current Structures in the Presence of a Heavy (Oxygen) Ions

Ion composition measurements in the lobes at distances $10R_E > X > 22R_E$ provided the evidence of the presence of relatively low energy (~10 eV–1 keV) heavy ions streaming anti-sunward (Delcourt *et al.*, 1996a; Vaisberg *et al.*, 1996; Sauvaud *et al.*, 2004). It was demonstrated that O^+ ions dominate both the pressure and density of the plasma sheet before the substorm onset and their contribution becomes even stronger after substorm expansion begins. Therefore, for modeling the current sheet of the Earth's magnetotail one should take into account its multicomponent composition including solar wind (H^+, He^{++}) and ionospheric (O^+, He^+) ions. Until recently the effects of such complicated composition both on the structure of kinetic quasi-equilibria and their dynamical characteristics were not well understood. Recently Kistler (2005) estimated the upper limit for the contribution of oxygen ions to the total cross-tail current to be as much as 10%. The estimates of the contributions of O^+ ions based on the semi-analytical kinetic model of magnetotail was made recently by Zelenyi *et al.* (2006), where the influence of oxygen constituents of the shape of the cross-tail current profile was analyzed.

Generally the basic principles of multicomponent model are the same as mentioned in Section 8. Oxygen ions, as well as protons, move in the magnetotail along nonadiabatic Speiser orbits. For comparable energy their Larmor radius is $\sqrt{m_{O^+}/m_{H^+}} \cong 4$ times larger than the proton one. Figure 12 illustrates the structure of magnetic field in TCS and particle trajectories (at the left) and the embedded

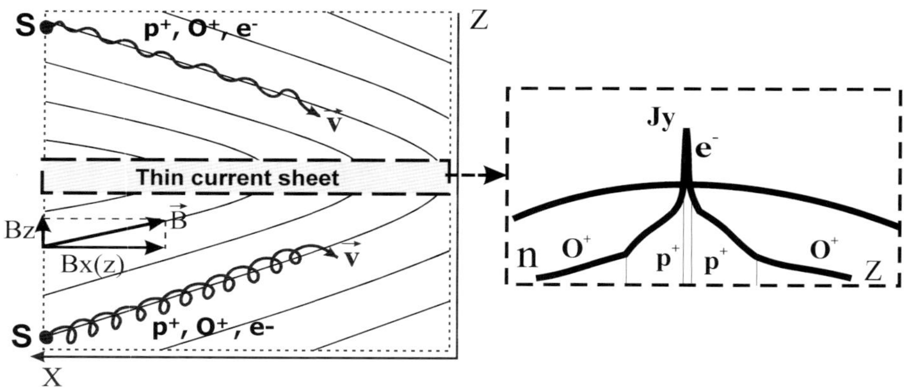

Fig. 12. The scheme of TCS model with three plasma components and embedded structure of plasma sheet. At the left: electrons and ions of hydrogen and oxygen moving from the edges of CS towards the central part. At the right: the schematic profiles of plasma density and current one are shown.

structure of current sheet (at the right). In the center of current sheet is located a narrow electron-dominated region, surrounded by larger proton-dominated one, which, in its turn, is embedded within substantially thicker oxygen-dominated domain. All these structures are embedded inside the much most thick plasma sheet. As in previous two-component calculation, the self-consistent solution can be found from the Ampere law for 1D current sheet:

$$\frac{d\vec{B}}{dz} = \frac{4\pi}{c}\left(\vec{j}_1 + \vec{j}_2 + \vec{j}_3\right). \tag{28}$$

Index 1 signifies the proton values, index 2 is for oxygen contribution, index "e" for the electron one. For protons H^+ and heavy ions O^+ the current density is defined as the moment of a distribution function $j_{1,2} = e\int v_y f_{1,2}(\vec{v})d^3v$, where $f_{1,2}(\vec{v})$ is the ion distribution function. Using abovementioned dimensionless variables as $\zeta = z\omega_{01}/\epsilon_1^{4/3}V_{D1}$, $w = \overleftarrow{v}/(v_{D1}\epsilon_1^{2/3})$, $\vec{x} = \vec{X}/(\omega_{01}v_{D1}\epsilon_1^{4/3})$, $I = I_z\epsilon_1^{2/3}\omega_{01}/(m_1 v_{T1})$, where $\epsilon_1 = \dfrac{v_{T1}}{v_{D1}}$, $\omega_{01} = \dfrac{eB_0}{m_1 c}$. $b = B/B_0$, $n = \tilde{n}/N_0$ we can rewrite the basic equation (28) in the form:

$$\frac{d(b)}{d\zeta} = \frac{4\epsilon_1}{\pi^{\frac{3}{2}}}\left(\frac{v_{D1}}{v_{A1}}\right)^2\left\{\frac{1}{1+\mathrm{erf}\left(\epsilon^{-1}\right)}j_1(\zeta) + \frac{e_r n_r}{v_{Tr}^3}\frac{1}{1+\mathrm{erf}(\epsilon_r^{-1}\epsilon_1^{-1})}j_2 + \frac{\epsilon_1\pi^{\frac{3}{2}}}{4}j_e\right\} \tag{29}$$

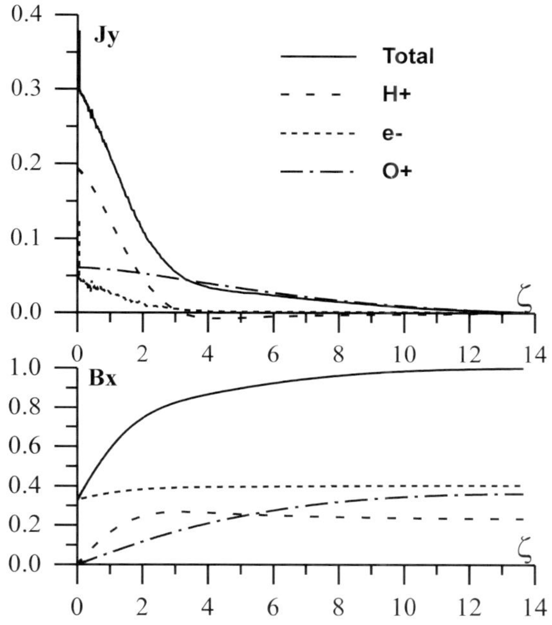

Fig. 13. Normalized current densities (at the top) and partial magnetic fields (at the bottom) for different components of plasma (dotted, dotted-dashed and dashed lines) and the total profiles (solid lines). The ratio of concentrations and temperatures of oxygen ions to protons are the equals: $n_2/n_1 = 1$, $T_2/T_1 = 1$.

This equation reflects the relation between the magnetic field and current density/ supported by three plasma components in TCS. The corresponding quasi-adiabatic integrals of particle motion acquire the form:

$$I_{1\pm} = \epsilon_1^{\frac{2}{3}} \tilde{I}_1 = \frac{2\epsilon_1^{\frac{2}{3}}}{\pi} \cdot \int_{\zeta_{0\pm}}^{\zeta_{1\pm}} \sqrt{w_0^2 - w_x^2 + \{\varphi(\zeta) - \varphi(\zeta')\} - \left\{\pm w_y - \epsilon_1^{\frac{2}{3}} \int_{\zeta'}^{\zeta} b(\zeta'')d\zeta''\right\}^2} \, d\zeta' \tag{30}$$

$$I_{2\pm} = \epsilon_1^{\frac{2}{3}} \tilde{I}_2 = m_r \frac{2\epsilon_1^{\frac{2}{3}}}{\pi} \cdot \int_{\zeta_{0\pm}}^{\zeta_{1\pm}} \sqrt{w_0^2 - w_x^2 + \frac{e_r}{m_r}\{\varphi(\zeta) - \varphi(\zeta')\} - \left\{\pm w_y - \frac{e_r}{m_r}\epsilon_1^{\frac{2}{3}} \int_{\zeta'}^{\zeta} b(\zeta'')d\zeta''\right\}^2} \, d\zeta'$$

with the limits of integration which can be found from the solution of the following

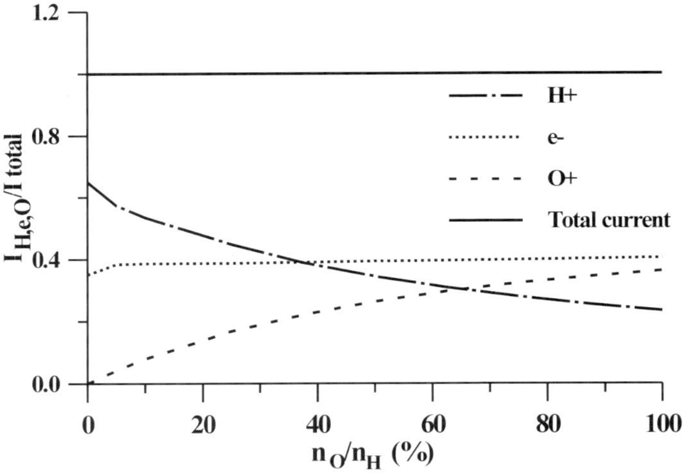

Fig. 14. Total cross tail current as a function of relative content of O^+ and H^+ ions for three component plasma model (the ratio n_O/n_H is estimated at the edges of CS).

equation:

$$w_0^2 - w_x^2 + \frac{e_r}{m_r}\{\varphi(\zeta) - \varphi(\zeta')\} - \left\{w_y - \frac{e_r}{m_r}\epsilon_1^{\frac{2}{3}} \int_{\zeta'}^{\zeta} b(\zeta'')d\zeta''\right\}^2 = 0 \tag{31}$$

The results of the solution of the system of Eqs. (28)–(31) are presented in Fig. 13, where one can see the current density profiles (total and partial ones) and the corresponding magnetic fields for multicomponent current sheet. Here one can clearly see the embedded current sheets dominated by electrons (in the center), protons and oxygen (at the peripheree). The contributions of oxygen ions to the magnetotail current, shown in Fig. 14 can be compared with the estimates of the oxygen currents in the magnetotail from the spacecraft observations. Estimates using Cluster measurements (Kistler, 2005) yield the relative density of oxygen ions to be of the order of 10%–15%. One should mention that our estimate of 30% is the upper limit, and it depends strongly on the oxygen flux at the source and on the oxygen temperature. In general our results provide estimates of the same order as the observations.

The principal conclusion is that the local currents of O^+ ions lead to an effective broadening of the total current profile and are dominant at the peripheree of the sheet. A combination of the contributions of the oxygen ions yields a multilayer structure of the current sheet resembling a "matreshka", a popular Russian nesting doll. Thus, a very thin electron layer $L \sim 0.05$–$0.1\rho_i$ is embedded into a thin proton sheet with $L \sim \rho_i$, which is in its turn embedded within a thick oxygen layer $L \sim 5$–$10\rho_i$. Further, the CS as whole is embedded inside the thicker plasma sheet, i.e., generally there are at least four levels of embedding (Fig. 12).

14 The Bifurcation and "Aging" of TCSs

In Section 12 we have considered the mechanism of current profile splitting due to the influence of electron drift currents. Unfortunately this mechanism could not explain the strong bifurcation of CS profile which is as a rule seen in experimental data. Here we consider another possible mechanism of bifurcation related with the nonadiabatic scattering of ions during current sheet crossings. In magnetic systems with $\kappa \leq 1$ (parameter κ should be sufficiently small to support a quasi-adiabaticity in the system) there is still some scattering of particle (for any finite value of κ) due to unavoidable jumps ΔI_z of quasiadiabatic invariant I_z at the moments of crossing the separatrix between two different types of motion (crossing plane $z = 0$ and non crossing it). The diffusion related with this scattering is quite slow for realistic magnetotail parameters and we disregard it in calculations above. However in this section we will consider its effects essential at the longer time scales. As a result of this diffusion the quasi-adiabatic current sheet might be gradually populated by quasi-trapped plasma (Section 6), which becomes trapped around CS plane due to the jumps of adiabatic invariant I_z. Gradual accumulation of quasi-trapped population within CS results in evolution of current profile and produces its increasing bifurcation. Such process might be considered as "aging" of CS (Zelenyi *et al.*, 2002). and finally could result in CS destruction.

Below we consider in general this mechanism of CS bifurcation. The contribution of quasi-trapped particles to the cross-tail current is different from one of transient particles. This is illustrated in Fig. 15, where the elements of different "serpentine" orbits in neutral plane are shown.

One can see that the local current of quasi-trapped particles is opposite to the current of transient ones. However the total current of quasitrapped particles is exactly equal to zero because such trajectories have a form of closed loops. Therefore, if the density of trapped particle population become too large, the primary current supporting the system, carried by Speiser's ions, should be redistributed without changes of its integral value. In the course of this redistribution the current maximum in the center should be displaced to the peripheries of current sheet, therefore we will see the local minimum in the center and two local off-center maxima of current density in the CS profile.

The self-consistent solutions of Ampere equation for current sheet with transient and quasitrapped ions (here for simplicity we consider electrons as a cold background) has the form

$$\frac{dB}{dz} = \frac{4\pi}{c}\left\{\int_{I_z<I'} v_y f(z,\vec{v})d\vec{v} + \int_{I'\leq I_z<I''} v_y \Psi(I(\vec{v}),\tau)d\vec{v}\right\} \tag{32}$$

where the first term in RHS is the current of transient particles having quasi-adiabatic invariant in the range of values $I_z < I' \equiv (m/\omega_0)v_0^2$. The second term in RHS describes the local current of quasi-trapped particles with quasi-adiabatic invariants belonging to the region $I' < I_z < I''$. Here I'' is the boundary value in a space of I_z invariants, separating the ions on "cucumber" and "ring" orbits (Savenkov and

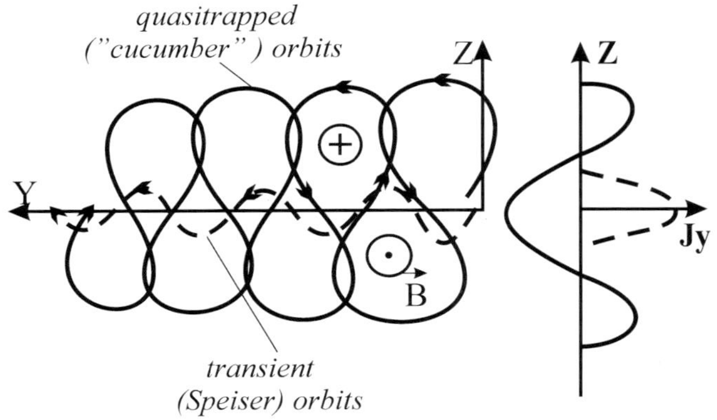

Fig. 15. The elements of quasi-adiabatic "cucumber" (solid line) and transient (dashed line) trajectories are presented. At the right the corresponding local current densities $J_y(z)$ supported by these group of ions are shown.

Zelenyi, 1996). The distribution function $\Psi = \Psi(I, \tau)(\tau = \omega_n t, \omega_n = eB_n/mc)$ might be found from the solution of the diffusion equation over the value of quasi-adiabatic invariants of motion I_z:

$$\frac{\partial \Psi}{\partial \tau} = \frac{\partial}{\partial I} D(I, \tau) \frac{\partial \Psi}{\partial I} \tag{33}$$

with the variable coefficient of diffusion D (Büchner and Zelenyi, 1989), corresponding to a small jumps ΔI_z in linear approximation:

$$D = \frac{3\pi^2}{16T_{ab}} \kappa^2 \left(1 - I^{\frac{4}{3}}\right) \tag{34}$$

For small value of κ the diffusion is slow and distribution function, obtained for different moments of time might be considered as instantaneous quasi-equilibrium, i.e., the system is assumed to relax very quickly to small variations of the distribution function. Therefore, the solutions of the diffusion equation (33) could be considered as snapshots during slow evolution of the diffusive ion distribution function. These temporal snapshots $\Psi = \Psi(I, \tau)$ were used to solve the self-consistent system of Eqs. (32)–(34). This allows to find the changing profiles of current densities and magnetic fields in TCS during the aging process, reflecting the accumulation of quasi-trapped plasma within TCS. Figure 16 demonstrates six self-consistent profiles of current density for different moments of time during TCS "aging". This figure shows that the evolution of current sheet profiles leads to the decrease of current density in the center of the sheet and redistribution of current maximums to the edges. At some critical moment when the minimum in the neutral plane reach almost zero the self-consistent solutions are absent. Whether current sheet will survive this strong bifurcation and will transform to the triple current sheet discussed in the next section

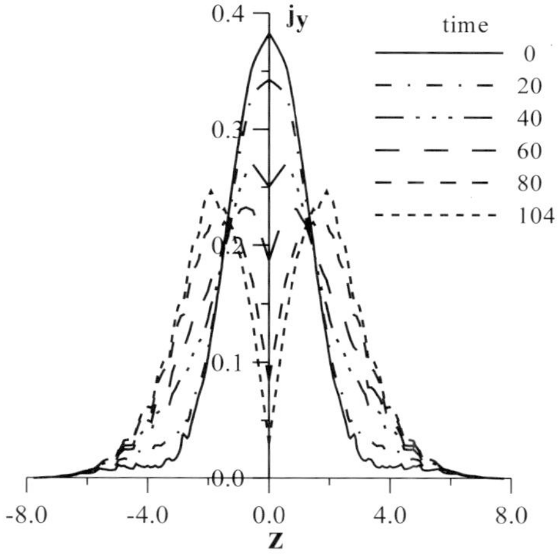

Fig. 16. The theoretical profiles of dimensionless current densities in TCS at six different moments of time $\tau = \omega_n t (\omega_n = eB_z/mc)$.

or will be destructed at the moment $j_y(z = 0) \to 0$ is still open question. Observations of triple sheets are very scarce and there is still theoretical problem to describe the transition to the case $j_y(z = 0) < 0$ (triple profile) because all approximations used above fail at the moment of such transition. Also it is clear that this bifurcation effect caused by "aging processes" could be very strong, contrary to bifurcation due to electrostatic effects, considered in Section 12.

The quasi-adiabatic diffusion is characteristic for magnetic configurations with small parameter of adiabaticity κ (see paragraph 6). Together with quasi-adiabatic diffusion plasma particles might reveal the nonadiabatic behavior and magnetic moment diffusion in configurations with $\kappa \geq 1$, i.e., when particle gyroradii are less than or about the curvature radius of magnetic lines. To investigate these scattering processes the heuristic Centrifugal Impulse Model (CIM) model of the nonadiabatic scattering of charged particles in a field reversal was developed by Delcourt *et al.* (1994). In this model, the nonadiabatic motion of magnetized ions is viewed as the result of an impulsive centrifugal force perturbing the particle gyromotion during a "critical" cyclotron turn near the magnetic field minimum. It was shown that such an interpretation framework is appropriate to characterize the $\kappa \approx 1$ regime in the near-Earth magnetotail. It leads to a three-branch pattern of magnetic moment variations (Delcourt *et al.*, 1996b). More specifically, assuming that the impulsive centrifugal force is $F(t) = (mV_{\|}^2/2R_{\mathrm{C}})\{1 - \cos(2\pi t/\tau)\}$ (m being the particle mass, $V_{\|}$ its parallel speed, R_{C} the field line curvature radius, and τ the impulse time scale) and solving the equation of motion in the gyration plane, the following expression can be obtained for the magnetic moment μ_f after the impulse, normalized to its initial value μ_o: $\mu_f/\mu_o \approx \lceil 1 + 1/\kappa_\alpha^2 \rceil$ (where $\kappa_\alpha = \kappa \sin^{1/2} \alpha_o / \cos \alpha_o$, α_o being the initial pitch angle of the particle at the field minimum and in the adiabatic limit, i.e., calculated assuming magnetic moment conservation). This expression yield a three-branch pattern of variations with (i) systematic μ enhancements at small pitch angles, (ii) either μ enhancement or damping depending upon gyrophase for intermediate pitch angles, and (iii) negligible μ change at large pitch angles. Parameter κ_α^2 provides a measure of the Lorentz force magnitude with respect to that of the centrifugal force. For small pitch angles ($\kappa_\alpha^2 < 1$), the centrifugal force dominates while, for large pitch angles, one has $\kappa_\alpha^2 > 1$, the Lorentz force dominates.

Though CIM was originally developed in the $\sim$1–3 κ range, it was shown that it can be applied also to limited $\kappa \ll 1$ domains. In particular, it reproduces well the energy resonances (Chen and Palmadesso, 1986) or, equivalently, the quasi-adiabatic behaviors (Büchner and Zelenyi, 1989) at specific values of κ. Delcourt *et al.* (2004) applied CIM to the case of double-hump current sheets that are observed in the magnetotail. For such current sheets the net nonadiabatic behaviour of the particles may be viewed as successions of elemental scattering sequences that can each be described using the CIM developed for single-hump current sheets. Each of the impulses leads to a three-branch pattern of magnetic moment variations, and the combination (controlled by gyration phase) of these elemental patterns lead to a wide variety of μ variations, with possibly quasi-adiabatic behavior (negligible μ change) at $\kappa \approx 1$. Such a result clearly contrasts with the single-hump situation where $\kappa \approx 1$ corresponds to enhanced μ scattering. As shown by Delcourt *et al.* (2006), this is of importance for instance for particles of ionospheric origin that intercept the magnetotail current sheet with $\kappa \approx 1$.

15 Triple Current Sheet Equilibrium

We have considered self-consistent solutions of semi-analytical TCS model and obtained two types of CS equilibrium configuration: bell-shape (one maximum of current density in the center and bifurcated (two maximums of current density at the edges). Here we present another current equilibrium in collisionless space plasma, which is difficult to find in a frame of analytical model. A hybrid numerical model of a one-dimensional current sheet in the Earth's magnetotail is used to calculate a new self-consistent equilibrium state with already double splitting of CS profile, i.e., consisting of three thin subsheets that are located close to each other and whose thickness are of the order of the ion gyroradius (Bykov *et al.*, 2008). The current in the central subsheet flows in the negative direction, opposite to the direction of the currents in the lateral subsheets.

We developed numerical self-consistent model with a modeling setup very close to the one of in analytical model described above. The model involves the Vlasov-Maxwell equations for the magnetic and electric fields, the equations of ion motion, and the equations of state for an equilibrium plasma. The electrons are described in the fluid approximation with an anisotropic pressure tensor. We assume that the magnetic field reversal occurs at $z = 0$ plane. As it is shown in Fig. 10, two plasma flows from two sources in the northern and southern portions of the magnetotail move downwards and upward toward the center of the TCS. Ion motion is modeled by the

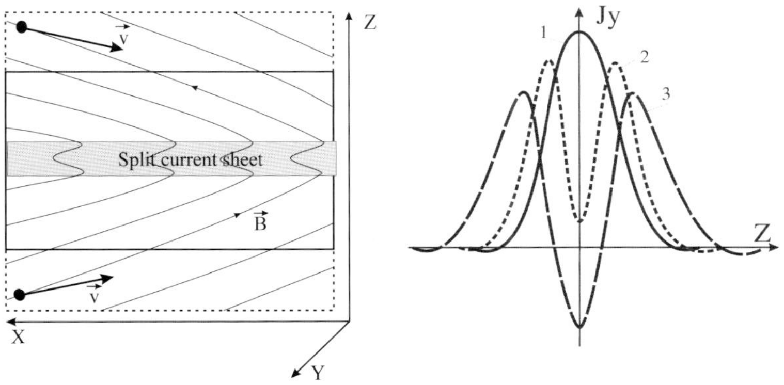

Fig. 17. Magnetic field lines in triple TCS and particles flying from the sources (at the left) and possible shape of profiles of current densities (at the right): 1—a bell-shaped and 2—a bifurcated (split) profile. For comparison, curve 3 shows the current density profile in the triply split current sheet considered in this study.

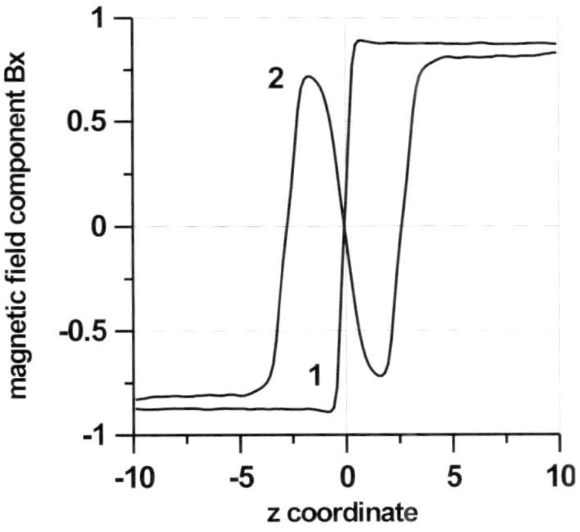

Fig. 18. Two self-consistent steady-state profiles of the magnetic field B_x, normalized to its value at infinity, in a TCS with one (curve 1) and three (curve 2) inversion planes.

macroparticle method. The electrons in the TCS are magnetized; they are assumed to be hot and fast enough to have time to maintain Boltzmann equilibrium along the magnetic field lines. We found that among other equilibrium solutions there exists an unusual one where current reverses its direction at the central plane $z = 0$.

The way initial conditions are set is an important factor which determines the particular type of emerging equilibria. For instance, in order to obtain a current sheet with a bell-shaped current density profile, it is sufficient to specify an initial classical current density profile and to switch on the initial normal magnetic field component.

In order to obtain more complicated triple CS equilibria, it is necessary to specify an initial magnetic field that changes its sign at three reversal planes (see the magnetic field profile in Fig. 17).

In this case, in a numerical macroparticle calculation, the plasma density and current density profiles $n(z)$, $j_y(z)$ relax to an equilibrium state at a certain time scale. In such self-consistent numerical simulations it was found that triple current sheet is stable against external magnetic field perturbations. By varying the input parameters of the system two types of steady equilibria were found, particularly, those with the one and with the three neutral planes. Figure 18 shows self-consistent steady-state profiles of the magnetic field for an unsplitted (curve 1) and for triply splitted (curve 2) current sheets. One can see that the split solution (curve 2) describes a current sheet consisting of three thin subsheets, each having a thickness on the order of the ion gyroradius, with alternating currents, positive at the edges of the sheet and negative at the sheet center. It should be noted that, although the current density distribution within the sheet is complicated, the total magnetic field drop across the sheet is supported by the total current of the Speiser particles and by electron drift currents.

As in the case of a usual current sheet, in a triple sheet ion trajectories could be classified into three main types: untrapped, quasi-trapped, and permanently trapped. In the entire plasma population, the fraction of quasi-trapped and trapped particles is small. In contrast to other particles, trapped ions permanently remain inside the sheet and do not cross the separatrices. Investigation of ion trajectories in a triply split TCS shows that untrapped particles can maintain such configuration self-consistently, so there is no need to add trapped plasma particles to the system (like in a case of "usual" sheet) in order to produce negative current density in the sheet center. Projections of characteristic trajectories of the transient particles to $\{x, y\}$ and $\{y, z\}$ planes are shown in Fig. 19. As one can see, in a triple sheet Speiser particle accomplishes more complicated meandering orbits, which support the split current configuration by carrying negative current in the center and positive currents at the CS edges.

Computer experiments show that strongly anisotropic split current sheet is stable, i.e., the thickness and position of each of the subsheets, as well as the magnetic field topology, remain practically unchanged when the parameters of the system are slowly varied or when the equilibrium state is deformed symmetrically. As the plasma drift velocity in the source is decreased (i.e., the parameter $\epsilon \equiv v_T/v_D$ is increasing), anisotropic current sheet configuration experience an abrupt transition from a splitted state to an unsplitted one. On the whole, a system with two equilibrium states behaves as a particle in a potential well with two minima.

The positions at the minima are stable, and there is an unstable local maximum in between. Both transitions are possible: from one stable state to another through an unstable state, and vice versa. As the anisotropy parameter ϵ is gradually increased to fall into the range $\epsilon < 0.5$, the current density changes in such a way that the negative current at the center of the triply splitted sheet becomes too small to maintain the splitted configuration. At a certain moment, the central subsheet is rapidly destroyed, and lateral subsheets, carrying positive currents, approach to each other and to merge

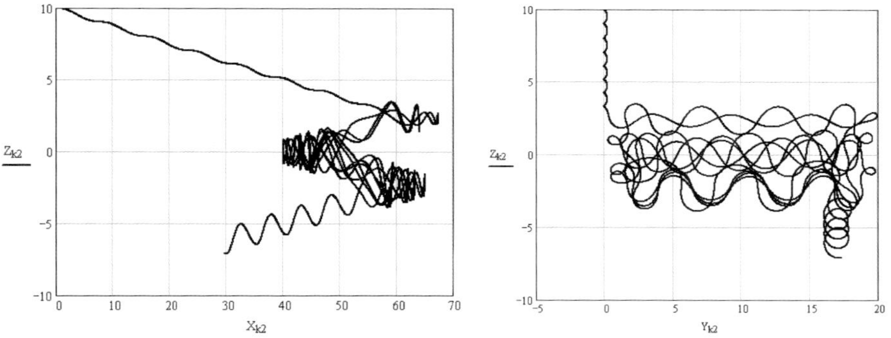

Fig. 19. Two projections of the trajectory of an untrapped ion in a triply split current sheet onto the planes $\{X\text{-}Z\}$ and $\{Y\text{-}Z\}$.

into a single wide current sheet with a positive current. The wide sheet is then rapidly thinning to become a classical bell-shaped thin CS.

In triply split TCSs transient particles support a negative current at the center and positive currents at the edges, thereby maintaining a jump in the magnetic field across the sheet. The equilibrium solution for a triply split TCS differs fundamentally from that for a conventional bell-shaped current sheet.

16 TCS Stability Dilemma and Possibility of Tearing Mode Growth

Tearing instability as the simplest explanation of substorm activity of Earth's magnetotail was proposed many years ago in a pioneering paper by Coppi *et al.* (1966). Authors considered a 1D Harris sheet equilibrium solution (the only one available at this time). Schindler (1974) realized that even small B_z component existing in a sheet completely destroys the resonant Landau damping. But strong stabilization effect of frozen electrons in presence of a finite normal component of magnetic field was the reason of decreased interest to tearing as potentially substorm related process (Lembege and Pellat, 1982). Even stochastic motion (Büchner and Zelenyi, 1989; Kuznetsova and Zelenyi, 1991) and pitch angles scattering (Coroniti, 1980) could not destroy strong stabilization effect of electrons (Pellat *et al.*, 1991). Recently Sitnov *et al.* used the idea of possible tearing destabilization due to transient electrons (Sitnov *et al.*, 1997, 2002) to suggest possible scenario of substorm activation. For many years A. T. Y. Lui and his coauthors advocate the approach relating the substorm onset to the fast growth of current driven instabilities (Lui, 1996 and references therein).

There are possibly many routes to substorm activization and all these mechanisms might find their place in the analysis of this complicated problem. We want to restore here the ion tearing mode as the one of most probable candidates for triggering of the substorm onset. The main confusion in many previous papers cited above was related with the fact that all earlier conclusions about tearing mode stability (Schindler,

1974; Lembege and Pellat, 1982; Kuznetsova and Zelenyi, 1991; Pellat *et al.*, 1991) were based on application of stability analysis to Harris-like current sheet with imposed normal component of magnetic field. Both the experiment (Hoshino *et al.*, 1996; Nakamura *et al.*, 2006; Runov *et al.*, 2006) and the theory discussed above indicate that CS structure is principally different from any of Harris like models and there exist serious problem in interpretation of CS observations without some mechanism capable to explain fast topological changes of magnetotail structure (Zhang *et al.*, 2002; Volwerk *et al.*, 2003; Sergeev *et al.*, 2004; Nakamura *et al.*, 2006). From the other hand last series of PIC modeling of collisionless reconnection (Pritchett, 2005a, b; Daughton *et al.*, 2006; Divin *et al.*, 2007; Daughton and Karimabadi, 2007) shows the appearance of strong anisotropy velocity distribution. There should exist, correspondingly, the additional factors controlling the stability of CS. Our point that the anisotropy is the factor should always be taken into account. Importance of anisotropy was also clarified in a number of recent numerical investigations of CS dynamics (Daughton, 2002; Karimabadi *et al.*, 2003; Daughton *et al.*, 2004; Sitnov *et al.*, 2004, 2006). These results make the problem of analytical stability analysis of anisotropic equilibrium much more actual. In this section we present the results of the evolution of eigenmodes (tearing and sausage) for thin anisotropic CS.

For this analysis of mode stability the theory by Schindler (2006) was modified to be applicable to a new 1D (but $B_z \neq 0$) anisotropic configuration. Thus the functional of the total energy could be presented in a form:

$$W = \int \frac{\mathbf{B}^2}{8\pi} d\mathbf{r} + \sum_j \int H_j f_j \, d\mathbf{v} d\mathbf{r} \tag{35}$$

Here $H_j = \{\mathbf{p} - (q_i/c)\mathbf{A}\}^2/2m_j + q_j\varphi$ is total (kinetic and potential) energy of particle with charge q_j and mass m_j. Tearing mode is a wave perturbation propagating in the direction of magnetic field: $f_{1j} = \hat{f}_{1j} \exp(ikx - i\omega t)$ and $\mathbf{A}_1 = \widehat{A}_1 \exp(ikx - i\omega t)$ contrary to kink like perturbation propagating along the equilibrium current ($\sim \exp(iky - i\omega t)$). Stationary equations for CS equilibrium discussed above could be considered as a "zero order" perturbation of Eq. (35). First order perturbation of Eq. (35) vanishes because for 1D CS $\int f_{1j}\, dx = 0$. Therefore at the next order we will have only second order perturbation of the total energy:

$$\delta W = \int \frac{\mathbf{B}_1^2}{8\pi} d\mathbf{r} + \sum_j \int \left(H_{0j} f_{2j} + H_{1j} f_{1j} + H_{2j} f_{0j} \right) d\mathbf{v} d\mathbf{r} \tag{36}$$

First and second order perturbations of the total particle energy can be presented in a Hamiltonian forms: $H_{2j} = (q_j/c)^2 \mathbf{A}_1^2/m_j$, $H_{1j} = -(q_j/cm_j)\mathbf{A}_1\{\mathbf{p} - (q_j/c)\mathbf{A}_0\} + q_j\varphi_1$.

The first order of velocity distribution perturbation could be found from the perturbed Vlasov equation:

$$\frac{df_{1j}}{dt} = -\frac{q_j}{cm_j} \left(c\mathbf{E}_1 + [\mathbf{v} \times \mathbf{B}_1] \right) \frac{\partial f_{0j}}{\partial \mathbf{v}} \tag{37}$$

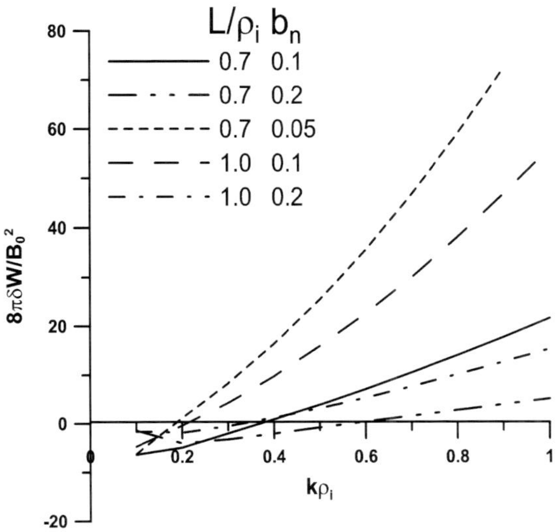

Fig. 20. Total energy of perturbation as a function of wave number for CS with different parameters.

Where d/dt is a full time derivative along particle trajectory, while $\mathbf{E}_1$, $\mathbf{B}_1$ can be obtained from Maxwell equation as some functional of f_{1j}.

The second order of velocity distribution perturbation provides the well known stabilization term (Galeev and Zelenyi, 1976; Lembege and Pellat, 1982; Sitnov *et al.*, 1997), which are due to perturbation of a scalar potential φ_1 (Zelenyi *et al.*, 2008) and which is the critical issue for all previous studies:

$$\sum_j \int H_{0j} f_{2j}\, d\mathbf{v} d\mathbf{r} = \frac{0.5 p_0 k^2 |A_{iy}|^2}{B_z^2} \tag{38}$$

Here p_0 is the total unperturbed particle pressure at the centre of CS.

Therefore Eq. (36) can be rewritten:

$$\delta W = \int \left\{ \left| \frac{dA_{1y}}{dz} \right|^2 + \left(k^2 - \frac{4\pi}{c} \frac{\partial j_{0y}}{\partial A_{0y}} + \frac{4\pi p_0 k^2}{B_z^2} \right) |A_{1y}|^2 \right\} d\mathbf{r} \tag{39}$$

We use Eq. (37) to change $\sum_j \int (H_{1j} f_{1j} + H_{2j} f_{0j}) d\mathbf{v} d\mathbf{r}$ to $\delta W = \int \frac{1}{2c} \frac{\partial j_{0y}}{\partial A_{0y}} |A_{1y}|^2 \, d\mathbf{r}$ and take into account the property of marginal stability ($\partial \ln(A_{1y})/\partial t \to 0$). To find CS parameters and wave number of perturbation k which corresponds to marginal stability one need to solve functional equation for minimizing δW. Once one will obtain such function A_{1y} for that $\delta W(A_{1y}) < 0$, then CS with the corresponding parameters becomes unstable. The total energy of perturbation at its minimum value is

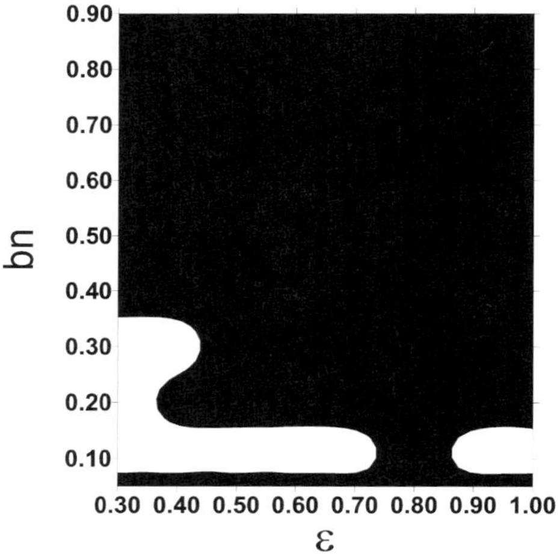

Fig. 21. Map of marginal tearing mode parameters. White region—instability, black region—stability.

shown in Fig. 20 as a function of wave number k (here we used estimate of thin CS thickness $L = \rho_i \epsilon^{1/3}$, where ρ_i is ion giroradius in the field B_0). As one can see there exist well defined region with $\delta W < 0$. Presence of this region signifies the long awaited fact that there are windows of ion tearing mode instability in the parameter space. To find the appropriate range of parameter region for each value of k we constructed a tearing mode stability map. At this parameter map (parameters ϵ and b_n are used as governing ones) white color is used for such points where $\delta W < 0$ at least for one value of k (Fig. 21). As shown at this map the most unstable CSs have strong anisotropic characteristics $\epsilon \ll 1$ and optimal value of parameter is $b_n \sim 0.1$–0.2. Thus thin anisotropic CS can be unstable to tearing instability contrary to isotropic CS with non zero normal component of magnetic field (Pellat *et al.*, 1991). This result gives a chance to revisit old but well forgotten paradigm associating substorm activity with the onset of tearing-type instability. Thinning of magnetotail CS with increasing anisotropy of ion streams can transfer thick isotropic (and stable) CS to a new unstable equilibrium where CS might experience topological changes enabling the process of fast reconnection.

17 General Analysis: Oblique Symmetric Low-frequency Electromagnetic Modes Stability Analysis for TCS

As was shown before TCS could become unstable for the electromagnetic modes propagating along magnetic field. But such mode with $\mathbf{k} = k\mathbf{e}_x$ is only a limiting case of more general situation when modes could propagate at arbitrary direction toward magnetic field (e.g., for $\mathbf{k} = k\mathbf{e}_y$ these are well-known kink and sausage modes (Lapenta and Brackbill, 1997; Büchner and Kuska, 1999; Daughton, 1999)), could be either symmetric or asymmetric with respect to $z = 0$ plane. Also as we will discuss below such modes could have different polarizations $\{A_{1x}, A_{1y}, 0\}$ or $\{0, A_{1y}, A_{1z}\}$. A lot of work have been done to analyze all these numerous instabilities both analytically (Lapenta and Brackbill, 1997; Daughton, 1999; Yoon and Lui, 2001) and numerically (Pritchett *et al.*, 1996; Zhu and Wingle, 1996; Büchner and Kuska, 1999; Karimabadi *et al.*, 2003). But all these studies were limited to the case of 1D Harris equilibrium with $B_z \neq 0$. As we saw in the previous section this is a drastic difference in stability properties for Harris and anisotropic TCS. So in this section we will revisit the general problem of stability analysis for oblique low frequency electromagnetic perturbations propagating in thin current sheet.

Before there were only few uncoordinated studies of this problem. Sitnov *et al.* (2004) studied LHD type perturbation propagating in Y-direction but practically ignored electron contribution limiting his analysis by the assumption that electrons provide only cold neutralizing background. According to our knowledge there was only one study (Silin *et al.*, 2002) devoted to general analysis of electromagnetic perturbation in CS, but it have been done also in a limit of infinitely thin 1D current sheet (so called Syrovatskys' sheet). As in the study by Silin *et al.* (2002) we assume $\mathbf{k} = k\cos\theta\mathbf{e}_x + k\sin\theta\mathbf{e}_y$ and $\tan\theta = k_y/k_x$ and using Coulomb gauge for vector potential div $\mathbf{A} = 0$ (where $\mathbf{A} = A_x\mathbf{e}_x + A_y\mathbf{e}_y$) we obtain: $A_{1y} = -A_{1y}\tan\theta$. For $\theta = \pi/2$ (sausage instability) component $A_{1x} \to \infty$ and existence of sausage-like perturbation becomes impossible. But on another hand another polarization could be used to study sausage instability $\mathbf{A}_1 = A_{1y}\mathbf{e}_y + A_{1z}\mathbf{e}_z$ (Lapenta and Brackbill, 1997; Büchner and Kuska, 1999). In this case the relation between perturbed vector potential components has a form $A_{1z} = -ik\sin\theta\int_{-\infty}^{z} A_{1y}(z')dz'$. In a limit $\theta \to 0$, $A_{1z} \to 0$ and these two different vector potential polarization coincide. However for the analysis of oblique perturbations we should use both potential sets (both satisfying, of course, the Coulomb calibration) to obtain the complete results.

In general case for oblique perturbations (contrary to tearing mode) mode frequency could have real oscillatory part and one should take into account inverse Landau damping resonant term $\omega = \omega_r + i\gamma$. In presence of normal magnetic field B_z electrons are frozen into magnetic field and can not participate in resonance interaction. Thus only ion resonance takes a place (Schindler, 1974) and corresponding mode is called ion mode. As was shown for field reversals where $B_x(0) \sim 0$ there exists central region with a width $\sim\sqrt{L\rho_\mathrm{i}}$ where ions "effectively" do not feel magnetic field (Dobrowolny, 1968). Here one could introduce the function $U_i(z)$:

$$U_i(z) = \begin{cases} 1, & |z| < \sqrt{L\rho_\mathrm{i}} \\ 0, & |z| > \sqrt{L\rho_\mathrm{i}} \end{cases} \tag{40}$$

Equation (35) can be rewritten for a general case (including resonant interactions and perturbation of the vector potential $\mathbf{A}_1$) by using

$$\delta W = \int \frac{\mathbf{B}_1^2}{8\pi} d\mathbf{r} - \frac{1}{2}\sum_j \int \frac{\tilde{f}_{1j}^2}{\dfrac{\partial f_{0j}}{\partial H_{0j}}} d\mathbf{r} d\mathbf{p} - \sum_j \frac{1}{2c}\int \frac{\partial \mathbf{j}_0}{\partial \mathbf{A}_0}\mathbf{A}_1^2 d\mathbf{r} d\mathbf{p}$$

$$-\sum_j \frac{q_j}{c}\int \mathbf{v}\mathbf{A}_1 f_{ij}^{\mathrm{res}} d\mathbf{r}d\mathbf{p} \tag{41}$$

Here $\tilde{f}_{1j} = f_{1j} - (\partial f_{0j}/\partial \mathbf{A}_0)\mathbf{A}_1$, and in assumption of frozen electrons, after using Swartz inequality one can obtain $-\frac{1}{2}\sum_j \int \frac{\tilde{f}_{1j}^2}{\partial f_{0j}/\partial H_{0j}} d\mathbf{r}d\mathbf{p} = p_0 k^2 |A_{1y}|^2/B_z^2$ (Galeev and Zelenyi, 1976; Lembege and Pellat, 1982). To find the characteristics of the possible wave modes (growth rate, real part of frequency and perturbations of vector potential) one need to minimize the functional δW: i.e., the variation of δW along $\mathbf{A}_1$ have to be zero. Thus there is a system of the linear differential equations for the vector potential components.

For the polarization of vector potential $\mathbf{A}_1$ in a form $\mathbf{A}_1 = A_{1x}\mathbf{e}_x + A_{1y}\mathbf{e}_y$ more convenient to consider one equation for magnitude $A_1 = \sqrt{A_{1x}^2 + A_{1y}^2}$ instead of the system of equations for each of these components. It is straight forward because A_{1x} and A_{1y} are linearly coupled: $A_{1x} = -A_{1y}\tan\theta$.

$$\frac{d^2 A_1}{dz^2} - k^2\left\{\left(1 + 4\pi p_0 B_z^{-2}\cos^4\theta\right) - 0.5c^{-1}\frac{\partial j_y}{\partial A_0}\cos^2\theta\right\} A_1 = -j^{\mathrm{res}}(z, \theta, A_1, t) \tag{42}$$

Expression for the resonant current $j^{\mathrm{res}}(z, \theta, A_{1y}, t)$ can be written in form:

$$j_y^{\mathrm{res}} = U_i\left(j_y^A\cos\theta + j_x^A\sin\theta + j_y^\varphi\cos\theta + j_x^\varphi\sin\theta + j_y^w\cos\theta + j_x^w\sin\theta\right) \tag{43}$$

All coefficients from (43) have integral form:

$$\begin{aligned}
j_y^A &= i v_{Ti} k_y \int_0^t Q(\mathbf{v}, t, t')dt' \int_{-\infty}^{+\infty} \frac{v_y^2}{n_0 v_{Ti}^2}\frac{\partial f_{0i}}{\partial A_{0y}} d\mathbf{v}\\
j_x^A &= i v_{Ti} k_y \int_0^t Q(\mathbf{v}, t, t')dt' \int_{-\infty}^{+\infty} \frac{v_y v_x}{n_0 v_{Ti}^2}\frac{\partial f_{0i}}{\partial A_{0y}} d\mathbf{v}\\
j_y^\varphi &= -i v_{Ti} k_x \int_0^t Q(\mathbf{v}, t, t')dt' \int_{-\infty}^{+\infty} \frac{v_y^2 v_x}{n_0 v_{Ti}^3}\frac{\partial f_{0i}}{\partial \varphi_0} d\mathbf{v}\\
j_x^\varphi &= -i v_{Ti} k_x \int_0^t Q(\mathbf{v}, t, t')dt' \int_{-\infty}^{+\infty} \frac{v_x^2 v_y}{n_0 v_{Ti}^3}\frac{\partial f_{0i}}{\partial \varphi_0} d\mathbf{v}\\
j_y^w &= 0.5 i v_{Ti} k_y \int_0^t Q(\mathbf{v}, t, t')dt' \int_{-\infty}^{+\infty} \frac{v_y^3}{n_0 v_{Ti}^3}\frac{\partial f_{0i}}{\partial \varphi_0} d\mathbf{v}\\
j_x^w &= 0.5 i v_{Ti} k_y \int_0^t Q(\mathbf{v}, t, t')dt' \int_{-\infty}^{+\infty} \frac{v_y^2 v_x}{n_0 v_{Ti}^3}\frac{\partial f_{0i}}{\partial \varphi_0} d\mathbf{v}
\end{aligned} \tag{44}$$

Where $Q(\mathbf{v}, t, t') = A_1(t')\exp(-i\mathbf{k}\mathbf{v}(t - t'))$ and $k_y = k\sin\theta$, $k_x = k\cos\theta$.

For another polarization $\mathbf{A}_1 = A_{1y}\mathbf{e}_y + A_{1z}\mathbf{e}_z$ we could solve only one equation for A_{1y}, like it was done by Lapenta and Brackbill (1997) for the sausage mode case,

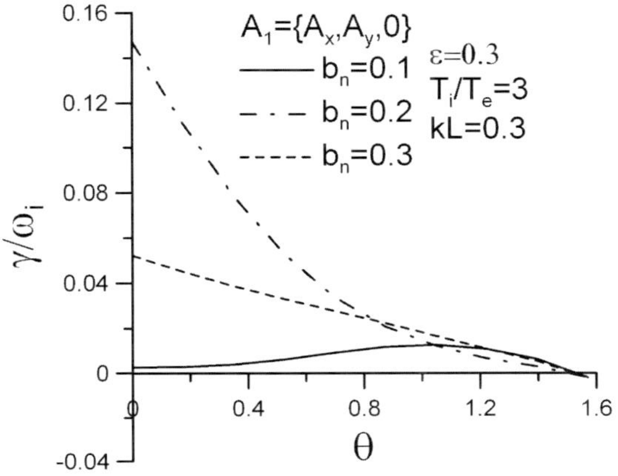

Fig. 22. Growth rate as a function of angle θ for TCS with different value of b_n for perturbation with polarization $\mathbf{A}_1 = A_{1x}\mathbf{e}_x + A_{1y}\mathbf{e}_y$.

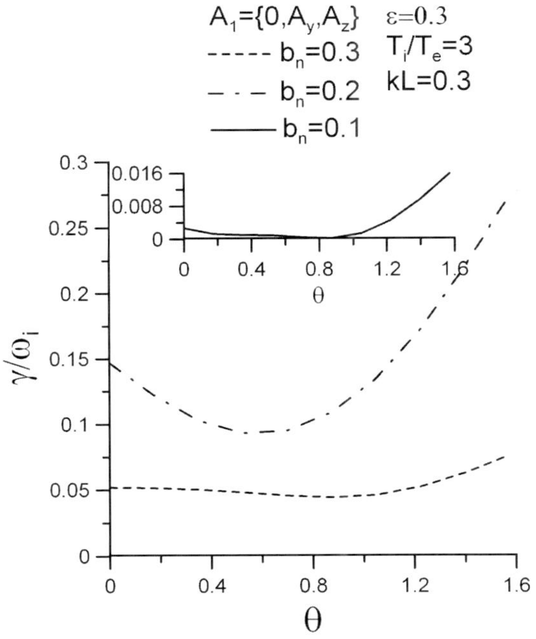

Fig. 23. Growth rate as a function of angle θ for TCS with different value of b_n for perturbation with polarization $\mathbf{A}_1 = A_{1x}\mathbf{e}_x + A_{1y}\mathbf{e}_y$.

but extending their results to an arbitrary angle propagation:

$$\frac{d^2 A_{1y}}{dz^2} - k^2 \left\{ \left(1 + 0.5 p_0 B_z^{-2} \cos^2\theta\right) - 0.5 c^{-1} \frac{\partial j_y}{\partial A_0} \right\} A_{1y} = -j^{\text{res}}(z, \theta, A_{1y}, t) \quad (45)$$

In such case the resonant current j^{res} is a function of only one component of perturbed vector potential A_{1y}:

$$j^{\text{res}} = \Theta U_i \left(j_y^A A_{1y} + j_y^{\varphi} A_{1y} + j_y^{w} A_{1y} \right) \quad (46)$$

By solving (41) and (44) using finite element method (Lapenta and Brackbill, 1997; Daughton, 1999; Sitnov *et al.*, 2002) one can obtain growth rate and real part of frequency as a function of angle θ and magnitude of k. As a spatial basis we use eigenfunctions of Eqs. (42) and (45) without time depended part at the right.

Growth rate γ as a function of angle θ for perturbation with polarization $\mathbf{A}_1 = A_{1x}\mathbf{e}_x + A_{1y}\mathbf{e}_y$ and different values of parameter $b_n = B_z/B_0$ is shown in Fig. 22. Maximum value of growth rate for tearing instability ($\theta = 0$) for TCS is achieved for $b_n = 0.2$. This result before was obtained recently for tearing studies (Zelenyi *et al.*, 2008). One the value of angle θ is increasing the value of growth rate becomes smaller. This result also corresponds to the previous investigation of Harris CS done by Silin *et al.* (2002).

If one consider another polarization: $\mathbf{A}_1 = A_{1y}\mathbf{e}_y + A_{1z}\mathbf{e}_z$ the growth rate as a function of θ will have significantly different behavior (Fig. 23). In this case the value of growth rate increases with the increase of θ and perturbation converges to the well known sausage mode (when $\theta \to \pi/2$). Also there exists different dependence of the growth rate on the value of parameter b_n.

Thus for different polarizations of perturbed vector potential we obtain different values of the growth rate as a function of angle θ. Also, as one can see from Figs. 22, 23, for each value of θ there exists mode of perturbation with positive growth rate. It means that simultaneously different types of wave modes could be at a neutral plane of magnetotail to form a complicated turbulent structure (Milovanov *et al.*, 2001). Zelenyi *et al.* (1998) suggested different approach to analyze this complicated turbulent ensemble of waves with different structure but already nonlinearly interacting with each other. Using the analysis based on fractal topology authors have shown that some universal scaling should exist so that the modes can be selfconsistently coupled with each other and magnetic field reversal of the current sheet.

18 From Theory to In-situ Measurement: Comparison of Theoretical and Experimental Data

We have considered above the properties of thin current sheets obtained particularly in a frame of the theoretical model of anisotropic current sheet based on quasi-adiabatic ion dynamics in the curved magnetic field, presented in series of publication by Zelenyi *et al.* (2000, 2002, 2004). In comparison with the usual isotropic models of current sheets this model seems to provide more adequate description real of magnetotail observation. The striking example of this inconsistency is shown in Fig. 22

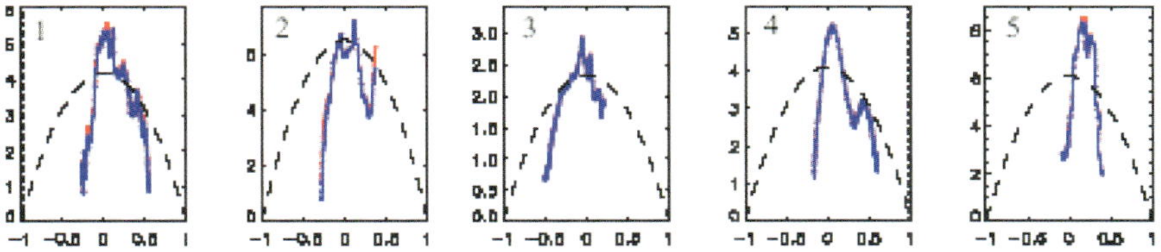

Fig. 24. Five current density profiles of thin current sheets. Corresponding current profiles, approximated by Harris model, are shown by dashed lines. (Adapted from Runov *et al.* (2006).)

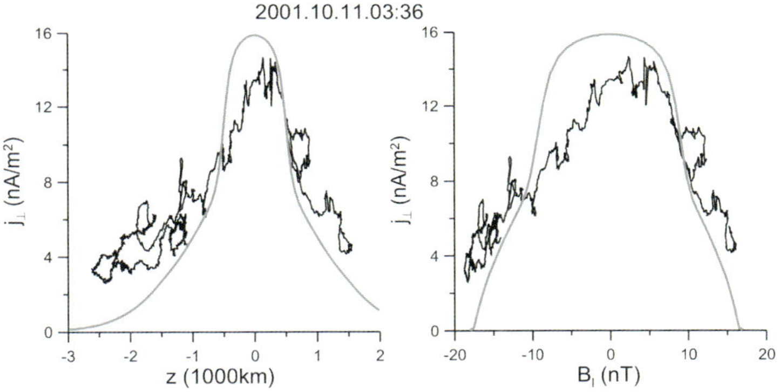

Fig. 25. Comparison of current density profiles versus coordinate along normal and magnetic field component B_l for experimental (black) and model (grey) case 2001.10.11.

(adapted from the paper by Runov *et al.*, 2006), where thin current sheet profiles are compared with Harris model (Fig. 24). The approximation is made accordingly to the data on magnetic field. One can clearly see that the observed thin current sheets are not Harris-like. Harris model can not describe these complicated profiles where very thin current sheets are embedded inside much thicker background plasma sheet.

Thus we have chosen the model by Zelenyi *et al.* (2004) with a set of its later modifications for comparison with corresponding experimental data. This model might be a useful tool to describe and explain peculiarities of a quite wide variety of the observed current structures. Interpretation of the observed sheet characteristics (and hence, applicability of model) may vary depending on the actual magnetic field geometry, local plasma properties and level of magnetospheric activity. Below we will show that the model of anisotropic current sheet might be successfully applied for comparison with to a specific subclass of Cluster current sheet observations, thin horizontal (with normal along z) single-peaked sheets.

Observational data were taken from several published investigations of fast current sheet crossings (Runov *et al.*, 2005, 2006). In particular the events during magne-

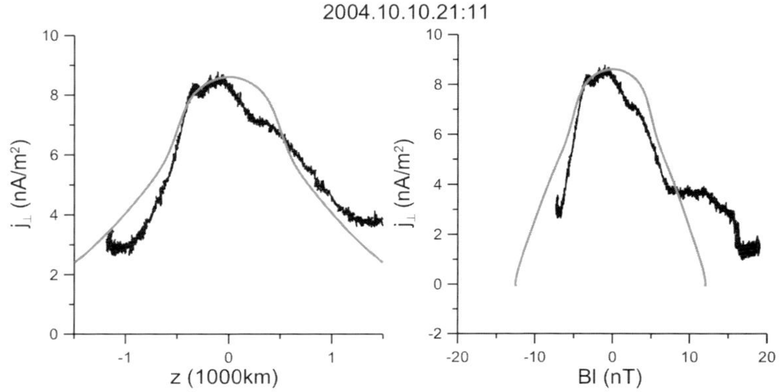

Fig. 26. Comparison of current density profiles versus coordinate along normal and magnetic field component B_l for experimental (black) and model (grey) case 2004.10.10.

totail seasons of 2001 and 2004 when Cluster spacecraft separation was about 1000–2000 km are considered and almost horizontal current sheets are chosen. Two characteristic examples of this comparison are presented in Figs. 25 and 26 contains current density profiles for the case 2001.10.11, plotted with respect to the spatial coordinate Z and magnetic field components along vector of maximum variation B_l accordingly to MVA method (Runov *et al.*, 2006). This example has a 15% of oxygen and therefore it is compared with the model with heavy ions. The example of current sheet in October 10, 2004 was compared with the model having two ion populations with two different temperatures, densities $n_i^{(2)}/n_i^{(1)} = 5.5$ and electrons with anisotropic pressure. Taking into account the hot ions helped us to describe current sheets rather wide in comparison with the first example. In both cases the difference between experimental data and theory (plasma density, maximum of current density, proton temperature and CS thickness) is quite good and does not exceed 30%.

As we have discussed in paragraph 16, thin anisotropic current sheets were found recently to be tearing unstable in parameter space $\{\epsilon, b_n\}(\epsilon = v_T/v_D, b_n = B_z/B_0)$ with the typical growth time of the order of one-two minutes (Zelenyi *et al.*, 2008). Figure 27 presents the map of instability of different current sheets, characterized by experimental points $\{B_z/B_0, B_{\text{ext}}/B_0\}$ (here we use two basic magnetic field values, i.e., B_0 is the magnetic field at the edge of thin current sheet and $B_{\text{ext}}^2 = 2\mu_0(n_pT_p + n_eT_e)|_{z=0}$ is the total (lobe equivalent) magnetic field). There exists approximate linear dependence between ratio B_{ext}/B_0 and parameter $\epsilon = v_T/v_D$ (Fig. 27), correspondingly we can replace $\{\epsilon, b_n\}$ instability map to $\{B_z/B_0, B_{\text{ext}}/B_0\}$ map. One can see that two points from the entire 14 one fall into the instability domain. Here 14 crossings from 2001 and 2004 years are considered. It is reasonable to assume that the observed sheets typically are in quasi steady state near the certain equilibrium and their parameters should belong to the stability domain of parameter space. Substorms are relatively rare events and if one will take the ratio of unstable/stable

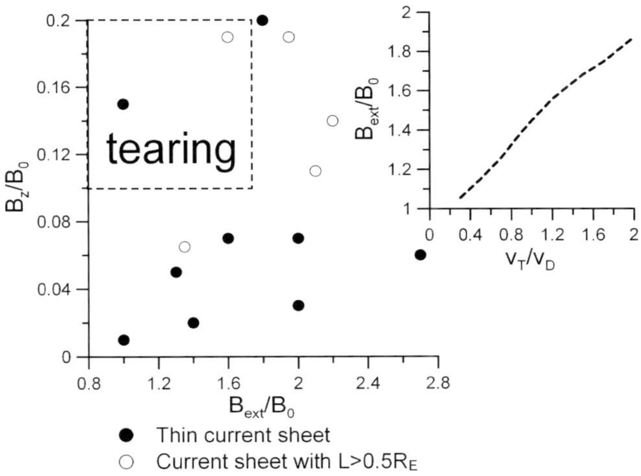

Fig. 27. Parametric map of instability region for modeling thin CS and experimental CS points from 2001 and 2004 years.

cases derived from our simple statistics ∼15% it seems to give a reasonable proxy to estimate the probability of substorm observation. Consequently one can conclude that experimental data are again not in a contradiction with the model.

Although statistics shown in Fig. 27 is unfortunately rather scanty today, we hope that observations that will be carried next decade by a new generation of magnetospheric spacecraft, will provide additional argument to the problem of magnetotail stability.

Abovementioned comparison with observations *in situ* confirm that considered anisotropic model is capable to reproduce the shape and maximum value of current density, including both ion and electron contributions as well as the important embedding property of current sheets.

19 Conclusions

In this review we tried to consider the specific properties of thin current sheets in the Earth's magnetotail, observed onboard Geotail, Interball and Cluster satellites and compare them with the corresponding characteristic obtained in a frame of the analytical "hybrid" model of quasiadiabatic current sheet. The basic idea of our model that we use two principally different description for ions (quasi-adiabatic approximation, based on assumption of I_z conservation) and electrons (guiding centre approximation with assumption about establishing of Boltzmann equilibrium along field lines). To avoid confusion we want to stress that ions which we called quasi-adiabatic usually (Speiser, 1965; Sonnerup, 1971; Buechner and Zelenyi, 1989) are called "non-adiabatic" (to emphasize the each understanding that magnetic moment μ is temporarily not conserved for them). Theoretically it was shown that there exists a class of equilibrium solutions of 1D self-consistent current sheet configuration where anisotropic electrons could provide a very characteristic strong and local contribution to the overall current structure. In the case of isotropic electron distribution the magnetic field profile is flattened in the center of the sheet (Zelenyi *et al.*, 2004), whereas in the anisotropic case it is steeper there. In other models it was also received that the electrons may carry a significant part of the cross-tail current and lead to a bifurcated structure of the current density profile (Asano, 2001; Asano *et al.*, 2003).

Recent CLUSTER data have shown (Nakamura *et al.*, 2004) that thin current sheets in the Earth's magnetotail during geomagnetically active periods become very dynamical and that the shape of the current density profile may substantially evolve with time, particularly exhibit double or even multiple peaks. It is not clear whether such structures represent successive phases of current sheet evolution or appear separately under specific external and/or internal conditions. So far the model presented here, which explicitly takes into account electron curvature drifts, is the only model that can explain more sophisticated cross-tail current profiles with triple peaks.

The experimental measurements in the Earth's magnetotail by Interball, Geotail, Cluster and other spacecraft revealed that characteristic dynamical properties of the magnetotail control the global substorm development. Theoretical arguments developed in this paper emphasize again the role of current sheet thinning in the course of substorm development, because thin current sheets could release an excess of the free energy and trigger the sequence of substorm steps (generate magnetic islands, produce x-line, initiate fast reconnection, produce well known auroral, mid and deep tail substorm manifestations). Tearing instability provides the natural mechanism for the spontaneous reconnection of magnetic field lines, which allows the appearance of magnetic X-line, which become the site of fast forced magnetic reconnection which accelerates and then dissipates plasma flows. Contrary to previous investigations of Harris-type current sheets which were found to be totally stable, our studies demonstrated the existence of certain "gaps" in parameter space where system could become unstable for tearing perturbation and thus launch the set of substorm events, discussed above. It was shown that the main domain of instability are concentrated near values $B_n/B_0 \sim 0.1$–0.2, and their width depends on both ion-to electron temperature ratio, average electron pitch-angles and the anisotropy of plasma flows from CS sources.

The experimental check of validity of these numerous theories indicates that the model satisfactorily describes non-Harris profiles and embedded maxima of current density and its stability criteria are consistent with some substorm features. Therefore we hope that model correctly captures basic physical effects important for the current sheet formation in the magnetotail and can be used for further analysis. The progress can be made adding new physical factors such as that related with magnetic field shear. For a more detailed comparison more information on actual ion and electron distribution functions should be obtained and incorporated to the theoretical model.

Acknowledgments. This work was supported in part by the RF Presidential Program for State Support of Leading Scientific Schools (project no. NSh-472.2008.2), the Russian Foundation for Basic Research (project nos. 08-02-00407, 06-02-72561 and 09-05-91008-ANF). One of the authors (L. M. Z.) is very grateful to organizers of Kyoto CAWSES symposium giving him a chance to enjoy hospitality at this very interesting and productive meeting.

References

Asano, Y., Configuration of the thin current sheet in substorms, Ph.D. thesis, Univ. Tokyo, 2001.

Asano, Y., T. Mukai, M. Hoshino, Y. Saito, H. Hayakawa, and T. Nagai, Evolution of the thin current sheet in a substorm observed by Geotail, *J. Geophys. Res.*, **108**(A5), doi:10.1029/2002JA009785, 2003.

Asano, Y., R. Nakamura, W. Baumjohann, A. Runov, Z. Voros, M. Volwerk, T. L. Zhang, A. Balogh, B. Klecker, and H. Reme, How typical are atypical current sheets?, *Geophys. Res. Lett.*, **32**(3), doi:10.1029/2004GL021834, 2005.

Ashour-Abdalla, M., J. Berchem, J. Büchner, and L. M. Zelenyi, Large and small scale structures in the plasma sheet: A signature of chaotic motion and resonance effects, *Geophys. Res. Lett.*, **18**(8), 1603–1606, 1991.

Ashour-Abdalla, M., L. M. Zelenyi, V. Peroomian, and R. L. Richard, Consequences of magnetotail ion dynamics, *J. Geophys. Res.*, **99**(A8), 14891–14916, 1994.

Baker, D. N., Magnetic reconnection during magnetospheric substorms, in *International Conference on Substorms, Proceedings of the 3rd International Conference held in Versailles*, edited by E. J. Rolfe and B. Kaldeich, ESA SP-389. Paris: European Space Agency, 365 pp., 1996.

Baumjohann, W. and R. Nakamura, What is Cluster telling us about magnetotail dynamics, *Proc. 34th COSPAR sci. assemb.*, Houston, TX, USA, 10–19 October, 2002, Proc. D3.1-0027-02, 1–9, 2002.

Biermann, L., Kometenschweife und solare Korpuskularstrahlung, *Zeitschrift fur Astrophysik*, **29**, 274, 1951.

Birn, J., R. Sommer, and K. Schindler, Open and closed magnetospheric tail configurations and their stability, *Astrophys. Space Sci.*, **35**(7), 389–402, 1975.

Birn, J., M. Hesse, and K. Schindler, Formation of thin current sheets in space plasmas, *J. Geophys. Res.*, **103**(A4), 6843–6852, 1998.

Birn, J., J. F. Drake, M. A. Shay, B. N. Rogers, R. E. Denton, M. Hesse, M. Kuznetsova, Z. W. Ma, A. Bhattacharjee, A. Otto, and P. L. Pritchett, Geospace Environmental Modeling (GEM) magnetic reconnection challenge, *J. Geophys. Res.*, **106**(A3), 3715–3720, 2001.

Borodkova, N. L., L. M. Zelenyi, V. N. Lutsenko, A. O. Fedorov, A. G. Yahnin, J. Hanasz, V. V. Klimenko, J. Manninen, R. Manninen, T. Mukai, R. H. W. Friedel, and J.-A. Sauvaud, Multi-point substorm observations of the outer plasma sheet dynamics on November 13, 1996, in *Interball in the ISTP Program: studies of the solar wind-magnetosphere-ionosphere interaction*, edited by D. G. Sibeck and K. Kudela, NATO science series, Series C, *Mathematical and physical sciences*, **537**, 11, Dordrecht, Boston: Kluwer, 1999.

Büchner, J. and L. M. Zelenyi, Regular and chaotic charged particle motion in magnetotaillike field reversals: 1. Basic theory of trapped motion, *J. Geophys. Res.*, **94**(10), 11821–11842, 1989.

Büchner, J. and J.-P. Kuska, Sausage mode instability of thin current sheets as a cause of magnetospheric substorms, *Ann. Geophys.*, **17**(5), 604–612, 1999.

Burkhart, G. R., J. F. Drake, P. B. Dusenbery, and T. W. Speiser, A particle model for magnetotail neutral sheet equilibria, *J. Geophys. Res.*, **97**(A9), 13799–13815, 1992.

Bykov, A. A., L. M. Zelenyi, and H. V. Malova, Triple splitting of a thin current shee—a new type of plasma equilibrium, *Plasma Physics Reports (Rus. Transl. Fizika Plasmy), 2008*, **34**(2), 128–134, Pleiades Publishing, Ltd., 2008.

Chen, J. and P. J. Palmadesso, Chaos and nonlinear dynamics of single-particle orbits in a magnetotaillike magnetic field, *J. Geophys. Res.*, **91**, 1499–1508, 1986.

Cheng, C. Z., Magnetospheric equilibrium with anisotropic pressure, *J. Geophys. Res.*, **97**(A2), 1497–1510, 1992.

Chew, G. F., M. L. Goldberger, and F. E. Low, The Boltzmann equation and the one-fluid hydromagnetic equations in the absence of particle collisions, *Proc. Roy. Soc. (London)*, **A236**, 112, 1956.

Coppi, B., G. Laval, and R. Pellat, Dynamics of the geomagnetic tail, *Phys. Rev. Lett.*, **16**(26), 1207–1210, 1966.

Coroniti, F. V., On the tearing mode in quasi-neutral sheets, *J. Geophys. Res.*, **85**, 6719–6728, 1980.

Daughton, W., The unstable eigenmodes of a neutral sheet, *Phys. Plasmas*, **6**(4), 1329–1343, 1999.

Daughton, W., Nonlinear dynamics of thin current sheets, *Phys. Plasmas*, **9**(9), 3668–3678, 2002.

Daughton, W. and H. Karimabadi, Collisionless magnetic reconnection in large-scale electron-positron plasmas, *Phys. Plasmas*, **14**(7), doi:10.1063/1.2749494, 2007.

Daughton, W., G. Lapenta, and P. Ricci, Nonlinear evolution of the lower-hybrid drift instability in a current sheet, *Phys. Rev. Lett.*, **93**(10), doi:10.1103/PhysRevLett.93.105004, 2004.

Daughton, W., J. Scudder, and H. Karimabadi, Fully kinetic simulations of undriven magnetic reconnection with open boundary conditions, *Phys. Plasmas*, **13**, 072101, 2006.

Delcourt, D. C., R. F. Martin, Jr., and F. Alem, A simple model of magnetic moment scattering in a field reversal, *Geophys. Res. Lett.*, **21**, 1543, 1994.

Delcourt, D. C., J.-A. Sauvaud, O. L. Vaisberg, L. A. Avanov, J. L. Burch, and J. H. White, Signatures of impulsive convection in the magnetospheric lobes, *Geophys. Res. Lett.*, **23**, 129, 1996a.

Delcourt, D. C., J.-A. Sauvaud, R. F. Martin Jr., and T. E. Moore, On the nonadiabatic precipitation of ions from the near-Earth plasma sheet, *J. Geophys. Res.*, **101**, 17409, 1996b.

Delcourt, D. C., H. V. Malova, and L. M. Zelenyi, Dynamics of charged particles in bifurcated current sheets: the $\kappa \approx 1$ regime, *J. Geophys. Res.*, **109**, A01222, doi:10.1029/2003JA010167, 2004.

Delcourt, D. C., H. V. Malova, and L. M. Zelenyi, Quasi-adiabaticity in bifurcated current sheets, *Geophys. Res. Lett.*, **33**, L06106, doi:10.1029/2005GL025463, 2006.

Divin, A. V., M. I. Sitnov, M. Swisdak, and J. F. Drake, Reconnection onset in the magnetotail: Particle simulations with open boundary conditions, *Geophys. Res. Lett.*, **34**(9), doi:10.1029/2007GL029292, 2007.

Dobrowolny, M., Instability of a neutral sheet, *Nuovo Cimento*, **LV**(2), 427–441, 1968.

Dungey, J. W., The steady state of the Chapman-Ferraro problem in two dimensions, *J. Geophys. Res.*, **66**, 1043, 1961.

Eastwood, J. W., Consistency of fields and particle motion in the "Speiser" model of the current sheet, *Planet Space Sci.*, **20**(10), 1555–1568, 1972.

Francfort, P. and R. Pellat, Magnetic merging in collisionless plasmas, *Geophys. Res. Lett.*, **3**(8), 433–436, 1976.

Galeev, A. A. and L. M. Zelenyi, Tearing instability in plasma configurations, *Zhurnal Eksperimental'noi i Teoreticheskoi Fiziki*, **70**(6), 2133–2151, 1976 (in Russian).

Gold, T., Motions in the magnetosphere of the Earth, *J. Geophys. Res.*, **64**(9), 1219–1224, 1959.

Greco, A., A. L. Taktakishvili, G. Zimbardo, P. Veltri, and L. M. Zelenyi, Ion dynamics in the near-Earth magnetotail: Magnetic turbulence versus normal component of the average magnetic field, *J. Geophys. Res.*, **107**(A10), doi:10.1029/2002JA009270, 2002.

Gringauz, K. I., V. V. Bezrukih, V. D. Ozerov, and R. E. Rybchinskyi, Investigation of interplanetary ionized gas and corpuscular radiation of the Sun with help of charged particle three-electrode traps at the Soviet space rocket, *Russ. J. Doklady AN SSSR*, **131**(6), 1301–1304, 1960.

Harris, E. G., On a plasma sheath separating regions of oppositely directed magnetic fields, *Nuovo Chimento*, **23**, 115–119, 1962.

Hesse, M., J. Birn, M. M. Kuznetsova, and J. Dreher, A simple model of core field generation during plasmoid evolution, *J. Geophys. Res.*, **101**(A5), 10797–10804, 1996.

Hoshino, M., A. Nishida, T. Mukai, Y. Saito, and T. Yamamoto, Structure of plasma sheet in magnetotail: double-peaked electric current sheet, *J. Geophys. Res.*, **101**, 24775–24786, 1996.

Kan, J. R., On the structure of the magnetotail current sheet, *J. Geophys. Res.*, **78**, 3773–3781, 1973.

Karimabadi, H., P. L. Pritchett, W. Daughton, and D. Krauss-Varban, Ion-ion kink instability in the magnetotail: 2. Three-dimensional full particle and hybrid simulations and comparison with observations, *J. Geophys. Res.*, **108**(A11), doi:10.1029/2003JA010109, 2003.

Kistler, L. M., C. Mouikis, E. Mebius, B. Klecker, J. A. Sauvaud, H. Reme, A. Korth, M. F. Marcucci, R. Lundin, G. K. Parks, and A. Balogh, Contribution of nonadiabatic ions to the cross-tail current in an O^+ dominated thin current sheet, *J. Geophys. Res.*, **110**(A6), doi:10.1029/2004JA010653, 2005.

Kojima, H., H. Matsumoto, S. Chikuba, S. Horiyama, M. Ashour-Abdalla, and R. R. Anderson, Geotail waveform observations of broadband/narrowband electrostatic noise in the distant tail, *J. Geophys. Res.*, **102**(A7), 14439–14456, 1997.

Kropotkin, A. P. and V. I. Domrin, Theory of a thin one-dimensional current sheet in collisionless space plasma, *J. Geophys. Res.*, **101**, 19893–19902, 1996.

Krall, N. A. and A. W. Trivelpiece, *Principles of Plasma Physics*, McGraw Hill, Inc., New York, 1993.

Kuznetsova, M. M. and L. M. Zelenyi, Magnetic reconnection in collisionless field reversals the universality of the ion tearing mode, *Geophys. Res. Lett.*, **18**(10), 1825–1828, 1991.

Kuznetsova, M. M., M. Hesse, and D. Winske, Ion dynamics in a hybrid simulation of magnetotail reconnection, *J. Geophys. Res.*, **101**(A12), 27351–27374, 1996.

Lapenta, G. and J. U. Brackbill, A kinetic theory for the drift-kink instability, *J. Geophys. Res.*, **102**(A12), 27,099–27,108, 1997.

Lembege, B. and R. Pellat, Stability of a thick two-dimensional quasineutral sheet, *Phys. Fluids*, **25**(11), 1995–2004, 1982.

Lui, A. T. Y., Inferring global characteristics of current sheet from local measurements, *J. Geophys. Res.*, **98**, 13423–13427, 1993.

Lui, A. T. Y., Current disruption in the Earth's magnetosphere: Observations and models, *J. Geophys. Res.*, **101**(A6), 13067–13088, 1996.

Lui, A. T. Y., Potential plasma instabilities for substorm expansion onsets, *Space Sci. Rev.*, **113**, 127–206, 2004.

Lui, A. T. Y., D. J. Williams, R. W. McEntire, S. Ohtani, L. J. Zanetti, W. A. Bristow, R. A. Greenwald, P. T. Newell, S. P. Christon, T. Mukai, K. Tsuruda, T. Yamamoto, S. Kokubun, H. Matsumoto, H. Kojima, T. Murata, D. H. Fairfield, R. P. Lepping, J. C. Samson, G. Rostoker, G. D. Reeves, A. L. Rodger, and H. J. Singer, Multipoint study of a substorm on February 9, 1995, *J. Geophys. Res.*, **103**(A8), 17333–17344, 1998.

Malova, H. V., M. I. Sitnov, L. M. Zelenyi, and A. S. Sharma, Self-consistent model of 1D current sheet: the role of drift, magnetization and diamagnetic currents, in *Proceedings of Chapman Conference: Magnetospheric Current Systems*, **118**, 313–322, 2000.

Matsumoto, H., H. Kojima, T. Miyatake, Y. Omura, M. Okada, I. Nagano, and M. Tsutsui, Electrotastic Solitary Waves (ESW) in the mag-netotail: BEN wave forms observed by GEOTAIL, *Geophys. Res. Lett.*, **21**, 2915–2918, 1994.

Milovanov, A. V., L. M. Zelenyi, and G. Zimbardo, Fractal structures and power law spectra in the distant Earth's magnetotail, *J. Geophys. Res.*, **101**, 19903–19910, 1996.

Milovanov, A. V., L. M. Zelenyi, P. Veltri, G. Zimbardo, and A. L. Taktakishvili, Geometric description of the magnetic field and plasma coupling in the near-Earth stretched tail prior to a substorm, *J. Atmos. Sol.-Terr. Phys.*, **63**(7), 705–721, 2001.

Mingalev, O. V., I. V. Mingalev, H. V. Malova, and L. M. Zelenyi, Numerical simulations of plasma equilibrium in a one-dimensional current sheet with a nonzero normal magnetic field component, *Plasma Phys. Rep.*, **33**(11), 942–955, 2007.

Mitchell, D. G., G. J. Williams, C. Y. Huang, L. A. Frank, and C. T. Russell, Current carriers in the near-Earth cross-tail current sheet during substorm growth phase, *Geophys. Res. Lett.*, **17**, 583–586, 1990.

Nagai, T., I. Shinohara, M. Fujimoto, M. Hoshino, Y. Saito, S. Machida, and T. Mukai, Geotail observations of the Hall current system: Evidence of magnetic reconnection in the magnetotail, *J. Geophys. Res.*, **106**(A11), 25929–25950, 2001.

Nakamura, R., W. Baumjohann, A. Runov, Y. Asano, A. Balogh, and H. Reme, Thin current sheets measured by Cluster at 200 km tetrahedron scale, The Second Workshop on Thin Current Sheets 19–21 April, 2004, College Park, Maryland, USA, 2004.

Nakamura, R., W. Baumjohann, Y. Asano, A. Runov, A. Balogh, C. J. Owen, A. N. Fazakerley, M. Fujimoto, B. Klecker, and H. Reme, Dynamics of thin current sheets associated with magnetotail reconnection, *J. Geophys. Res.*, **111**(A11), doi:10.1029/2006JA011706, 2006.

Ness, N. F., The Earth's magnetic tail, *J. Geophys. Res.*, **70**, 2989, 1965.

Parker, E. N., Interaction of the solar wind with the geomagnetic field, *Phys. Fluid.*, **1**, 171–187, 1958.

Pellat, R., F. V. Coroniti, and P. L. Pritchett, Does ion tearing exist?, *Geophys. Res. Lett.*, **18**, 143–146, 1991.

Peroomian, V., L. M. Zelenyi, and D. Schriver, Imprints of small-scale nonadiabatic particle dynamics on large-scale properties of dynamical magnetotail equilibria, *Adv. Space Res.*, **30**(12), 2657–2662, 2002.

Petrukovich, A. A., V. A. Sergeev, L. M. Zelenyi, T. Mukai, T. Yamamoto, S. Kokubun, K. Shiokawa, C. S. Deehr, E. Y. Budnick, J. Büchner, A. O. Fedorov, V. P. Grigorieva, T. J. Hughes, N. F. Pissarenko, S. A. Romanov, and I. Sandahl, Two spacecraft observations of a reconnection pulse during an auroral breakup, *J. Geophys. Res.*, **103**(A1), 47–60, 1998.

Petrukovich, A. A., T. Mukai, S. Kokubun, S. A. Romanov, Y. Saito, T. Yamamoto, and L. M. Zelenyi, Substorm-assotiated pressure variations in the magnetotail plasma sheet and lobe, *J. Geophys. Res.*, **104**(3), 4501–4513, 1999.

Petrukovich, A. A., W. Baumjohann, R. Nakamura, A. Balogh, T. Mukai, K.-H. Glassmeier, H. Reme, and B. Klecker, Plasma sheet structure during strongly northward IMF, *J. Geophys. Res.*, **108**(A6), SMP 22-1, CiteID 1258, doi:10.1029/2002JA009738, 2003.

Pritchett, P. L., Collisionless magnetic reconnection in a three-dimensional open system, *J. Geophys. Res.*, **106**(A11), 25961–25978, 2001.

Pritchett, P. L., Externally driven magnetic reconnection in the presence of a normal magnetic field, *J. Geophys. Res.*, **110**(A5), doi:10.1029/2004JA010948, 2005a.

Pritchett, P. L., The "Newton Challenge": Kinetic aspects of forced magnetic reconnection, *J. Geophys. Res.*, **110**(A10), doi:10.1029/2005JA011228, 2005b.

Pritchett, P. L. and F. V. Coroniti, Formation and stability of the self-consistent one-dimensional tail current sheet, *J. Geophys. Res.*, **97**, 16773–16787, 1992.

Pritchett, P. L. and F. V. Coroniti, Formation of thin current sheets during plasma sheet convection, *J. Geophys. Res.*, **100**, 23551–23565, 1995.

Pritchett, P. L., F. V. Coroniti, and V. K. Decyk, Three-dimensional stability of thin quasi-neutral current sheets, *J. Geophys. Res.*, **101**(A12), 27,413–27,429, 1996.

Pulkkinen, T. I., D. N. Baker, C. J. Owen, J. T. Gosling, and N. Murthy, Thin current sheets in the deep geomagnetotail, *Geophys. Res. Lett.*, **20**, 2427–2430, 1993.

Pulkkinen, T. I., D. N. Baker, D. G. Mitchell, R. L. McPherron, C. Y. Huang, and L. A. Frank, Thin current sheets in the magnetotail during substorms: CDAW 6 revisited, *J. Geophys. Res.*, **99**, 5793–5804,1994.

Rich, F. J., V. M. Vasyliunas, and R. A. Wolf, On the balance of stresses in the plasma sheet, *J. Geophys. Res.*, **77**, 4670–4676, 1972.

Runov, A., R. Nakamura, W. Baumjohann, T. I. Zhang, and M. Volverk, Cluster observation of a bifurcated current sheet, *Geophys. Res. Lett.*, **30**, 1036, doi:10.1029/2002GL016136, 2003a.

Runov, A, R. Nakamura, W. Baumjohann, R. A. Treumann, T. L. Zhang, M. Volwerk, Z. Voros, A. Balogh, K.-H. Glassmeier, B. Klecker, H. Reme, and L. Kistler, Current sheet structure near magnetic X-line observed by Cluster, *Geophys. Res. Lett.*, **30**, 1579, doi:10.1029/2002GL016730, 2003b.

Runov, A., V. A. Sergeev, W. Baumjohann, R. Nakamura, S. Apatenkov, Y. Asano, M. Volwerk, Z. Vörös, T. L. Zhang, A. Petrukovich, A. Balogh, J.-A. Sauvaud, B. Klecker, and H. Réme, Electric current and magnetic field geometry in flapping magnetotail current sheets, *Ann. Geophys.*, **23**(4), 1391–1403, 2005.

Runov, A., V. A. Sergeev, R. Nakamura, W. Baumjohann, S. Apatekov, Y. Asano, T. Takada, M. Volwerk, Z. Voros, T. L. Zhang, J.-A. Sauvaud, H. Reme, and A. Balogh, Local structure of the magnetotail current sheet: 2001 Cluster observations, *Ann. Geophys.*, **24**, 247–262, 2006.

Sanny, J., R. L. McPherron, C. T. Russell, D. N. Baker, T. I. Pulkkinen, and A. Nishida, Growth phase thinning of the near-Earth current sheet during the CDAW-6 substorm, *J. Geophys. Res.*, **99**, 5805–5816, 1994.

Sauvaud, J.-A. and R. A. Kovrazhkin, Two types of energy-dispersed ion structures at the plasma sheet boundary, *J. Geophys. Res.*, **109**, A12213, 2004.

Savenkov, B. V. and L. M. Zelenyi, Application of separatrix crossing theory to nondiffusion model of current sheet resonance, *Geophys. Res. Lett.*, **23**, 3255–3259, 1996.

Schindler, K., A theory of the substorm mechanism, *J. Geophys. Res.*, **79**(19), 2803–2810, 1974.

Schindler, K., *Physics of Space Plasma Activity*, Cambridge University Press, ISBN: 9780521858977, doi:10.2277/0521858976, 2006.

Sergeev, V. A., D. G. Mitchell, C. T. Russell, and D. J. Williams, Structure of the tail plasma/current sheet at $\sim 11 R_{\rm E}$ and its changes in the course of a substorm, *J. Geophys. Res.*, **98**(A10), 17345–17366, 1993.

Sergeev, V., V. Angelopoulos, C. Carlson, and P. Sutcliffe, Current sheet measurements within a flapping plasma sheet, *J. Geophys. Res.*, **103**(A5), 9177–9188, 1998.

Sergeev, V., A. Runov, W. Baumjohann, R. Nakamura, T. L. Zhang, M. Volwerk, A. Balogh, H. Reme, J. A. Sauvaud, M. Andre, and B. Klecker, Current sheet flapping motion and structure observed by Cluster, *Geophys. Res. Lett.*, **30**(6), doi:10.1029/2002GL016500, 2003.

Sergeev, V., A. Runov, W. Baumjohann, R. Nakamura, T. L. Zhang, A. Balogh, P. Louarnd, J.-A. Sauvaud, and H. Reme, Orientation and propagation of current sheet oscillations, *Geophys. Res. Lett.*, **31**(5), doi:10.1029/2003GL019346, 2004.

Silin, I., J. Büchner, and L. Zelenyi, Instabilities of collisionless current sheets: theory and simulations, *Phys. Plasm.*, **9**(4), 1104–1112, 2002.

Sitnov, M. I., H. V. Malova, and A. T. Y. Lui, Quasi-neutral sheet tearing instability induced by electron preferential acceleration from stochasticity, *J. Geophys. Res.*, **102**(A1), 163–174, 1997.

Sitnov, M. I., L. M. Zelenyi, H. V. Malova, and A. S. Sharma, Thin current sheet embedded within a thicker plasma sheet: Self-consistent kinetic theory, *J. Geophys. Res.*, **105**, 13029–13044, 2000a.

Sitnov, M. I., L. M. Zelenyi, A. S. Sharma, and H. V. Malova, Distinctive features of forced current sheets: electrostatic effects, *Proc. Int. Conf. Substorm-5, St.-Petersburg, Russia, 16–20 May*, 197–200, 2000b.

Sitnov, M. I., A. S. Sharma, P. N. Guzdar, and P. H. Yoon, Reconnection onset in the tail of Earth's magnetosphere, *J. Geophys. Res.*, **107**(A9), doi:10.1029/2001JA009148, 2002.

Sitnov, M. I., A. T. Y. Lui, P. N. Guzdar, and P. H. Yoon, Current-driven instabilities in forced current sheets, *J. Geophys. Res.*, **109**, A03205, doi:10.1029/2003JA010123, 2004.

Sitnov, M. I., M. Swisdak, P. N. Guzdar, and A. Runov, Structure and dynamics of a new class of thin current sheets, *J. Geophys. Res.*, **111**(A8), doi:10.1029/2005JA011517, 2006.

Sonnerup, B. U. Ö., Adiabatic particle orbits in a magnetic null sheet, *J. Geophys. Res.*, **76**, 8211–8222, 1971.

Soop, M. and K. A. Schindler, Self-consistent two-dimensional approach to magnetospheric structures, *Astrophys. Space Sci.*, **20**(2), 287–305, 1973.

Speiser, T. W., Particle trajectories in model current sheets; 1. Analytical solutions, *J. Geophys. Res.*, **70**, 4219–4226, 1965.

Syrovatsky, S. I., On the appearance of current sheet in plasma with frozen-in strong magnetic field, *Sov. Phys. JETP*, **60**, 1727, 1971.

Tsyganenko, N. A. and D. G. Suslikov, Large-scale magnetic effects of field-aligned currents in the magnetosphere, *Magnitosfernye Issledovaniia*, **1**, 60–64, 1982 (in Russian).

Van Allen, J. A. and L. A. Frank, Radiation around the Earth to a radial distance of 107,400 km, *Nature*, **183**, 430, 1959.

Vaisberg, O. L., L. A. Avanov, J. L. Burch, and J. H. Waite, Jr., Measurements of plasma in the magnetospheric tail lobes, *Adv. Space Res.*, **8**(8), 63, 1996.

Vernov, S. N., N. L. Grigorov, Iu. I Logachev, and A. E. Chudakov, Cosmic radiation measured on the second artificial satellite, *Soviet Phys. Doklady*, **3**, 617, 1958.

Volwerk, M., K.-H. Glassmeier, A. Runov, W. Baumjohann, R. Nakamura, T. L. Zhang, B. Klecker, A. Balogh, and H. Réme, Kink mode oscillation of the current sheet, *Geophys. Res. Lett.*, **30**(6), doi:10.1029/2002GL016467, 2003.

Voronina, V. A. and J. R. Kan, A kinetic model of the plasma sheet—Isotropic nonuniform plasma temperature, *J. Geophys. Res.*, **98**(A8), 13,395–13,402, 1993.

Whipple, E. C., M. Rosenberg, and M. Brittnacher, Magnetotail acceleration using generalized drift theory—A kinetic merging scenario, *Geophys. Res. Lett.*, **17**, 1045–1048, 1990.

Yermolaev, Yu. I., Observations of the plasma sheet: INTERBALL project, *Cosmic Res.*, **36**(3), 256, 1998.

Yin, L., D. Winske, S. P. Gary, and J. Birn, Hybrid and Hall-magnetohydrodynamics simulations of collisionless reconnection: Effect of the ion pressure tensor and particle Hall-magnetohydrodynamics, *Phys. Plasmas*, **9**, 2575, 2002.

Yoon, P. H. and A. T. Y. Lui, On the drift-sausage mode in one-dimensional current sheet, *J. Geophys. Res.*, **106**(A2), 1939–1947, 2001.

Zhang, T. L., W. Baumjohann, R. Nakamura, A. Balogh, and K.-H. Glassmeier, A wavy twisted neutral sheet observed by CLUSTER, *Geophys. Res. Lett.*, **29**(19), 1899, doi:10.1029/2002GL015544, 2002.

Zelenyi, L. M., A. V. Milovanov, and G. Zimbardo, Multiscale magnetic structure of the distant tail: self-consistent fractal approach, in *New Perspectives on the Earth's Magnetotail*, Geophysical Monograph, **105**, edited by A. Nishida, D. N. Baker, and S. W. H. Cowley, ISBN 0-87590-088-7, American Geophysical Union, Washington D.C., 321 pp., 1998.

Zelenyi, L., M. I. Sitnov, H. V. Malova, and A. S. Sharma, Thin and superthin ion current sheets, quasiadiabatic and nonadiabatic models, *Nonlinear Proc. Geophys.*, **7**, 127–139, 2000.

Zelenyi, L. M., D. C. Delcourt, H. V. Malova, and A. S. Sharma, "Aging" of the magnetotail thin current sheets, *Geophys. Res. Lett.*, **29**, 1608, doi:10.1029/2001GL013789, 2002.

Zelenyi, L. M., H. V. Malova, and V. Yu. Popov, Splitting of thin current sheets in the Earth's magnetosphere, *J. Exp. Theor. Phys. Lett.*, **78**(5), 296–299, 2003.

Zelenyi, L. M., H. V. Malova, V. Yu. Popov, D. Delcourt, and A. S. Sharma, Nonlinear equilibrium structure of thin currents sheets: influence of electron pressure anisotropy, *Nonlinear Proc. Geophys.*, **11**, 1–9, 2004.

Zelenyi, L. M., H. V. Malova, V. Y. Popov, D. C. Delcourt, N. Y. Ganushkina, and A. S. Sharma, "Matreshka" model of multilayered current sheet, *Geophys. Res. Lett.*, **33**, L05105, doi:10.1029/2005GL025117, 2006.

Zelenyi, L., A. Artemyev, H. Malova, and V. Popov, Marginal stability of thin current sheets in the Earth's magnetotail, *J. Atmos. Sol.-Terr. Phys.*, **70**, 325–333, 2008.

Zimbardo, G., A. Greco, A. L. Taktakishvili, P. Veltri, and L. M. Zelenyi, Magnetic turbulence and particle dynamics in the Earth's magnetotail, *Ann. Geophys.*, **21**(9), 1947–1953, 2003.

Zhu, Z. and R. M. Winglee, Tearing instability, flux ropes, and the kinetic current sheet kink instability in the Earth's magnetotail: A three-dimensional perspective from particle simulations, *J. Geophys. Res.*, **101**(A3), 4885–4897, 1996.

Climate and Weather of the Sun-Earth System (CAWSES): Selected Papers from the 2007 Kyoto Symposium,
Edited by T. Tsuda, R. Fujii, K. Shibata, and M. A. Geller, pp. 171–199.

Simulating solar 'climate'

Mausumi Dikpati

*High Altitude Observatory, National Center for Atmospheric Research,**
3080 Center Green, Boulder, Colorado 80301, USA
E-mail: dikpati@hao.ucar.edu

With the progress of observational and theoretical knowledge, global-scale solar dynamo models have reached the stage that it is possible to calibrate them for the Sun by adjusting only a few tunable parameters. The important ingredients in a dynamo model that is most plausibly calibrated are differential rotation (Omega-effect), helical turbulence (alpha-effect), meridional circulation and turbulent diffusion. The meridional circulation works as a conveyor belt for transporting the magnetic flux and governs the dynamo cycle period. Meridional circulation and magnetic diffusivity together govern the memory of the Sun's past magnetic fields. After describing the physical processes involved in this class of dynamo, we will show that a predictive tool can be built from it to predict the solar 'climate', namely the mean solar cycle features, particularly cycle amplitude and duration, by assimilating magnetic field data from previous cycles. We will discuss the successes, limitations and future prospects for such predictive solar dynamo models.

1 Introduction

Among various manifestations of solar activity cycle, the most popular one, known to us for more than 200 years now, is the appearance and variation in the number of sunspots with an 11-year periodicity. The annual sunspot number as a function of solar cycle is shown in Fig. 1; it reveals that the amplitude of a cycle varies from cycle to cycle as does the duration of a cycle, ranging from 8 to 14 years. In 1908 G. E. Hale discovered that the sunspots are actually the sights of strong magnetic fields.

Most likely a magnetohydrodynamic dynamo operating inside the Sun is responsible for producing solar activity cycle. Through its outputs of magnetic fields and energetic particles, the solar activity cycle influences the terrestrial systems in various ways: posing hazards to satellites, disrupting powergrids, causing blackouts in radio and telecommunication systems, even affecting the astronauts in space. Solar cycle variations in visible and UV output (Foukal *et al.*, 2006; Wenzler *et al.*, 2006) influence the density of the upper atmosphere, affecting the lifetime of satellites. Such variations also affect the global dynamics of the stratosphere and above, and may have some influence on the climate.

*The National Center for Atmospheric Research is sponsored by the National Science Foundation.

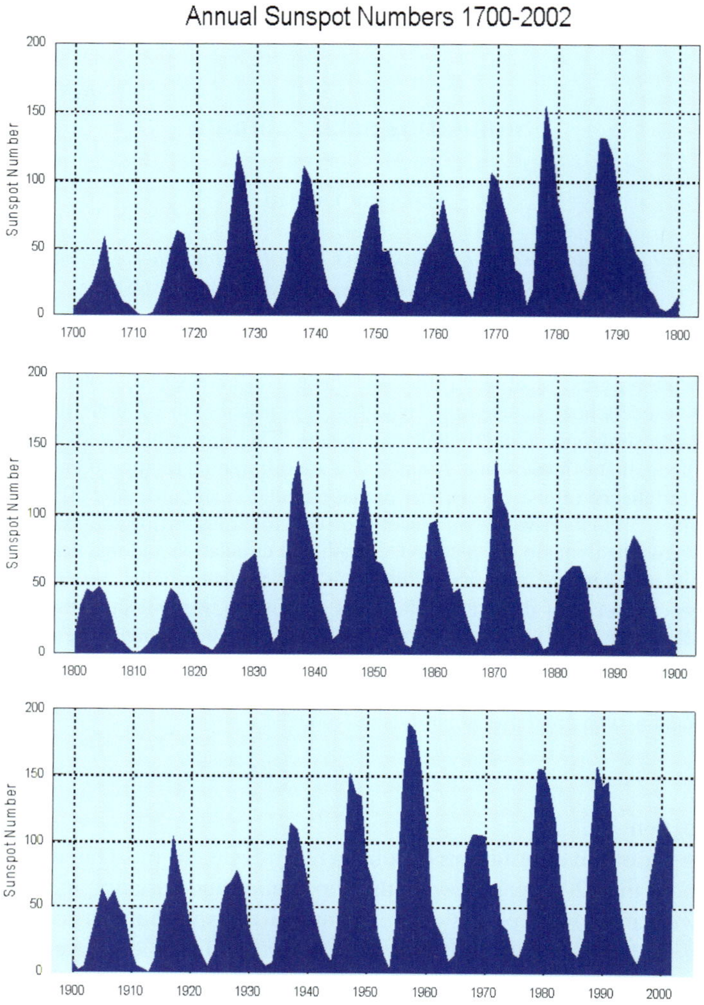

Fig. 1. Annual mean sunspot number plotted as function of time during 1700–2006. (Adopted from National Geophysical Data Center; URL: www.ngdc.noaa.gov/stp/SOLAR/SSN/image/annual.gif.)

There has been substantial interest in predicting future solar cycles for the past forty years (Ohl, 1966). In the current era of extensive use of the high atmosphere and neighboring interplanetary medium by man, such predictions have considerable practical value. However, the approach to solar cycle forecasts has been mostly statistical, using the empirical fit formulae. There is obvious need for predictions based on physical models. Atmospheric and oceanic weather and climate prediction mod-

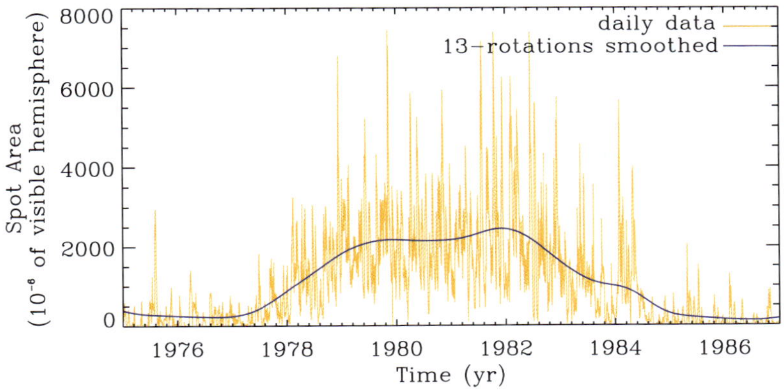

Fig. 2. Daily spot area in a typical solar cycle is plotted in yellow curve as function of time while the 13-rotation average spot area in blue curve.

els have been built over the past 50 years starting from their corresponding calibrated models; only very recently the physical models for solar 'climate' predictions have been developed (Dikpati and Gilman, 2006; Dikpati *et al.*, 2006)—these models integrate forward in time the induction equation, as is done in meteorology, with the specified initial conditions derived from observations. These models can be called solar 'climate' prediction models rather than weather prediction models, because the predictions have so far been made about long-term (decadal time-scale) mean features of solar cycle, as shown by blue-filled curves in Fig. 1 and in blue curve in Fig. 2, but not short-term (daily or monthly) features which can be seen in yellow curve of Fig. 2.

The skill of such models should depend on how well they can be calibrated, how well the initial conditions are known and also on the quality and quantity of observational data to be assimilated in these models. In solar science this is a particularly important time for developing such models that will assimilate data, since so much data are being collected by various ground- and space-based instruments. And in fact, since the first data-assimilation model by Dikpati *et al.* (2006), the use of similar methodology in physical models has been growing (Cameron and Schüssler, 2007; Choudhuri *et al.*, 2007), not only in solar cycle research but also has been applied in flare prediction (Bélanger *et al.*, 2007).

A question that immediately arises is: what solar cycle features are predictable? The answer depends on the current status of the solar cycle modeling. Given the complexity of the physics of the convection zone and adjacent domains of the sun, it is a formidable challenge yet to develop a unified global dynamo model for the solar cycle that includes all physical processes occuring with a wide range of length and time scales. We focus in this paper on a class of large-scale solar dynamo models called flux-transport dynamos that have now reached a stage that it is possible to calibrate them. That means these models are able to simulate important solar cycle features of a 'typical' solar cycle such as the 'butterfly diagram', and the phase relation in time between toroidal (sunspot) fields and poloidal (polar) fields.

In the following sections we describe the physical processes involved in a flux-transport dynamo, the validity of its calibration, development of predictive tool and data-assimilation procedures, skill-tests of such a predictive tool, and results, in the form of forecasts of certain properties of the upcoming solar cycle 24. We close with the discussion of some skepticisms expressed about solar cycle prediction models as well as future prospects for such predictive research.

Figure 2 shows the daily variation in the spot area (yellow curve) within a cycle as well as the spot area averaged over 13-rotations (blue curve). Solar cycle modeling research has reached the stage that the prediction of the overall strength of a cycle, its mean duration, rise and fall patterns seen in the blue curve in Fig. 2 is possible, but not the details shown in the yellow curve. This is because the current global solar cycle models are based on mean-field electrodynamics; automatically there is a scale selection so that the solar cycle features that occur in the time-scale longer than six months have a chance to be successfully simulated.

2 Development of Solar Cycle Prediction Model

2.1 A brief historical background

Flux-transport dynamos are just the so-called α-Ω dynamos with meridional circulation. The very first α-Ω solar dynamo model was built by Parker (1995) who included two basic processes in the dynamo mechanism—the shearing of the poloidal magnetic fields to produce toroidal fields by the Sun's differential rotation (the Ω-effect), and the regeneration of poloidal fields by displacing and twisting the toroidal flux tubes by helical motions (the so-called α-effect). The equatorward propagating dynamo wave solution was used to explain the equatorward migration of the sunspot zones with the progress of the solar cycle.

With further progress in solar observations many other solar cycle features were discovered. Early magnetograms (Babcock and Babcock, 1955) revealed that weak, diffuse fields outside the sunspots which arise from the decay of active regions drift poleward with the solar cycle (in contrast to the equatorward migration of the sunspot zones) and cause the reversal in the Sun's polar field in every 11 year. Helioseismic observations revealed details about the Sun's flow fields, namely the Sun's azimuthal flow primarily contains the latitudinal differential rotation in the bulk of the convection zone, while the radial differential rotation exists in a thin layer at the base of the solar convection zone (Brown *et al.*, 1989; Dziembowski *et al.*, 1989). The meridional circulation was detected by various observations to have a poleward surface flow of speed 10–20 m s^{-1} (Duvall, 1979; Hathaway *et al.*, 1996) with the appearance and disappearance of a high-latitude reverse cell from time to time (Haber *et al.*, 2002; Basu and Antia, 2003; Ulrich and Boyden, 2005). An equtorward return flow somewhere near the base of the convection zone is not observed yet, but it should exist, because piling up of plasma near the pole has not been seen.

In the 1990's these detailed findings about the Sun's large-scale flow fields as well as the magnetogram observations of decay of active regions gave a new twist

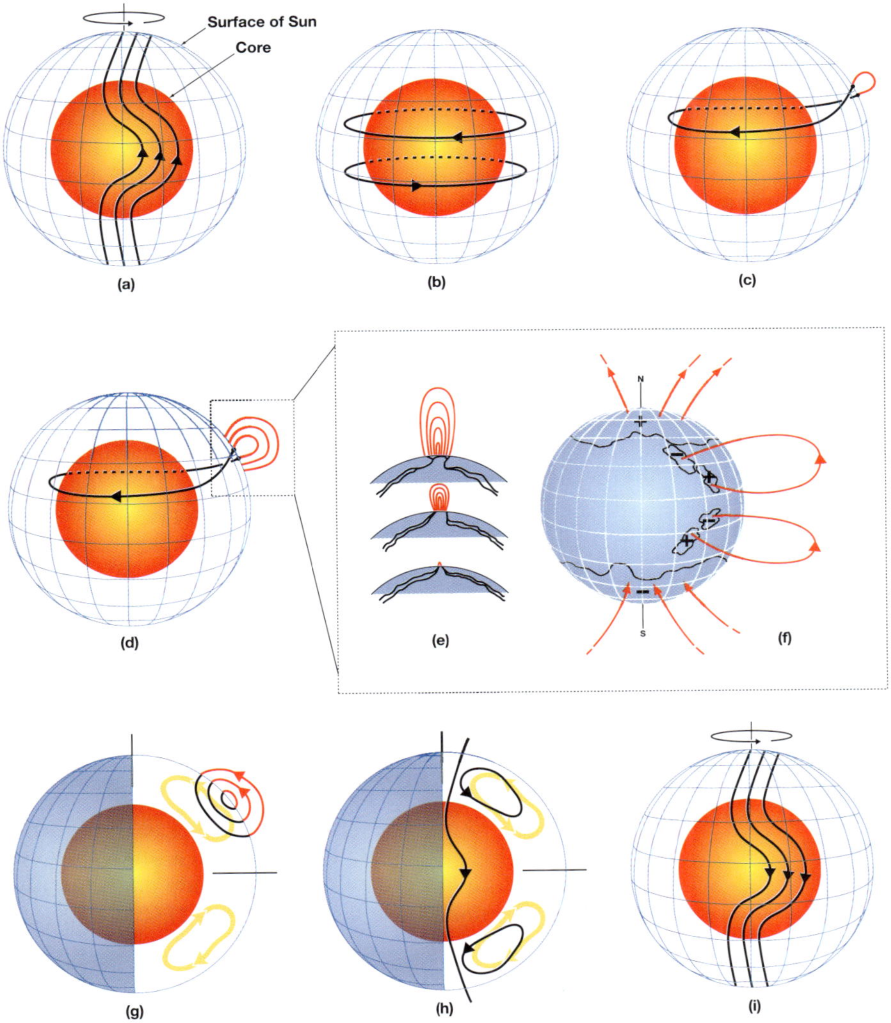

Fig. 3. Schematic of solar flux-transport dynamo processes. Red inner sphere represents the Sun's radiative core and blue mesh the solar surface. In between is the solar convection zone where dynamo resides. (a) Shearing of poloidal field by the Sun's differential rotation near convection zone bottom. The Sun rotates faster at the equator than the pole. (b) Toroidal field produced due to this shearing by differential rotation. (c) When toroidal field is strong enough, buoyant loops rise to the surface, twisting as they rise due to rotational influence. Sunspots (two black dots) are formed from these loops. (d, e, f) Additional flux emerges (d, e) and spreads (f) in latitude and longitude from decaying spots (as described in figure 5 of Babcock (1961)). (g) Meridional flow (yellow circulation with arrows) carries surface magnetic flux poleward, causing polar fields to reverse. (h) Some of this flux is then transported downward to the bottom and towards the equator. These poloidal fields have sign opposite to those at the beginning of the sequence, in frame (a). (i) This reversed poloidal flux is then sheared again near the bottom by the differential rotation to produce the new toroidal field opposite in sign to that shown in (b).

to the modeling of the solar dynamo. Wang *et al.* (1991) first showed, using a simplified rectangular geometry, that the meridional circulation is an essential ingredient in the solar dynamo mechanism. The large-scale poloidal fields, produced from the decay of active regions, drift towards the pole due to transport by the meridional circulation and supergranular diffusion. Eventually these poloidal fields cause polar reversal when enough new-cycle's fields have been transported to the polar regions. The equatorward meridional flow component near the base of the solar convection zone help transport the toroidal magnetic flux, which is actually the spot-producing flux, towards the equator. This class of the dynamo models is successful in explaining the equatorward migration of the sunspot zones, poleward drift of the large-scale diffuse magnetic fields which eventually cause the polar reversals and the correct phase-relationship in time between the above two components of the solar magnetic fields. These models are called 'flux-transport' dynamos.

During the past two decades many flux-transport dynamo models have been built (Choudhuri *et al.*, 1995; Durney, 1995; Dikpati and Charbonneau, 1999; Küker *et al.*, 2001; Bonanno *et al.*, 2002; Guerrero and Muñoz, 2004; Rempel, 2006; Jouve and Brun, 2007) —these models show that the meridional circulation works as a conveyor belt by transporting flux from the surface to the bottom of the convection zone where the new spot-producing flux is generated. Figure 3 shows a sequence of schematic diagrams that depict qualitatively the succession of processes contained in a flux-transport dynamo solution that leads to cyclic magnetic fields of the correct dominant symmetry about the equator.

2.2 Predictive power of a flux-transport dynamo

In the kinematic regime, flux-transport dynamos include four major ingredients: (i) differential rotation, (ii) meridional circulation, (iii) Babcock-Leighton poloidal source and (iv) magnetic diffusivity. The dynamo cycle period and the Sun's magnetic memory are primarily governed by the meridional flow speed in this class of models. For a typical solar-like meridional circulation pattern with a single flow-cell with maximum surface flow-speed of 15 m s^{-1} the flow-particles are plotted in blue dots in Fig. 4 along a selected streamline in one year intervals. This plot reveals: (i) the latitude-zone of spot-producing fields drift from mid-latitudes to equator in $\sim$11 years, indicating how the average length of a sunspot cycle becomes 11 year, (ii) the surface poloidal fields are transported from the high-latitudes at the surface to the mid-latitudes at the bottom in $\sim$20 years, indicating how the meridional circulation plays a crucial role in determining the Sun's memory about its past magnetic fields. The variations in the meridional flow-speed should cause the variations in the cycle-length from cycle to cycle.

The property of magnetic memory rendered by the meridional circulation provides the predictive power to this class of models. The solar meridional circulation is not a unique circulation in the nature. Analogous latitudinal circulations exist in the terrestrial system, such as Hadley cell, polar cell, Ferrel cell in the tropopause, and which are known to have influence in weather forecast. The great ocean conveyor belt which is a thermohaline circulation is driven primarily by the formation and sinking of cold water (from around 1500 m to the Antarctic bottom water over-

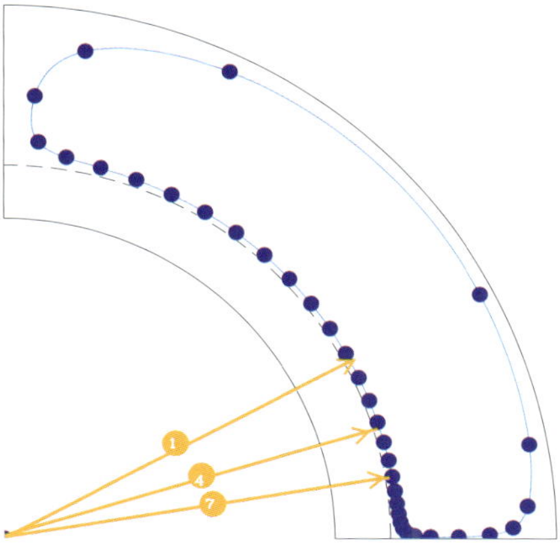

Fig. 4. The blue flow-dots are plotted on a streamline in one-year interval. The number of dots from mid-latitudes at the surface to the mid-latitudes at the base of the convection zone indicates the Sun's magnetic memory length in years and that from mid-latitudes to the equator indicates how this circulation of ~15–20 years.

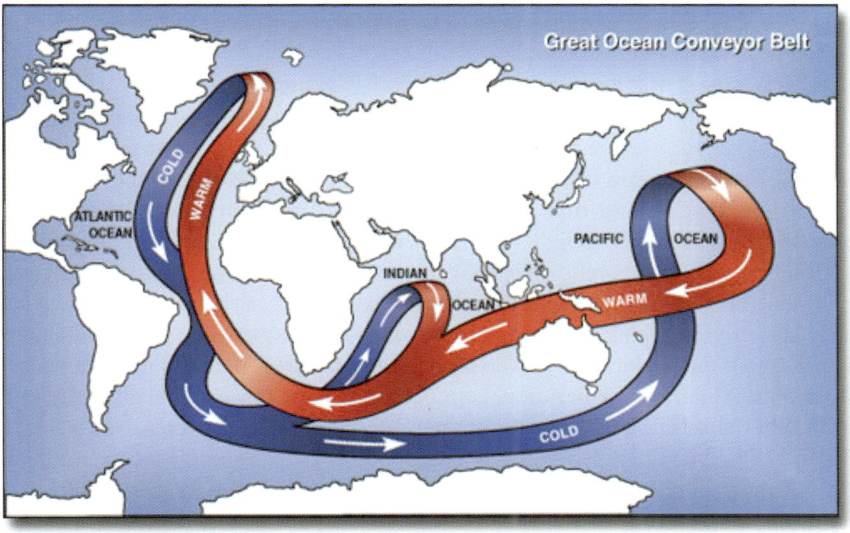

Fig. 5. The great ocean conveyor belt is created due to thermo-haline forcing.

lying the bottom of the ocean) in the Norwegian Sea. This circulation is thought to be responsible for the large flow of upper ocean water from the tropical Pacific to the Indian Ocean through the Indonesian Archipelago. The two counteracting forcings operating in the North Atlantic control the conveyor belt circulation: (1) the thermal forcing (high-latitude cooling and low-latitude heating) which drives a polar southward flow; and (2) haline forcing (net high-latitude freshwater gain and low-latitude evaporation) which moves in the opposite direction (see Fig. 5).

The common features between the Sun's meridional flow conveyor belt and that of the ocean are: both are persistent slow flow in the turbulent medium and both carry the surface forcing (surface magnetic flux and thermal patterns respectively in the case of the Sun and the ocean) with long memory. Warm North Atlantic meridional overturning circulation is linked to excessive rain-fall over Sahel and western India, while variations in overturning flow in eastern Pacific determine the timing, amplitude and duration of an El Nino event. Similarly the variations in the meridional flow amplitude and profile determine the timing and shape of a cycle, and amplitude to some extent.

The theory for the Sun's meridional circulation is still under development (Rempel, 2005; Miesch *et al.*, 2008). Meridional flow is the second-order effect of convection in a rotating sphere. The detailed physics behind this theory is quite complicated. In brief, Coriolis force acting on convection creates Reynolds stresses that transport angular momentum towards the equator. Thus equatorial acceleration is produced. Meridional circulation is then driven by the Coriolis force from the differential rotation—outward radial velocity created from large rotational flow at low latitudes make the fluid particles flow upward from the base of the convection zone and then poleward at the surface.

2.3 Development of a flux-transport dynamo-based predictive tool

Given the complexity of the physics of the convection zone and adjacent domains of the sun, it is a formidable challenge to take up the development of a global dynamo model for the solar cycle that includes all physical processes occuring with a wide range of length scale and time scale. How then is such prediction possible? To understand how it can be done, one must consider a wide range of approaches to modelling the solar dynamo, and what constraints need to be applied. Any such model must be able to simulate the important features of a 'typical' solar cycle such as the 'butterfly diagram', and the phase relation in time between toroidal (sunspot) fields and poloidal (polar) fields. The model must run with dynamo ingredients such as differential rotation and meridional circulation that are close to that observed in the sun. In other words, the dynamo model must be 'calibrated' with solar observations (Dikpati *et al.*, 2004; Dikpati and Gilman, 2007a). This process also constrains the choice of more poorly known physical parameters, such as the magnetic diffusivity. Once calibrated, the model can be used to simulate and predict departures from the typical cycle, caused by variations in input data, such as observed magnetic fields from previous times, as well as such dynamo ingredients as the meridional flow.

To be calibrated, a solar dynamo model must contain sufficient realism to allow inclusion of ingredients known from solar observations. For example, a spherical shell configuration is necessary in order to incorporate the observed differential rotation and meridional circulation. In order to incorporate the observed Babcock-Leighton type poloidal source in the most realistic way, one needs to consider departures from axisymmetry, because this poloidal source arises from the decay of bipolar active regions which are tilted in latitude and longitude. For simplicity, averaging this poloidal source in longitude can be a good starting point, and the amplitude and time-dependence of this source can be derived from observations. The magnetic diffusivity is the least known ingredient, and we have to rely on theoretical arguments. Kinematic dynamo models that solve only certain forms of the induction equation, can be calibrated to many solar cycle features, but adding an equation of motion would allow calibration of the model with the so-called torsional oscillations, which are clearly associated with the solar cycle.

At the same time, there are practical limits to the realism that is possible to include for predictive purposes. Dynamo action in the Sun no doubt occurs on many space and timescales, from the global down to granulation scales (10^{-4} of the solar radius), involving many turbulent processes. Capturing all or even most of these scales and processes in one numerical model is not possible yet. Current 3D global MHD models for solar differential rotation, convection and magnetic fields are truncated at larger spatial scales than supergranulation (10^{-2} of the solar radius), so they must parameterize all smaller-scale turbulent processes. It is therefore not possible to do direct numerical simulations (DNS) for the solar convection zone. Such models show dynamo action, but no evidence yet of such basic features as global field reversals (Browning *et al.*, 2006) so it is not possible yet to calibrate them. These simulations are probably so expensive in supercomputing power as to be impractical yet for multiple cycle simulation and predictions. The earliest such simulation attempts (Gilman, 1983), done with much coarser resolution, did show field reversals but the resulting butterfly diagrams were not solar-like. Toroidal fields produced in them migrated toward the poles rather than the equator.

As a consequence of the above considerations, the large-scale flux-transport dynamos can be formulated in the following way starting from the induction equation:

$$\frac{\partial \mathbf{B}}{\partial t} = \nabla \times (\mathbf{U} \times \mathbf{B} - \eta \nabla \times \mathbf{B}), \tag{1}$$

in which $\mathbf{B}$ is the vector magnetic field, $\mathbf{U}$ is the specified velocity, and η is the magnetic diffusivity. Then applying mean field theory (Moffat, 1978) to Eq. (1) in the standard way, we obtain the dynamo equation (2), written as

$$\frac{\partial \mathbf{B}}{\partial t} = \nabla \times (\mathbf{U} \times \mathbf{B} + \alpha \mathbf{B} - \eta \nabla \times \mathbf{B}), \tag{2}$$

in which $\mathbf{B}$ is now the mean magnetic field, $\mathbf{U}$ the mean differential rotation and meridional circulation (for simplicity we omit the averaging overbars in (2) to distinguish it from the original equation (1)), α a parametric form of production of one component of magnetic field from another by lifting and twisting, such as by kinetic helicity of the turbulent flow, and η is now a turbulent magnetic diffusivity.

To simplify the problem further, we assume axisymmetry for the mean fields and decompose them into toroidal ($B_\phi(r, \theta, t)\,\hat{\mathbf{e}}_\phi$) and poloidal ($\nabla \times A(r, \theta, t)\,\hat{\mathbf{e}}_\phi$) and separate the vector components of Eq. (2) into equations for A and B_ϕ:

$$\frac{\partial A}{\partial t} + \frac{1}{r\sin\theta}(\mathbf{u}.\nabla)(r\sin\theta A) = \eta\left(\nabla^2 - \frac{1}{r^2\sin^2\theta}\right)A + S(r,\theta,B_\phi) + \alpha B_\phi, \quad (3)$$

$$\frac{\partial B_\phi}{\partial t} + \frac{1}{r}\left[\frac{\partial}{\partial r}(ru_r B_\phi) + \frac{\partial}{\partial\theta}(u_\theta B_\phi)\right] =$$

$$r\sin\theta(\mathbf{B}_p.\nabla)\Omega - \hat{\mathbf{e}}_\phi \,.\, \left[\nabla\eta \times \nabla \times B_\phi\hat{\mathbf{e}}_\phi\right] + \eta\left(\nabla^2 - \frac{1}{r^2\sin^2\theta}\right)B_\phi. \quad (4)$$

In Eq. (3) for the poloidal field we have added a surface poloidal field source term $S(r,\theta,B_\phi)$, which represents the production of new poloidal field near the outer boundary that originates from toroidal field near the bottom that rises to the surface by magnetic buoyancy to form active regions. In Eq. (4) $\Omega(r,\theta,t)$ is the specified differential rotation and in both equations $\mathbf{u} = u_r\,\hat{\mathbf{e}}_r + u_\theta\,\hat{\mathbf{e}}_\theta$ is the specified meridional circulation. In Eq. (3) α now is an α-effect that arises from instabilities in the overshoot tachocline near the bottom of the domain. The details of the tachocline α-effect we have used, such as its latitudinal profile, are given in Dikpati and Gilman (2001). Both the surface poloidal source and the α-effect near the bottom of the domain are "quenched" above a certain field strength, so that the solutions do not grow without bound. This is the only nonlinearity in the system.

Dikpati *et al.* (2006) and others have focussed on building dynamo-based predictive tools starting from the above formulation of a mean-field kinematic large-scale flux-transport dynamo model which has enough physics in it to be calabratible with global solar observations, while at the same time using modest computing resources. These models generally parameterize key turbulent processes occurring on scales smaller than the resolution of the model, such as the turbulent transport of magnetic flux, captured by a turbulent magnetic diffusivity. The turbulent lifting and twisting of magnetic field lines, is represented by one or more 'α-effects'. The α-effects in the convection zone or in the tachocline regions cannot be observed yet, but the Babcock-Leighton type α-effect can be derived from observations at the surface. These models also omit others, such as formation and buoyant lifting of magnetic flux tubes. Successful comparison of the simulated solar cycle features with appropriate observations is necessary but not necessarily a sufficient condition for the validity of the model.

The Dikpati *et al.* (2004) flux-transport dynamo model included dynamo ingredients such as differential rotation, meridional circulation, Babcock-Leighton poloidal source taking guidence from observations, as well as the magnetic diffusivity profile and tachocline α-effect from theoretical considerations. Calibration is done by tuning the unknown ingredients, primarily the magnetic diffusivity profile and amplitude. The differential rotation does not change much with observations. The meridional flow pattern is not known at the lower half of the convection zone. But it has been

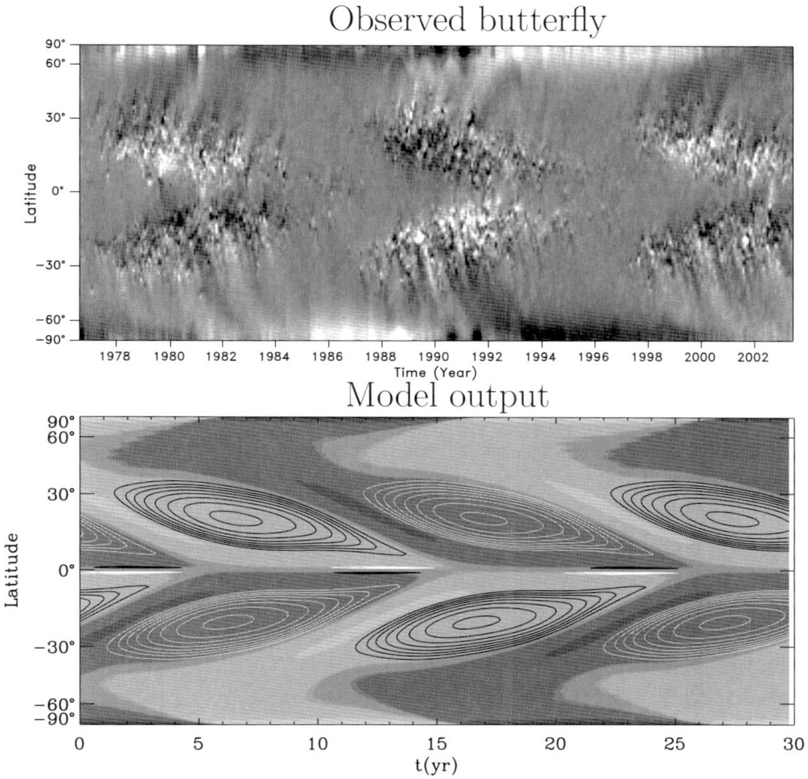

Fig. 6. Top frame: NSO synoptic map of observed longitude-averaged photospheric magnetic fields. Equatorward propagating branch represents the evolution of bipolar spots as a function of solar cycle; poleward branch represents polar field evolution. Bottom frame: Gray-scale map of $B_\phi|_{r=0.725R}$ (primarily confined below 35° latitude) is superimposed on that of surface radial field in time-latitude plane. Bright shades represent positive fields, which means that in low latitudes, toroidal fields run from positive to negative along the direction of rotation and hence follower spots will erupt with positive polarity; in mid-latitudes and high latitudes bright shades denote radial fields that are radially outward. Dark shades in low and high latitudes respectively represent toroidal field lines opposite to the direction of rotation and inward radial fields. Innermost shade in low-latitude has a value of 100 kG and 3 contours cover an order of magnitude field strength. Radial fields are 3–4 orders of magnitude weaker. (Adopted from figures 8 and 9 of Dikpati *et al.* (2004). Reproduced by permission of the AAS.)

widely shown that the flux-transport type dynamos are not too sensitive to the variations in streamlines in terms of how dense or rare they are near the equator or pole, near the surface or base of the convection zone, as long as the flow is a singe cell in each hemisphere. The model becomes sensitive when the flow becomes multicell pattern (Jouve and Brun, 2007). Therefore by selecting a single cell flow and tuning the diffusivity profile Dikpati *et al.* (2004, see also Dikpati and Gilman, 2007a) performed the validity test of calibration by comparing the time-latitude diagram derived from model-output with observations. Figure 6 shows that the model-output is in good agreement with observations.

The variations in solar cycle features from cycle to cycle are actually the departure from the mean and these departures are what are predicted. Therefore the time-variations in the dynamo ingredients need to be incorporated in order to simulate the departures from the mean features. As a simple example, incorporating the observed time-variation in the meridional flow and in the polar field sources in cycle 23, Dikpati *et al.* (2004) hindcast the unusually slow polar reversal in cycle 23. The physical causes for this delay in polar reversal were that the reduction in meridional flow speed by $\sim$50% during 1996–2003 and the weaker poloidal field source in cycle 23 compared to that in cycle 22. Due to the latter, already much weaker poloidal fields were created and the diffusion acting on them more efficiently in the presence of a slow advective transport weakened them further—consequently it took a longer time to cancel all the old-polarity flux and to establish the new polarity.

2.4 Cycle 24 onset-timing and amplitude forecasts

In order to forecast, the model-run needs to be extended into the future. Therefore some assumptions regarding the time-variations in the dynamo ingredients are necessary. As mentioned earlier in the introduction section, the initial conditions are important in a prediction model. Similarly plausible assumptions about the behaviour of the input in future would contribute to the skill of the forecast. In order to clarify, we start with a specific example—the cycle 24 onset-timing prediction.

We know that the cycle length depends primarily on the meridional circulation flow speed in flux-transport dynamos, and also we know from observations that the flow speed decreased by $\sim$50% during 1996–2003, the whole ascending phase (1996–2001) of cycle 23 and a little bit of the declining phase. At that time (say 2003), if we intend to forecast the onset of cycle 24, we have to make some assumptions regarding how the meridional flow would vary after 2003 into the future until 2010 or so. From a reduced flow speed, there could be two plausible assumptions: (i) the flow will continue to slow down, (ii) the flow speed will increase. Dikpati (2004, see also Dikpati, 2005) ran the flux-transport dynamo since 1996 until 2010 first with a steady flow with the speed as in 1996, and then incorporating the above two assumed time-variations. If the flow would have remained steady then the onset of cycle 24 would have indeed happened in the middle of 2006, confirming the cycle-length of 10.5 year for the cycle 23. This is the mean cycle length for the last four cycles. The red curve in Fig. 7 denotes this mean feature. The departure is coming from the combination of actual observation during 1996–2003, and the two assumptions of flow speed variations during 2004–2010. These two cases are represented respectively by the blue and green curves in Fig. 7. The simulation results indicate that the onset of cycle 24 will be delayed by 1 to 2 years, in the case (i) the cycle 24 setting in by late 2007 and in the latter case by early 2008.

This prediction of a late minimum between cycles 23 and 24 and hence delayed onset of cycle 24 is now supported by a variety of observational evidence. For example, the observational butterfly diagram from http://solarscience.msfc.nasa.gov/images/bfly.gif clearly indicates the late onset of cycle 24 compared to past 3 cy-

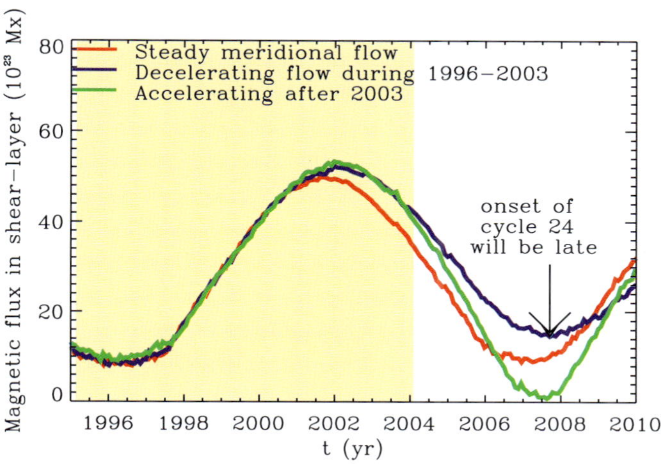

Fig. 7. Time trace of integrated toroidal flux in bottom shear layer for simulations from end of 1995 with 3 different meridional flow variations with time: red curve denotes the case with steady flow, blue with flow decreasing during 1996–2010, and green with flow decreasing during 1996–2003 and then increasing.

cles's onset. The observed large-scale corona also indicates the same—the corona at total eclipse of March 29, 2006 looked very much like the corona of November 1994 (see Dikpati and Gilman, 2007b), not the dipolar corona of early 1996, during the last minimum. The first spot of new cycle 24 has been reported to appear in December 13, 2007 (www.sidc.be).

While variations in the meridional flow speed instantaneously influence the cycle length, the amplitude of a cycle is governed by the Sun's magnetic memory. As we have noted in Fig. 4, the surface forcing of the Sun's large-scale magnetic flux is carried down to the bottom of the convection zone by the meridional circulation conveyor belt in about 20 years, and the seed for the future cycles are produced there. Therefore for forecasting the amplitude of a future cycle, we have to run the model with past cycles' data so that the model builds the magnetic memory.

Thus Dikpati *et al.* (2006) and Dikpati and Gilman (2006) started from the calibrated flux transport dynamo and made certain modifications. First, in Eq. (3) for the poloidal field in terms of vector potential A, the link between the toroidal field in the dynamo at the bottom and the poloidal field at the surface is replaced by a surface forcing term that is derived from the waxing and waning of the observed poloidal fields created by the decay of tilted bipolar active regions in previous solar cycles. This represents a form of 2D (time-latitude) data assimilation in the model, a very common practice for meteorological forecast models, but perhaps the first time it has been used in its most simplified form (which can be called the 'data-nudging') in a solar forecasting problem.

As a result of the above modification, no quenching of the Babcock-Leighton poloidal source and no buoyancy mechanism are included in the model. However the nonlinear quenching in the α-effect near the base of the convection zone is retained. Since this α-effect is much smaller compared the Babcock-Leighton surface α-effect, in effect the model is mathematically changed from being a self-contained nonlinear system to a primarily linear system that is forced at the upper boundary by a measure of the sun's past surface fields. The resulting equation for A is given by

$$\frac{\partial A}{\partial t} + \frac{1}{r\sin\theta}(\mathbf{u}.\nabla)(r\sin\theta A) = \eta\left(\nabla^2 - \frac{1}{r^2\sin^2\theta}\right)A + \mathcal{F}(r,\theta,t) + \alpha\frac{B_\phi}{\left[1+\left(B_\phi/B_0\right)^2\right]}, \tag{5}$$

in which, B_0 is the quenching field strength, set to 30 kGauss.

The inclusion of the surface forcing function, $\mathcal{F}$, representing the time variation of past magnetic fields of the Sun, imposes on the model the (variable) period of past solar cycles. But flux-transport dynamos already have their own intrinsic period, determined primarily by the strength of the meridional flow. We do not know directly from velocity observations prior to 1996 what the amplitude of this meridional flow was. In our initial approach, we estimate an average amplitude of meridional flow from the average period of as many past cycles as we have high quality sunspot data for, which takes us back to cycle 12. From this data we get an average period of 10.75 years. A more sophisticated approach might involve an ensemble of 12 cycle simulations with different or time-varying meridional circulation, but that remains for the future.

If the meridional flow amplitude is determined in this way, but the model is forced with data that has the observed period of each sunspot cycle, a phase mismatch in the link between the toroidal and poloidal fields grows to an unacceptably large value over several cycles' time integration. This makes simulation and forecast of cycle peaks impossible. To avoid this problem, Dikpati *et al.* (2006) stretched and compressed in time the observed data so that each cycle in the surface forcing had the same period, 10.75 years, as the intrinsic period of the dynamo, so that no such destructive phase mismatches would occur. As a result, this method cannot be used to simultaneously simulate the period correctly, as all the cycles have the same length. Figure 8(a) compares the stretched and compressed sunspot data with the original data. Representiing the latitude distribution of the poloidal source within each cycle by a Gaussian with half-width 6° latitude (Frame 8(b)) whose peak matches with spot area curve for that cycle, the surface forcing is migrated toward the equator at a fixed rate. Clearly this representation is not unique, and other possibilities should be explored.

Strictly speaking, the surface poloidal forcing should come from synoptic observations of the photospheric magnetic flux that originates from the decay of active regions, averaged over longitude to fit the axisymmetric mean-field formulation, but such measurements exist only from 1976. Photospheric magnetic flux and sunspot

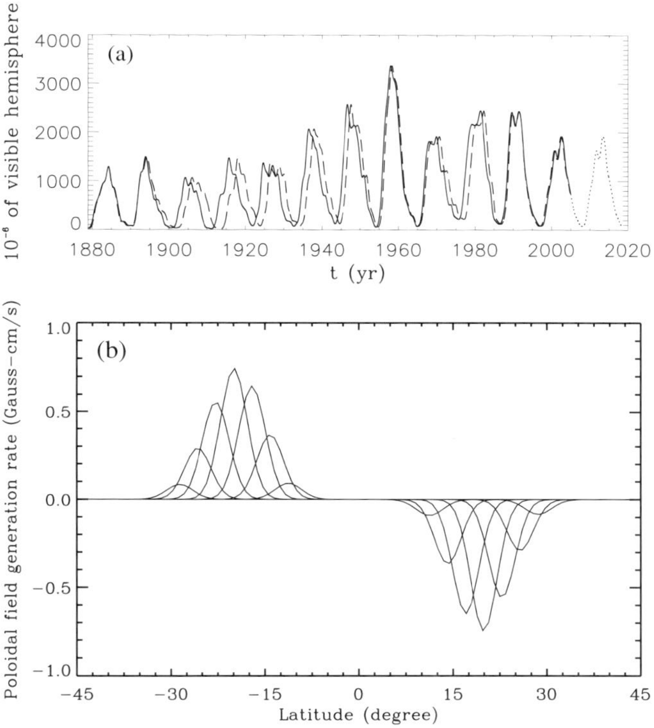

Fig. 8. (a) Sunspot area data stretched and compressed (dashed curve) in time to equalize all solar cycle periods to 10.75 years, compared to original data (solid curve); (b) Gaussian profiles used to represent latitude width of sunspot zone at various phases of a sunspot cycle.

area averaged over several rotations correlate very well (Dikpati *et al.*, 2006), so Dikpati *et al.* (2006) were able to use the much longer sunspot area record to create the forcing.

Starting from a fully converged calibrated model solution, and initializing the model from the beginning of solar cycle 12, Dikpati *et al.* (2006) ran it for the next 13 cycles extending through the upcoming cycle 24. The results of the simulations for the cycle-peaks are shown in Figs. 9(a) and (b), adapted from Dikpati *et al.* (2006). In Fig. 9(a) the observed solar cycles (cycle periods all stretched or compressed to 10.75 years) are compared with the sequence of simulated peaks. It is clear from Fig. 9(a) that the correspondence between observed and simulated cycles is very good, particularly beyond cycle 15. For the first few cycles the agreement is not as good, because the model is still loading the meridional circulation 'conveyer belt' with fields from previous cycles. The simulations show two curves for the upcoming cycle 24. The solid curve represents the case for which the meridional circulation remains fixed at a value that produces the 10.75 year dynamo period, and shows cy-

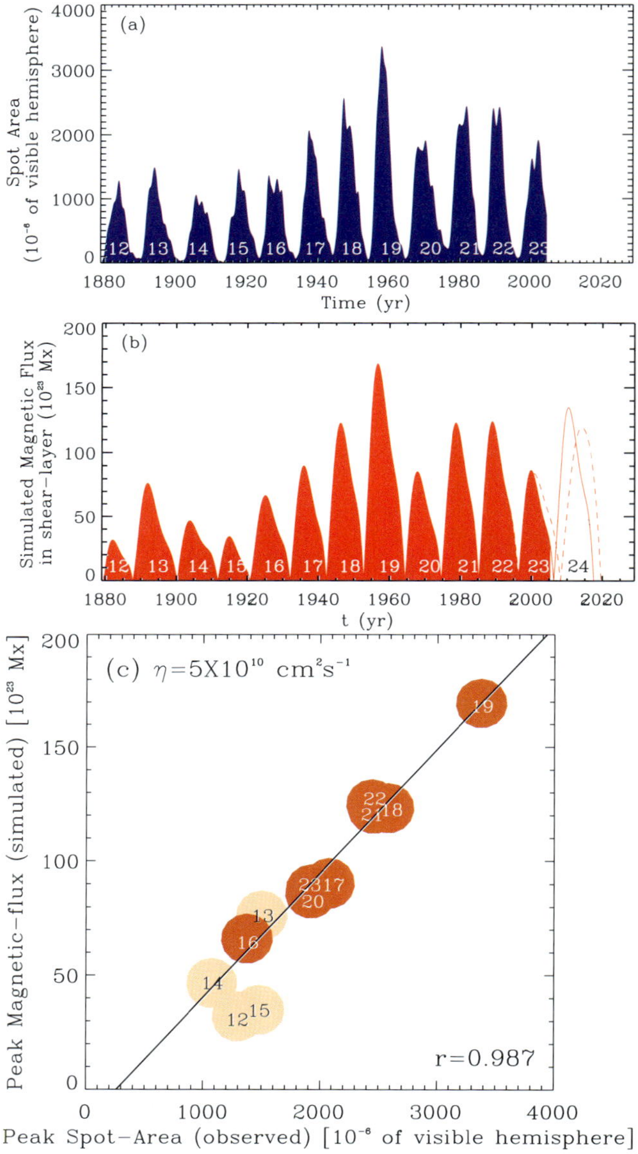

Fig. 9. (a) Observed sunspot area data for cycles 12–23; (b) integral from 0–45° latitude of simulated toroidal magnetic flux in bottom shear layer for cycles 12–24, plus two forecasts for cycle 24; (c) scatterplot of sunspot area peaks vs peak of toroidal magnetic flux integral.

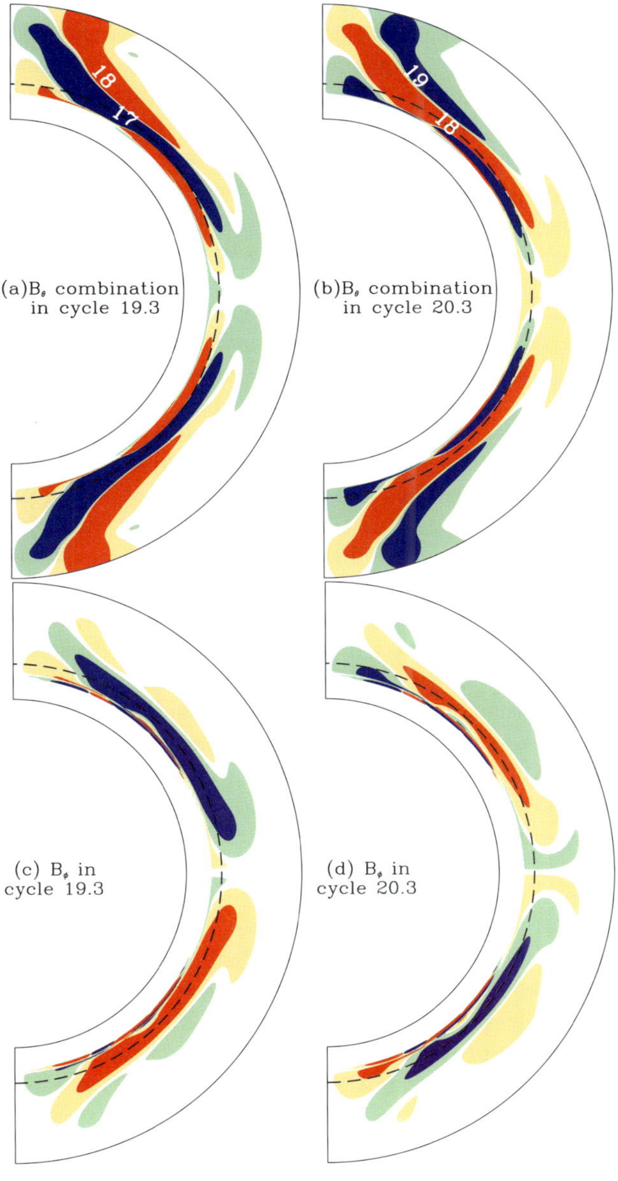

Fig. 10. (a), (b) latitudinal component of simulated poloidal field (red and blue opposite signs) at phase 0.3 of cycles 19 and 20; (c), (d) same for simulated toroidal field.

cle 24 should have a ~50% higher peak than cycle 23. The dashed curve represents a simulation that takes account of time variations in observed meridional circulation since 1996, as described earlier while doing onset-timimg prediction. For this case, cycle 24 is forecast to be ~30% higher than cycle 23. Thus, variations in meridional circulation have some influence in determining the cycle amplitude also. As of today, there exists many other forecasts for cycle 24, ranging from very low (Svalgaard *et al.*, 2005) to extremely high (Tsirulnik *et al.*, 1997) amplitude.

The skill of the forecast model is demonstrated more clearly in the scatterplot shown in Fig. 9(c), which shows the observed peaks for cycles 12–23 against the peaks of the toroidal flux integral computed for the tachocline between the equator and 45°. The correlation coefficient for these points is extremely high, 0.987 for cycles 16–23, for a particular value of the magnetic diffusivity in the bulk of the convection zone, namely $5 \times 10^{10}\,\mathrm{cm^2\,s^{-1}}$, but in Dikpati and Gilman (2006) similar levels are reached for a wide range of diffusivity values, illustrating the robustness of the model.

How the model works to achieve its forecast skill is discussed in detail in Dikpati and Gilman (2006). A key element of its workings is shown in Fig. 10, in which is plotted the latitudinal poloidal field and the toroidal field at similar phases in cycles 19 and 20. The frames in Fig. 10 are snapshots from a continuous video animation for cycles 16–23, in which the path followed by new poloidal field of a particular sign in B_θ can be tracked until it disappears near the base of the convection zone near the equator 2–3 cycles later. The numbers within the plot denote the cycle of origin of each field tracked in this way through all of cycles 16–23. Figure 10 then saves a sample from this tracking, illustrating that there is a memory at high latitudes of the poloidal field from previous cycles, with up to 3 successive cycles' fields accumulated at high latitudes by the meridional flow. The latitudinal differential rotation shearing these fields produces the toroidal fields shown. The equatorward-flowing bottom branch of the meridional circulation then sweeps both poloidal and toroidal fields toward the equator to put in place the toroidal field that creates the peak of the next cycle.

3 Skepticism

Some skepticism have been expressed regarding solar cycle predictions using a flux-transport dynamo. Bushby and Tobias (2007) have argued that predicting solar cycle amplitudes using a mean-field dynamo is impossible. They have shown, using dynamo models different than flux-transport dynamos, that either modest stochasticity or nonlinear effects should destroy the prediction skill. They raised other issues regarding assumptions in the inputs, as well as how such a slow meridional flow present in the vigorously turbulent convection zone play important role in prediction.

Tobias *et al.* (2006) also claim that since the dynamo equations are extremely nonlinear, chaos intrinsic to the system makes prediction much more difficult, particularly for longer time periods. Within a solar cycle, there are short-term turbulent features which are probably governed by a variety of nonlinear processes including the buoyant rise of initially toroidal flux tubes.

Using a 1D surface transport model similar to that of Leighton (1969) and Wang and Sheeley (1991), Cameron and Schüssler (2007) showed that a new solar cycle predictor they proposed, the so-called cross-equatorial flux (CEF), achieved nearly the same skill level as the toroidal flux integral of Dikpati *et al.* (2006), provided they used the same 'migrating Gaussian' representation of the surface poloidal flux forcing. But when they introduced what they characterized as 'more realistic' representations, with a variable migration rate or even actual distributions of spot areas with no Gaussian fit (which is really beyond the limit of the approximations of mean field models), the skill of their predictor went down substantially, to essentially insignificant skill levels. They suggested that the same might happen in the Dikpati *et al.* (2006) model. This possibility can not be tested properly in the flux-transport dynamo-based prediction scheme until time variations in meridional circulation is known for the past 12 cycles.

Furthermore Cameron and Schüssler (2007) have speculated that the Waldmeier effect (the anti-correlation between the peak in sunspot number of a cycle, and the time from minimum to reach that peak) may be responsible for much of the skill found by Dikpati *et al.* (2006) in forecasting the peak of a solar cycle from the record of past surface magnetic flux. But Dikpati *et al.* (2008b) showed recently that the Waldmeier effect is not present in the sunspot area data which were actually used in Dikpati *et al.*'s (2006) prediction scheme. Figure 11 (adapted from figure 2 of Dikpati *et al.* (2008b)) demonstrates that the Waldmeier effect may be specific to only Wolf sunspot number. Previuosly Hathaway *et al.* (2002) showed that the Waldmeier effect is not present in the group sunspot number.

Some responses to those issues can be found in a series of papers Dikpati, de Toma and Gilman have written in 2007 and 2008 (Dikpati and Gilman, 2007a, b; Dikpati *et al.*, 2007, 2008a, b). In brief, using a different model ("B") than a Babcock-Leighton flux-transport dynamo ("A"), it is logically impossible to disprove the skill of model ("A"). While it is true that all the inputs are not known yet from solar observations, but most of them are constrained by observations, and unknown ingredients have been tuned to calibrate the model, as discussed earlier.

We have already mentioned in the introduction how a slow meridional flow can play an important role, as does the slow oceanic flow in the El Nino prediction. In this context, we note that the atmospheric and oceanic weather and climate prediction models went through similar periods of skepticisms during the early years of the development of those models. Adopting from Cane *et al.* (1986) we present in Fig. 12 the comparison between the predicted and the observed El Nino event that is now known to occur due to Pacific Ocean overturning circulation and also quote from Cane's url the following: "Making public forecast did not receive universal approval at that time, as even the notion of climate predictability on the time-scale of a year was not generally accepted". Comparing the top and bottom frames of Fig. 12, we quote again from Cane's url that "Differences in timing and other details showed that the prediction scheme is far from perfect, while the mere appearance of El Nino proved it successful". Eventually the ENSO model correctly forecasted the major events in 1972, 76, 82, 86, 91, a year or more ahead, but had little skill in predicting smaller

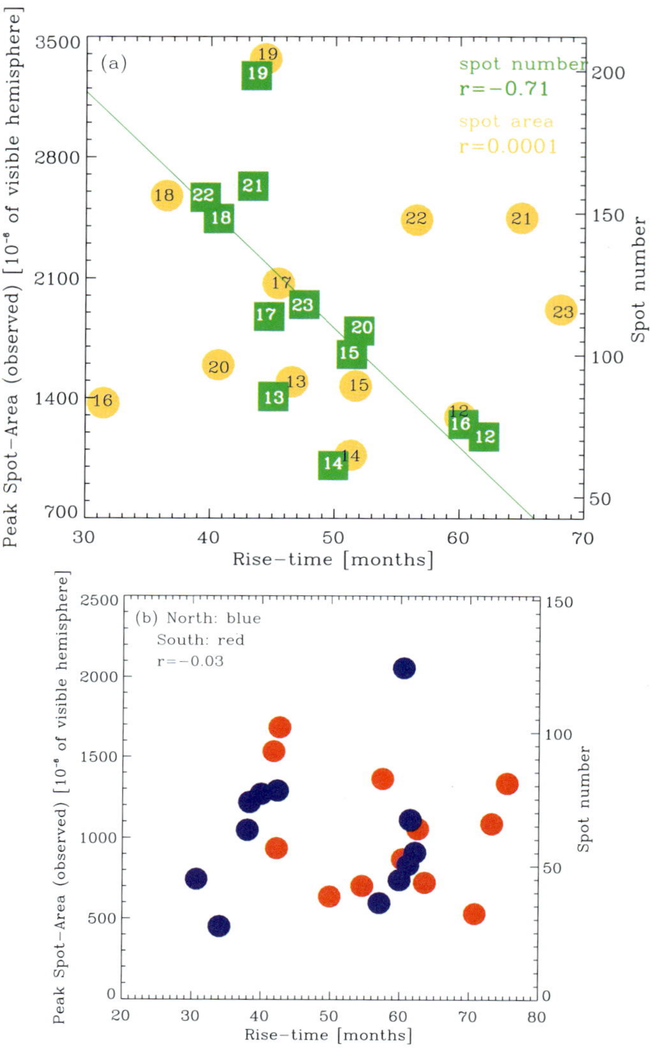

Fig. 11. (a) Yellow circle show scatterplot of correlations between observed peak spot area and cycle rise-time; green squares show that between Zurich sunspot number peak and cycle rise-time. (b) Scatterplot between peak spot area and rise time for North (blue) and South (red) hemispheres. (From figure 2 of Dikpati *et al.* (2008b). Reproduced by permission of the AAS.)

fluctuations.

Using another version of a flux-transport dynamo model, Choudhuri *et al.* (2007) recently forecasted a weak cycle 24, in contrast to the strong cycle forecast in Dikpati *et al.* (2006). The principal differences between the two approaches is that Choudhuri

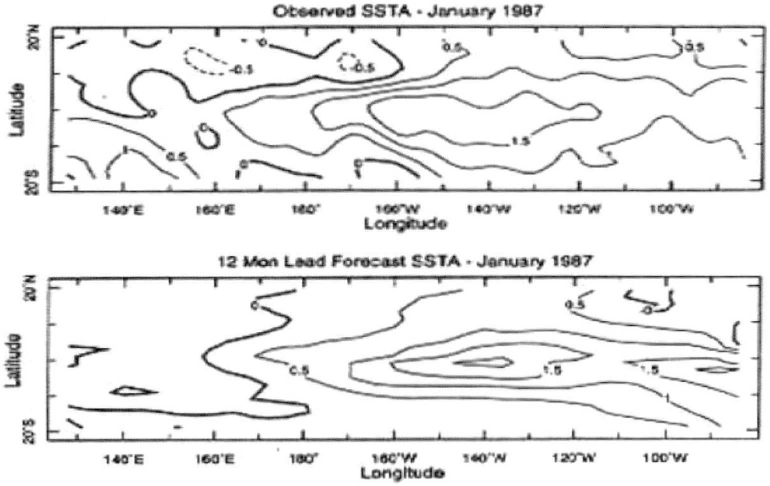

Fig. 12. Observed El Nino event and an early model-based forecast. (Adopted from Cane (1997), http://ccb.colorado.edu/ijas/ijasno2/cane4.jpg.)

et al. (2007) compute solutions with magnetic diffusivities in the bulk of the convection zone that are almost two orders of magnitude higher than in Dikpati *et al.* (2006), and they treat the input of surface poloidal very differently from us. As pointed out in Dikpati *et al.* (2005) the high diffusivity ($>3 \times 10^{11}$ cm^2 s^{-1}) puts a flux-transport dynamo model with solar-like meridional flow into a diffusion rather than advection dominated regime, in which meridional circulation plays very little role, because the high diffusivity moves surface magnetic flux to the bottom in only 1–2 years, very short compared to a sunspot cycle. In addition, such a dynamo run without surface forcing gives a cycle period much shorter than that of the sun, if it sustains a dynamo at all. With such high diffusivity, the model retains almost no memory of its past, so no true 'forecasts' are possible, only what meteorologists call 'nowcasts'. The reason Choudhuri *et al.* (2007) can say anything about the peak of the next solar cycle is because they delay introduction of the surface polar fields to the end of the previous cycle. In the Sun, of course, polar fields at the surface exist for all cycle phases and are being continually reprocessed and are continuously participating in the dynamo. Finally, Dikpati *et al.* (2007) have shown that within flux transport dynamos of lower diffusivity, the surface polar field produced in the dynamo correlates very well with the cycle peak that precedes it, not the one that follows it. One-dimensional models also show that polar fields are the follower of a cycle rather than precursor to it (see Cameron and Schüssler, 2007). So by that test polar fields are not good predictors of the next cycle, even though the original motivation for picking polar fields as a predictor came from concepts of the dynamo (Schatten *et al.*, 1978).

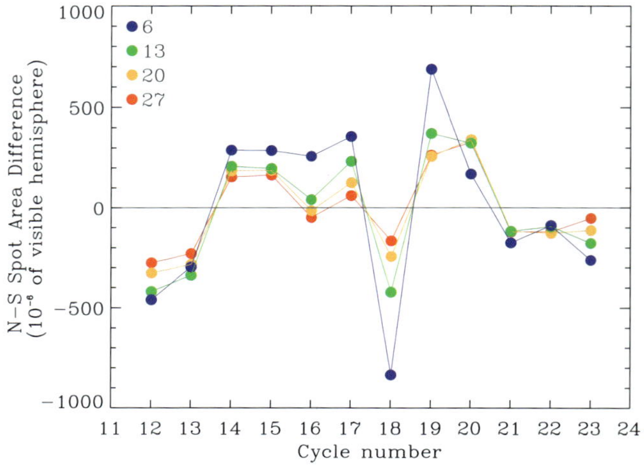

Fig. 13. N–S difference in sunspot cycle peaks for cycles 12–23, for averaging lengths listed.

4 Predicting N/S Asymmetry in Solar Cycle

Observations indicate (see Fig. 13) that there exists North-South asymmetries in amplitude which may persist for several cycles, but there is no regular switching in strength between North and South. Figure 14 shows that the North and the South hemispheres of the Sun are closer in phase at minima than in maxima. The difference in rise times between the two hemispheres is always a year or more.

Separate predictions for the North and South hemispheres of the Sun are important. The Earth's North pole is tilted towards the Sun in June and the South pole in December, whereas the North solar pole is titled towards the Earth in september and the South solar pole in March. The N/S asymmetry in solar magnetic fields changes seasonal modulation in the geo-effectiveness in global sense.

By incorporating the asymmetry in the time-varying meridional flow in the North and South hemispheres of the Sun, Dikpati *et al.* (2007) have predicted that the amplitude of the cycle 24 is likely to be stronger in the South than in the North. The results are shown in Fig. 15. Frames (a), (b), (c) show three different combinations of meridional flow amplitude with time (the profile with latitude and radius is kept fixed for simplicity). Figure 15(a) (Case 1) is closest to the observations for the period 1996–2004, in which the meridional flow toward the poles declined in both hemispheres starting around 1995 but with different rates, and appeared to be leveling off by 2004. The North declined somewhat more. In this case the flow was assumed to increase again in both hemispheres, reaching the 1995 values by about 2013. In Fig. 15(b) (Case 2) we assume the meridional circulation in the South decreases to the minimum seen in the North, reaching that level in 2007, whereupon it increases again to the 1995 level, but reaching it later, in 2016. In Fig. 15(c) (Case 3), the meridional flow in the South was assumed to remain slow, asymptoting to a minimum value in

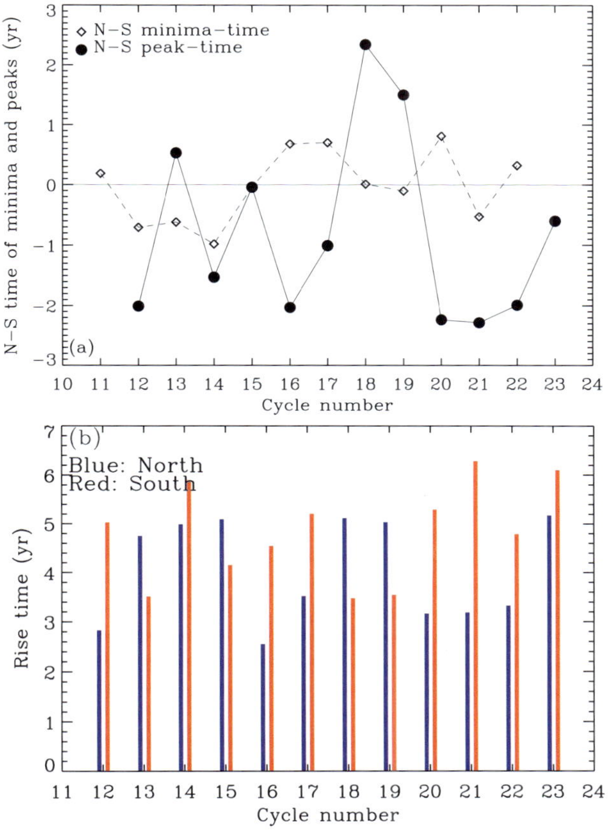

Fig. 14. (a) Diamonds denote the difference between North and South in the timing of the minimum at the end of each numbered sunspot cycle; filled circles give the corresponding difference in the timing of the sunspot area peak for each numbered cycle. (b) The "rise-time" or time from the previous minimum to the maximum of sunspot area for each numbered is shown for north (blue) and south (red) hemispheres.

2012.

Figures 15(d), (e), (f) show the simulated spot-producing toroidal flux integral from separate calculations for North and South hemispheres for cases 1, 2, 3 respectively with the corresponding flow variations shown in Figs. 15(a), (b), and (c). The blue curves are for the North, the red for the South. All the differences between North and South prior to 1995 are due to different surface poloidal flux sources in North and South. After 1995, both these sources and differences in meridional flow contribute to the differences between hemispheres in toroidal flux integral. From Figs. 15(d), (e), (f) we see that in all cases the model predicts that in cycle 24 both hemispheres will be stronger than in cycle 23, by 30% or more. In Case 1 (Fig. 15(d)) whose meridional flow is closest to known observations, the South is predicted to have a

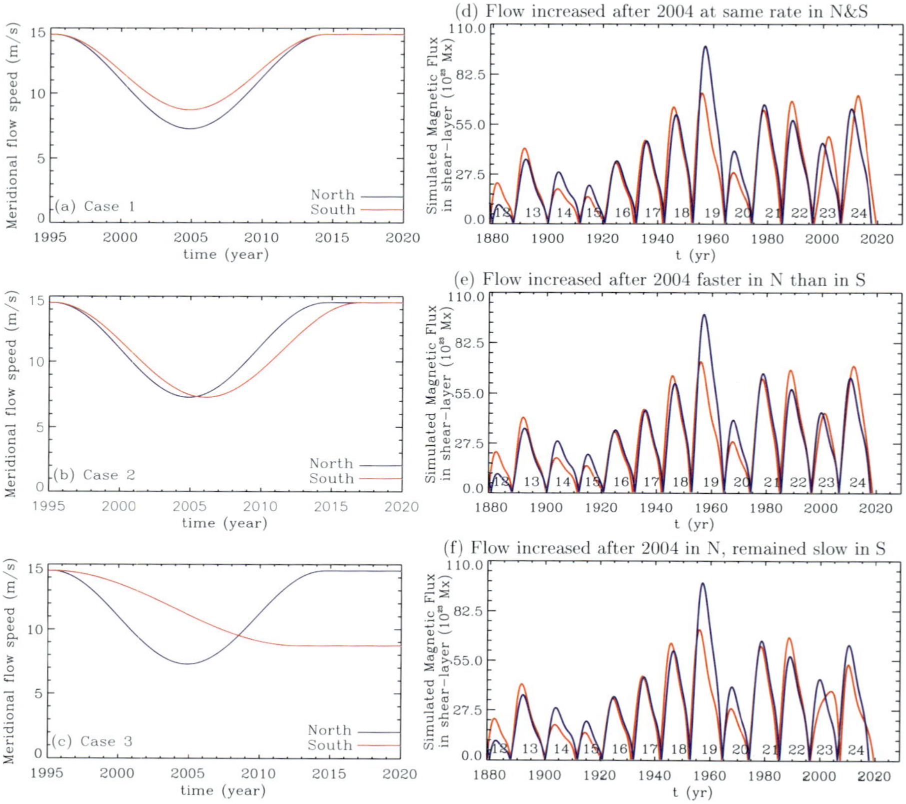

Fig. 15. Predictions of the peak of sunspot area in North (blue curves) and South (red curves) hemispheres for the upcoming sunspot cycle 24, using the spot-producing toroidal flux integral from the flux-transport dynamo-based prediction model of DGdT, for various combinations of time-varying meridional flow amplitude in each hemisphere; these time variations begin in 1995. Case 1, closest to the observed meridional circulation changes, is shown in (a) and (d); Case 2, with slower rise of meridional flow after 2004 in the South, is shown in (b) and (e); Case 3, in which the meridional circulation remains slow in the South after 2004, is shown in (c) and (f). The time varying meridional flow in the North is the same in all three cases, with a decline until 2005 followed by a rise until about 2014.

higher peak than the North, by 10–15%. Case 2 gives results similar to case 1, with the South slightly larger. Only when the meridional flow in the South is kept low (case 3) does the peak in the South fall below that in the North.

5 Discussions and Future Goals

The flux-transport dynamo-based predictive tool we have described above has shown skill in the hindcast and forecast of solar cycle peaks; this implies that the progress has been made, but it does not imply that the solar dynamo problem has

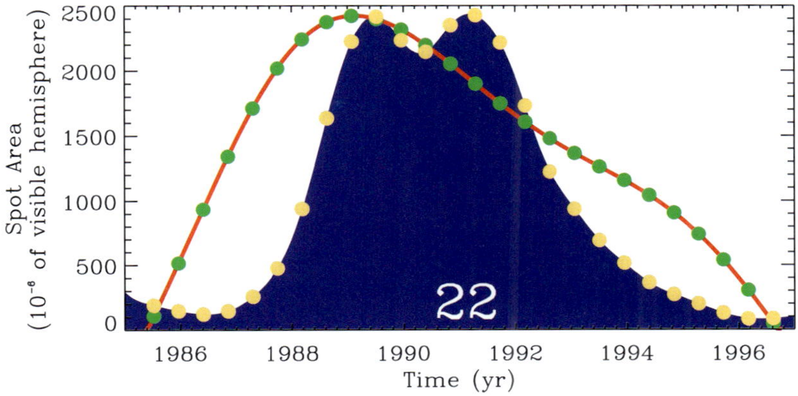

Fig. 16. Blue-filled curve shows observed spot area data and superimposed on that in red curve is the simulated tachocline toroidal flux. Solid green and yellow circles at 5.3 months' intervals respectively on red and blue-filled curves denote corresponding phases of simulated and observed solar cycles.

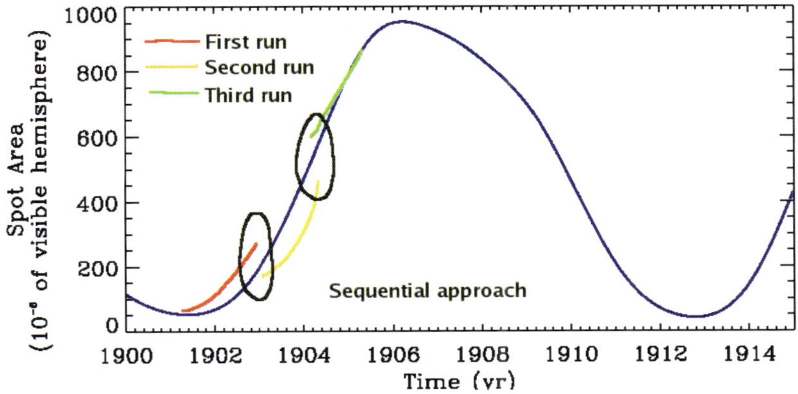

Fig. 17. A sequential data-assimilation diagram with three successive runs; meridional flow is adjusted in the model after each run until the match between model-output and observation is good.

been solved. Figure 16 illustrates how much of the solar cycle pattern is reproducible within a solar cycle. The cycle 22 has been chosen for comparing the simulated and observed phases in each 5.3 months' interval time. The correlation between the simulated and observed phases is clearly not as good as that between simulated and observed cycle peaks for past 12 cycles.

There are various reasons for why the detailed features within a cycle are not possible to reproduce by using a mean-field dynamo. One important reason is that there is a process missing in the two-dimensional large-scale dynamo calculation, namely the formation of concentrated toroidal bands and flux tubes and their eruption to the surface. Recently Isik *et al.* (2007) have studied the influence of emergence latitudes in the advective-diffusive transport of magnetic flux using a surface flux-transport model. In order to investigate the details within a solar cycle, it is necessary to include such flux-emergence physics in the two-dimensional flux-transport dynamo simulations.

Another reason for much degraded correlation between the simulated and observed cycle phases within a cycle shown in Fig. 16 is that the physics behind the rise and fall patterns of a cycle is not clearly known. This is partly because the transport rate of the magnetic flux is related to the shape of a cycle in this class of models, and the details of the ingredients, the meridional circulation and the magnetic diffusivity, that govern the transport processes, are not known.

The question arises whether we should proceed forward to simulate the shape of a cycle, or we should wait until observations give us more guidance about the unknown dynamo ingredients. One plausible answer perhaps lies in further development of data-assimilation in solar physics (see Brun, 2007). This is also an appropriate time for that because of the accumulation of large solar data from various ground- and space-based instruments, and data-assimilation techniques have begun to be applied not only in solar cycle prediction models, but also in other solar research topics, such as flare predictions (Bélanger *et al.*, 2007).

So far a very simplified data-assimilation technique has been applied in solar cycle prediction models, which can actually be called the 'data-nudging'. Thus the technique was restricted to predict the cycle-timing and cycle-amplitude in separate calculations. In order to be able to simultaneously predict the timing, amplitude and shape of a cycle, more sophisticated 'sequential' and/or 'variational' data-assimilation techniques need to be built. An example for a sequential data-assimilation in solar cycle model is presented in Fig. 17, which demonstrates that the surface magnetic data is assimilated in the model and the model output is compared with observations after a certain run time; if the match is not good, the unknown ingredients such as meridional flow, and/or diffusivity is adjusted and then the next run is performed. Proceeding in this way until the end of calculation one could also reconstruct the meridional flow and eventually could make an ensemble average from many such runs. These profiles can be compared in future with observed flow when SDO is launched.

Acknowledgments. The author gratefully acknowledges the CAWSES 2007 symposium organizers for the support for her trip to Kyoto. This work is partially supported by the NASA grant NNH05AB521.

References

Babcock, H. W., The topology of the Sun's magnetic field and the 22-YEAR cycle, *Astrophys. J.*, **133**, 572, doi:10.1086/147060, 1961.

Babcock, H. W. and H. D. Babcock, The Sun's magnetic field, 1952–1954, *Astrophys. J.*, **121**, 349, doi:10.1086/145994, 1955.

Basu, S. and H. M. Antia, Changes in solar dynamics from 1995 to 2002, *Astrophys. J.*, **585**, 553, doi:10.1086/346020, 2003.

Bélanger, E., A. Vincent, and P. Charbonneau, Predicting solar flares by data assimilation in avalanche models, *Sol. Phys.*, **245**, 141, doi:10.1007/s11207-007-9009-3, 2007.

Bonanno, A., D. Elstner, G. Rüdiger, and G. Belvedere, Parity properties of an advection-dominated solar $\alpha^2\Omega$-dynamo, *Astron. Astrophys.*, **390**, 673, doi:10.1051/0004-6361:20020690, 2002.

Brown, T. M., J. Christensen-Dalsgaard, W. A. Dziembowski, P. R. Goode, D. O. Gough, and C. A. Morrow, Inferring the sun's internal angular velocity from observed p-mode frequency splittings, *Astrophys. J.*, **343**, 526, doi:10.1086/167727, 1989.

Browning, M. K., M. S. Miesch, A. S. Brun, and J. Toomre, Dynamo action in the solar convection zone and tachocline: pumping and organization of toroidal fields, *Astrophys. J.*, **648**, L157, doi:10.1086/507869, 2006.

Brun, A. S., Toward using modern data assimilation and weather forecasting methods in solar physic, *Astron. Nachr.*, **328**, 329, doi:10.1002/asna.200610739, 2007.

Bushby, P. J. and S. M. Tobias, On predicting the solar cycle using mean field models, *Astrophys. J.*, **661**, 1289, doi:10.1086/516628, 2007.

Cameron, R. and M. Schüssler, Solar cycle prediction using precursors and flux transport models, *Astrophys. J.*, **659**, 801, doi:10.1086/512049, 2007.

Cane, M. A., ENSO and its prediction: How well can we forecast it?, *Internet Journal of African Studies*, http://ccb.colorado.edu/ijas/ijasno2/ijasno2.html, 1997.

Cane, M. A., S. E. Zebiak, and S. C. Dolan, Experimental forecasts of El-Nino, *Nature*, **321**, 827, doi:10.1038/321827a0, 1986.

Choudhuri, A. R., P. Chatterjee, and J. Jiang, Predicting solar cycle 24 with a solar dynamo model, *Phys. Rev. Lett.*, **98**, 131103, doi:10.1103/PhysRevLett.98.131103, 2007.

Choudhuri, A. R., M. Schüssler, and M. Dikpati, The solar dynamo with meridional circulation, *Astron. Astrophys.*, **303**, L29, 1995.

Dikpati, M., Global MHD theory of tachocline and the current status of large-scale solar dynamo, in *Helio- and Asteroseismology: Towards a Golden Future*, edited by D. Dansey, ESA-SP 559, 233, 2004.

Dikpati, M., Solar magnetic fields and the dynamo theory, *Adv. Space Res.*, **35**, 322, doi:10.1016/j.asr.2005.04.061, 2005.

Dikpati, M., Predicting solar 'climate' by assimilating magnetic data into a flux-transport dynamo, *Astron. Nachr.*, **328**, 1092, doi:10.1002/asna.200710885, 2007.

Dikpati, M. and P. Charbonneau, A Babcock-Leighton flux transport dynamo with solar-like differential rotation, *Astrophys. J.*, **518**, 508, doi:10.1086/307269, 1999.

Dikpati, M. and P. A. Gilman, Flux-transport dynamos with α-effect from global instability of tachocline differential rotation: A solution for magnetic parity selection in the Sun, *Astrophys. J.*, **559**, 428, doi:10.1086/322410, 2001.

Dikpati, M. and P. A. Gilman, Simulating and predicting solar cycles using a flux-transport dynamo, *Astrophys. J.*, **649**, 498, doi:10.1086/506314, 2006.

Dikpati, M. and P. A. Gilman, Steps for building a calibrated flux-transport dynamo for the Sun, *Sol. Phys.*, **241**, 1, doi:10.1007/s11207-007-0380-x, 2007a.

Dikpati, M. and P. A. Gilman, Global solar dynamo models: simulations and predictions of cyclic photospheric fields and long-term non-reversing interior fields, *New J. Phys.*, **9**, 297, doi:10.1088/1367-2630/9/8/297, 2007b.

Dikpati, M., G. de Toma, P. A. Gilman, C. N. Arge, and O. R. White, Diagnostics of polar field reversal in solar cycle 23 using a flux transport dynamo model, *Astrophys. J.*, **601**, 1136, doi:10.1086/380508, 2004.

Dikpati, M., M. Rempel, P. A. Gilman, and K. B. MacGregor, Comments on "Full-sphere simulations of circulation-dominated solar dynamo: Exploring the parity issue", *Astron. Astrophys.*, **437**, 699, doi:10.1051/0004-6361:20042435, 2005.

Dikpati, M., G. de Toma, and P. A. Gilman, Predicting the strength of solar cycle 24 using a flux-transport dynamo-based tool, *Geophys. Res. Lett.*, **33**, L05102, doi:10.1029/2005GL025221, 2006.

Dikpati, M., P. A. Gilman, G. de Toma, and S. S. Ghosh, Simulating solar cycles in northern and southern hemispheres by assimilating magnetic data into a calibrated flux-transport dynamo, *Sol. Phys.*, **245**, 1, doi:10.1007/s11207-007-9016-4, 2007.

Dikpati, M., G. de Toma, and P. A. Gilman, Polar flux, cross-equatorial flux, and dynamo-generated tachocline toroidal flux as predictors of solar cycles, *Astrophys. J.*, **675**, 920, doi:10.1086/524656, 2008a.

Dikpati, M., P. A. Gilman, and G. de Toma, The Waldmeier effect: An artifact of the definition of Wolf sunspot number?, *Astrophys. J.*, **673**, L99, doi:10.1086/527360, 2008b.

Durney, B. R., On a Babcock-Leighton dynamo model with a deep-seated generating layer for the toroidal magnetic field, *Sol. Phys.*, **160**, 213, doi:10.1007/BF00732805, 1995.

Duvall, T. L., Jr., Large-scale solar velocity fields, *Sol. Phys.*, **63**, 3, doi:10.1007/BF00155690, 1979.

Dziembowski, W. A., P. R. Goode, and K. G. Libbrecht, The radial gradient in the sun's rotation, *Astrophys. J.*, **337**, L53, doi:10.1086/185377, 1989.

Foukal, P., C. Fröhlich, H. Spruit, and T. M. L. Wigley, Variations in solar luminosity and their effect on the Earth's climate, *Nature*, **443**, 161, doi:10.1038/nature05072, 2006.

Gilman, P. A., Dynamically consistent nonlinear dynamos driven by convection in a rotating spherical shell. II—Dynamos with cycles and strong feedbacks, *Astrophys. J. Suppl.*, **53**, 243, doi:10.1086/190891, 1983.

Guerrero, G. A. and J. D. Muñoz, Kinematic solar dynamo models with a deep meridional flow, *MNRAS*, **350**, 317, doi:10.1111/j.1365-2966.2004.07655.x, 2004.

Haber, D. A., B. W. Hindman, J. Toomre, R. S. Bogart, R. M. Larsen, and F. Hill, Evolving submerged meridional circulation cells within the upper convection zone revealed by ring-diagram analysis, *Astrophys. J.*, **570**, 855, doi:10.1086/339631, 2002.

Hathaway, D. H., P. A. Gilman, J. H. Harvey, F. Hill, R. F. Howard, H. Jones, J. Kasher, J. Leibacher, J. Pintar, and G. Simon, GONG observations of solar surface flows, *Science*, **272**, 1306, doi:10.1126/science.272.5266.1306, 1996.

Hathaway, D. H., R. M. Wilson, and E. J. Reichmann, Group sunspot numbers: Sunspot cycle characteristics, *Sol. Phys.*, **211**, 357, 2002.

Isik, E., M. Schüssler, and S. Solanki, Magnetic flux transport on active cool stars and starspot lifetimes, *Astron. Astrophys.*, **464**, 1049, doi:10.1051/0004-6361:20066623, 2007.

Jouve, L. and A. S. Brun, On the role of meridional flows in flux transport dynamo models, *Astron. Astrophys.*, **474**, 239, doi:10.1051/0004-6361:20077070, 2007.

Küker, M., G. Rüdiger, and M. Schültz, Circulation-dominated solar shell dynamo models with positive alpha-effect, *Astron. Astrophys.*, **374**, 301, doi:10.1051/0004-6361:20010686, 2001.

Leighton, R. B., A magneto-kinematic model of the solar cycle, *Astrophys. J.*, **156**, 1, doi:10.1086/149943, 1969.

Miesch, M. S., A. S. Brun, M. L. De Rosa, and J. Toomre, Structure and evolution of giant cells in global models of solar convection, *Astrophys. J.*, **673**, 557, doi:10.1086/523838, 2008.

Moffatt, H. K., *Magnetic field generation in electrically conducting fluids*, 353 p, Cambridge University Press, Cambridge, England, 1978.

Ohl, A. I., Forecast of sunspot maximum number of cycle 20, *Soln. Dann.*, **12**, 84, 1966.

Parker, E. N., Hydromagnetic dynamo models, *Astrophys. J.*, **122**, 293, doi:10.1086/146087, 1955.

Rempel, M., Solar differential rotation and meridional flow: the role of a subadiabatic tachocline for the Taylor-Proudman balance, *Astrophys. J.*, **622**, 1320, doi:10.1086/428282, 2005.

Rempel, M., Flux-transport dynamos with Lorentz force feedback on differential rotation and meridional flow: saturation mechanism and torsional oscillations, *Astrophys. J.*, **647**, 662, doi:10.1086/505170, 2006.

Schatten, K. H., P. H. Scherrer, L. Svalgaard, and J. M. Wilcox, Using dynamo theory to predict the sunspot number during solar cycle 21, *Geophys. Res. Lett.*, **5**, 411, doi:10.1029/GL005i005p00411, 1978.

Svalgaard, L., E. W. Cliver, and Y. Kamide, Sunspot cycle 24: smallest cycle in 100 years?, *Geophys. Res. Lett.*, **32**, L01104, doi:10.1029/2004GL021664, 2005.

Tobias, S. M., D. E. Hughes, and N. O. Weiss, Letter to the editor, *Nature*, **442**, 26, 2006.

Tsirulnik, L. B., T. V. Kutnetsova, and V. N. Oraevsky, Forecasting the 23rd and 24th solar cycles on the basis of MGM spectrum, *Adv. Space Res.*, **20**, 2369, doi:10.1016/S0273-1177(97)00909-5, 1997.

Ulrich, R. K. and J. E. Boyden, The solar surface toroidal magnetic field, *Astrophys. J.*, **620**, L123, doi:10.1086/428724, 2005.

Waldmeier, M., *The Sunspot Activity in Years, 1610–1960*, pp.171, Schulthess, Zurich, 1961.

Wang, Y. -M. and N. R. Sheeley, Jr., Magnetic flux transport and the sun's dipole moment—New twists to the Babcock-Leighton model, *Astrophys. J.*, **375**, 761, doi:10.1086/170240, 1991.

Wang, Y. -M., N. R. Sheeley, Jr., and A. G. Nash, A new solar cycle model including meridional circulation, *Astrophys. J.*, **383**, 431, doi:10.1086/170800, 1991.

Wenzler, T., S. K. Solanki, N. A. Krivova, and C. Fröhlich, Reconstruction of solar irradiance variations in cycles 21-23 based on surface magnetic fields, *Astron. Astrophys.*, **460**, 583, doi:10.1051/0004-6361:20065752, 2006.

Climate and Weather of the Sun-Earth System (CAWSES): Selected Papers from the 2007 Kyoto Symposium,
Edited by T. Tsuda, R. Fujii, K. Shibata, and M. A. Geller, pp. 201–216.

Evidence for solar forcing: Some selected aspects

Jürg Beer[1] and Ken McCracken[2]

[1]*Eawag, CH-8600 Duebendorf, Switzerland*
E-mail: beer@eawag.ch
[2]*Institute for Physical Sciences and Technology,*
University of Maryland, College Park,
Maryland, USA

It is believed that the global warming since the mid-20th century is primarily the result of the combustion of fossil fuel. The fact that the climate also changed in the past during periods of rather constant atmospheric greenhouse gas concentrations points to additional factors such as solar and volcanic forcing. The Sun is by far the most important source of energy for Earth and direct satellite based observations during the past 30 years show that the solar constant (total solar irradiance TSI) changes in phase with the solar magnetic activity. The past 30 years are characterized by a high, rather constant mean level of activity, however, during the last 2 years the minima in TSI, IMF (interplanetary magnetic field), NM (neutron monitor count rate), and Φ (solar modulation function) have clearly deviated from the earlier minima, suggesting that TSI is now decreasing in response to a lower level of solar magnetic activity. Unfortunately our knowledge of past solar activity is very limited, the longest record available being the sunspot record going back to 1610. The record can be extended from centuries to millennia by using the cosmogenic radionuclides which are primarily produced by the galactic cosmic rays. Their intensity is modulated by the open solar magnetic and the geomagnetic field. Removing the geomagnetic effects results in the solar modulation function Φ which can be reconstructed for the past 10,000 years, as can the strength of the interplanetary magnetic field. The comparison of Φ with selected climate records provides strong evidence that solar forcing was important in the past and will possibly play a role in the future. Confirmation of the synchronous declines in TSI and IMF will allow the reconstructed IMF to be used to estimate TSI for the past 10,000 years.

1 Introduction

In the recently published IPCC report (Solomon *et al.*, 2007) the authors conclude that the available evidence is now strong enough to state that "Most of the observed increase in global average temperatures since the mid-20th century is very likely due to the observed increase in anthropogenic greenhouse gas concentrations". This means that the global change observed during the past few decades is outside the range of expected natural variability, or in other words, it cannot be explained as a result of natural forcings. It is important to note that this statement does not mean that without anthropogenic forcing the climate would stay unchanged in the future. The climate has always changed and will continue to do so in the future. As a consequence the predictions of the impact caused by anthropogenic activities on future climate change must allow for the natural variability.

The climate is a dynamic non-linear system of large complexity that varies on time scales from months to millions of years. It is an open system and interacts with space through electromagnetic radiation, matter in the form of galactic and solar cosmic ray particles, and magnetic fields. By far the strongest interaction takes place with the Sun which is the most important source of energy. The power from cosmic ray particles is in the order of 10^9 W which is comparable to the power received by the stars during the night. However, the total power of solar radiation arriving at the top of the atmosphere is about 8 orders of magnitude larger and amounts to $2\ 10^{17}$ W. Processes of reflection, absorption, distribution, and emission control the flow of solar radiation, with the climate system attempting to equilibrate the temperatures and to reach an equilibrium between incoming short wave radiation and out going long wave radiation. As a consequence the conditions at the Earth's surface depend strongly on the amount of incoming solar radiation (Total Solar Irradiance TSI), its spectral distribution (Spectral Solar Irradiance SSI), the atmospheric composition (greenhouse gases, aerosols), the albedo (clouds, ice and snow, vegetation), and the internal variability caused by the transport processes redistributing the energy (ocean and atmospheric circulation and latent heat transport). Some of these have long time constants with the consequence that the responses to external forcings may be long delayed.

The fact that the climate is a complex non-linear system makes it difficult to obtain a quantitative understanding of its temporal and spatial variability in the past. This makes it especially difficult to make reliable predictions about the future. A good example of the difficulties we face when dealing with the climate system is provided by the modern weather forecast. Even today, with an almost unlimited amount of information from weather stations and satellites, and with the most advanced general circulation models (GCM's), it is impossible to make reliable predictions beyond about seven days. In spite of all our impressive technological progress, the chaotic properties of the weather system will always prevent us from making detailed and accurate long-term predictions. One could therefore come to the conclusion that understanding climate change is hopeless. However, it seems that many of the short-term chaotic climate fluctuations are averaged out when going to time scales of decades and larger and that a limited set of parameters exists which if known well enough will enable us to make useful predictions.

We are therefore faced with two distinctly different problems. (1) To understand how the climate system works, and to determine the parameters, that best determine its secular changes, and (2) to be able to predict the magnitude of the natural and anthropogenic forcings in the future. Probably the best way to address the first problem is to improve our understanding of the longer-term dynamics of the climate system by studying the history of past natural forcings and the corresponding responses of the climate system. Based on this information more reliable predictions about future

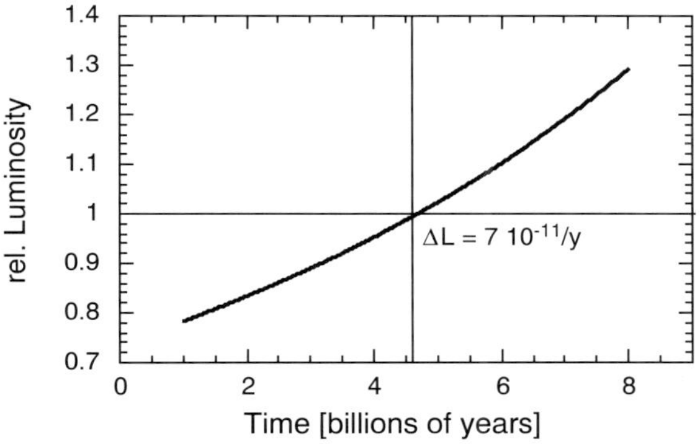

Fig. 1. Increase in solar luminosity relative to the present ($L = 1$). Note that a billion years after the formation of the solar system the luminosity was more than 20% smaller than today.

natural forcing changes can be made. In the case of the second problem, the anthropogenic forcings scenarios of e.g. greenhouse gas emissions can be constructed which are based on certain assumptions about the future development of the world's population and its economy and technology. Even though the increase in greenhouse gases is going to be the dominant factor in climate forcing during the coming decades, natural forcing will continue to play a role. In the following we will focus on some aspects related to solar forcing.

2 The Source of the Sun's Emissions

Since the Sun is by many orders of magnitude the most important source of energy it is quite reasonable to assume that any change in TSI and SSI will affect the climate on Earth. Such changes can have very different causes and may occur on very different time scales.

The fusion of hydrogen to helium takes place in the core of the Sun where every second some 4.2 Million tons of mass are turned into electromagnetic radiation. According to the standard solar model this process is very stable but increases monotonically from $L \approx 0.8$ three billion years ago to $L \approx 1.3$ in 3.5 billion years time (Fig. 1), when the sun will run out of hydrogen and first turn into a red giant and then into a white dwarf. The change in luminosity at present is only $7\ 10^{-11}$ per year and therefore completely irrelevant for climate changes on time scales of centuries and millennia. Nevertheless, the dramatic change on a billion year time scale raises the interesting question how planet Earth avoided becoming a "snowball" in its young age. This question is often called the "faint young sun paradox" (Sagan and Chyba, 1997).

On its way to the Sun's surface the electromagnetic radiation is repeatedly absorbed and reemitted which steadily shifts the wavelength towards longer values which in turn increases the probability for absorption. At about 2/3 of the Sun's radius radiative transport becomes so inefficient that the thermal gradient gets very large and convection sets in. Most of the solar power is radiated into space by the photosphere, a layer of thickness about 500 km, at a temperature of 5770 K. The spectrum of the emitted light can be approximated in a first order by the Planck spectrum of a black body at this temperature.

To date, not much attention has been paid to the question whether the energy transport from the core to the sun's surface is subject to variability. Although there are ideas under development how fluctuations in the convection zone might occur (Kuhn *et al.*, 1988; Sofia and Li, 2004; Steiner and Ferriz-Mas, 2006) so far no evidence is available and many arguments are put forward against such fluctuations (Foukal *et al.*, 2006). It is important to note, however, that even small changes in the solar diameter could lead to a significant change in luminosity (Sofia and Li, 2006).

3 Emission from the Surface

The emission from the photosphere is what we see from Earth. For a long time the total emission was considered to be constant and it is still often called the "solar constant". However, there is a long history of investigations to determine whether the solar constant is really constant (Langley, 1903; Abbot, 1910). Unfortunately, fluctuations in the atmospheric opacity prevented these pioneering investigators from proving that the solar constant is not constant. The ability to observe the Sun from satellites ultimately yielded the precision necessary to detect changes in TSI smaller than 0.1%.

Figure 2 shows the compilation of TSI measurements which has been carefully put together from different satellites after correcting for various effects such as degradation of the instruments with time (see Fröhlich, this volume). Many attempts have been made to explain the observed fluctuations of TSI. The principle of most approaches is to separate the solar disc into several components such as a background component considered as constant, a negative component given by the dark sunspots including umbra and penumbra and a positive component consisting of the bright faculae and the magnetic network. By weighing these different components accordingly with the so-called filling factor the measured TSI fluctuations can be reproduced surprisingly well for the period 1980–2004 (Wenzler *et al.*, 2006; Krivova *et al.*, 2007). In spite of this success this approach has the inherent disadvantage that it is basically a regression model, which is based to a large extent on observations (filling factors) that are only available for the recent past. In particular, it provides no information on the variability of the background component on decadal to centennial time scales; these being most relevant as far as climate forcing is concerned.

It has been long recognized that the variability of TSI must be closely linked to the magnetic activity of the Sun (Baliunas and Jastrow, 1990). The factor of three reduction in solar forcing in the IPCC4 report was primarily due to a reassessment of the long-term changes in properties of the solar magnetic fields between the Maunder Minimum and the present. The strength of the interplanetary field near Earth is closely related to the magnetic fields on the Sun (Wang *et al.*, 2000), and is plotted in Fig. 2. Both TSI and IMF exhibit 11 year variations, a 0.4 W m^{-2} change in TSI corresponding to a 1.0 nT change in IMF. While the strength of the IMF at the three

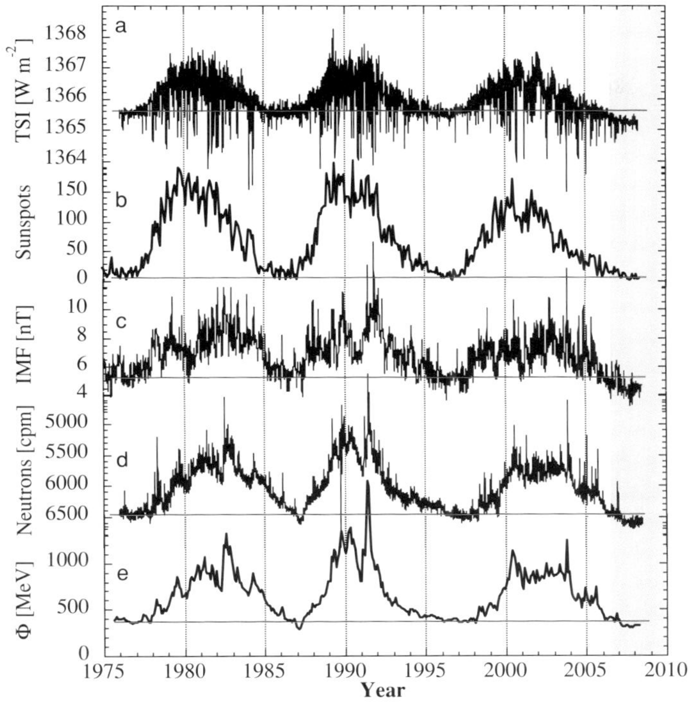

Fig. 2. (a) Composite of TSI measurements during the period 1978 until 2008 compiled by C. Fröhlich (daily data). (b) Sunspots (monthly data) (c) Interplanetary magnetic field (daily data) (d) count rate of the neutron monitor from Oulu (daily data on a reversed *y*-axis) (e) Solar modulation function Φ derived from neutron monitor data (monthly data).

previous sunspot minima was ~5.2 nT, we note that it decreased through that value in the first half of 2006 and decreased steadily to a mean of ~4.25 nT in late 2007. In exactly the same manner, TSI decreased below the previous minimum values in early 2006, and was ~0.4 W m^{-2} lower by late 2007 probably pointing to a long-term change in the background contribution.

While the first direct measurements of the IMF started in the 1960s, three methods have been used to extrapolate the present day values to the past, these being used to some extent in IPCC4 as a proxy for TSI. The cosmogenic nuclides are the basis of one of those methods, and are unique in their ability to provide estimates of the IMF over the previous millennia, offering the possibility to estimate TSI far into the past.

4 The Cosmogenic Nuclides and the IMF as Proxies for TSI

The production rate of the cosmogenic nuclides (e.g. ^{10}Be and ^{14}C) is modulated by the open magnetic field which is carried away from the Sun by the solar wind (see below) (McCracken and Beer, 2007). ^{10}Be from ice cores and ^{14}C from tree rings,

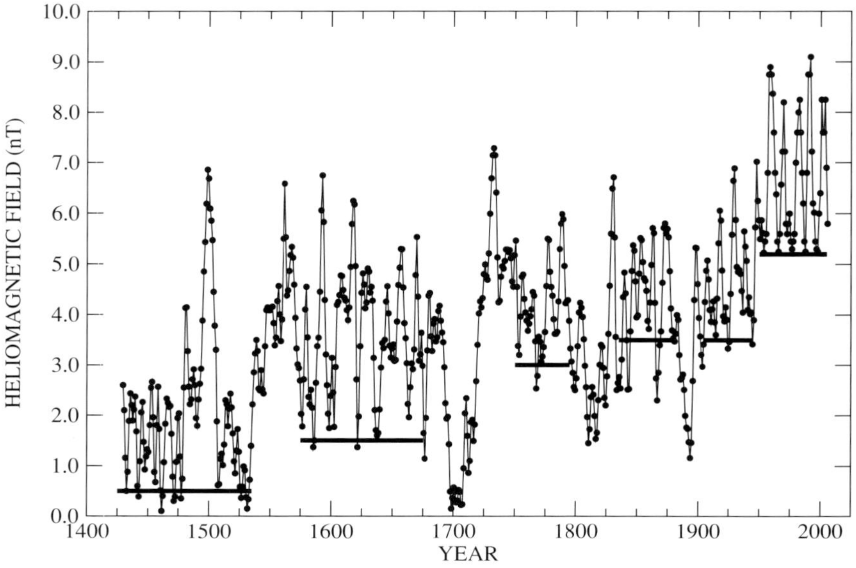

Fig. 3. IMF derived from ^{10}Be for the past 600 years. (From McCracken, K. G., Heliomagnetic field near Earth, 1428–2005, *J. Geophys. Res.—Space Phys.*, **112**, A09106, 2007. Copyright 2007 American Geophysical Union. Reproduced by permission of American Geophysical Union.)

together with other cosmogenic radionuclides, provide our only continuous record of the long-term variability of solar activity prior to the commencement of the sunspot record in 1610. They also constitute a cosmic magnetometer that provides estimates of the strength of the IMF for the past 10000 years. As an example, Fig. 3 presents the strength of the IMF for the past six centuries derived from the ^{10}Be data (McCracken, 2007). Note the estimated field strengths of ≤ 1 nT for the Spoerer and Maunder Minima. Note also the $\sim$85 year variability (the Gleissberg Cycle). It appears possible that the decreasing field after 2006 in Fig. 2 may be the commencement of another period of reduced field strength similar to $\sim$1815 or $\sim$1900, with implications for a lower TSI.

Cosmogenic radionuclides are produced by nuclear interactions of the galactic cosmic rays (GCR) with atoms (N, O, Ar) in the atmosphere. To reach the atmosphere the GCR have to propagate through the heliosphere which forms a bubble with a radius of about 150 Astronomical Units (AU) around the Sun that is filled with solar plasma carrying magnetic fields (Fig. 4). The propagation of the cosmic rays is described by the transport equation derived by Parker (Parker, 1965). It is difficult to use the transport equation to parameterize the intensity of the GCR, however the so-called force field approximation (Gleeson and Axford, 1967) has proven to be a good approximation near Earth. This approximation describes the modulation effect of the Sun on the energy spectrum of the GCR in terms of a parameter Φ called the

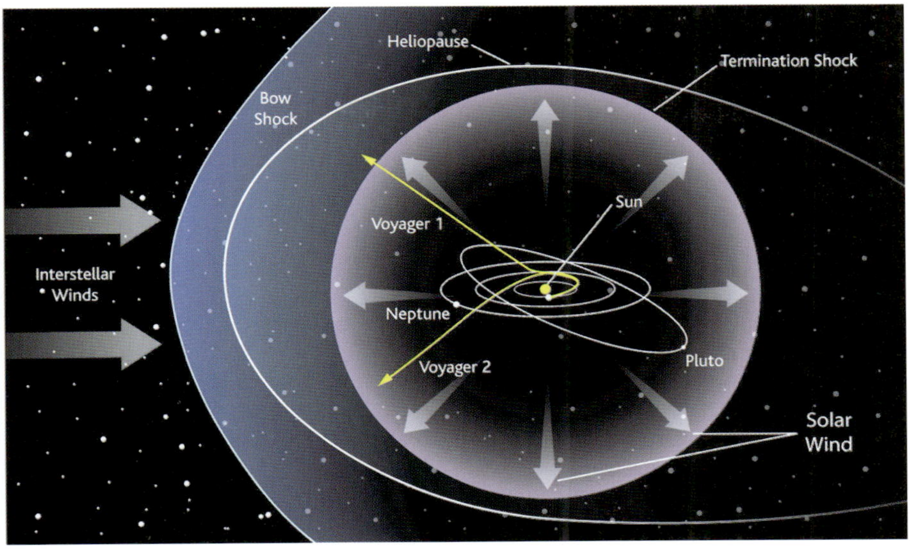

Fig. 4. Voyagers 1 and 2 have flown on different trajectories past the outer planets of the solar system since 1977, and Voyager 1 crossed the termination shock of the solar wind at 94 AU from the Sun in December 2004. Voyager 2 did likewise in 2007. The solar wind is a supersonic flow, and a shock—the termination shock—is required for the wind to decelerate and merge with the local interstellar medium that bounds the solar system. The solar wind and interstellar gas do not merge easily, so further out beyond the termination shock, there is a thick boundary region between the solar wind and the interstellar medium: the heliosheath. Further out still, if the solar system is itself moving supersonically relative to the interstellar medium, there may be a large bow shock. (From Fisk, L. A., Journey into the unknown beyond, *Science*, **309**, 2016–2017, 2005. Reprinted with permission from AAAS.)

solar modulation function. Φ basically corresponds to the average energy lost by a cosmic ray proton on its way to the Earth.

Figure 5 shows the differential energy spectrum of the GCR proton flux for different levels of solar activity. $\Phi = 0$ MeV corresponds to the local interstellar spectrum outside the heliosphere (Fig. 4). This spectrum is an estimate because no space probe has left the heliosphere yet and actually measures this spectrum. Voyager 1 and 2 have crossed the termination shock and are passing through the heliosheath (Fig. 4). Figure 5 shows that the shielding effects of the solar open magnetic field are most pronounced at the low energy end of the spectrum. As a consequence GCR particles above about 20 GeV are hardly affected by the heliospheric magnetic fields.

Before reaching Earth the cosmic ray particles have to overcome a second barrier, the geomagnetic field. This field prevents particles with too low a rigidity (momentum per unit charge) from reaching the top of the atmosphere. In a first approximation the geomagnetic field is considered as a dipole and in this case the cut-off rigidity depends only on the angle of incidence and the geomagnetic latitude. At low latitudes the cut-off rigidity for vertical incidence is presently $\sim$14.9 GV. This means that a cosmic ray

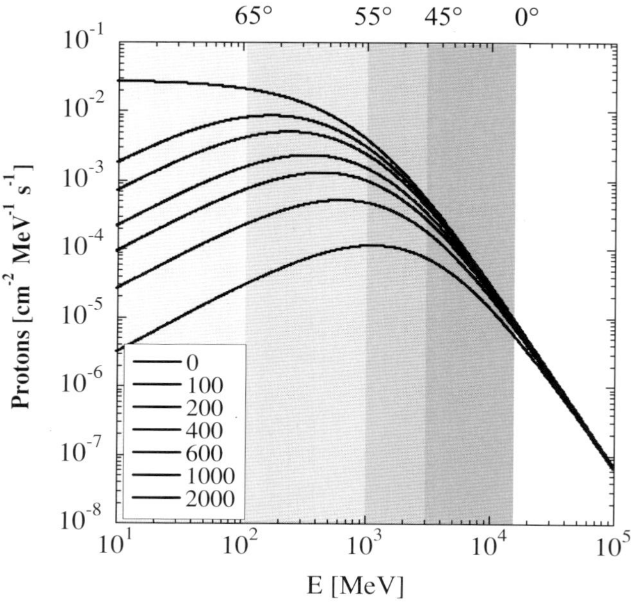

Fig. 5. Differential GCR proton fluxes for different levels of solar activity ranging from $\Phi = 0$ MeV corresponding to the local interstellar spectrum arriving at Earth without any solar influence, to $\Phi = 2000$ MeV which corresponds to a very active Sun. There are similar curves for cosmic ray alpha particles and heavier nuclides. The vertical bands illustrate the effect of the geomagnetic field which cuts of all protons approaching vertically with an energy below about 100 MeV for a geomagnetic latitude of 65°; below 1 GeV for 55°, and below 3 GeV for 45°. At 0° the cut-off energy is 13.9 GeV for the present geomagnetic field.

proton needs a kinetic energy of at least 13.9 GeV to reach the top of the atmosphere (see shaded bands in Fig. 5). The solar modulation is a monotonic decreasing function of particle energy (Fig. 5) and consequently the modulation is small near the equator (~14 GeV) and large at high latitudes which are accessible to the strongly modulated energies near 1 GeV.

If a primary cosmic ray particle makes its way through the heliosphere and the geomagnetic field and enters the atmosphere it will interact quickly with an atom of oxygen, nitrogen, or argon. Since the energies of incoming particles are generally very high, only part of their kinetic energy is transferred to the first atom they hit. They continue their travel and hit a few more atoms until their energy is dissipated. Each collision results in the generation of secondary particles covering the full spectrum of hadrons and leptons, which either decay or interact with other atoms of the atmosphere. In this way a cascade of secondary particles develops which can be simulated using the Monte Carlo technique (Masarik and Beer, 1999, 2009).

The simulations show that the majority of the secondaries are neutrons followed by protons. Both, in turn, collide with atmospheric atoms initiating spallation reactions (Masarik and Beer, 1999; Webber and Higbie, 2003; Masarik and Beer, 2009), that generate the cosmogenic nuclides that are archived for us in ice (^{10}Be, ^{36}Cl) or tree rings (^{14}C). In addition, the cosmic ray produced neutrons have been monitored continuously since 1951 by so-called neutron monitors. In Fig. 2(d) the count rate of the Oulu neutron monitor clearly shows the modulation of the GCR by the 11-y Schwabe cycle (Fig. 2(b)). Whenever the magnetic activity is high (large sunspot numbers) the shielding is strong and the neutron flux is low. As we discussed above the solar modulation of the GCR can be described by the modulation function Φ which is shown in panel e of Fig. 2. Many studies have shown that the 11 yr and longer-term variations are faithfully reproduced in the cosmogenic data, and they and the neutron monitor data have been inter-calibrated to yield a continuous cosmic ray record for the past 10,000 years (McCracken and Beer, 2007; Steinhilber *et al.*, 2008).

In practice, the cosmogenic data contain substantial statistical variations, and some residual atmospheric effects. The quality of the solar signal can be improved by combining different ^{10}Be records from different sites, together with the ^{14}C record from tree rings. ^{14}C is produced almost identically as ^{10}Be, but behaves geochemically in a completely different manner. It forms $^{14}CO_2$ which exchanges between atmosphere, biosphere, and ocean. These large reservoirs cause a considerable attenuation of the high frequency production changes and delays while ^{10}Be is removed from the atmosphere quickly within 1–2 years. Combined together, however, the two cosmogenic nuclides provide a result that is largely devoid of atmospheric or other "system effects".

The cosmogenic radionuclides record the cosmic ray intensity with a relatively low temporal resolution of 1 year compared to a few minutes for a neutron monitor and furthermore, a relatively low signal to noise ratio. However, they have the unique advantage that at present they are the only "neutron monitor" capable of recording the cosmic ray flux on Earth for the past 10,000 years compared to the almost 60 years of modern neutron monitors. This is another example of nature providing its own solution to an engineering problem long before mankind even was aware of the problem.

5 The Long-term Solar Variability Record

In the following we describe how the long-term solar variability record is derived, and from it, the estimated strength of the interplanetary magnetic field near Earth. Some of its spectral properties are then discussed, and finally they are compared with some selected examples of climate change, pointing to a significant role of the Sun in the past.

As discussed above the ^{10}Be record reflects changes in the open magnetic field filling the heliosphere, in the geomagnetic dipole field, and to some extent in the transport of ^{10}Be from the atmosphere where it is produced to the ice sheet where it is stored. GCM models show that the transport effects were relatively stable during the climatic conditions prevailing during the Holocene (the last 10,000 years), so to a first approximation they can be neglected. This is not the case for the geomagnetic field which exhibits significant long-term changes (Muscheler *et al.*, 2005; Vonmoos *et al.*, 2006).

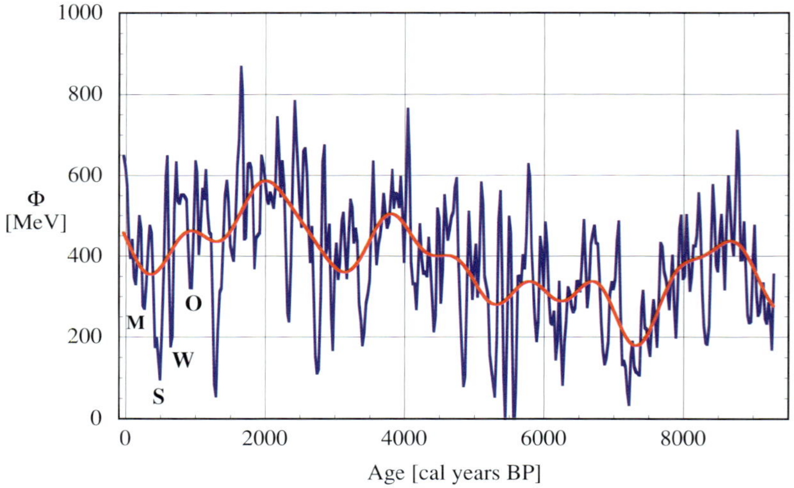

Fig. 6. Solar modulation function Φ from the present (0 BP corresponds to 1950) back to 9350 BP (Steinhilber *et al.*, 2008). The blue curve data is low-pass filtered with a cut-off of 150 years, the red one with 1000 years. The most recent solar minima are indicated: M: Maunder; S: Spoerer; W: Wolf, and O: Oort.

Using our Monte Carlo simulations (Masarik and Beer, 1999, 2009), the effects of secular changes of the geomagnetic dipole field have been removed and we are left with the solar modulation function Φ (Fig. 6). The GRIP ice core record is limited to the period from 1640 to 9300 BP and has recently be complemented by the most recent 360 years which are a composite of Φ derived from neutron monitor data and those from a shallow ice core (Steinhilber *et al.*, 2008). The data of Fig. 6 have been low-pass filtered with a 150 y cutoff. The most striking features of the Φ record are the many distinct minima which correspond to grand solar minima such as the Maunder (M), Spoerer (S), Wolf (W), and Ort (O). The fact that Φ never reaches zero means that there is always some residual open magnetic flux; in other words the solar dynamo seems to weaken from time to time, but it never stops.

The maxima are less pronounced. It is interesting to note that the present level of solar activity is comparatively high although there were earlier periods with similar or possibly even higher activity around 2000, 4000, and 9000 BP. There is also a clear long-term trend indicated by the thick line that is low-pass filtered with a cut-off of 1000 years.

For a more detailed analysis we calculate the power spectrum using wavelet analysis (Grinsted, 2002–2004). Figure 7 shows the wavelet spectrum of Φ. There are several distinct periodicities some of which are listed in Table 1. Since the time scales for ^{10}Be in ice cores are not as easily established as those for ^{14}C in tree rings we also give the corresponding periodicities for Δ^{14}C (Reimer *et al.*, 2004) and Q^{14}C calculated for almost the same time interval (1750–9300 BP). Q^{14}C is the ^{14}C production

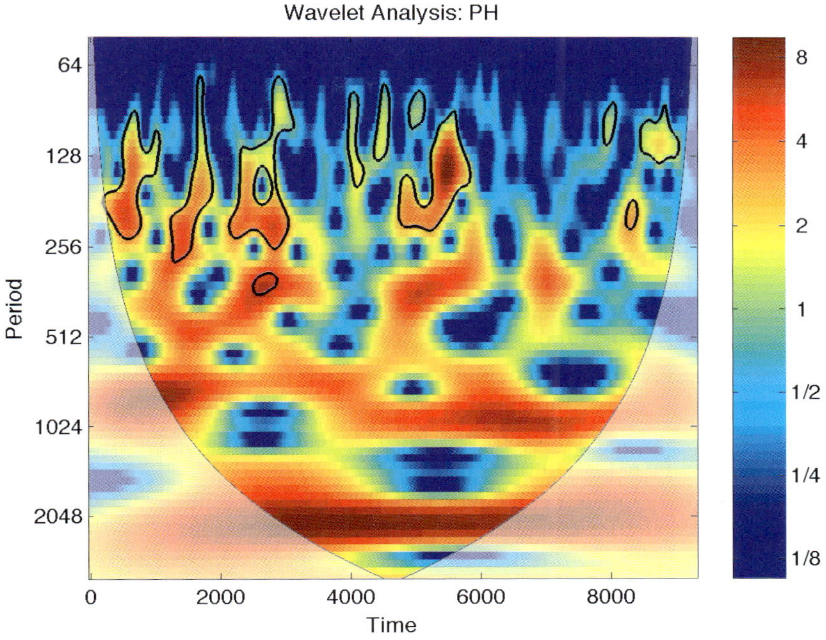

Fig. 7. Wavelet analysis (Grinsted, 2002–2004) of the Φ data from Fig. 6.

Table 1.

Cycle/Period	Φ	$\Delta^{14}C$	$Q^{14}C$
Hallstatt	2194	2275	2424
	982	984	957
DeVries, Suess	207	208	208
	352	350	350
	704	714	713
	497	512	512
	105	105	105
Gleissberg	86	87.9	87.0

rate which was calculated using the Intcal04 calibration curve and the Siegenthaler Oeschger carbon cycle model (Oeschger *et al.*, 1975). An interesting feature of Fig. 7 is that the cycles wax and wane during the Holocene. There are periods when most cycles show large amplitudes (between 2000 and 3000, and between 5000 and 6000 BP) and times when the amplitudes are generally low (between 4000 and 5000 BP).

6 Solar Variability and Past Climate Change

Finally we address the question whether the reconstructed solar variability correlates with the known climate change in the past, and whether it has the potential to contribute to a better understanding of future climate change. An unequivocal attribution of an observed climate change to a reconstructed change in solar activity is a difficult task. First of all we do not yet know the precise quantitative change in solar forcing in W m^{-2}. Secondly the response of the climate system is non-linear and can therefore be attenuated, deformed, and delayed with respect to the forcing signal. In addition the climate models show that the response is generally very heterogeneous. Finally the information on a past climate change is based on paleodata usually derived from natural archives such as ice cores, sediments, stalagmites, and tree rings. The usual climate parameters (temperature, precipitation rate) are not directly available but have to be derived from so-called proxies such as the oxygen isotopic ratio $^{18}O/^{16}O$ which, in the case of precipitation, basically measures the temperature at the site where water vapor condenses and forms water droplets. Most proxies are also dependent to some extent on other parameters and have to be calibrated. Therefore the reconstructed climate parameters are subject to uncertainties regarding the climate parameter they represent, but also regarding the time scale. Nevertheless the temporal and spatial resolution of the records is continuously increasing and due to improvements of the existing and the development of new analytical techniques the uncertainties of the data is decreasing.

With those caveats in mind, we now examine two examples where it appears that solar forcing has played an important role in climate change long before the recent increase in anthropogenic forcing. We should mention that the literature of such examples is quickly growing and there would be many more examples and maybe more convincing ones. However, we believe that these two illustrate how the modern climate models, together with the reconstructed solar activity, will allow the climate models to be refined, and the proxies such as the modulation function and the strength of the IMF to be calibrated in terms of TSI.

The first example concerns the extensions of alpine glaciers. It is a well-known phenomenon that as a result of the present global warming most of the glaciers on the globe are shrinking. Using radiocarbon dating of trees that were killed by an advancing glacier it is possible to reconstruct the history of the glacier's extension over the past few millennia (Fig. 8) (Holzhauser *et al.*, 2005). Similar observations were made elsewhere (Denton and Karlén, 1973; Hormes *et al.*, 2006).

The size of a glacier is mainly related to winter precipitation and summer temperature and integrates over several years to decades. This makes it insensitive to individual weather events and delays its response by a few decades. Figure 8 shows the history of the great Aletsch glacier in Switzerland, the largest glacier in the Alps. The reconstruction shows that it has retreated by more than 3 km since about 1850, and will probably continue to do so. But the figure also shows that the present retreat distance is not unique. It retreated similar distances in the medieval warm period, and at the end of the Roman era, and each time advanced back to where it was in the "little ice-ages". Comparison with the solar modulation function Φ shows that the advances

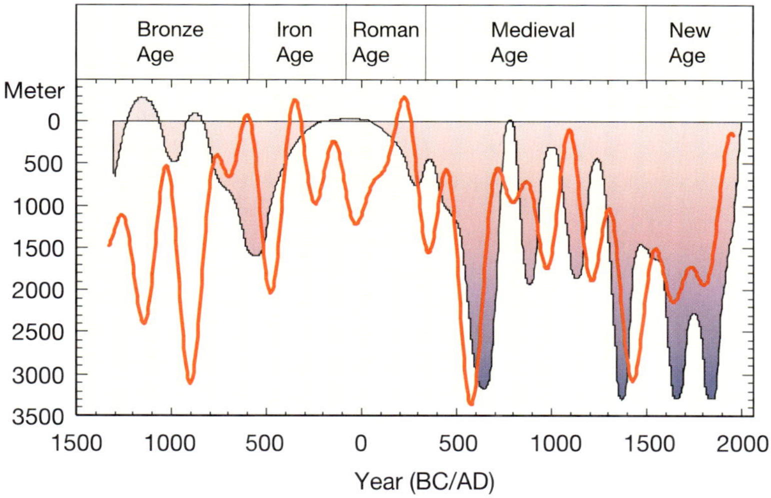

Fig. 8. Extension of the "great Aletsch glacier" in the Swiss Alps. Photographic records show that the glacier has retreated by more than 3 km since the 19th century. However, the present retreat distance is not unique; similar retreats occurred in Medieval and Roman times. The changes in extension are compared with the Φ curve (Fig. 6) and it is clear that low solar activity corresponded to large extensions of the glacier.

correspond in general to low Φ, and retreats to high Φ. It should be mentioned that a relatively small number of tree samples means that the timing of the glacier dynamics is not very well constrained. The lag of the glacier in response to climate change has been taken into account.

The second example concerns $\delta^{18}O$ measurements in a stalagmite from the Chinese Dongge cave (Wang *et al.*, 2005). Stalagmites consist of $CaCO_3$ which precipitates from the drip water when the pressure is reduced. The authors have shown that $\delta^{18}O$ in this stalagmite is a proxy for the local precipitation rate. The stalagmite was dated using the U/Th technique and some tuning (<50 years) was done to match the $\Delta^{14}C$ curve. In Fig. 9 the $\delta^{18}O$ and the Φ record are both low-pass filtered with a 100 years cutoff. Again, there is clear evidence for a solar signal in the data. For example, the grand solar minimum around 2700 BP, one of the largest minima during the Holocene, shows up very clearly. This 2700 BP grand minimum is associated with evidence of climate change all over the globe (van Geel and Renssen, 1998). A spectral analysis of the data reveals the same periodicities that were found in the Φ record (Table 1). This is another indication that the solar signal is imprinted in the precipitation rate.

7 Summary and Conclusions

The Earth is an open system which is driven by energy coming almost exclusively from the Sun. Modern space based measurements of TSI show that the energy supply

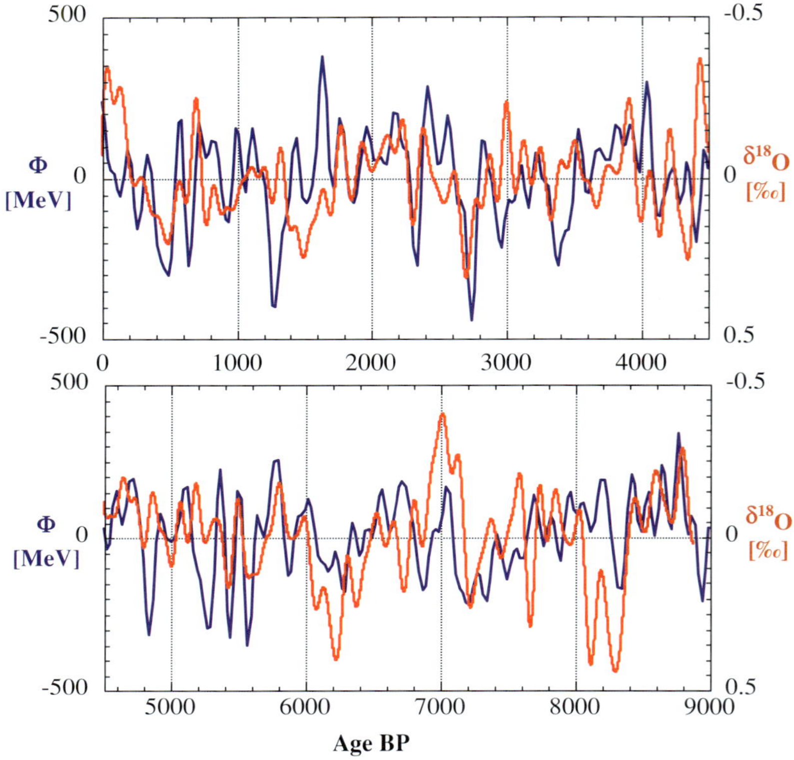

Fig. 9. Comparison of the $\delta^{18}O$ measurements on a stalagmite from the Dongge cave in China (Wang *et al.*, 2005) with the Φ record from Fig. 6. Low Φ values corresponding to low solar activity generally agree with high $\delta^{18}O$ values interpreted as increased precipitation. Both data was first detrended by a polynomial of degree 3 and then low-pass filtered with a cut-off of 100 years and are shown as deviations from the long-term mean.

from the Sun is subject to small changes (0.1%) which seem to be related to variations in the magnetic field of the Sun. This raises the important question to what extent solar variability affects the climate on Earth. Cosmogenic radionuclides provide the unique opportunity to reconstruct the history of the variability of the Sun, and its magnetic fields, over at least the past 10,000 years. The reconstruction of the solar modulation function is characterized by long-term changes as well as short-term cyclicity (11-y Schwabe cycle). Other typical periodicities are 2200, 208, and 87 years. A special feature of solar variability are the so-called grand minima, periods when the solar activity is strongly reduced which leads to an almost complete absence of sunspots (only confirmed for the Maunder Minimum).

A comparison of the solar modulation function Φ with climate records points to a relationship between solar variability and climate forcing. Taking into account that the Sun is just one forcing factor among others (volcanic, internal, greenhouse since the 20th century); that the climate response is spatially heterogeneous; and that there is some uncertainty in the time scales, the case for solar forcing looks promising. Evidence is strengthened by new high-precision records, together with GCM modeling which shows that the climate system is very sensitive to solar forcing (Ammann *et al.*, 2007).

The modern space age has shown us that there is a close relationship between the strength of the IMF, and TSI. Further still, we now have the ability to invert the cosmogenic data itself, or the modulation function Φ, to provide estimates of the time dependence of the IMF far into the past. Such reconstructions indicate that the sunspot minimum value of the IMF near Earth has been lower (to $\sim$1 nT) and higher (to $\sim$8 nT) compared to 5.2 nT in 1976, 1968, and 1997. That is, the means now exists to (1) extrapolate the observed relationship between TSI and IMF, together with the estimated dependence of the IMF, to provide TSI as a function of time for the past 10,000 years for input to climate models, and (2) use the same process to test non-linear relationships between TSI and IMF.

Acknowledgments. This work was supported by the International Space Science Institute (ISSI) and the Swiss National Science Foundation.

References

Abbot, C. G., The solar constant of radiation, *Annual Report*, 319 pp., Smithonian Institution, 1910.

Ammann, C. M. *et al.*, Solar influence on climate during the past millennium: Results from transient simulations with the NCAR Climate System Model, *Proc. Natl. Acad. Sci. USA*, **104**, 3713–3718, 2007.

Baliunas, S. and R. Jastrow, Evidence for long-term brightness changes of solar-type stars, *Nature*, **348**, 520–523, 1990.

Denton, G. H. and W. Karlén, Holocene climatic variations—their pattern and possible cause, *Quat. Res.*, **3**, 155–205, 1973.

Fisk, L. A., Journey into the unknown beyond, *Science*, **309**, 2016–2017, 2005.

Foukal, P. *et al.*, Variations in solar luminosity and their effect on the Earth's climate, *Nature*, **443**, 161–166, 2006.

Gleeson, L. J. and W. I. Axford, Cosmic rays in the interplanetary medium, *Astrophys. J.*, **149**, L115–L118, 1967.

Grinsted, A., Continuous Wavelet Transform, http://www.pol.ac.uk/home/research/waveletcoherence/, 2002–2004.

Holzhauser, H. *et al.*, Glacier and lake-level variations in west-central Europe over the last 3500 years, *Holocene*, **15**, 789–801, 2005.

Hormes, A. *et al.*, A geochronological approach to understanding the role of solar activity on Holocene glacier length variability in the Swiss Alps, *Geogr. Ann. Ser.-A Phys. Geogr.*, **88A**, 281–294, 2006.

Krivova, N. A. *et al.*, Reconstruction of solar total irradiance since 1700 from the surface magnetic flux, *Astron. Astrophys.*, **467**, 335–346, 2007.

Kuhn, J. R. *et al.*, The surface-temperature of the Sun and changes in the solar-constant, *Science*, **242**, 908–911, 1988.

Langley, S. P., The "solar constant" and related problems, *Astrophys. J.*, **17**, 89–99, 1903.

Masarik, J. and J. Beer, Simulation of particle fluxes and cosmogenic nuclide production in the Earth's atmosphere, *J. Geophys. Res.*, **104**, 12,099–12,111, 1999.

Masarik, J. and J. Beer, Simulation of particle fluxes and cosmogenic niclide production in the Earth's atmopshere revisited, *J. Geophys. Res.*, 2009 (in press).

McCracken, K. G., Heliomagnetic field near Earth, 1428–2005, *J. Geophys. Res.—Space Phys.*, **112**, A09106-09101–09109, 2007.

McCracken, K. G. and J. Beer, Long-term changes in the cosmic ray intensity at Earth, *J. Geophys. Res.—Space Phys.*, **112**, 1428–2005, 2007.

Muscheler, R. *et al.*, Geomagnetic field intensity during the last 60,000 years based on ^{10}Be & ^{36}Cl from the Summit ice cores and ^{14}C, *Quat. Sci. Rev.*, **24**, 1849–1860, 2005.

Oeschger, H. *et al.*, A box diffusion model to study the carbon dioxide exchange in nature, *Tellus*, **27**, 168–192, 1975.

Parker, E. N., The passage of energetic charged particles through interplanetary space, *Planet. Space Sci.*, **13**, 9–49, 1965.

Reimer, P. J. *et al.*, IntCal04 terrestrial radiocarbon age calibration, 0–26 cal kyr BP, *Radiocarbon*, **46**, 1029–1058, 2004.

Sagan, C. and C. Chyba, The early faint Sun paradox: organic shielding of ultraviolet-labile greenhouse gases, *Science*, **276**, 1217–1221, 1997.

Sofia, S. and L. H. Li, Solar variability caused by structural changes of the convection zone, *Sol. Variability Eff. Clim.*, **141**, 15–31, 2004.

Sofia, S. and L. H. Li, Mechanisms for global solar variability, *Sol. Variability Earth. Clim.*, **76**, 768–772, 2006.

Solomon, S. *et al.*, *IPCC, 2007: Summary for Policymaker*, Cambridge, 2007.

Steiner, O. and A. Ferriz-Mas, The deep roots of solar radiance variability, *Sol. Variability Earth. Clim.*, **76**, 789–793, 2006.

Steinhilber, F. *et al.*, Solar modulation during the Holocene, *Astrophys. Space Sci. Trans.*, **4**, 1–6, 2008.

van Geel, B. and H. Renssen, Abrupt climate change around 2,650 BP in North-West Europe: evidence for climatic teleconnections and a tentative explanation, in *Water, Environment and Society in Tiomes of Climatic Change*, edited by A. S. Issar and N. Brown, pp. 21–41, Kluwer Academic Publishers, The Netherlands, 1998.

Vonmoos, M. *et al.*, Large variations in Holocene solar activity: Constraints from Be-10 in the Greenland Ice Core Project ice core, *J. Geophys. Res.—Space Phys.*, **111**, 2006.

Wang, Y. J. *et al.*, The Holocene Asian monsoon: Links to solar changes and North Atlantic climate, *Science*, **308**, 854–857, 2005.

Wang, Y. M. *et al.*, The long-term variation of the Sun's open magnetic flux, *Geophys. Res. Lett.*, **27**, 505–508, 2000.

Webber, W. R. and P. R. Higbie, Production of cosmogenic Be nuclei in the Earth's atmosphere by cosmic rays: Its dependence on solar modulation and the interstellar cosmic ray spectrum, *J. Geophys. Res.*, **108**, 1355–1365, 2003.

Wenzler, T. *et al.*, Reconstruction of solar irradiance variations in cycles 21–23 based on surface magnetic fields, *Astron. Astrophys.*, **460**, 583–595, 2006.

Climate and Weather of the Sun-Earth System (CAWSES): Selected Papers from the 2007 Kyoto Symposium,
Edited by T. Tsuda, R. Fujii, K. Shibata, and M. A. Geller, pp. 217–230.

Total solar irradiance variability: What have we learned about its variability from the record of the last three solar cycles?

Claus Fröhlich

Physikalisch-Meteorologisches Observatorium Davos, World Radiation Center, CH–7260 Davos Dorf, Switzerland
E-mail: cfrohlich@pmodwrc.ch

Since November 1978 a set of total solar irradiance (TSI) measurements from space is available, yielding a time series of almost 30 years. Presently, there are three TSI composites available, called PMOD, ACRIM and IRMB, which are all constructed from the same original data, but use different procedures to correct for sensitivity changes. The PMOD composite is the only one which also corrects the early HF data for degradation. The origin of the differences between the three composite are discussed by comparison with the record of ERBE. For the discussion of the similarities and differences of the three cycles the PMOD composite is used. The most interesting feature is the low value of TSI at the present minimum which cannot be seen in proxies such as $F_{10.7}$, CaK and MgII indices. Thus, we see for the first time an effect on TSI which is not due to the direct effect of the superficial magnetic structures of solar activity.

1 Introduction

Since late 1978 total solar irradiance (TSI) measurements were made by different radiometers in space, HF on NIMBUS 7, ACRIM I on SMM, ERBE on ERBS, ACRIM II on UARS, SOVA on EURECA, VIRGO on SOHO, ACRIM III on ACRIM-Sat and since 2003 TIM on SORCE. Figure 1 shows these original time series and it is clear that not only the absolute values are quite different, especially at the beginning of the series, but there are also important differences between the series. This is e.g. obvious from comparison of early ACRIM-I with HF and is due to the fact that the original data from the HF radiometer cannot be corrected for degradation by internal means. These time series can be used for the construction of a TSI composite by shifting each series to a common level and merging them together. Presently there are three composites available, the first one was presented in 1997 at the IAU General Assembly in Kyoto by Fröhlich and Lean (1998b) and is now called PMOD composite (for the newest version see Fröhlich, 2006). A few months later the ACRIM composite was published by Willson (1997) which has been updated by Willson and Mordvinov (2003). Somewhat later a third composite, called IRMB, was presented

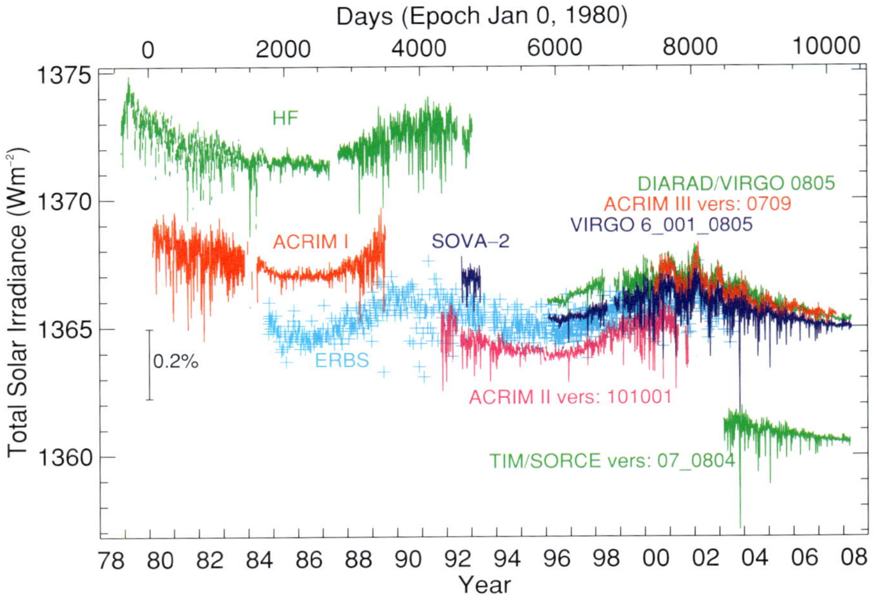

Fig. 1. Compared are daily averaged values of the Sun's total irradiance from radiometers on different space platforms as published by the instrument teams since November 1978. Note, that the VIRGO TSI is determined from both VIRGO radiometers (PMO6V and DIARAD), whereas the DIARAD TSI is only based on this one alone.

by Dewitte *et al.* (2004).[1]

2 Construction of TSI Composites

The ACRIM and IRMB composite use the data as published, whereas the PMOD composite introduces corrections of effects not considered in the original data sets. Already in the first versions of the PMOD composite corrections for the HF degradation were introduced (Fröhlich and Lean, 1998a, b). Similarly, the degradation of ACRIM-I during its first year was corrected for the effect of the rather short exposure time during the spin mode which was not taken into account in the original treatment by Willson and Hudson (1991). The two time series with the corrections mentioned were the basis for the composite of Fröhlich and Lean (1998a) during the period before the end of ACRIM-I in 1989 and remained unchanged up to version d40_61_0502. Since then the HF corrections have been re-determined (Fröhlich, 2006): after removal of all glitches throughout the mission a new HF time series was produced which is now corrected for the early increase, the degradation and the

[1] The composites are available from: PMOD at http://www.pmodwrc.ch/pmod.php?topic=tsi/composite/SolarConstant, ACRIM at http://www.acrim.com/Data%20Products.htm and IRMB at http://remotesensing.oma.be/solarconstant/sarr/SARR.txt.

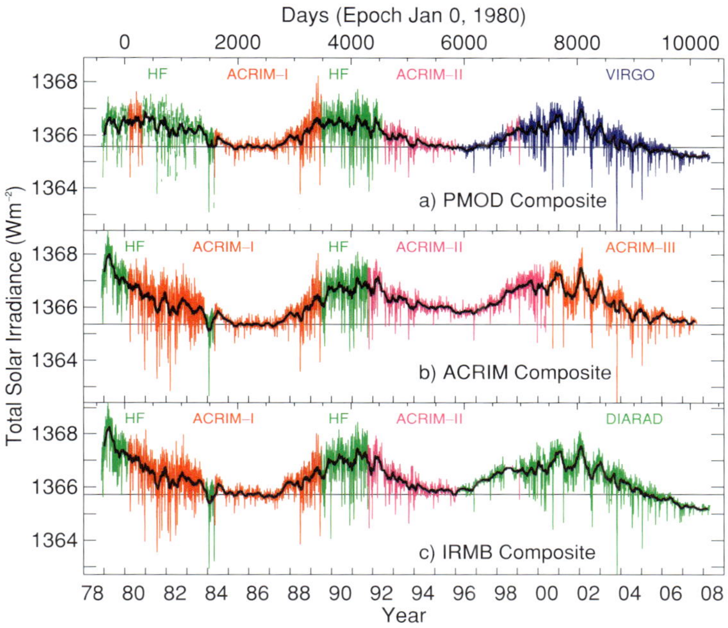

Fig. 2. Shown are the three composites: (a) PMOD (Fröhlich and Lean,1998b; Fröhlich, 2006), (b) ACRIM (Willson, 1997; Willson and Mordvinov, 2003) and (c) IRMB (Dewitte *et al.*, 2004).

long-term sensitivity increase over the full period of the NIMBUS-7 mission from November 1978 until January 1993.

Another problem for all composite constructions is how to bridge the so-called ACRIM gap between the end of ACRIM-I and the start of ACRIM-II, from June 1989 to October 1991. During this period daily values from HF and some 70 data points from ERBE with a sampling every 14 days are available. Already Lee III *et al.* (1995) revealed two slips in the HF data resulting in a total change of -0.68 W m^{-2} over this period which was confirmed by comparison with a model from the San Fernando group (Chapman *et al.*, 1996). Fröhlich (2000) re-analized this period and the overall change was also confirmed, but with a slightly different value of -0.58 W m^{-2}. With the corrected HF time series from Fröhlich (2006), there is no longer a need to treat the ACRIM gap separately, but the construction of the composite has become straightforward and internally consistent. The ACRIM composite neglects the corrections of the HF during the gap and this is the main reason for the claimed upward trend of TSI over the last three cycles (Willson and Mordvinov, 2003). The IRMB composite traces ACRIM-II to I via ERBE which results in an insignificant difference between the two minima. As IRMB uses ERBE data to trace ACRIM-I to ACRIM-II it yields a similar result for the difference between the minima in 1986 and 1996 as the PMOD composite. However, the IRMB times series during the ACRIM gap shows a strange

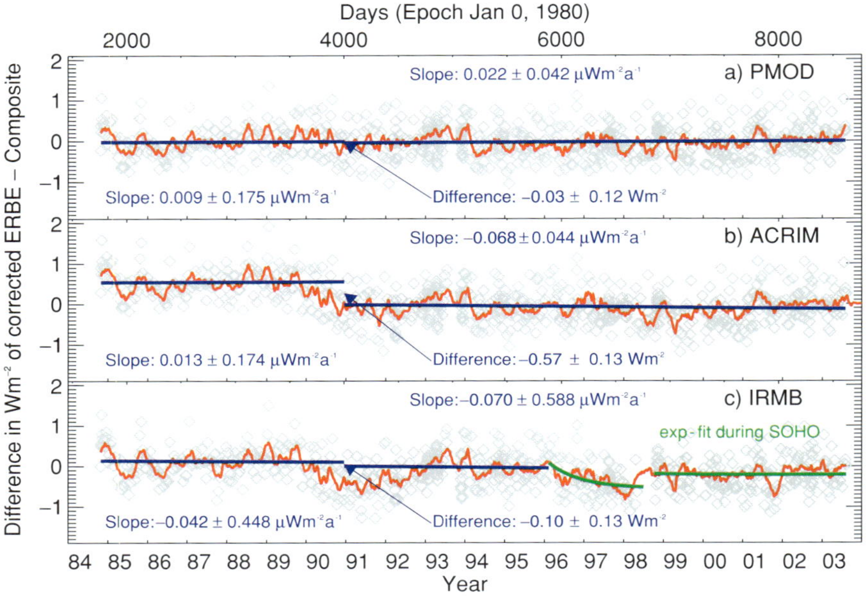

Fig. 3. Comparison of the three composites with the ERBE time series. ERBE is corrected for the early increase with the model parameters determined for ACRIM-I.

variation which comes from the fact that IRMB determines the composite as daily averages of the SARR[2] referred values of all available radiometers at that day.

After 1996 the PMOD uses version 6_001_0805 of the VIRGO data, which is updated to end of April 2008. How to derive the VIRGO data set is described in Fröhlich (2003) and in the online document[3]. The ACRIM composite uses for the same period a combination of ACRIM-II and III, whereas the IRMB composite uses the DIARAD/VIRGO data. In contrast to the official VIRGO TSI, the analysis of DIARAD/VIRGO is based on the measurements of DIARAD alone and the result is quite different. This is mainly due to the fact that their analysis implicitly neglects the non-exposure dependent increase of sensitivity which can only be determined by comparison with an independent radiometer.

The three composites are shown in Fig. 2. In terms of absolute scale the PMOD composite is referred to SARR, the ACRIM composite is on the scale of ACRIM-III and the IRMB refers first all data to SARR and then combines them without changing their level and thus the absolute scale. In order to get an independent view of the

[2]Space Absolute Radiometric Reference as defined by Crommelynck *et al.* (1995).

[3]http://www.pmodwrc.ch/pmod.php?topic=tsi/virgo gives some details about the corrections and presents links to the hourly and daily TSI data from VIRGO. Together with the official VIRGO TSI also the level 1.8 and 2 data for PMO6V and DiARAD are included, as well as the DIARAD/VIRGO data as determined by IRMB independently.

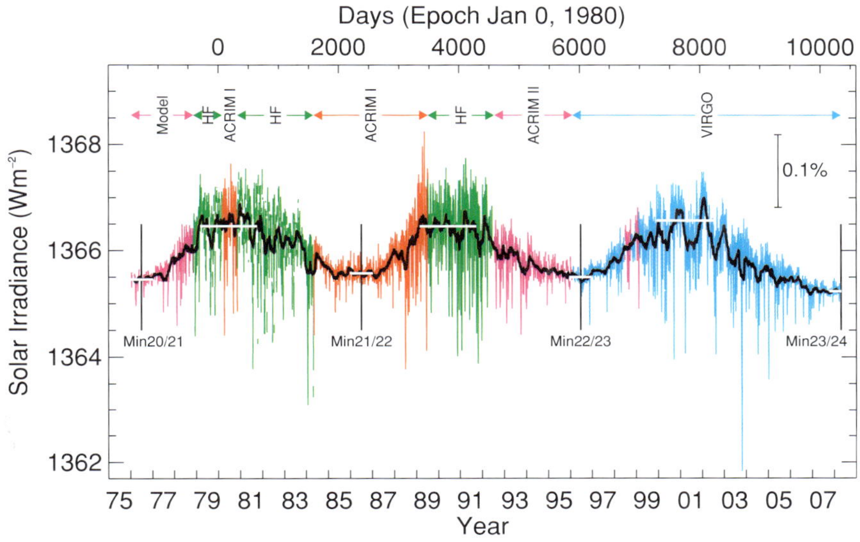

Fig. 4. Shown is the PMOD composite extended back to 1975 with the result of a 3-component proxy model as described in Section 4.

three composite we compare them with the ERBE data set, shown Fig. 3. As the ERBE radiometer is an ACRIM type and as the total exposure time is only 4 days we have applied the early increase correction determined for ACRIM-I (as described in Fröhlich, 2006). The most important result is that there is essentially no trend between the corrected ERBE time series and the PMOD composite, nor is there a significant step over the ACRIM gap. In the case of the IRMB composite the comparison with ERBE shows a similar exponential increase of sensitivity which has been identified during the evaluation of the VIRGO radiometry (Fröhlich, 2003) and is used in the evaluation of VIRGO TSI.

The above discussion of the three composites shows that the PMOD composite may possibly be the closest representative of the solar variability over the last 30 years. In order to have a more complete view of the behaviour of the Sun during the last three cycles the PMOD composite has been extended back to the minimum of 1975 with the result of a calibrated 3-component model as will be described in Section 4. This extended time series is shown in Fig. 4 and will be used in the following discussion.

3 Behaviour of TSI during the Last Three Solar Cycles

The changes of the maxima and minima from cycle to cycle are listed in Table 1. The other differences and similarities of the three cycles are more related to rotational variations. Cycle 21 and and 22 are similar, but quite different from cycle 23. The latter has less sunspots, as also shown by the lower maximum sunspot number of

Table 1. Listed are the values for the minima and maxima of the solar cycles 21–23 and the trend of the minima as change over the cycle relative to its amplitude. The amplitude is calculated from the difference between the maximum value and the mean of the two adjacent minima. The standard deviation are formal uncertainties of the corresponding average of the irradiance. For the maximum the average is over about 1000 days, for the minima over 350 days with the exception of the last minimum which covers yet only about 200 days.

		Cycle 21		Cycle 22		Cycle 23		
	Min	Max	Min	Max	Min	Max	Min	
Maxima		1366.49		1366.46		1366.39		$W\,m^{-2}$
Amplitude		0.980		0.921		1.029		$W\,m^{-2}$
StdDev		0.019		0.020		0.017		$W\,m^{-2}$
Minima	1365.46		1365.57		1365.50		1365.23	$W\,m^{-2}$
Dev. from Mean	0.015		0.131		0.060		−0.206	$W\,m^{-2}$
StdDev	0.006		0.006		0.008		0.006	$W\,m^{-2}$
Trend		11.8		7.7		−25.8		%

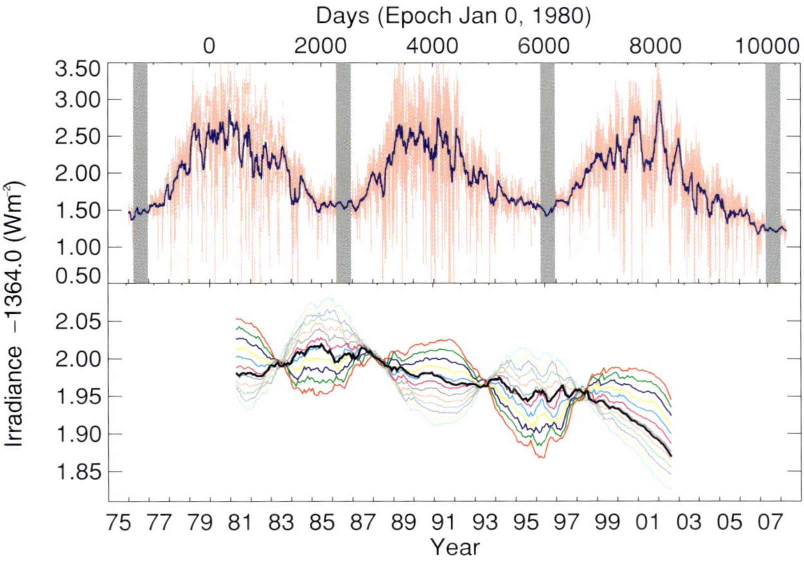

Fig. 5. The top panel shows the extended PMOD composite and the lower panel the cycle averaged long-term changes as in Lockwood and Fröhlich (2007). The red to light cyan lines show running means over intervals of 9 to 13 years. The black line gives the average length between the nodes (for details, see Lockwood and Fröhlich, 2007). Note the steeper decrease during the last cycle indicates that the general trend due to the diminishing amplitudes of the cycles is accentuated by the downward trend of TSI between the last minima.

about 115, compared to about 150 for the earlier cycles. On the other hand, this cycle has as many active regions as the others which is manifested by more peaks from the facular brightening which are not compensated by sunspot darkening. This is also the reason for the highest amplitude of this cycle relative to the others. Another point related to the sunspot numbers is the long-term change observed in cycle averaged TSI as shown in Lockwood and Fröhlich (2007, 2008) and updated in Fig. 5.

An important question is whether the recent downward trend is real and not an artefact of the analysis of the VIRGO data. For this we compare the PMOD composite with the ACRIM-II and III data over the whole cycle 23 and with TIM on SORCE since its launch in January 2003. Figure 6 shows that the comparison with ACRIM is quite noisy and shows some oscillatory behaviour with e.g. a 1-year periodicity, but with an overall trend of only 6.5 ppm/dec. The comparison with TIM is less noisy and shows a trend over the period 2003–2008 of 100 ppm/dec. It is interesting to note that the trend shown in this figure is from TIM version 8, whereas the corresponding trend of version 7 was only 6 ppm/dec. This may be partly due to the fact that an exponential function is fitted to the ratio of the operational TIM radiometer to the back-up ones, which is now no longer reproducing the steep decrease at the beginning of the mission. Moreover, the preliminary results from the re-analysis of

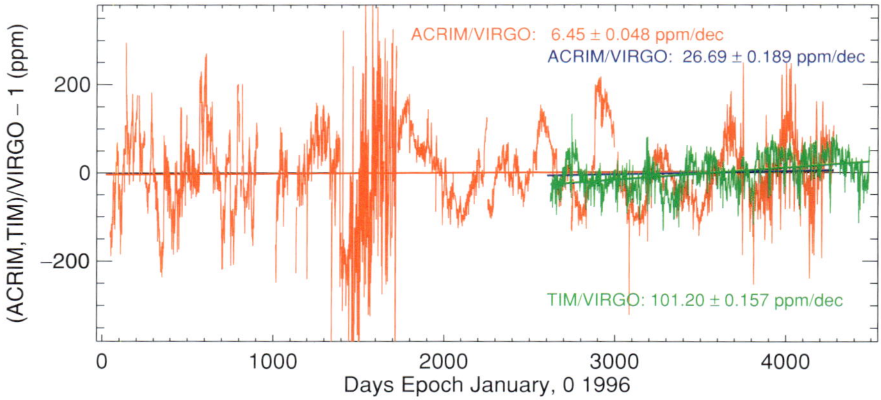

Fig. 6. Shown is the comparison of VIRGO TSI with ACRIM-II/III and TIM. TIM seems to indicate that there is a problem with the downward trend. For details see text, the conclusion is that this comparison may indicate that the trend as seen by VIRGO may be somewhat too large, maybe not 26% of the cycle amplitude, but still certainly more than 18%.

the VIRGO radiometry indicate that the cycle amplitude may be somewhat underestimated which would also reduce the slope of the ratio TIM/VIRGO. The trend of the ratio to ACRIM over the same period as for TIM is also increased, but amounts only to about one quarter of the one seen by TIM. Taking these results as an indication of the uncertainty we may attribute to the observed trend an uncertainty of about 50 ppm/dec. Thus the downward trend during cycle 23 maybe somewhat overestimated by the VIRGO data, but it is still around 18%. So, we are confident that the downward trend is real and we have to find a physical explanation, which must be different from changes of the upper atmosphere due to surface magnetism.

4 Proxy Models and Reconstructions

Proxy models are based on the photometric sunspot index (PSI) and some proxy for the influence of faculae and network. PSI was first described by Hudson *et al.* (1982) and is based on time-dependent information about sunspot areas and locations determined from ground-based white light images made from 1882 to 1976 by the Greenwich Observatory, and most recently by the U.S. Air Force operational Solar Observation Optical Network (SOON) sites.[4] Here we use an improved version of the same basic algorithm Fröhlich *et al.* (1994) together with a size dependent contrast from Brandt *et al.* (1994). The explicit calculation of a facular index PFI analogous to PSI is also possible (Lean *et al.*, 1998) but more difficult because faculae have lower contrasts in the visible spectrum and more fragmented areas than sunspots. This motivates the use of alternative facular indices such as the Mg-II index, which consists

[4]The observations of the sunspot regions are available from the National Geophysical Data Center (NGDC) operated by the National Oceanographic and Atmospheric Administration (NOAA) at Boulder, Colorado, http://www.ngdc.noaa.gov/stp/SOLAR/ftpsunspotregions.html.

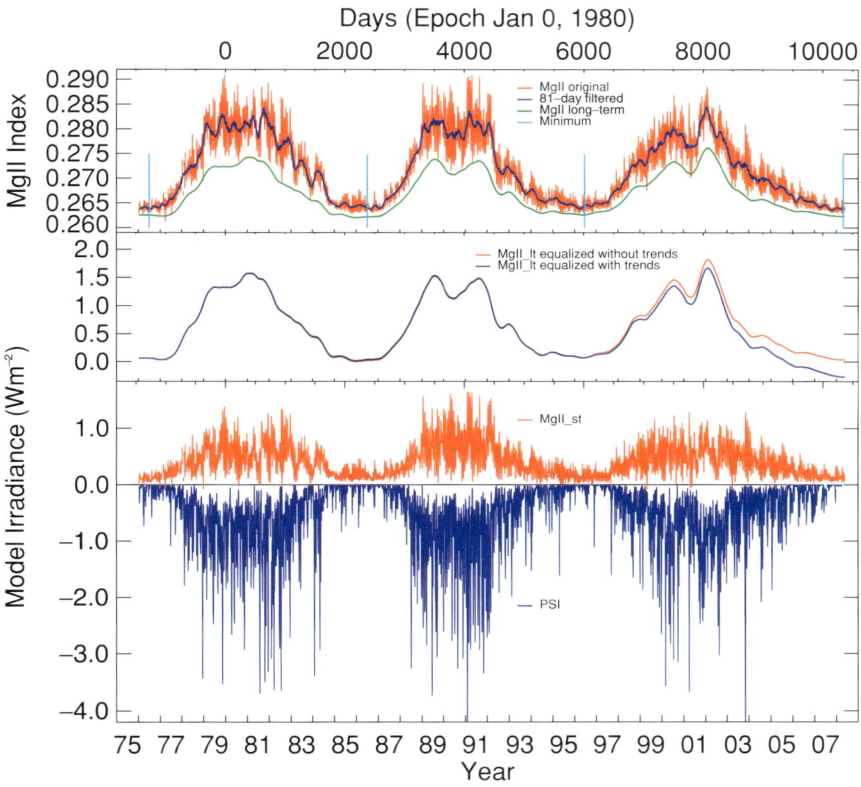

Fig. 7. The top panel shows the composite MgII index (red) and the long-term part determined as a lower envelope of short-term variation mainly due to the passage of faculae. The lower two panels show the calibrated indices: long-term MgII, short-term MgII together with PSI. The vertical lines on the top panel indicate the time of the solar cycle minima.

of the ratio of the core and wing fluxes and minimizes the influence of instrumental drifts. The line cores originate in the chromosphere where bright features have greater contrast than their photospheric counterparts (a factor of 2 versus a few percent for an active region), and the flux measurements automatically integrate emission over the full disk. One disadvantage is that chromospheric and photospheric radiances have different center-to-limb dependence of emission and contrast. By following an active region passage without a sunspot we observe in TSI two maxima at about 4–5 days before and after the central-meridian passage with a minimum at the central meridian, and in contrast the MgII index shows one peak at the central meridian. A method to take this difference into account has been presented earlier by Fröhlich (2004), but here we make no correction for this effect because the improvement is only marginal for the present discussions. The composite MgII index from Viereck *et al.* (2004)

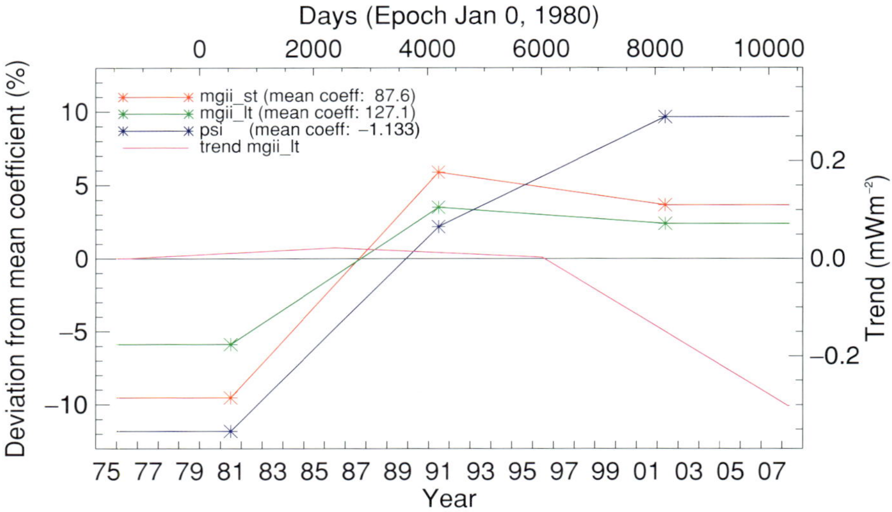

Fig. 8. Shown are the changes of the coefficients determined for each cycle relative to there mean value. These changes are then used to 'correct' the long and short-term MgII and PSI for further analysis. Also shown are the trends needed to be introduced to account for the changes in TSI not represented by the MgII index, nor PSI.

is used which has been recently corrected during cycle 23 and updated until 2008 by Viereck *et al.* (2008). This composite has been extended back to spring 1976 with HeI equivalent width data, calibrated against MgII index over the rest of cycle 21. The MgII index is shown in the top panel of Fig. 7. From comparison of the MgII and TSI time series it is clear that the recent downward trend cannot be explained by a model based on such proxies. This is not only true for MgII index, also CaK index, the radio flux at 10.7 cm and similarly the sunspot number do not show any substantial changes of the values at solar minima. For $F_{10.7}$ this is true back to the beginning of the measurements in the late forties and for the CaK index back to 1915. This means that the surface manifestations of activity such as sunspots, faculae and network, which change the temperature structure of the upper solar atmosphere, explain only the cycle variations of TSI, but not the trends from minimum to minimum. These trends may still be related to activity, by e.g. temperature changes of the photosphere. TSI is with 0.07%/K quite sensitive to such changes whereas the for UV radiation at e.g. 200 nm a change of 0.22%/K is negligible compared to the effect of chromospheric changes during a cycle.

Moreover, there may be a difference of the effect influencing the variability over a full solar cycle and over only solar rotation which is directly linked to the passage of active regions. So, the MgII index is separated into a long-term part which is determined as the lower envelope of the index as shown in Fig. 7. The short-term is then obtained by subtracting the so determined long-term from the observed MgII index.

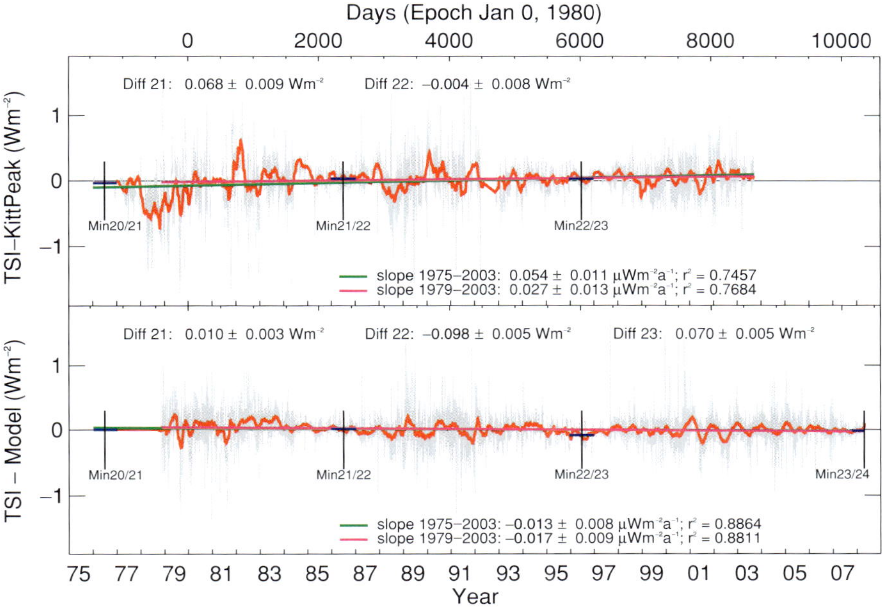

Fig. 9. The top panel shows the comparison between TSI and the reconstruction by Wenzler *et al.* (2006) and the bottom panel the one between TSI and corrected 3-component proxy model with trends for each cycle added.

The long-term on the other hand is varying slowly and may be due to changes of the network inside and outside active regions in the course of a cycle. Multiple regression of the three components against TSI is used to 'calibrate' the model. However, as the MgII index has essentially no change of the minima, we need to add a fourth component to account for trends between the minima of each cycle. In a first step each cycle is fitted separately in order to account for possible changes of e.g. the contrast of sunspots and faculae. The results are shown in Fig. 8. As in earlier analysis (e.g. Fröhlich and Lean, 2004) we find different coefficients for the long- and short-term MgII index which may be explained by a different specific contrast of faculae and network (Ortiz, 2005) and thus may indicate that the solar-cycle modulation is indeed due to changes of the network and not due to the faculae and sunspots alone. The differences between the three cycles show a steady increase of all parameters relative to their mean—especially between cycle 21 and 22 and somewhat less between 22 and 23. Albeit the coefficients change the ratios of the short-term to the long-term coefficients stay about constant at 1.46 ± 0.05. Also the contrasts for sunspots and faculae seem to change from cycle to cycle, but again not so much their ratio. Both observations are by itself very interesting and may help to understand better the influence of surface magnetism. Now we adjust the short and long-term MgII and PSI to the mean coefficients from Fig. 8, add the trends and redo the calibration against TSI. The result of this adjusted model is shown as comparsion with TSI in the bottom panel of Fig. 9. The correlation is extremely high and it explains 86% of the variance. This indicates that not only the effect of faculae, network and sunspots are quite different during the three cycles, but also that different trends for each cycle are necessary.

Another way to reconstruct TSI is based on the identification of sunspot (umbra, penumbra and pores) pixels from white light images and of quiet, network and facular pixels from the magnetograms (see e.g. Wenzler *et al.*, 2005). Together with a contrast (see e.g. Krivova *et al.*, 2003) the radiance of each pixel is determined and summed up to yield the irradiance. The pixels from the sunspot and pores are subtracted from the magnetograms and the remaining pixels represent faculae and network for which a filling factor is defined which depends linearly on the magnetic field in the pixel up to a saturation field after which it is set to one. This saturation field is the only free parameter in this type of reconstruction which is determined by calibration against the PMOD composite as 340 G for the Kitt-Peak magnetograms (Wenzler *et al.*, 2005, 2006). This reconstruction covering the period of the composites from the beginning of the measurements of TSI up to the maximum of solar cycle 23 can also be used to compare the three composites. As with ERBE the best agreement is found for the PMOD composite explaining 77% of the variance (see Fig. 9, top panel). This is a rather high correlation for the simple assumptions with only one free parameter on which the reconstruction is based. On the other hand there also are important deviations which explain the lower correlation than the one achieved by the corrected 3-component model with trends. They come mainly from the fact that the saturation field is determined from the data of cycle 23 and obviously the level of agreement is much lower for the ascending parts of cycle 21 and 22 than for the ascending part of cycle 23. Part of this discrepancy may be removed if each cycle would be calibrated separately as in the proxy model. These reconstructions fail also to explain minimum-to-minimum trends, similar to the other chromospheric proxies. The reason why this reconstruction fits the TSI composite quite well is due to the fact that there was only a small trend over cycle 22 and that the comparison is limited to the period from maximum of cycle 21 to the maximum of 23.

5 Conclusions

The most important result is that TSI shows trends from minimum to minimum which cannot be reproduced by proxies for the brightening due to magnetic surface structures, such as the 10.7 cm radio flux, CaK and MgII indices or the ones deduced from magnetograms. They vary from minimum to minimum at most by a few percent whereas TSI changed during cycle 23 by 18% or more. As the reconstructions and proxy models represent very well the irradiance changes during a cycle of TSI and the UV radiation they cannot explain the long-term trends. This in turn means also that the UV radiation has probably no long-term trend even back to the Maunder Minimum, only changes of the cycle amplitude which can most probably be well represented by the amplitude of the sunspot number. On the other hand TSI shows a long-term trend which could be due to a change of the temperature of the photosphere, which changes TSI but not the UV radiation. This temperature change may also be related to the magnetic field emerging during a cycle and thus may well be related to the long-term changes of the amplitude of the sunspot number also. So, a long-term trend for TSI on top of the amplitude changes of the individual cycles is highly probable. But the same long-term trend cannot be applied to the UV irradiance, which shows only an important variation related to the cycle amplitude.

Acknowledgments. Continued support of solar irradiance research at PMOD/WRC by the Swiss National Science Foundation is greatly acknowledged. Thanks are extended to the VIRGO team for supporting the experiment on SOHO, a cooperative ESA/NASA mission, and to the ACRIM and SORCE teams for still unpublished data.

References

Brandt, P. N., M. Stix, and H. Weinhardt, Modelling solar irradiance variations with an area dependent photometric sunspot index, *Sol. Phys.*, **152**, 119–124, 1994.

Chapman, G. A., A. M. Cookson, and J. J. Dobias, Variations in total solar irradiance during solar cycle 22, *J. Geophys. Res.*, **101**, 13541–13548, 1996.

Crommelynck, D., A. Fichot, R. B. Lee III, and J. Romero, First realisation of the space absolute radiometric reference (SARR) during the ATLAS 2 flight period, *Adv. Space Res.*, **16**(8), 17–23, 1995.

Dewitte, S., D. Crommelynck, S. Mekaoui, and A. Joukoff, Measurement and uncertainty of the long-term total solar irradiance trend, *Sol. Phys.*, **224**, 209–216, doi:10.1007/s11207-005-5698-7, 2004.

Fröhlich, C., Observations of irradiance variability, *Space Sci. Res.*, **94**, 15–24, 2000.

Fröhlich, C., Long-term behaviour of space radiometers, *Metrologia*, **40**, 60–65, 2003.

Fröhlich, C., Solar irradiance variability, in *Geophysical Monograph 141: Solar Variability and its Effect on Climate*, Chapter 2: Solar Energy Flux Variations, 97–110, AGU, Washington DC, USA, 2004.

Fröhlich, C., Solar irradiance variability since 1978: Revision of the PMOD composite during solar cycle 21, *Space Sci. Res.*, **125**, 53–65, doi:10.1007/s11214-006-9046-5, 2006.

Fröhlich, C. and J. Lean, The sun's total irradiance: Cycles and trends in the past two decades and associated climate change uncertainties, *Geophys. Res. Lett.*, **25**, 4377–4380, 1998a.

Fröhlich, C. and J. Lean, Total solar irradiance variations: The construction of a composite and its comparison with models, in *IAU Symposium 185: New Eyes to See Inside the Sun and Stars*, edited by F. L. Deubner, J. Christensen-Dalsgaard, and D. Kurtz, 89–102, Kluwer Academic Pub., Dordrecht, The Netherlands, 1998b.

Fröhlich, C. and J. Lean, Solar radiative output and its variability: Evidence and mechanisms, *Astron. Astrophys. Rev.*, **12**, 273–320, doi:10.1007/s00159-004-0024-1, 2004.

Fröhlich, C., J. M. Pap, and H. S. Hudson, Improvement of the photometric sunspot index and changes of disk-integrated sunspot contrast with time, *Sol. Phys.*, **152**, 111–118, 1994.

Hudson, H. S., S. Silva, M. Woodard, and R. C. Willson, The effects of sunspots on solar irradiance, *Sol. Phys.*, **76**, 211–218, 1982.

Krivova, N. A., S. K. Solanki, M. Fligge, and Y. C. Unruh, Reconstruction of solar total and spectral irradiance variations in cycle 23: is solar surface magnetism the cause?, *Astron. Astrophys.*, **399**, L1–L4, 2003.

Lean, J. L., J. Cook, W. Marquette, and A. Johannesson, Magnetic sources of the solar irradiance cycle, *Astrophys. J.*, **492**, 390–401, 1998.

Lee III, R. B., M. A. Gibson, R. S. Wilson, and S. Thomas, Long-term total solar irradiance variability during sunspot cycle 22, *J. Geophys. Res.*, **100**, 1667–1675, 1995.

Lockwood, M. and C. Fröhlich, Recent oppositely directed trends in solar climate forcing and the global mean surface air temperature, *Proc. R. Soc. A*, **463**, 2447–2460, doi:10.1098/rspa.2007.1880, 2007.

Lockwood, M. and C. Fröhlich, Recent oppositely-directed trends in solar climate forcings and the global mean surface air temperature: II. different reconstructions of the total solar irradiance variation and dependence on response timescale, *Proc. R. Soc. A*, **464**, 1367–1385, doi:10.1098/rspa.2007.0347, 2008.

Ortiz, A., Solar cycle evolution of the contrast of small photospheric magnetic elements, *Adv. Space Res.*, **35**, 350–360, doi:10.1016/j.asr.2005.03.014, 2005.

Viereck, R. A., L. E. Floyd, P. C. Crane, T. N. Woods, B. G. Knapp, G. Rottman, M. Weber, L. C. Puga, and M. T. DeLand, A composite Mg II index spanning from 1978 to 2003, *Space Weather*, **2**, S10005, doi:10.1029/2004SW000084, 2004.

Viereck, R. A. *et al.*, Update of the composite Mg II index, 2, 2008 (personal communication, will be published).

Wenzler, T., S. K. Solanki, and N. A. Krivova, Can surface magnetic fields reproduce solar irradiance variations in cycles 22 and 23?, *Astron. Astrophys.*, **432**, 1057–1061, doi:10.1051/0004-6361:20041956, 2005.

Wenzler, T., S. K. Solanki, N. A. Krivova, and C. Fröhlich, Reconstruction of solar irradiance variations in cycles 21–23 based on surface magnetic fields, *Astron. Astrophys.*, **460**, 583–595, doi:10.1051/0004-6361:20065752, 2006.

Willson, R. C., Total solar irradiance trend during solar cycles 21 and 22, *Science*, **277**, 1963–1965, 1997.

Willson, R. C. and H. S. Hudson, The Sun's luminosity over a complete solar cycle, *Nature*, **351**, 42–44, 1991.

Willson, R. C. and A. V. Mordvinov, Secular total solar irradiance trend during solar cycles 21–23, *Geophys. Res. Lett.*, **30**, 1199, doi:10.1029/2002GL016038, 2003.

Climate and Weather of the Sun-Earth System (CAWSES): Selected Papers from the 2007 Kyoto Symposium,
Edited by T. Tsuda, R. Fujii, K. Shibata, and M. A. Geller, pp. 231–256.

Mechanisms for solar influence on the Earth's climate

Joanna D. Haigh

Imperial College London
E-mail: j.haigh@imperial.ac.uk

Solar radiation is the fundamental energy source for the atmosphere and the global average equilibrium temperature of the Earth is determined by a balance between the energy acquired by the solar radiation absorbed and the energy lost to space by the emission of heat radiation. The interaction of this radiation with the climate system is complex but it is clear that any change in incoming solar irradiance has the potential to influence climate. There is increasing evidence that changing solar activity, on a wide range of time scales, influences the Earth's climate although details of the mechanisms involved remain uncertain. This article provides a brief review of the observational evidence and an outline of the mechanisms whereby rather small changes in solar radiation may induce detectable signals in the lower atmosphere.[1]

1 Introduction

In the past, although much had been written on relationships between sunspot numbers and the weather, the topic of solar influences on climate has often been disregarded by meteorologists. This was due to a combination of factors of which the key was the lack of any robust measurements indicating that the solar radiation incident on Earth did indeed vary. There was also mistrust of the statistical validity of much of the supposed evidence and, importantly, no established scientific mechanisms whereby the apparent changes in the Sun might induce detectable signals near the Earth's surface. Another concern of the emergent meteorological profession in the nineteenth century was to distance itself from the popular "astrometeorology" movement (Anderson, 1999; see e.g. Fig. 1). This estrangement of studies of solar-climate links from mainstream scientific endeavour lasted until the late 20th century when the necessity of attributing causes to global warming meant increased international effort into distinguishing natural from human-induced factors. Nowadays, with improved measurements of both solar and climate parameters, evidence of a solar signal in the climate of the lower atmosphere has begun to emerge from the noise and computer models of the atmosphere are providing a route to understanding the sometimes complex and subtle processes involved.

[1] It is not possible to review here all potential mechanisms for solar-climate links; I restrict this article to those relating to solar irradiance and thus neglect any involving energetic particles.

Fig. 1. "Murphy the Dick-Tater, Alias the Weather Cock of the Walk" (1837). (Reproduced with permission from Guildhall Library, City of London.)

2 Solar Influences on the Earth's Lower Atmosphere

2.1 Measurements and reconstructions

Assessment of climate variability and climate change depends crucially on the existence and accuracy of records of meteorological parameters. Ideally records would consist of long time series of measurements made by well-calibrated instruments located with high density across the globe. In practice, of course, this ideal cannot be met. Measurements with global coverage have only been made since the start of the satellite era about thirty years ago. Instrumental records have been kept over the past few centuries at a few locations in Europe. For longer periods, and in remote regions, records have to be reconstructed from the indirect indicators of climate often referred to as proxy data.

Proxy indicators provide information about weather conditions at a particular location through records of a physical, biological or chemical response to these conditions. Some proxy datasets provide information dating back hundreds of thousands of years which make them particularly suitable for analysing long term climate variations and their correlation with solar activity. One well established technique for providing proxy climate data is dendrochronology, or the study of climate changes by comparing the successive annual growth rings of trees (living or dead). Much longer records of temperature have been derived from analysis of oxygen isotopes in ice cores obtained from Greenland and Antarctica and evidence of very long term temperature variations can also be obtained from ocean sediments.

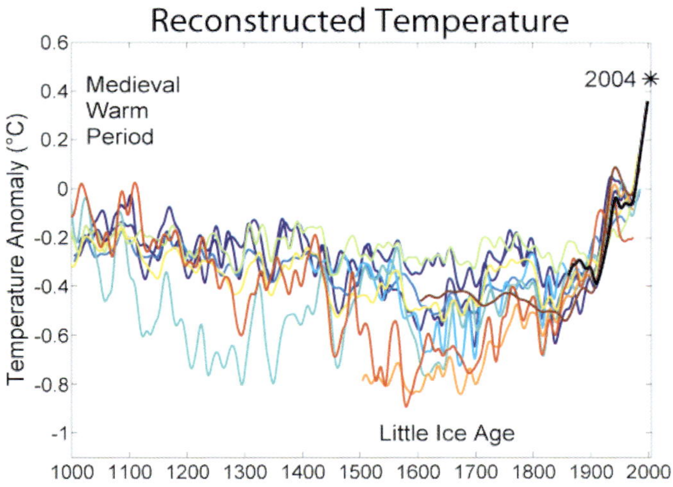

Fig. 2. Northern Hemisphere surface temperature (5-year running-mean) over the past millennium. The thick black curve (1856-present) is from measurements; the other curves are reconstructions by various authors based on proxy data (see http://www.globalwarmingart.com/wiki/Image:1000_Year_Temperature_Comparison_png for further details).

Figure 2 presents reconstructions of the Northern Hemisphere surface temperature record produced using a variety of proxy datasets. There are some large differences between them, especially in long-term variability, but there is general agreement that current temperatures are higher than they have been for at least the past 2 millennia. Other climate records suggesting that the climate has been changing over the past century include the retreat of mountain glaciers, sea level rise, thinner Arctic ice sheets and an increased frequency of extreme precipitation events. A key concern of contemporary climate science is to attribute cause(s) to these changes, including the contribution of solar variability.

2.2 Solar signals in climate records

Many different approaches have been adopted in the attempt to identify solar signals in climate records. Probably the simplest has been spectral analysis, in which cycles of 11 (or 22 or 90, etc.) years are assumed to be associated with the Sun. Alternatively records of observational data are correlated with time series of solar activity. This can be developed to extract the response in the measured parameter to a chosen solar activity forcing factor. A further sophistication allows a multiple regression, in which the responses to other factors are simultaneously extracted along with the solar influence. Each of these approaches gives more certainty than the previous one that the signal extracted is actually due to the Sun and not to some other factor, or to random fluctuations in the climate system, but it should be remembered that such detection is based only on statistics and not on any understanding of how the presumed solar influence takes place.

2.2.1 Millennial, and longer, timescales

On timescales of many millennia the amount of solar radiation received by the Earth is modulated by variations in its orbit around the Sun. The distance between the two bodies varies during the year due to the ellipticity of the orbit which varies with periods of around 100,000 and 413,000 years due to the gravitational influence of the Moon and other planets. At any particular point on the Earth the amount of radiation striking the top of the atmosphere also depends on the tilt of the Earth's axis to the plane of its orbit, which varies cyclically with a period of about 41,000 years, and on the precession of the Earth's axis which varies with periods of about 19,000 and 23,000 years (see Fig. 3). Averaged over the globe the solar energy flux at the Earth depends only on the ellipticity but seasonal and geographical variations of the irradiance depend on the tilt and precession. These are important because the intensity of radiation received at high latitudes in summer determines whether the winter growth of the ice cap will recede or whether the climate will be precipitated into an ice age. Thus changes in seasonal irradiance can lead to much longer-term shifts in climatic regime. Cyclical variations in climate records with periods of around 19, 23, 41, 100 and 413 kyr are generally referred to as Milankovitch cycles after the geophysicist who made the first detailed investigation of solar-climate links related to orbital variations.

Ocean sediments have been used to reveal a history of temperature in the North Atlantic by analysis of the minerals believed to have been deposited by drift ice (Bond

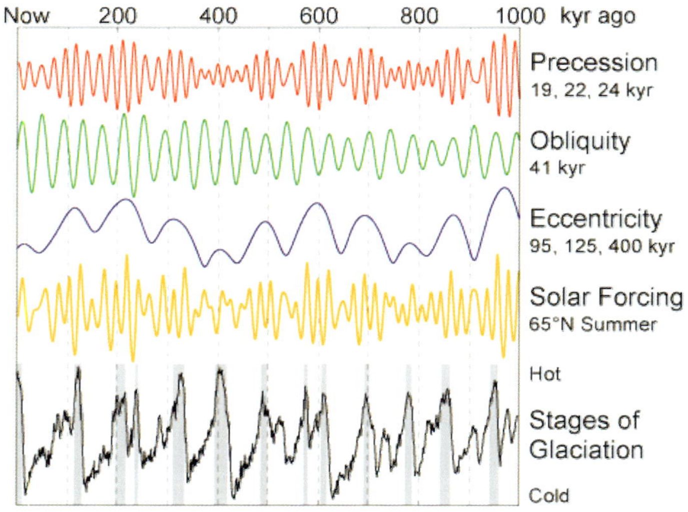

Fig. 3. Top 3 curves: variations in parameters determining the Earth's orbit. 4th curve: the resulting changes in solar energy flux at high latitude in summer. Lowest curve: global temperature deduced from ^{18}O records in deep ocean cores, grey bands indicate inter-glacial periods (see http://www.globalwarmingart.com/wiki/Image:Milankovitch_Variations_png for further details).

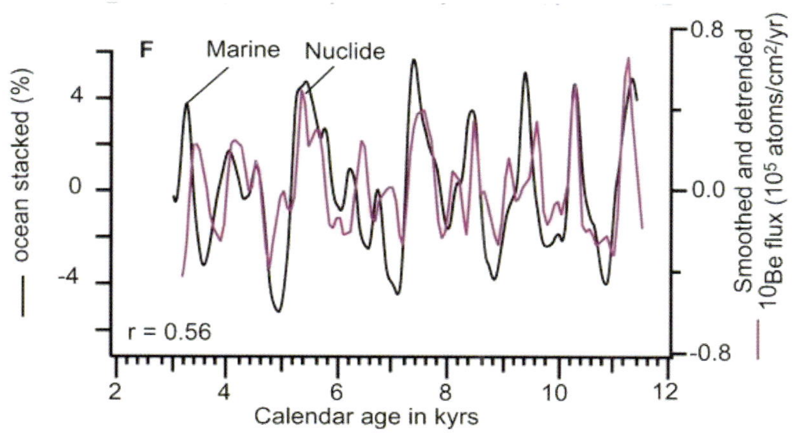

Fig. 4. Records of ^{10}Be (lighter curve) and ice-rafted minerals (darker curve) extracted from ocean sediments in the North Atlantic. (From Bond *et al.*, Persistent solar influence on north Atlantic climate during the Holocene, *Science*, **294**, 2130–2136, 2001. Reprinted with permission from AAAS.)

et al., 2001). In colder climates the rafted ice propagates further south where it melts, depositing the minerals. These materials also preserve information on cosmic ray flux, and thus solar activity, in isotopes such as ^{10}Be and ^{14}C. Thus simultaneous records of climate and solar activity may be retrieved. An example is given in Fig. 4 which shows fluctuations on the 1,000 year timescale well correlated between the two records, suggesting a long-term solar influence on North Atlantic temperatures.

2.2.2 Century scale

On somewhat shorter timescales it has frequently been remarked that the Maunder Minimum in sunspot numbers in the second half of the seventeenth century coincided with what is sometimes referred to as the "Little Ice Age" during which western Europe experienced significantly cooler temperatures. Figure 5 shows decadal trends in winter temperatures averaged over the period 1684–1738 as solar activity emerged from its Grand Minimum state. This pattern is symptomatic of the positive phase of the North Atlantic Oscillation indicating that changes in solar activity may influence climate on a regional scale (Luterbacher *et al.*, 2004). There is also some evidence (see Fig. 2) that the temperature averaged over the whole Northern Hemisphere was cooler during the 17th century, hence this period is sometimes referred to as "The Little Ice Age", but cooler temperatures are not found in all records for the same period across the globe so attribution of the temperature anomalies to solar variability remains uncertain. Crowley (2000) points out that the higher levels of volcanism prevalent during the 17th century, as well as lower solar irradiance, may contribute to the cooler northern hemisphere experienced at that time.

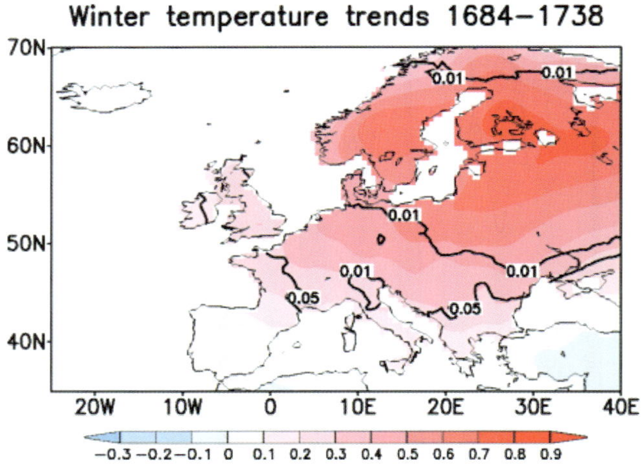

Fig. 5. Winter temperature trends (°C per decade) from 1684 to 1738. The thick solid lines represent the 95% and 99% confidence level. Except for the Mediterranean area, the warming trends are statistically significant over the whole of Europe. (From Luterbacher *et al.*, European seasonal and annual temperature variability, trends, and extremes since 1500, *Science*, **303**, 1499–1503, 2004. Reprinted with permission from AAAS.)

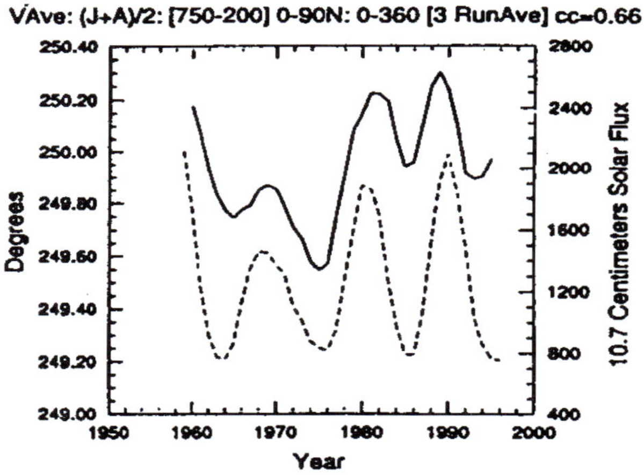

Fig. 6. Time series of the mean temperature of the summer northern hemisphere upper troposphere (750–200 hPa) (solid line) and the solar 10.7 cm flux (dashed line). (From van Loon and Shea, A probable signal of the 11-year solar cycle in the troposphere of the northern hemisphere, *Geophys. Res. Lett.*, **26**, 2893–2896, 1999. Copyright 1999 American Geophysical Union. Reproduced by permission of American Geophysical Union.)

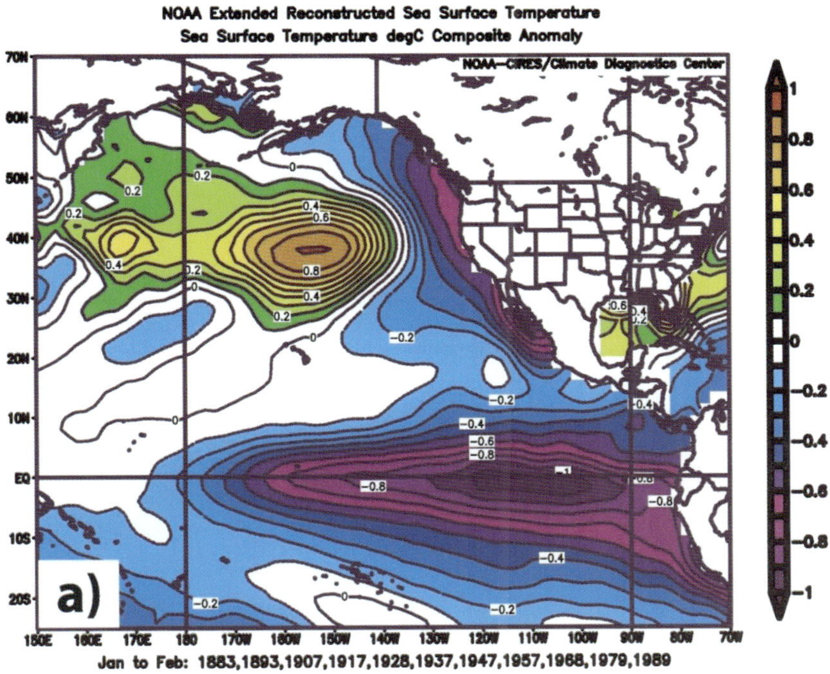

Fig. 7. Anomalies of sea surface temperature (°C) in January–February of solar cycle peak years relative to mean values 1871–1998. (From van Loon *et al.*, Coupled air-sea response to solar forcing in the Pacific region during northern winter, *J. Geophys. Res.—Atmos.*, **112**, D02108, 2007. Copyright 2007 American Geophysical Union. Reproduced by permission of American Geophysical Union.)

2.2.3 Solar cycle

Many studies have purported to show variations in meteorological parameters in phase with the "11-year" solar cycle. Some of these are statistically not robust and some show signals that appear over a certain interval of time only to disappear, or even reverse, over another interval. There is, however, considerable evidence that solar variability on decadal timescales does influence climate. An example is shown in Fig. 6 which presents the mean summer time temperature of the upper troposphere (between about 2.5 and 10 km) averaged over the whole northern hemisphere. This parameter varies in phase with the solar 10.7 cm index with an amplitude of 0.2–0.4 K.

The hemispheric average, however, hides the fact that the solar signal is not uniformly distributed. Solar signals detected in sea surface temperatures (SSTs) by White *et al.* (1997) show that SSTs do not respond uniformly: at higher solar activity

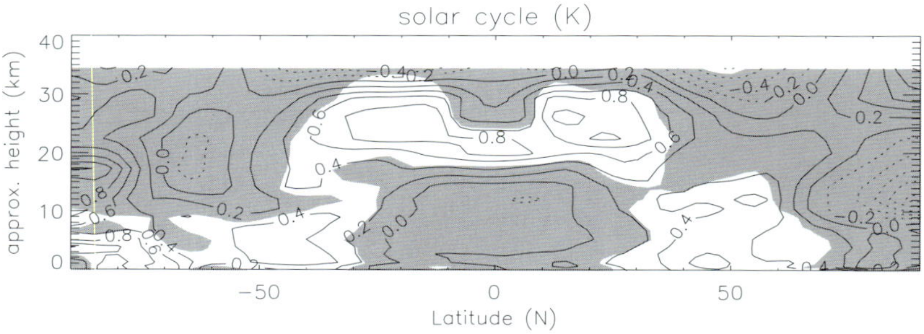

Fig. 8. Maximum difference between peaks and troughs of solar cycles in zonal mean temperatures 1979–2002, grey areas are not significant at 95% level. Log pressure (approximately linear altitude) scale for ordinate. From multiple regression analysis of Haigh (2003).

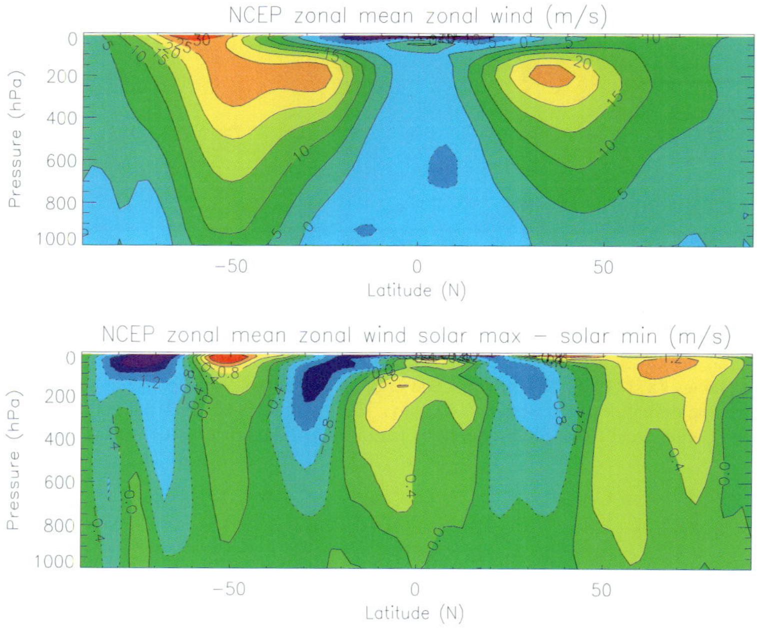

Fig. 9. Top: Zonal mean zonal wind (m s^{-1}) averaged over 1979–2002. Bottom: Maximum difference between peaks and troughs of solar cycles in this period. Linear pressure scale for ordinate. (From Haigh *et al.* (2005).)

the pattern shows latitudinal bands of warming and cooling with warmer temperatures in the tropics lagging the peak in solar forcing by 1–2 years. Van Loon *et al.* (2007) suggest a solar cycle signal in surface temperatures in the Pacific Ocean which bears a close relationship to the negative phase of the ENSO cycle (Fig. 7).

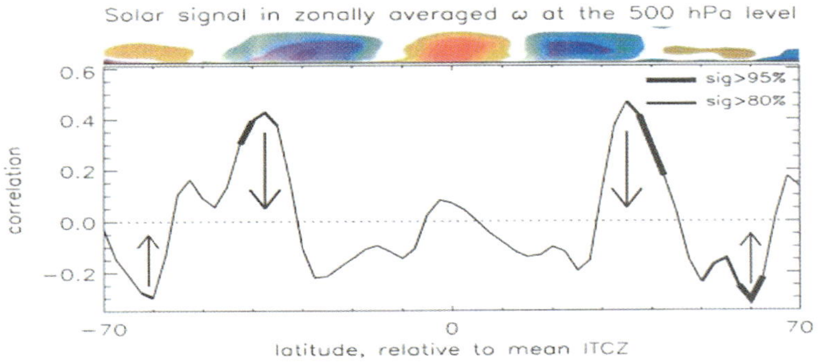

Fig. 10. Correlations between solar $F_{10.7\,\mathrm{cm}}$ index and zonally (relative to the mean ITCZ) averaged pressure velocity (ω) at the 500 hPa level. The climatology of ω is qualitatively outlined at the top of the panels, to indicate regions of upward (shades of yellow and red) and downward (blue and green) atmospheric motions (NB ω is positive downwards). The arrows show the effects on ω by increased solar activity. (From Gleisner and Thejll, Patterns of tropospheric response to solar variability, *Geophys. Res. Lett.*, **30**, 17129, 2003. Copyright 2003 American Geophysical Union. Reproduced by permission of American Geophysical Union.)

Latitudinal variations have also been found for the solar response in air temperatures. Haigh (2003) presented results of a multiple regression analysis of NCEP/NCAR reanalaysis (Kalnay *et al.*, 1996) zonal mean temperatures in which data for 1978–2002 were analysed simultaneously for ten signals: a linear trend, El Niño-Southern Oscillation (ENSO), North Atlantic Oscillation (NAO), solar activity, stratospheric aerosol from volcanic eruptions, Quasi-Biennial Oscillation (QBO) and the amplitude and phase of the annual and semi-annual cycles. The patterns of response for each signal are statistically significant and separable from the other patterns. The solar response (Fig. 8) shows largest warming in the stratosphere and bands of warming, of >0.4 K, throughout the troposphere in mid-latitudes. Associated with the temperature response is a signal in zonal mean zonal wind (Haigh *et al.*, 2005, see Fig. 9) which shows an 11-year solar cycle influence in which the effect of increasing solar activity is to weaken the westerly jets and to move them slightly polewards. These temperature and zonal wind patterns are consistent with a situation in which the tropical Hadley cells are broader at solar maximum, as suggested by an analysis of vertical velocity data by Gleisner and Thejll (2003, see Fig. 10). Kodera (2004) finds a suppression of tropical convection when the Sun is more active, and Kodera *et al.* (2007) relate this to a change in location of the downward portion of the Walker cell, accompanied by non-linear interactions with the phase of ENSO. However, the response of tropical circulations to solar activity is not at all well established: van Loon *et al.* (2004) find an intensification of the Hadley and Walker circulations at higher solar activity.

3 Variations in Solar Irradiance and Mechanisms for Influence on Climate

3.1 Total solar irradiance and radiative forcing of climate change

Direct measurements of total solar irradiance (TSI) made outside the Earth's atmosphere began with the launch of satellite instruments in 1978. Previous surface-based measurements could not provide the accuracy required to show solar cycle variations as they were subject to uncertainties and fluctuations in atmospheric absorption that swamped the small solar variability signal. The data from each of half a dozen satellite instruments, however, show consistent variations of approximately 0.08% ($\sim$1.1 W m^{-2}) in TSI over three 11-year cycles. Nevertheless, significant uncertainties remain with regard to the absolute value of TSI. In particular, data from the newest instrument, the Total Irradiance Monitor (TIM) on the SORCE satellite, is giving values approximately 5 W m^{-2} lower than other contemporaneous instruments which disagree among themselves by a few W m^{-2}. This uncertainty, related to the calibration of the instruments and their degradation over time, is a serious problem underlying current solar-climate research and makes it particularly difficult to establish the existence of any underlying trend in TSI over the past 2 cycles. Fröhlich (2006) reviews the problems encountered in compositing the available satellite data to provide a longer-term record; his assessment shows essentially no difference in TSI values between the cycle minima occurring in 1986 and 1996. The results of Willson and Mordinov (2003), however, show an increase in irradiance of 0.045% between these dates. Such a trend implies an increase in radiative forcing of about 0.1 W m^{-2} per decade which is about one-third that due to the increase in concentrations of greenhouse gases (decadal trend averaged over the past 50 years). Data for the current solar minimum, however, suggests that TSI is now lower than in 1996 so that there is clearly no upward secular trend in TSI.

To assess the potential influence of the Sun on the climate on centennial timescales it is necessary to know TSI further back into the past than is available from satellite data so proxy indicators of solar variability have been used to produce an estimate of its temporal variation over the past centuries. There are several different approaches taken to "reconstructing" the TSI, all employing a substantial degree of empiricism and in all of which the proxy data (such as sunspot number) are calibrated against the recent satellite TSI measurements, despite the problems outlined above. Some examples of TSI time series reconstructed over the past four centuries are given in Fig. 11. The estimates diverge as they go back in time due to the different assumptions made about the state of the Sun during the Maunder Minimum. This uncertainty is a major problem for understanding the role of the Sun in recent global warming.

In assessing the role played by different factors in causing climate change a useful parameter is Radiative Forcing (RF). This gives a measure of the imbalance in top-of-atmosphere radiation budget caused by the introduction of a given factor (e.g. an increase in the concentration of greenhouse gases would trap outgoing heat radiation, implying a positive RF; an increase in global albedo would result in greater reflection of solar radiation to space implying a negative RF). Solar RF may be deduced from anomalies in TSI but it is necessary to scale the TSI values by a factor of 0.18 to take account of global averaging and global albedo. Thus the range of TSI values of 0.4 to

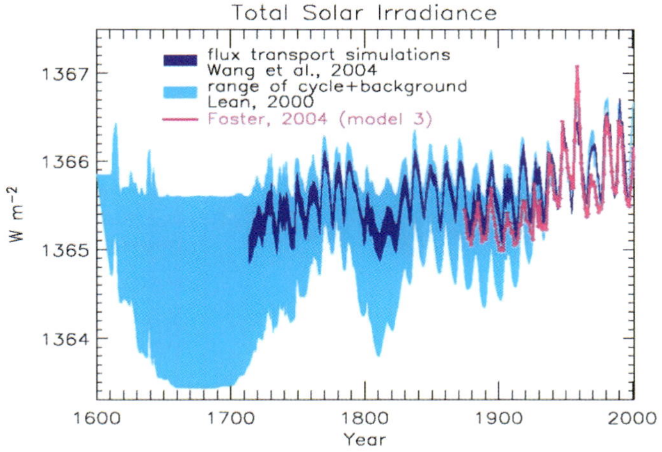

Fig. 11. Reconstructions of total solar irradiance based on sunspot numbers and different values for secular increase since Maunder Minimum (Lean, 2000), calibrated against MDI images (Foster, 2004), magnetic flux evolution from model (Wang *et al.*, 2005). (Figure courtesy Judith Lean.)

1.5 W m^{-2} shown in Fig. 11 as the increase since 1750 would translates into a range of RFs of 0.07 to 0.27 W m^{-2}. The Intergovernmental Panel on Climate Change in its 2007 report shows solar radiative forcing since 1750 as 0.12 W m^{-2} (see Fig. 12) which at the low end of the estimated range. It should be noted that the solar value included in this well-publicised figure might be considerably larger, or smaller, if a different year were assumed for the pre-industrial start-date.

Simulations with computer models of the global circulation of the atmosphere (GCMs) have been carried out to represent the climate from ~1860 to 2000 with time-evolving natural (solar and volcanic) and anthropogenic (greenhouse gases, sulphate aerosol) forcings. The GCMs are generally able to reproduce, within the bounds of observational uncertainty and natural variability, the temporal variation of global average surface temperature over the twentieth century with the best match to observations obtained when all the above forcings are included. Separation of the effects of natural and anthropogenic forcing suggests that the solar contribution is particularly significant to the observed warming over the period 1900–1940. However, uncertainties remain with the solar effect, particularly regarding the impact of the choice of TSI reconstruction. Rind (2000) suggests, given some anthropogenic warming and large natural variability, that solar forcing is not necessarily involved in early 20th century warming but the analyses of Stott *et al.* (2000) and Meehl *et al.* (2003) do detect a solar influence.

Historically many authors have suggested, based on analyses of observational data, that the solar influence on climate is larger than would be anticipated based on radiative forcing arguments alone. The problems with these studies (apart from any question concerning the statistical robustness of their conclusions) is that (i) they

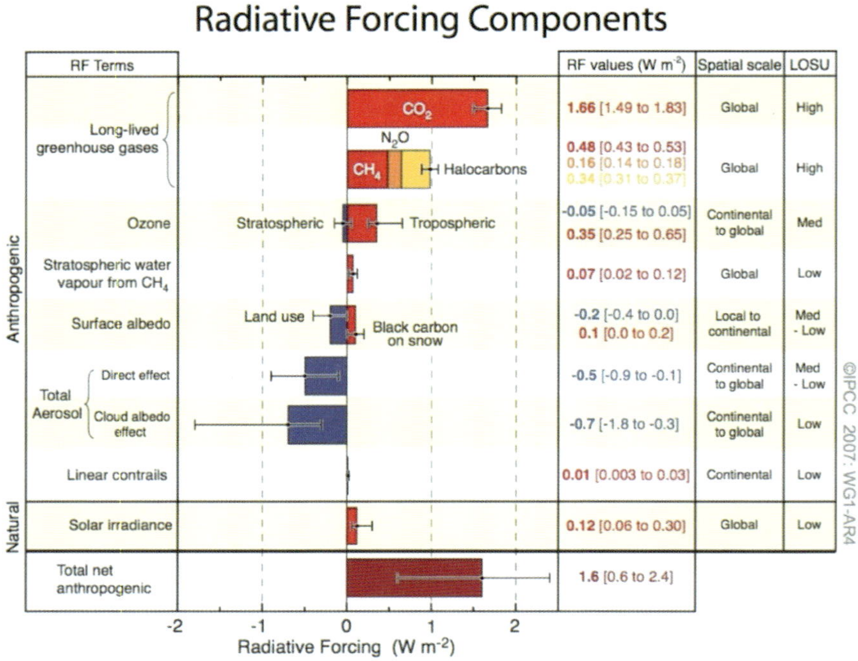

Fig. 12. Radiative forcing estimates 1750–2005 as compiled by IPCC (2007).

frequently apply only to certain locations, and (ii) they do not offer any advances in understanding of how the supposed amplification takes place. Recently some interesting developments have been made to address these problems based on two different approaches: the first uses detection/attribution techniques to compare model simulations with observations. These studies (North and Wu, 2001; Stott *et al.*, 2003) show that the amplitude of the solar response, derived from multiple regression analysis of the data using model-derived signal patterns and noise estimates, is larger than predicted by the model simulations (by up to a factor 4). This technique does not offer any physical insight but suggests the existence of deficiencies in the models and also shows how the solar signal may be spatially distributed. The second approach proposes feedbacks in the climate system (that may already exist in GCMs) and uses these to explain features found both in model results and in observational data. The proposed mechanisms are generally either concerned with radiative and thermodynamic processes, water vapour feedback and clouds or involve dynamical coupling between upper and lower layers of the atmosphere.

One suggestion (Meehl *et al.*, 2003) is that, because solar heating of the tropical sea surface is larger where there is less cloud, there may be changes in atmospheric circulation associated with anomalies in horizontal temperature gradient. Kodera

(2004) and Kodera *et al.* (2007) suggest that changes in the stratosphere influence static stability and tropical convection. Van Loon and Meehl (2008) also invoke changes in the stratosphere but find an intensification of tropical precipitation at high levels of solar activity, which appears to be at odds with the Kodera results. The reason for the similarity of the apparent sea surface temperature response to the cold phase of the ENSO cycle (Fig. 7) is also contentious. Van Loon and Meehl (2008) see two separate processes resulting in similar patterns. Emile-Geay *et al.* (2007), however, find that ENSO is actually modulated by solar activity, through a change in ocean circulations induced by longitudinally asymmetric changes in sea surface temperatures, although this takes several years to become established so may not apply on solar cycle timescales.

3.2 Solar spectral irradiance and photochemical effects in the stratosphere

The peak in the solar spectrum occurs at visible wavelengths but significant energy also resides in the ultraviolet region (Fig. 13(a)). UV radiation is absorbed in the middle and upper atmosphere with shorter wavelengths tending to be absorbed at high altitudes. Uncertainties in the absolute value of spectral irradiance imply uncertainties in the amount of energy absorbed in the atmosphere: Fig. 14(a) shows profiles of heating rates calculated due to absorption of radiation at wavelengths in the range 200–320 nm using two available solar spectra datasets, differences of up to 15% are apparent. This has implications for assessments of middle atmosphere temperatures: Fig. 14(b) shows large differences in the upper stratospheric and mesospheric temperature fields calculated in a general circulation model (GCM) between runs using the two spectra.

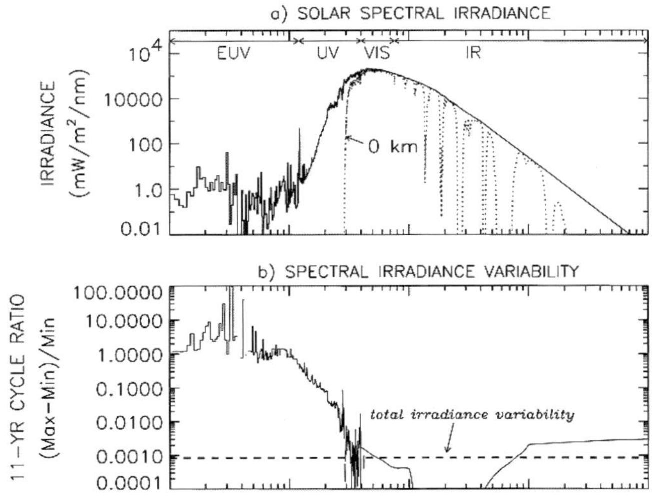

Fig. 13. Top: solar spectrum. Bottom: Fractional difference in solar spectral irradiance between maximum and minimum of the 11-year cycle. (From Lean and Rind (1998).)

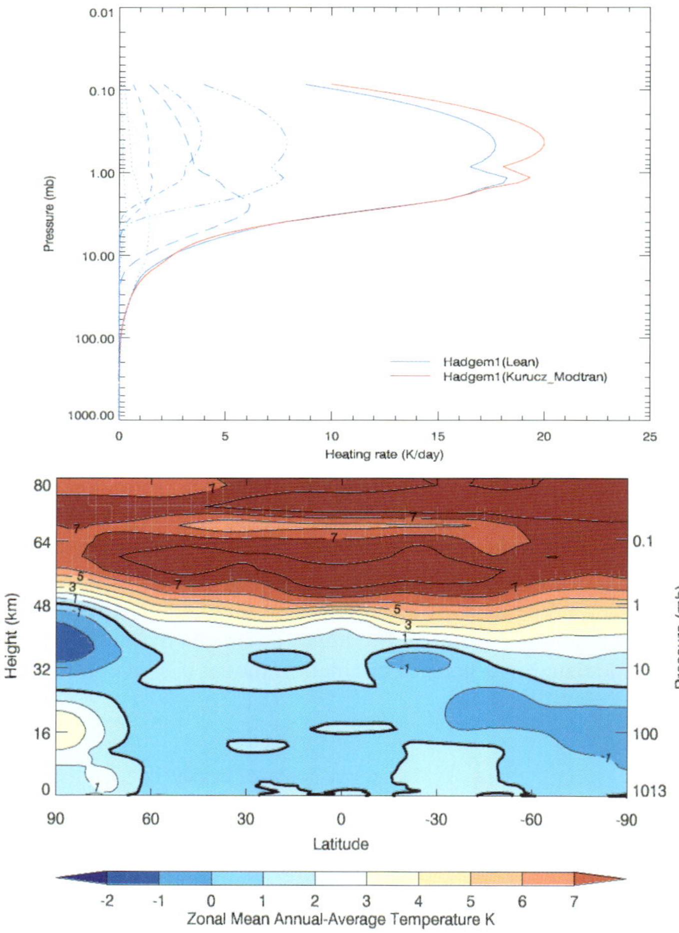

Fig. 14. Top: UV (200–320 nm) heating rates for mid-latitude summer conditions calculated using a line-by-line radiative code with two different spectral data files. Bottom: Impact on zonal mean temperatures simulated in GCM of replacing UV spectrum. (From Zhong *et al.*, Influence of the prescribed solar spectrum on calculations of atmospheric temperature, *Geophys. Res. Lett.*, **35**, L22813, 2008. Copyright 2008 American Geophysical Union. Reproduced by permission of American Geophysical Union.)

Measurements of solar spectral irradiance in the visible and ultraviolet show that the amplitude of solar cycle variability is greater at shorter wavelengths (Fig. 13(b)). The result of this is that the response of atmospheric temperatures to solar variability is large in the upper atmosphere, with, for example, variations of 400 K being typical at 300 km over the 11-year cycle, reflecting the large modulation of far and extreme

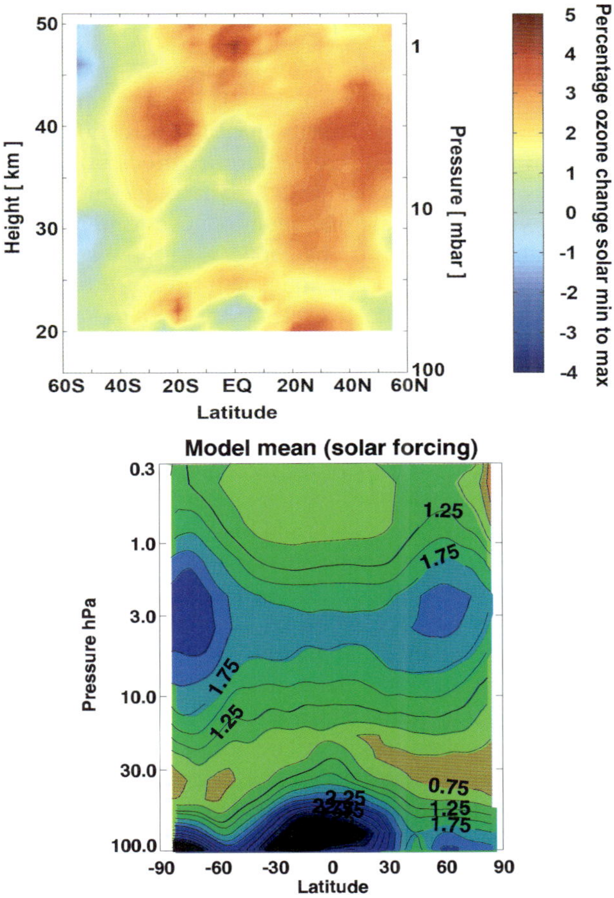

Fig. 15. Top: Meridional cross section of solar cycle regression fit of the SAGE I and II data (% per 100 units $F_{10.7}$ radio flux. (From Randel and Wu, A stratospheric ozone profile data set for 1979–2005: Variability, trends, and comparisons with column ozone data, *J. Geophys. Res.—Atmos.*, **112**, D06313, 2007. Copyright 2007 American Geophysical Union. Reproduced by permission of American Geophysical Union.) Bottom: Modelled annually averaged ozone solar cycle response, as a function of latitude and pressure, averaged over 8 chemistry-climate models. Units as above, contour interval is 0.25%. (From Austin *et al.*, Coupled chemistry climate model simulations of the solar cycle in ozone and temperature, *J. Geophys. Res.*, **113**, D11306, 2008. Copyright 2008 American Geophysical Union. Reproduced by permission of American Geophysical Union.)

ultraviolet radiation in that region. At lower altitudes the response is smaller; measurements made from satellites suggest an increase of up to about 1 K in the upper stratosphere at solar maximum; a minimum, or possibly even a negative change, in the mid-low stratosphere with another maximum, of a few tenths of a degree, below.

However, precise values, as well as the position (or existence) of the negative layer, vary between datasets.

Ozone is produced by short wavelength solar ultraviolet radiation and destroyed by radiation at somewhat longer wavelengths. This means that ozone production is more strongly modulated by solar activity than its destruction and this leads to a higher net production of stratospheric ozone during periods of higher solar activity. Observational records show approximately 2% higher values in vertically-integrated ozone amount at 11-year solar cycle maximum relative to minimum. The vertical distribution, as with the temperature signal, appears to show a minimum in the low latitude middle stratosphere (see Fig. 15, top panel) but as the observational data are only available over less than two solar cycles there remains some doubt about the statistical robustness of the signals derived from them.

Nevertheless, it is clear that the temperature and ozone responses are intimately linked and Austin *et al.* (2008) show that some coupled chemistry-climate GCMs are now able to reproduce the lower stratosphere maxima in temperature and ozone (Fig. 15, lower panel) although details of the mechanism whereby they occur remain uncertain.

3.3 Vertical coupling through the middle and lower atmosphere

3.3.1 Northern hemisphere winter polar stratosphere

During the winter the high latitude stratosphere becomes very cold and a polar vortex of strong westerly winds is established. The date in spring when this vortex finally breaks down is very variable, particularly in the northern hemisphere, but plays a key role in the global circulation of the middle atmosphere. Because variations in solar UV input change the latitudinal temperature gradient in the upper stratosphere, the evolution of the winter polar vortex may be affected. Satellite data suggest that the vortex strengthens in November and December in response to enhanced solar activity. This positive perturbation to zonal mean zonal winds then propagates polewards and downwards, until by February it is replaced by an easterly anomaly (Kodera *et al.*, 1990; Kodera, 1995).

Planetary-scale waves induce a large-scale meridional circulation (Haynes *et al.*, 1991) that is strengthened when the winter polar vortex is more disturbed. The meridional circulation is therefore a prime route for winter polar events to influence the lower stratosphere, not only in polar latitudes but throughout the winter hemisphere and even the equatorial and summer subtropical latitudes, through a modulation of the strength of equatorial upwelling. A dynamical feedback via the meridional circulation would serve to amplify the direct radiative solar influence since the less disturbed early winter conditions in solar maximum lead to a weaker meridional circulation, weaker equatorial upwelling and hence a warmer equatorial lower stratosphere at solar maximum than solar minimum (Kodera and Kuroda, 2002; see their schematic reproduced in Fig. 16). Recent advances in modelling these phenomena have been made (Matthes *et al.*, 2004) but there are still several aspects that are not adequately reproduced or understood, including the apparent modulation of the solar signal by the quasi-biennial oscillation in tropical stratospheric zonal winds (Labitzke, 1987; Gray *et al.*, 2004).

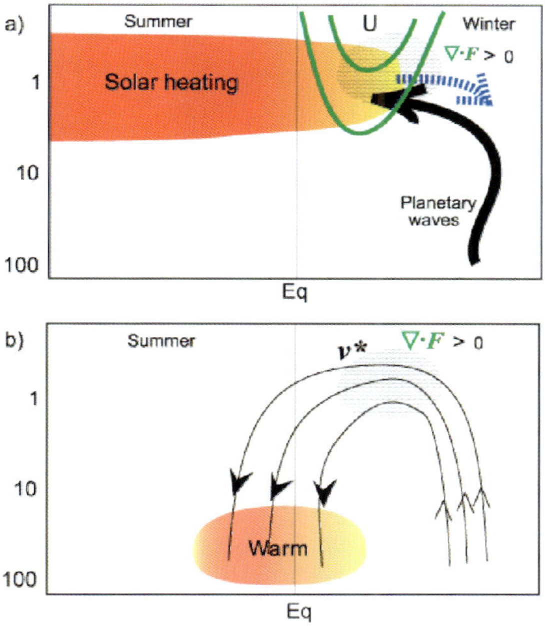

Fig. 16. Schematic illustration of the solar influence on the lower stratosphere. (a) Stronger stratospheric jet due to increased solar forcing during the maximum phase deflects planetary waves from the subtropics (dashed arrow), which creates anomalous divergence of the planetary wave flux (stippled). (b) Decrease in wave forcing results in reduces mean meridional circulation (arrows) and warming in the tropical lower stratosphere (shading). (From Kodera and Kuroda, Dynamical response to the solar cycle, *J. Geophys. Res.—Atmos.*, **107**, 4749, 2002. Copyright 2002 American Geophysical Union. Reproduced by permission of American Geophysical Union.)

3.3.2 Stratosphere-troposphere coupling

Patterns similar to the solar effects seen in the NCEP temperature and wind data (Fig. 8 and Fig. 9) have been reproduced in GCM simulations of the effects of increased solar UV variability (Haigh, 1996, 1999; Larkin *et al.*, 2000; Matthes *et al.*, 2004): see e.g. Fig. 17. Haigh (1999) showed that the solar response was enhanced if solar-induced ozone changes were included, implying that the additional UV heating in these runs, especially in the lower stratosphere, were important. Shindell *et al.* (2006), using a GCM with coupled ocean, has also found an enhanced solar effect in the troposphere if stratospheric chemistry is allowed to respond; Fig. 18, from that study, shows a reduced solar impact on tropical precipitation when the chemistry is omitted. The GCM studies also predict a weakening and expansion of the tropical Hadley cells in response to solar activity (see Fig. 19) not dissimilar to that found in observations (Fig. 10).

The similarity of the signals found in the observational data and model runs is intriguing and, by the nature of the model experiments, suggests that changes in the

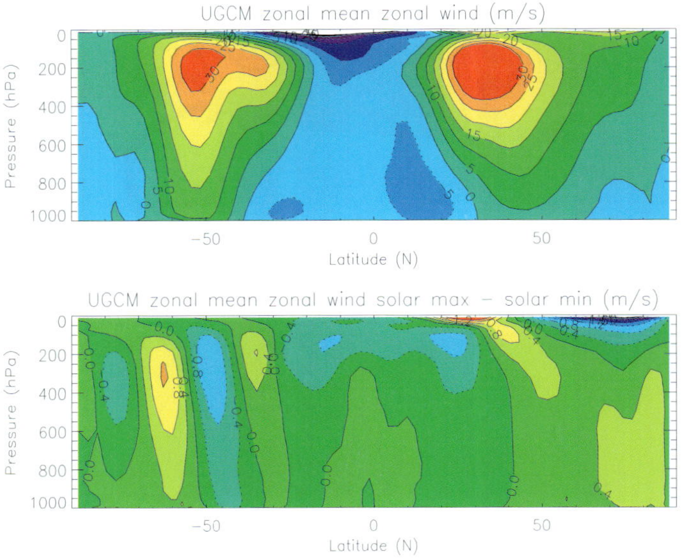

Fig. 17. Top: Average January zonal mean zonal wind (m s^{-1}) from a global climate model. Bottom: Difference in zonal mean zonal wind calculated for solar-cycle changes in solar UV radiation and ozone. (From Haigh, The impact of solar variability on climate, *Science*, **272**, 981–984,1996. Reprinted with permission from AAAS.)

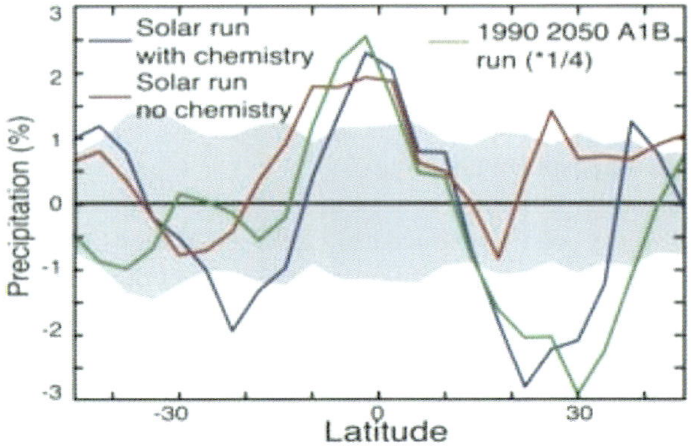

Fig. 18. Zonal mean changes in precipitation (%) changes from simulations of: red curve—solar irradiance change typical of a solar cycle, without stratospheric chemistry response; blue curve—solar change with coupled stratospheric chemistry; green curve—climate change 1880–2003. For further details see Shindell *et al.* (2006). (From Shindell *et al.*, Solar and anthropogenic forcing of tropical hydrology, *Geophys. Res. Lett.*, **33**, L24706, 2006. Copyright 2006 American Geophysical Union. Reproduced by permission of American Geophysical Union.)

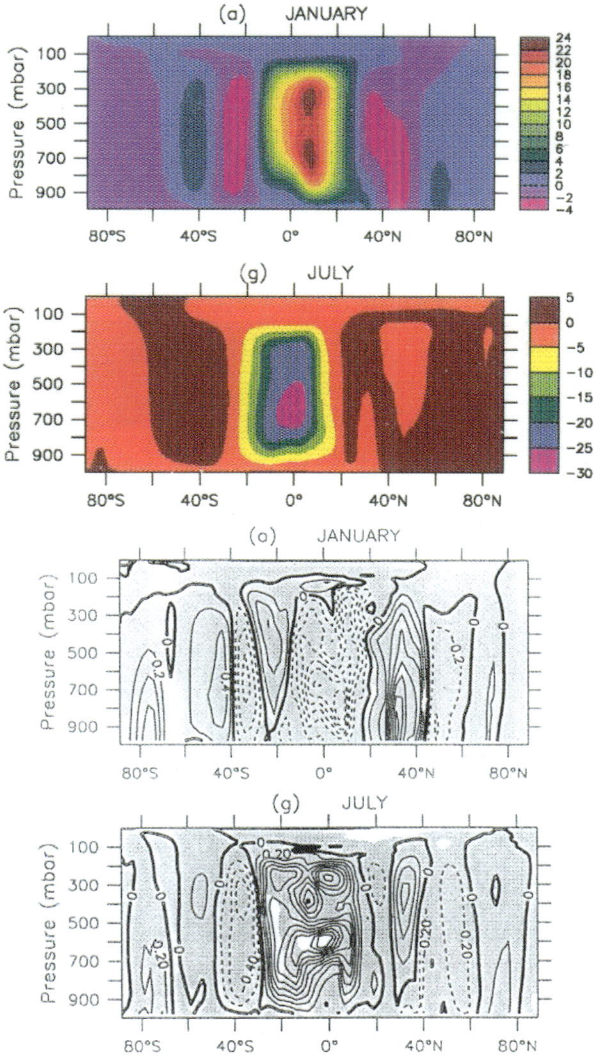

Fig. 19. Zonal mean tropospheric meridional circulation from GCM simulations. Coloured panels show January and July mean fields, positive/negative values indicate clockwise/counterclockwise circulation (major Hadley cell in winter hemisphere). Monochrome panels indicate the difference between solar maximum and minimum simulations, solid/dashed contours indicate positive/negative values; in both seasons the Hadley cell is weaker and broader at solar max. (With kind permission from Springer Science+Business Media: *Space Sci. Rev.*, The effect of solar UV irradiance variations on the Earth's atmosphere, **94**, 2000, 199–214, Larkin *et al.*, figure 4.)

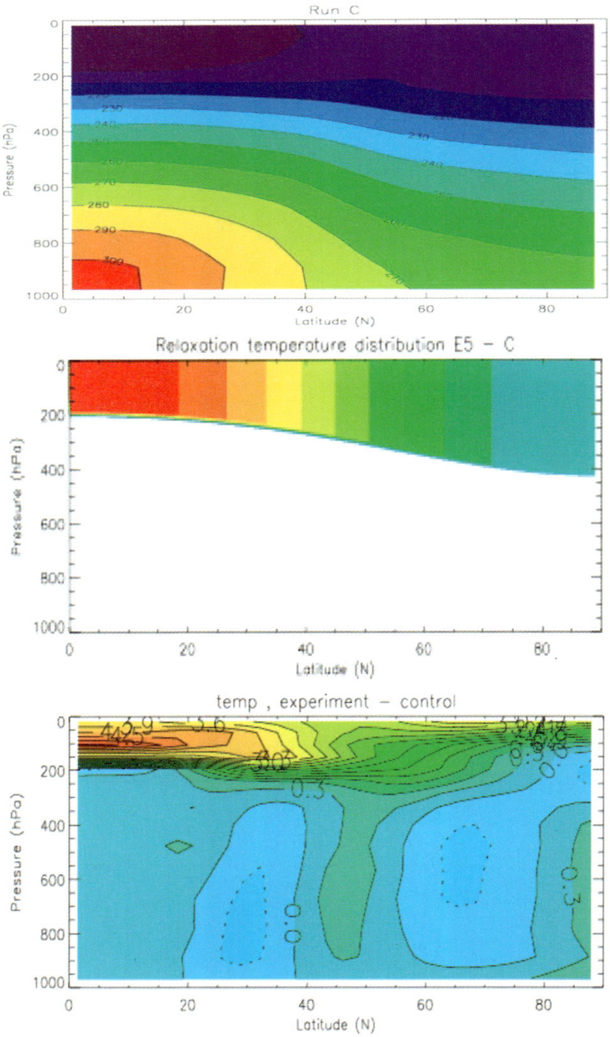

Fig. 20. Zonal mean temperature fields (K) in one hemisphere from a simplified GCM. Top: model climatology, middle: experimental change in stratospheric radiative equilibrium temperature. Bottom: difference between experiment and control; note that, despite forcing only being imposed in the stratosphere, a temperature response is seen throughout the lower atmosphere. (From Haigh *et al.* (2005).)

stratosphere, introduced by modulation of solar ultraviolet radiation and ozone, are key. It does not, however, explain the mechanisms whereby such a change in tropospheric circulation is brought about by thermal perturbations to the stratosphere. Haigh *et al.* (2005) have carried out some experiments with a simplified GCM de-

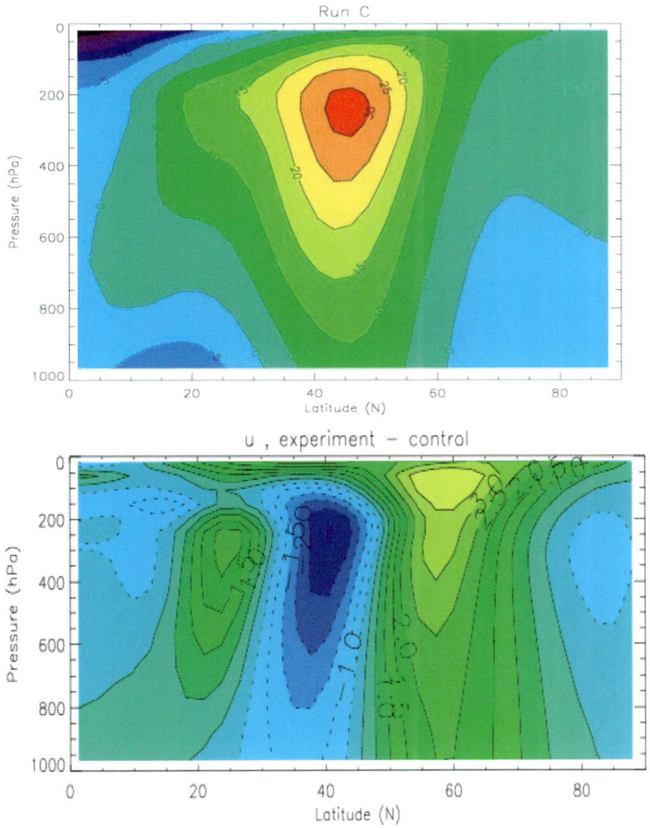

Fig. 21. Top: simple GCM climatology of zonal mean zonal wind (m s^{-1}). Bottom: difference between experiment (defined in Fig. 20) and control. (From Haigh *et al.* (2005).)

signed to elucidate some of these mechanisms. The model, using the same basic framework as the dynamical core outlined by Held and Suarez (1994), includes a full representation of the fluid dynamics but diabatic processes are represented by a simple Newtonian relaxation back to an equilibrium state. This means that solar heating is not explicitly included but can be represented by changes in the equilibrium temperature field. The advantages of using such a model are that the dynamical responses are easier to understand and also, because it is much less computationally demanding, that it is feasible to carry out a range of forcing and diagnostic runs. Similar models have been used to investigate stratosphere-troposphere coupling processes, particularly in the context of polar modes of variability (e.g. Polvani and Kushner, 2002).

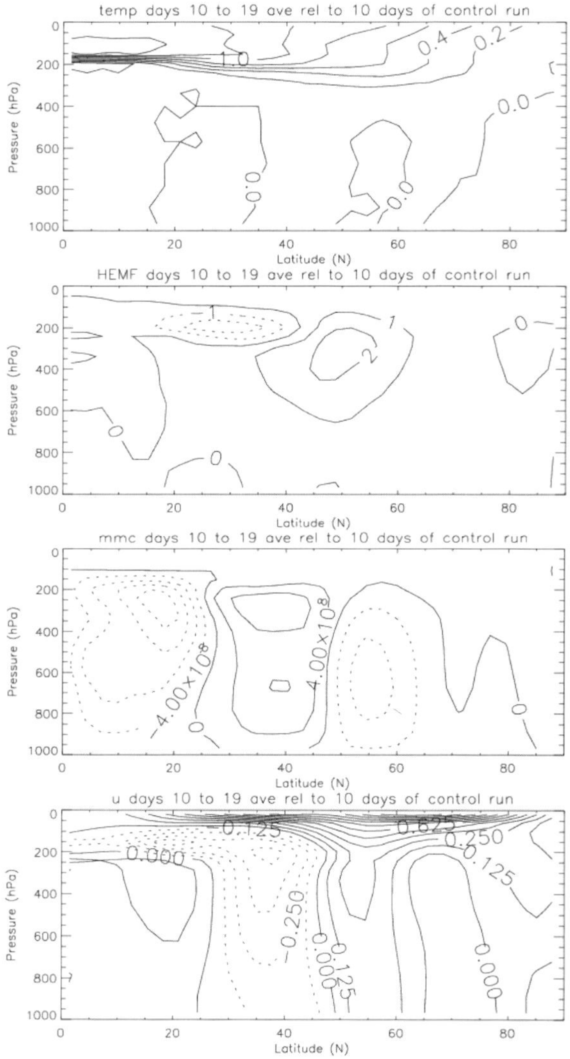

Fig. 22. Zonal mean fields from an ensemble of spin-up experiments of the simple GCM. Days 10–19 average anomaly of, from top to bottom: temperature (K), horizontal eddy momentum flux (m^2 s^{-2}), mean meridional circulation (kg s^{-1}), zonal wind (m s^{-1}). (From Simpson *et al.* (2009).)

In the Haigh *et al.* (2005) study perturbations to the stratospheric equilibrium temperature were imposed and the response investigated. Figure 20 presents an example; the perturbation consists of 5 K at the equator decreases by $\cos^2$(latitude) to zero at the poles, chosen to resemble coarsely the pattern found in the stratosphere

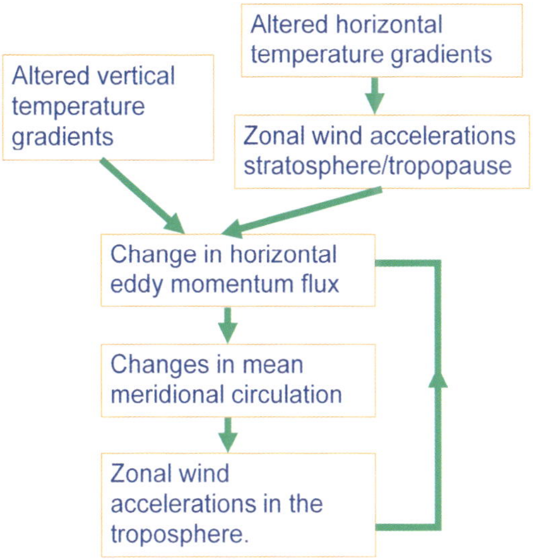

Fig. 23. Outline of mechanism proposed by Simpson *et al.* (2009) for transmission of a (solar) thermal perturbation in the stratosphere to the climate of the troposphere.

response to solar variability (Fig. 8), although with larger magnitude. A response is seen, not only in the stratosphere but with vertical bands of warming throughout the troposphere—qualitatively similar to the observed solar signal. Figure 21 shows the response in zonal wind, the weakening and broadening of the mid-latutude jets is again markedly similar to the solar signal shown in observational data in Fig. 9 and in the full GCM experiment of Fig. 17.

From these experiments it was concluded that imposed changes in the lower stratospheric temperature forcing lead to coherent changes in the latitudinal location and width of the mid-latitude jetstream and its associated storm-track, and that wave/mean-flow feedbacks are crucial to these changes. The stratospheric warming, and an associated lowering of the tropopause, weakens the jet and storm-track eddies and it was also found that equatorial stratospheric warming displaces the jet polewards while uniform warming displaces it markedly equatorwards. It was concluded that the observed climate response to solar variability is brought about by a dynamical response in the troposphere to heating predominantly in the stratosphere, that the effect is small, and frequently masked by other factors, but not negligible in the context of the detection and attribution of climate change. The results also suggested that, at the Earth's surface, the climatic effects of solar variability will be most easily detected in the sub-tropics and mid-latitudes.

Further details of the mechanisms involved in the transfer of the effects of the stratospheric forcing to the troposphere have been revealed by "spin-up" experiments in which the perturbation is imposed on the unperturbed atmosphere and the evolution of the response diagnosed (Haigh and Blackburn, 2006; Simpson *et al.*, 2009). 10-day averages from an ensemble of such spin-up experiments is shown for temperature, horizontal eddy momentum flux, mean meridional circulation and zonal wind in Fig. 22. What happens in the sGCM, and by implication also in the real atmosphere, although other factors are bound to complicate the process, is as follows: changes to the temperature structure in the tropopause region influence the wind field there and thus the propagation of synoptic-scale waves and the deposition of zonal momentum; the meridional velocity, and by continuity the vertical velocity then respond, transferring the response to the lower troposphere and influencing the zonal winds there. Finally, and crucially, the changes in low level wind affect the propagation of waves and thus provide a feedback on the tropopause level effects. An outline of this mechanism is presented schematically in Fig. 23 providing an indication of how a solar (or indeed any other) perturbation to the lower stratosphere may influence tropospheric climate.

4 Summary

Radiation from the Sun ultimately provides the only energy source for the Earth's atmosphere and thus changes in solar activity clearly have the potential to affect climate. There is now statistical evidence for solar influence on various meteorological parameters on a wide range of timescales, although extracting the signal from the noise in a naturally highly variable system remains a key problem. Many questions remain concerning details of the mechanisms which determine to what extent, where and when the solar impacts are felt but advances in understanding are being made through the use of global climate models. It is only by further investigation of the complex interactions between radiative, physical, chemical and dynamical processes in the atmosphere that these difficult questions will be answered.

Acknowledgments. The studies at Imperial College London were supported by the UK Natural Environment Research Council.

References

Anderson, K., The weather prophets: science and reputation in Victorian meteorology, *Hist. Sci.*, **37**, 179–216, 1999.

Austin, J. *et al.*, Coupled chemistry climate model simulations of the solar cycle in ozone and temperature, *J. Geophys. Res.*, **113**, D11306, doi:10.1029/2007JD009391, 2008.

Bond, G., B. Kromer, J. Beer, R. Muscheler, M. N. Evans, W. Showers, S. Hoffmann, R. Lotti-Bond, I. Hajdas, and G. Bonani, Persistent solar influence on north Atlantic climate during the Holocene, *Science*, **294**, 2130–2136, 2001.

Crowley, T. J., Causes of climate change over the past 1000 years, *Science*, **289**, 270–277, 2000.

Emile-Geay, J., M. Cane, R. Seager, A. Kaplan, and P. Almasi, El Nino as a mediator of the solar influence on climate, *Paleoceanogr.*, **22**, 2007.

Foster, S., Reconstructions of solar irradiance variations, Ph.D thesis, University of Southampton, UK, 2004.

Fröhlich, C., Solar irradiance variability since 1978: Revision of the PMOD Composite during Solar Cycle 21, *Space Sci. Rev.*, **125**, 53–65, doi:10.1007/s11214-006-9046-5, 2006.

Gleisner, H. and P. Thejll, Patterns of tropospheric response to solar variability, *Geophys. Res. Lett.*, **30**, 17129, 2003.

Gray, L. J., S. A. Crooks, C. Pascoe, and S. Sparrow, Solar and QBO influences on the timings of stratospheric sudden warmings, *J. Atmos. Sci.*, **61**, 2777–2796, 2004.

Haigh, J. D., The impact of solar variability on climate, *Science*, **272**, 981–984,1996.

Haigh, J. D., A GCM study of climate change in response to the 11-year solar cycle, *Quart J. Roy. Meteorol. Soc.*, **125**, 871–892, 1999.

Haigh, J. D., The effects of solar variability on the Earth's climate, *Phil. Trans. Roy. Soc.*, **A361**, 95–111, 2003.

Haigh, J. D. and M. Blackburn, Solar influences on dynamical coupling between the stratosphere and troposphere, *Space Sci. Rev.*, **125**, 331–344, 2006.

Haigh, J. D., M. Blackburn, and R. Day, The response of tropospheric circulation to perturbations in lower stratospheric temperature, *J. Clim.*, **18**, 3672–3691, 2005.

Haynes, P. H., C. J. Marks, M. E. McIntyre, T. G. Shepherd, and K. P. Shine, On the 'downward control' of extratropical diabatic circulations by eddy-induced mean zonal forces, *J. Atmos. Sci.*, **48**, 651–678, 1991.

Held, I. M. and M. J. Suarez, A proposal for the intercomparison of the dynamical cores of atmospheric general circulation models, *Bull. Am. Meteor. Soc.*, **75**, 1825–1830, 1994.

IPCC, Climate Change 2007: The Physical Science Basis, CUP, 2007.

Kalnay, E. and co-authors, The NCEP/NCAR 40-year reanalysis project, *Bull. Am. Meteor. Soc.*, **77**, 437–471, 1996.

Kodera, K., On the origin and nature of the interannual variability of the winter stratospheric circulation in the northern-hemisphere, *J. Geophys. Res.*, **100**, 14077–14087, 1995.

Kodera, K., Solar influence on the Indian Ocean Monsoon through dynamical processes, *Geophys. Res. Lett.*, **31**, 2004.

Kodera, K. and Y. Kuroda, Dynamical response to the solar cycle, *J. Geophys. Res.—Atmos.*, **107**, 4749, 2002.

Kodera, K., K. Yamazaki, M. Chiba, and K. Shibata, Downward propagation of upper stratospheric mean zonal wind perturbation to the troposphere, *Geophys. Res. Lett.*, **17**, 1263–1266, 1990.

Kodera, K., K. Coughlin, and O. Arakawa, Possible modulation of the connection between the Pacific and Indian Ocean variability by the solar cycle, *Geophys. Res. Lett.*, **34**, L03710, 2007.

Labitzke, K., Sunspots, the QBO and the stratospheric temperature in the north polar region, *Geophys. Res. Lett.*, **14**, 535–537, 1987.

Larkin, A., J. D. Haigh, and S. Djavidnia, The effect of solar UV irradiance variations on the Earth's atmosphere, *Space Sci. Rev.*, **94**, 199–214, 2000.

Lean, J., Evolution of the sun's spectral irradiance since the Maunder Minimum, *Geophys. Res. Lett.*, **27**, 2425–2428, 2000.

Lean, J. and D. Rind, Climate forcing by changing solar radiation, *J. Clim.*, **11**, 3069–3094, 1998.

Luterbacher, J., D. Dietrich, E. Xoplaki, M. Grosjean, and H. Wanner, European seasonal and annual temperature variability, trends, and extremes since 1500, *Science*, **303**, 1499–1503, 2004.

Matthes, K., U. Langematz, L. J. Gray, K. Kodera, and K. Labitzke, Improved 11-year solar signal in the Freie Universitat Berlin climate middle atmosphere model (FUB-CMAM), *J. Geophys. Res.*, doi:10.1029/ 2003/D004012, 2004.

Meehl, G. A., W. M. Washington, T. M. L. Wigley, J. M. Arblaster, and A. Dai, Solar and greenhouse forcing and climate response in the twentieth century, *J. Clim.*, **16**, 426–444, 2003.

North, G. R. and Q. Wu, Detecting climate signals using space-time EOFs, *J. Clim.*, **14**, 1839–1863, 2001.

Polvani, L. M. and P. J. Kushner, Tropospheric response to stratospheric perturbations in a relatively simple general circulation model, *Geophys. Res. Lett.*, **29**, doi:10.1029/2001GL014284, 2002.

Randel, W. J. and F. Wu, A stratospheric ozone profile data set for 1979–2005: Variability, trends, and comparisons with column ozone data, *J. Geophys. Res.—Atmos.*, **112**, D06313, 2007.

Rind, D., Relating paleoclimate data and past temperature gradients: Some suggestive rules, *Quat Sci. Rev.*, **19**, 381–390, 2000.

Shindell, D. T., G. Faluvegi, R. L. Miller, G. A. Schmidt, J. E. Hansen, and S. Sun, Solar and anthropogenic forcing of tropical hydrology, *Geophys. Res. Lett.*, **33**, L24706, 2006.

Simpson, I. R., M. Blackburn, and J. D. Haigh, The role of eddies in driving the tropospheric response to stratospheric heating perturbations, *J. Atmos. Sci.*, 2009 (in press).

Stott, P. A., S. F. B. Tett, G. S. Jones, M. R. Allen, J. F. B. Mitchell, and G. J. Jenkins, External control of 20th century temperature by natural and anthropogenic forcings, *Science*, **290**, 2133–2137, 2000.

Stott, P. A., G. S. Jones, and J. F. B. Mitchell, Do models underestimate the solar contribution to recent climate change?, *J. Clim.*, **16**, 4079–4093, 2003.

van Loon, H. and D. J. Shea, A probable signal of the 11-year solar cycle in the troposphere of the northern hemisphere, *Geophys. Res. Lett.*, **26**, 2893–2896, 1999.

van Loon, H. and G. A. Meehl, The response in the Pacific to the sun's decadal peaks and contrasts to cold events in the Southern Oscillation, *J. Atmos. Sol.-Terr. Phys.*, **70**, 1046–1055, 2008.

van Loon, H., G. A. Meehl, and J. M. Arblaster, A decadal solar effect in the tropics in July-August, *J. Atmos. Sol.-Terr. Phys.*, **66**, 1767–1778, 2004.

van Loon, H., G. A. Meehl, and D. J. Shea, Coupled air-sea response to solar forcing in the Pacific region during northern winter, *J. Geophys. Res.—Atmos.*, D02108, **112**, 2007.

Wang, Y. M., J. L. Lean, and N. R. Sheeley, Modeling the sun's magnetic field and irradiance since 1713, *Astrophys. J.*, **625**, 522–538, 2005.

White, W. B., J. Lean, D. R. Cayan, and M. D. Dettinger, Response of global upper ocean temperature to changing solar irradiance, *J. Geophys. Res.*, **102**, 3255–3266, 1997.

Willson, R. C. and A. V. Mordinov, Secular total solar irradiance trend during solar cycles 21 and 22, *Geophys. Res. Lett.*, **30**, 1199–1202, 2003.

Zhong, W., S. M. Osprey, L. J. Gray, and J. D. Haigh, Influence of the prescribed solar spectrum on calculations of atmospheric temperature, *Geophys. Res. Lett.*, **35**, L22813, doi:10.1029/2008GL035993, 2008.

Climate and Weather of the Sun-Earth System (CAWSES): Selected Papers from the 2007 Kyoto Symposium,
Edited by T. Tsuda, R. Fujii, K. Shibata, and M. A. Geller, pp. 257–278.

Variability in the stratosphere: The sun and the QBO

Karin Labitzke and Markus Kunze

Institute for Meteorology, Free University of Berlin, Germany
E-mail: karin.labitzke@met.fu-berlin.de

Large effects of solar variability related to the 11-year sunspot cycle (SSC) are seen in the stratosphere, but only if the data are grouped according to the phase of the QBO (Quasi-Biennial Oscillation). New results based on an extended, 66-year long data set fully confirm earlier findings and suggest a significant effect of the SSC on the occurrence of the Major Midwinter Warmings (MMWs) over the Arctic as well as on the strength of the stratospheric polar vortex and on the mean meridional circulation. By means of teleconnections the dynamical interaction between the Arctic and the Tropics in the stratosphere and in the troposphere is shown for the whole data set and compared with the anomalies of single events. The results suggest strongly that during the northern winter the teleconnections between the Arctic and the Tropics were determined by the MMWs and the undisturbed, cold winters, respectively. These events in the stratosphere depend, however, on the 11-year SSC and on the QBO. The stratosphere is least disturbed during the northern summer when the interannual variability is small. And if the different phases of the QBO are introduced, a large solar signal is found in the eastphase of the QBO (more than two standard deviations). It is shown that the QBO not only modulates the solar signal on the decadal scale, but that the QBO is itself modulated by the solar variability.

1 Introduction

The large interannual variability of the Arctic winters in the stratosphere depends on the development of Major Midwinter Warmings (MMWs), but as well on the frequently occurring very cold winters which are connected with a very strong polar vortex (e.g., Labitzke, 1982). As discussed before (Labitzke, 1987; Labitzke and van Loon, 1988; van Loon and Labitzke, 2000; Crooks and Gray, 2005), these events depend strongly on the SSC and on the respective phase of the QBO (Quasi-Biennial Oscillation).

By means of teleconnections the dynamical interaction between the Arctic and the Tropics in the stratosphere and in the troposphere is shown for the whole data set and compared with the anomalies of single events. The results suggest strongly that during the northern winter the teleconnections between the Arctic and the Tropics were determined by the MMWs and the COLD winters, respectively. These events in the stratosphere depend, however, on the 11-year SSC and on the QBO.

The stratosphere is dynamically least disturbed during the northern summer and the interannual variability is low, Fig. 1. And if the different phases of the QBO are

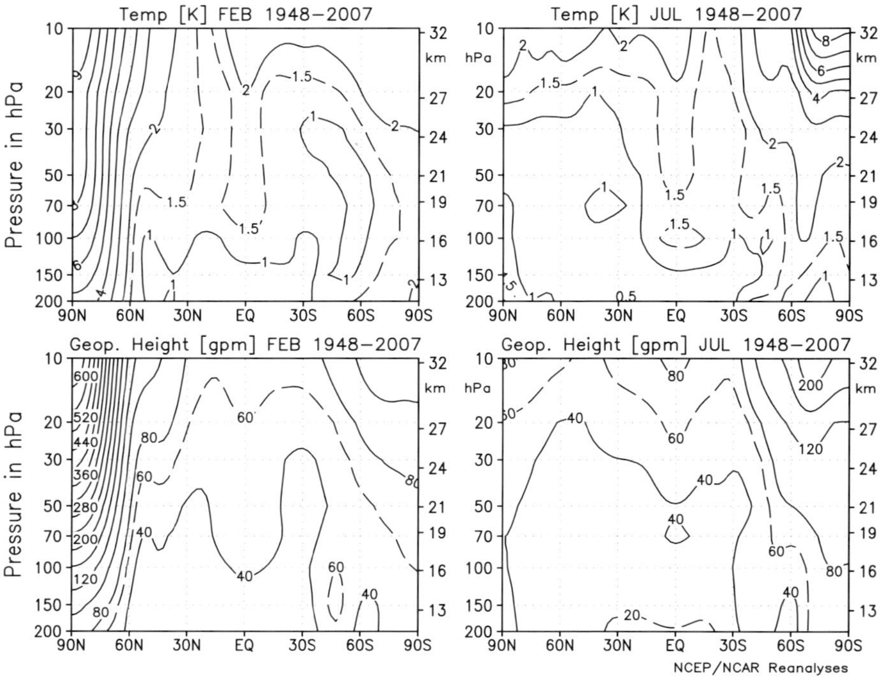

Fig. 1. Vertical meridional sections of the standard deviations for February (left) and July (right) of the zonal mean monthly mean temperatures (K), upper panels, and of the zonal mean monthly mean geopotential heights (geopot.m), lower panels, for the period 1948–2007, n (number of years) = 60 (NCEP/NCAR re-analyses) (Labitzke *et al.*, 2006, updated).

introduced, a large solar signal is found also in summer.

2 Data and Methods

For this study, some data are available for the period 1942–2007 (n = number of years = 66), as decribed in Labitzke *et al.* (2006). These are mainly the NCEP/NCAR re-analyses 1948 till 2007 (Kalnay *et al.*, 1996) and the statistically reconstructed 30-hPa heights at the North Pole (1942 till 1947) from Brönnimann *et al.* (2005). 29 MMWs occurred in Dec., Jan., or Feb., i.e. during 44% of all winters (Labitzke *et al.*, 2006). The QBO is an oscillation in the atmosphere which is best observed with radiosonde data in the stratosphere above the equator, where the zonal winds change between east and west with time. The period of the QBO varies in space and time, with an average value near 28 months at all levels, see reviews by Naujokat (1986) and Baldwin *et al.* (2001).

Because the QBO modulates the solar signal in the stratosphere, and in turn is modulated by the sun (e.g., Soukharev and Hood, 2001; Labitzke, 2005), it is necessary to stratify the data into years for which the equatorial QBO in the lower stratosphere (at about 45 hPa, e.g., Holton and Tan, 1980) was in its westerly or easterly phase (QBO data set (starting 1953) in: Labitzke and Collaborators, 2002).

Information on the phase of the QBO before 1953 was extended back to 1942 based on historical pilot balloon wind data (Labitzke *et al.*, 2006). Note that this extension is only tentative as the historical upper-level wind data are very sparse and of uncertain quality. But meanwhile Brönnimann *et al.* (2007) confirmed our QBO analysis.

Monthly mean values of the 10.7 cm solar flux are used as a proxy for variations through the SSC. The flux values are expressed in solar flux units: 1 s.f.u. $= 10^{-22}$ W m^{-2} Hz^{-1}. This is an objectively measured radio wave intensity, highly and positively correlated with the 11-year SSC and particularly with the UV part of the solar spectrum (Hood, 2003).

For the earliest years, before the regular measurements of the 10.7 cm solar flux became available in 1947, we derived the solar flux from a regression between the sunspot numbers and the flux (see Labitzke *et al.*, 2006).

For the range of the SSC, the mean difference of the 10.7 cm solar flux between solar minima (about 70 units) and solar maxima (about 200 units) is used, i.e., 130 units. Any linear correlation can be represented also by a regression line with $y = a + bx$, where x in this case is the 10.7 cm solar flux and b is the slope. This slope is used here, multiplied by 130, in order to get the differences between solar minima and maxima (Labitzke, 2003).

The significance of our results depends on the number of solar cycles available. We have reached 6.5 solar cycles and we can now safely say that the results for the northern winters, especially in the west phase of the QBO, with r reaching 0.7, are highly significant; ($r = 0.5 \sim 95\%$; $r = 0.66 \sim 99\%$).

3 Variability of the Stratospheric Winters

3.1 Standard deviations for February and July

The Arctic stratosphere reaches its highest variability in winter. Figure 1 gives an example of the variability of the stratosphere during the northern winter (February) and summer (July). It is remarkable that in the lower and middle stratosphere the standard deviations in the Arctic winter are three to four times larger than those in the Antarctic winter. This is due to the fact that Major Mid-Winter Warmings which create the large variability of the Arctic, do usually not penetrate to the lower stratosphere over the Antarctic. But the variability is large in the upper stratosphere over the Antarctic, where so-called Minor Mid-Winter Warmings occur frequently (Labitzke and van Loon, 1972).

When the Antarctic westerly vortex breaks down in spring (September–November), the middle stratosphere varies so much from one spring to another that the standard deviation at the South Pole in October (Labitzke and van Loon, 1999, their figure 2.11) approaches that at the North Pole in January and February. In summer the variability is low in both hemispheres. A relative maximum of variability is observed on the equator due to the QBO.

3.2 Time series over the Arctic

To give an impression of the large interannual variability of the Arctic stratosphere in winter, time series over the North Pole of the 30-hPa temperatures (°C) in January and of the 30-hPa heights (geopot.dam) in February are given in Fig. 2(a) and 2(b), respectively. The variability is largest during this part of the winter, with standard deviations (sigma) of 8.5 K in January and 45.5 gpdam in February.

It is of great interest to note that the overall trend of the 30-hPa temperatures in January is slightly positive, 0.8 K per decade (84% significance) and that of the 30-hPa heights in February is practically zero (0.23 geopot.dam per decade)—(see, e.g., discussions about temperature trends in the Arctic by Pawson and Naujokat (1999) and Labitzke and Kunze (2005)).

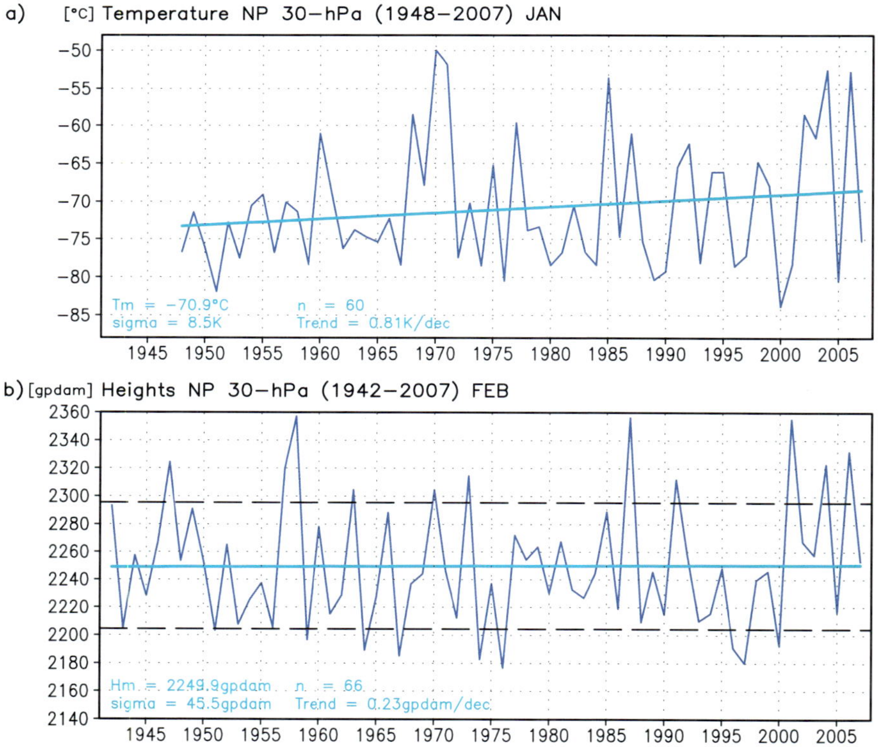

Fig. 2. (a) Time series of the monthly mean 30-hPa temperatures (°C) over the North Pole in January, 1948 through 2007. A trend line is given for the whole period (NCEP/NCAR re-analyses). (b) Time series of the monthly mean 30-hPa heights (geopot.dam) over the North Pole in February, 1942 through 2007, with a trend line and the + and −1 sigma lines (dashed). n = number of years. (NCEP/NCAR re-analyses and reconstructions, see Labitzke *et al.*, 2006.)

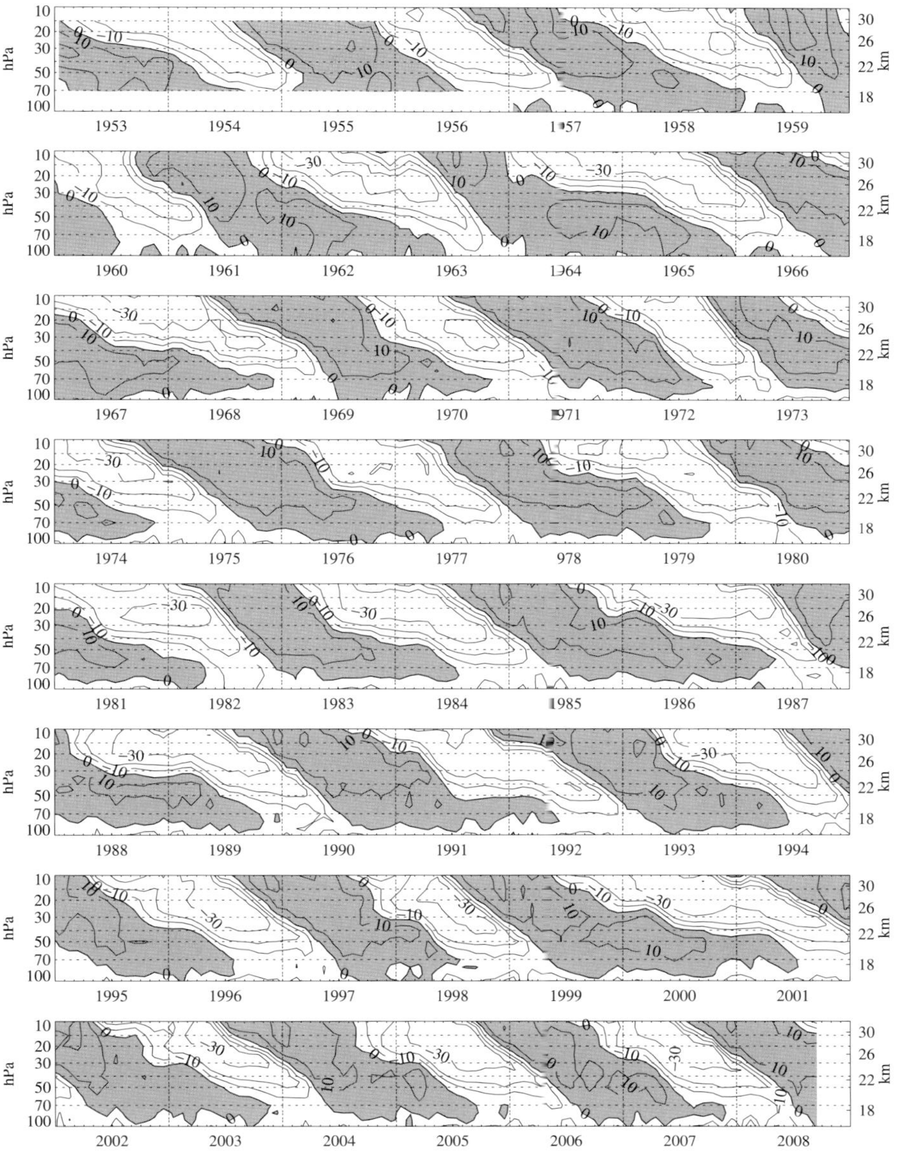

Fig. 3. Time height section of monthly mean zonal winds (m/s) at equatorial stations: Canton Island, 3°S/172°W (Jan 1953–Aug 1967), Gan/Maledive Islands, 1°S/73°E (Sep 1967–Dec 1975) and Singapore, 1°N/104°E (since Jan 1976). Isopleths are at 10 m/sec intervals; winds from the west are grey, updated from Naujokat (1986).

3.3 Factors responsible for the variability

The state of the Arctic westerly vortex in northern winter is influenced by several factors (van Loon and Labitzke, 1993):

- The QBO, Fig. 3, consists of downward propagating west and east winds in the stratosphere with an average period of about 28 month; this pattern is centered on the equator. A historical review and the present explanation of the QBO can be found in Labitzke and van Loon (1999). The QBO modulates the Arctic and also the Antarctic polar vortex (Labitzke, 2004a), but this modulation changes sign depending on the phase in the solar cycle, see Section 3.4.

- Another quantity whose effect is felt in the stratosphere is the Southern Oscillation. The SO is defined as a "see-saw" in atmospheric mass (evidenced by sea-level pressure) between the Pacific Ocean and the Australian-Indian region, (see, e.g., Labitzke and van Loon, 1999). Its influence is global and reaches into the stratosphere. The anomalies in the lower stratosphere associated with extremes of the SO are described in van Loon and Labitzke (1987), where they are discussed in terms of other influences such as the QBO and volcanic eruptions. In the warm extremes (WE) of the SO (i.e., El Ninos) the stratospheric temperatures and heights at Arctic latitudes are most of the time well above normal (about 1 standard deviation), and conversely in the cold extremes (CE).

- The stratosphere is also influenced by different types of waves which penetrate under certain conditions from the troposphere to the stratosphere. These are the very large scale planetary waves. Further, different types of the so-called gravity waves reach the stratosphere where they deposit their momentum, depending on the vertical profiles of the zonal winds.

- Tides are very important for the dynamics in the upper stratosphere and mesosphere. They develop in these regions mainly through thermal forcing of the rotating earth's atmosphere by the sun.

- Yet another influence on the stratosphere is the solar variability which until recently received little attention. The influence of the 11-year SSC will be discussed in the next section.

3.4 Connections between the Arctic stratosphere, the sun and the QBO during winter

Based on results published in 1982, Labitzke (1987) found that a signal of the 11-year SSC emerged when the Arctic stratospheric temperatures and geopotential heights were grouped into two categories determined by the direction of the equatorial wind in the stratosphere (QBO). Figure 4 shows in two scatter diagrams the correlations between the 30-hPa heights over the North Pole in February (i.e., the data of Fig. 2(b)) and the SSC, with the data grouped according to the QBO. The data span 6.5 solar cycles, see Fig. 5.

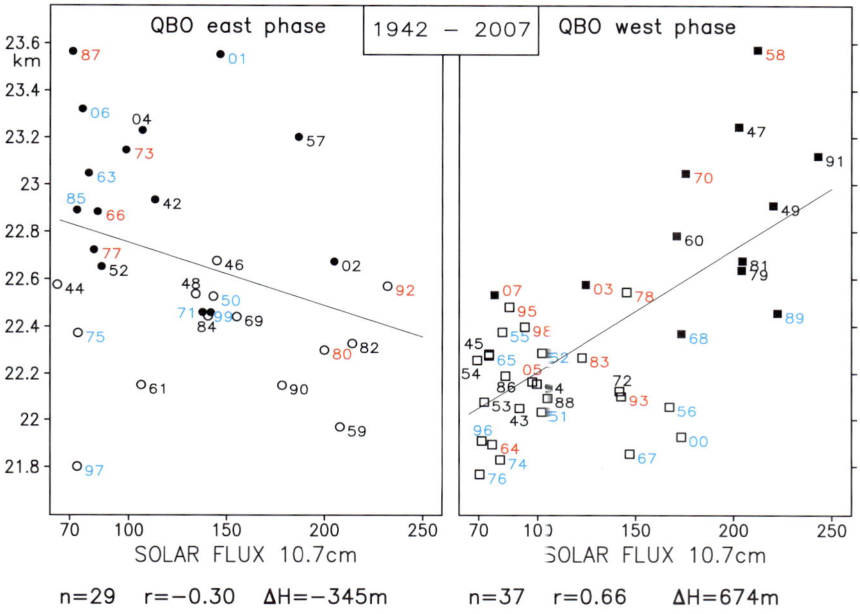

Fig. 4. Scatter diagrams of the monthly mean 30-hPa geopotential heights (geopot.km) in February at the North Pole (1942 till 2007), plotted against the 10.7 cm solar flux. Left: Circles: years in the east phase of the QBO ($n = 29$). Right: Squares: years in the west phase ($n = 37$). The numbers indicate the respective years, with WE in red and CE in blue. Filled symbols denote MMWs. r = correlation coefficient; ΔH gives the mean difference of the heights (geopot.m) between solar maxima (200 s.flux units) and solar minima (70 s.flux units). (Reconstructions: 1942 till 1947; NCEP/NCAR re-analyses: 1948 till 2007.) (van Loon and Labitzke (1994), updated.)

It is obvious (Fig. 4) that the correlations are opposite in the two different phases of the QBO: while they are significantly positive in the QBO-west group ($r = 0.66$, significance about 99%), they are negative in the QBO-east group. The filled symbols indicate the occurrence of MMWs, which took place preferably during solar maxima in the west phase and during solar minima in the east phase. There is a tendency for warm events to occur together with MMWs in solar maximum/west phase, but in solar minimum with the east phase.

The average height difference (ΔH in Fig. 4) between solar maxima and minima is very large in the west phase winters, reaching 67.4 geopot.dam which is 1.5 standard deviations of the interannual variability (cf. Fig. 2(b)).

Figure 5 presents the SSC based on the 10.7 cm solar flux in January and February for the period 1942–2007. It is indicated whether a winter (January/February) belonged to the west (squares) or east (circles) phase of the QBO. Large filled symbols indicate the occurrence of the 27 MMWs in January and February. n = number of years.

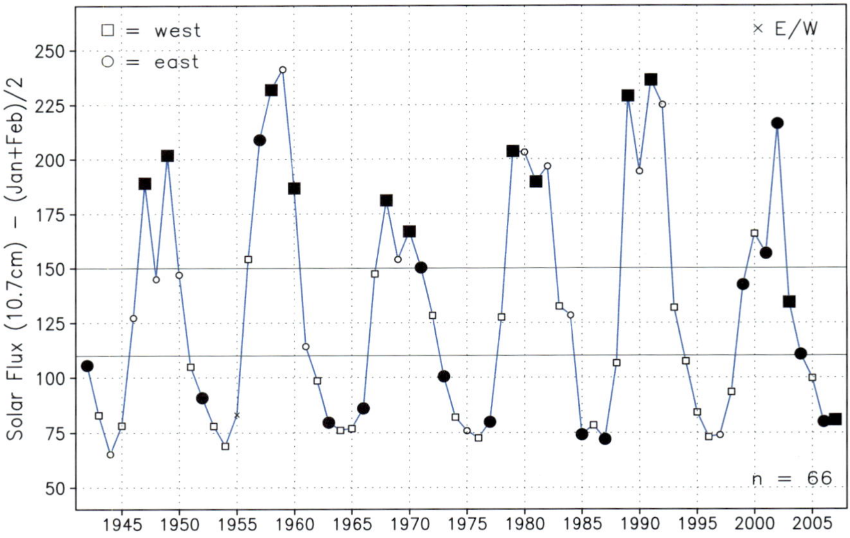

Fig. 5. Time series of the 10.7 cm solar flux, 1942 through 2007, (January+February)/2. Squares denote winters in the west phase of the QBO, circles winters in the east phase. For the classification, the QBO-winds in January and February are used at the 40 and 50 hPa levels, divided by 4. Large filled symbols indicate the occurrence of Major Midwinter Warmings (MMWs) (Labitzke and van Loon, 1990, updated).

While the data shown in Fig. 4 are only for the lower stratosphere (30 hPa) above the North Pole, Fig. 6 shows on the left hand side the correlations of the 30-hPa heights for the whole Northern Hemisphere, with the winters in the east phase of the QBO in the upper part of the figure, and the winters in the west phase of the QBO in the lower part. Again, the pattern of correlations is clearly very different in the two groups, with negative correlations over the Arctic in the east phase and large positive correlations there in the west phase. Outside of the Arctic the correlations are positive and strong in the east phase, but very weak in the west phase.

The respective height differences between solar maxima and minima are given on the right hand side of Fig. 6. In the east phase of the QBO the heights tend to be below normal over the Arctic in solar maxima (about one standard deviation, see Fig. 1); they are above normal towards the equator, indicating an intensification of the Aleutian High. In the west phase, the Arctic heights tend to be well above normal (about 1.5 standard deviations) in solar maxima; there are only very small height differences outside the Arctic.

Figure 7 shows on the left hand side vertical meridional sections of the correlations between the solar 10.7 cm flux and zonally averaged temperatures, and on the right hand side the corresponding temperature differences between solar maxima and minima. When all years are used in February, the correlations and the corresponding

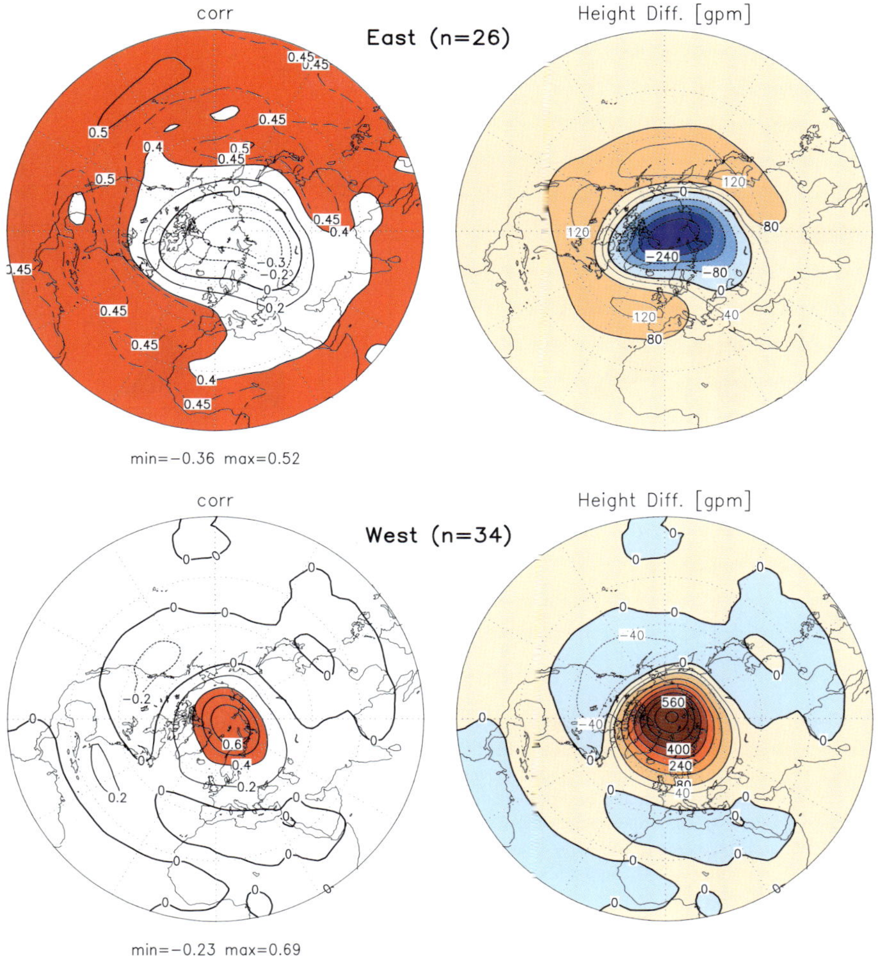

Fig. 6. Left: Correlations between the 10.7 cm solar flux (the 11-year solar cycle) and 30-hPa heights in February, shaded for emphasis where the correlations are above 0.4: upper panel: years in the east phase of the QBO; lower panel: years in the west phase of the QBO. Right: Respectively, height differences (geopot.m) between solar maxima and minima. NCEP/NCAR re-analyses, period: 1948–2007, n = 60); (Labitzke, 2002, updated).

temperature differences (top left and right, respectively) are very small. But, in the east phase of the QBO, the correlations of the zonally averaged temperatures with the solar cycle are positive from 60°N to the South Pole in the summer hemisphere, and negative north of 60°N, in the winter hemisphere. On the right hand side in the middle panel are the zonally averaged temperature differences between solar maxima and

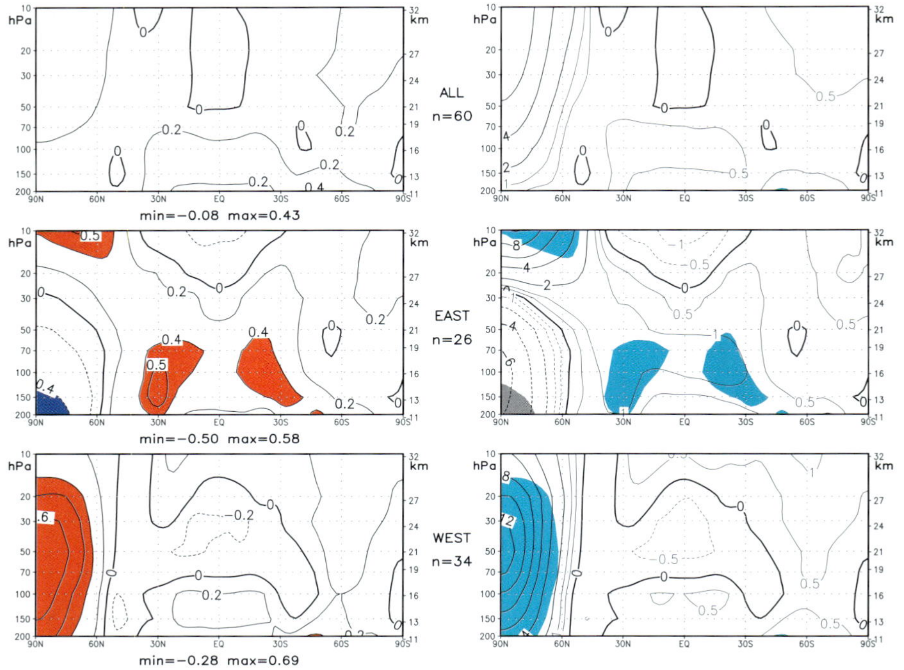

Fig. 7. Vertical meridional sections between 200 and 10 hPa of, on the left, the correlations between the detrended zonally averaged temperatures for February and the 10.7 cm solar flux (shaded for emphasis where the correlations are larger than 0.4), and, on the right, the respective temperature differences (K) between solar maxima and minima, shaded where the corresponding correlations on the left hand side are above 0.4. The upper panels show all years, the middle panels only years in the east phase of the QBO, and the lower panels only years in the west phase of the QBO (NCEP/NCAR re-analyses, 1948–2007) (Labitzke, 2002, updated).

minima in the east phase of the QBO which correspond to the correlations on the left side; the shading is the same as that in the correlations where it denotes correlations above 0.4.

In the west phase of the QBO (Fig. 7, bottom), the correlations with the 10.7 cm solar flux are highly positive over the Arctic ($r = 0.7$) and near zero or weakly negative elsewhere. The large positive correlations are associated with the frequent MMWs which occur when the QBO is in the west phase at solar maxima (e.g., van Loon and Labitzke, 2000). The Arctic temperatures and heights in the stratosphere are then determined by strong subsidence. Outside the Arctic the lower latitudes are expected to warm at solar maximum but, because of the subsidence and warming in the Arctic, the warming to the south is dynamically counter-balanced by a rising motion and cooling, well into the southern (summer) hemisphere.

The height and temperature changes shown in Figs. 6 and 7 indicate that the solar cycle influences the "Mean Meridional Circulation (MMC)", also called "Brewer-Dobson Circulation (BDC)". Forced by planetary waves the MMC regulates wintertime polar temperatures through downwelling and adiabatic warming (e.g., Kodera and Kuroda, 2002; Kuroda and Kodera, 2002; Hood and Soukharev, 2003; Labitzke, 2003; Hood, 2004; Salby and Callaghan, 2004, 2006; Matthes *et al.*, 2006).

During the west phase of the QBO the MMC is intensified during solar maxima (and *vice versa* during solar minima), with large positive anomalies over the Arctic (intensified downwelling and warming), and concurrent weak anomalies (anomalous upwelling/adiabatic cooling) over the Tropics and Subtropics, as shown in the lower maps in Fig. 6 and the lowest panels in Fig. 7. During the east phase the MMC is weakened in solar maxima, with reduced downwelling (anomalous upwelling/cooling) and negative anomalies over the Arctic in solar maxima, and concurrent anomalous downwelling with positive anomalies over the Tropics and Subtropics.

4 Teleconnections in Winter

The strong signal of the SSC in combination with the QBO provokes the question whether there are related signals outside the Arctic and maybe also in the troposphere. One way to investigate this question is using teleconnections (here: correlations) which showed interesting results (e.g., Shea *et al.*, 1991; van Loon and Labitzke, 1998). The opposite temperature changes between high and low latitudes during stratospheric warmings were already shown with early rocket and satellite data (Labitzke, 1972; Labitzke and Barnett, 1973).

4.1 North pole versus equator

To look for the connections in the lower stratosphere itself, we correlated in (Jan+Feb)/2 the 70-hPa temperatures at the North Pole with the zonal mean 70-hPa temperatures over the equator, for the period 1948 till 2007, Fig. 8. (The tropical tropopause is near 70 hPa.) The correlation is negative, as expected, because warming in the Arctic stratosphere is accompanied by cooling over the Tropics and Subtropics, and *vice versa*. But the correlation of -0.65 is very high, and considering the large number of years ($n = 60$), this correlation is very significant (better than 99.9%), as it is independent from the solar cycle and the QBO. This figure shows nicely the direct connection between high and low latitudes.

The variability of the temperatures over the equator is dominated by the QBO, and the east phase is generally colder than the mean and the west phase warmer. The lowest temperatures over the equator (Fig. 8) occurred with MMWs in the east phase, e.g., in 1966, 1973, 2004 and 2006. Three eruptions of tropical volcanoes lead to significant warming of the tropical stratosphere, marked here with A for Agung (April 1963), CH for Chichon (April 1982) and P for Pinatubo (June 1991). While Agung and Chichon belong to the west phase of the QBO and the temperatures should already be higher than the mean, Pinatubo belongs to the east phase and temperatures should be lower than the mean. Therefore the warming after Pinatubo in Jan/Feb 1992 is as strong as that after the two other volcanoes (e.g., Labitzke and van Loon, 1996). And after the QBO changed to west (Jan/Feb 1993 (P + 1)) the warming due

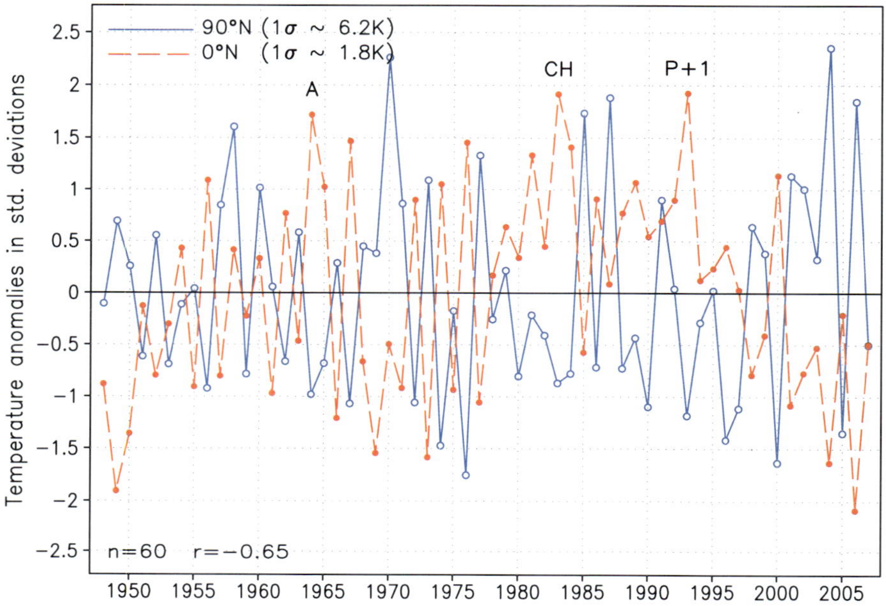

Fig. 8. Time series of the anomalies (in standard deviations) in (Jan+Feb)/2 of the 70-hPa temperatures over the North Pole (solid line) and around the equator (dashed line), 1948–2007; A, CH and P indicate the eruptions of three tropical volcanoes. $n = 60$; correlation $r = -0.65$ (NCEP/NCAR re-analyses).

to Pinatubo is fully comparable to that of the other volcanoes.

4.2 Teleconnections versus the five strongest Major Midwinter Warmings and the five coldest winters

4.2.1 The 5 strongest Major Midwinter Warmings

Figure 9(a) shows the global distribution of the correlations (i.e., the teleconnections) between the 70-hPa temperatures at the North Pole and the whole map, in (Jan+Feb)/2.

Clearly, the region south of about 60°N is negatively correlated with the Arctic. This is consistent with the idea of an intensification of the Mean Meridional Circulation (MMC) over the Arctic, when a MMW with anomalous warming, connected with downwelling, governs the Arctic and anomalous cooling (upwelling) is observed south of the polar region, far into the southern hemisphere (Kodera and Kuroda, 2002; Kuroda and Kodera, 2002; Labitzke, 2003; Salby and Callaghan, 2004, 2006; Kodera, 2006). (The difference of the correlations between the two different phases of the QBO is small, not shown).

Figure 9(b) gives for the 5 strongest MMWs the deviations of the 70-hPa temperatures in (Jan+Feb)/2 from the long-term mean 1968–2007. We choose MMWs of the QBO east/solar min group: 1977; 1985; 1987; 2004; 2006. But the anomalies of

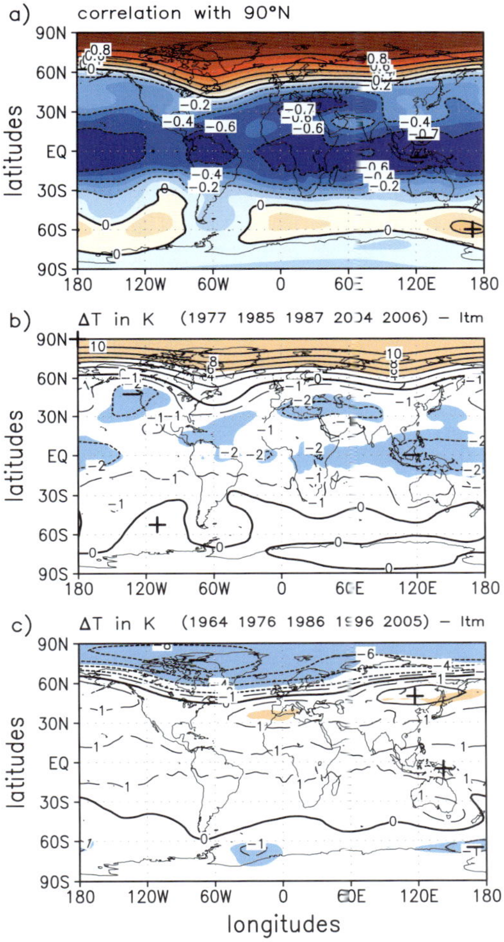

Fig. 9. (a) Teleconnections (i.e. correlations) at the 70-hPa level between the North Pole and the rest of the map for the temperatures in (Jan+Feb)/2. (NCEP/NCAR re-analyses, 1948–2007; $n = 60$). (b) Deviations of the 70-hPa temperatures (K) in (Jan+Feb)/2 of the five strongest MMWs (in solar min/east) from the long-term mean (ltm) (1968 through 2007). Shaded are anomalies larger than 1 standard deviation. (NCEP/NCAR re-analyses.) (c) As (b), but for the five coldest winters (in solar min/west).

the QBO west/solar max group are very similar (not shown).

The pattern of the anomalies can be directly compared with the correlations in Fig. 9(a) and shows the strong interaction between high and low latitudes connected with the MMWs. The influence of the northern winters reaches as far as 40°S, as observed already with early satellite data (Fritz and Soules, 1970; Labitzke and Barnett,

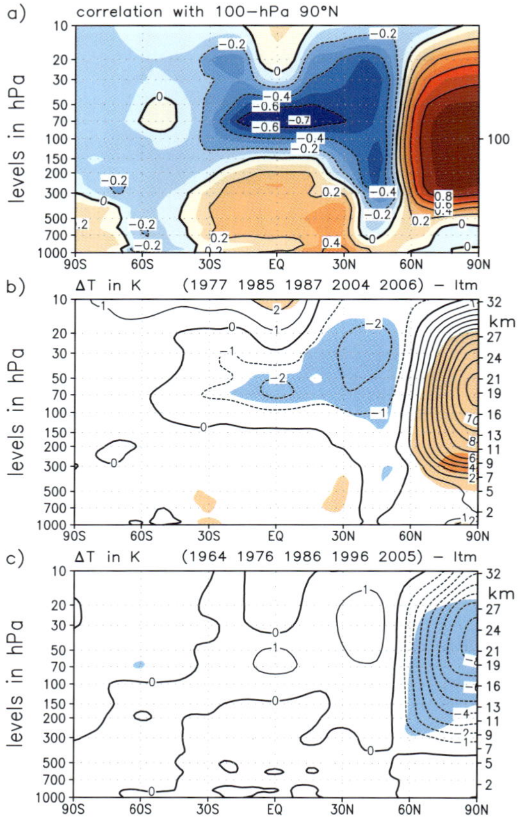

Fig. 10. (a) Vertical meridional section from 1000 to 10 hPa of the teleconnections (i.e., correlations) between the 100-hPa temperature at the North Pole and the rest of the globe in (Jan+Feb)/2. (NCEP/NCAR re-analyses, period 1968–2007). (b) Vertical meridional section from 1000 to 10 hPa of the deviations of the zonal mean temperatures (K) in (Jan+Feb)/2 of the five strongest MMWs from the long-term mean (1968 through 2007). Light shading: anomalies larger than 1 standard deviation, heavy shading: anomalies larger than 2 standard deviations. (c) As (b), but for the five coldest winters. (NCEP/NCAR re-analyses.)

1973).

Considering the large signals of the MMWs and the interaction between high and low latitudes, it is of great interest to see if this signal penetrates to the troposphere. Figure 10(a) shows a vertical meridional section of the teleconnections between the 100-hPa level over the Arctic and the rest of the atmosphere from the ground to 10 hPa. The interaction between high and low latitudes is very clearly visible with a maximum of the negative correlation over the equator at the 70-hPa level which is the tropopause region here, (cf. Fig. 8). And over the Arctic the positive cor-

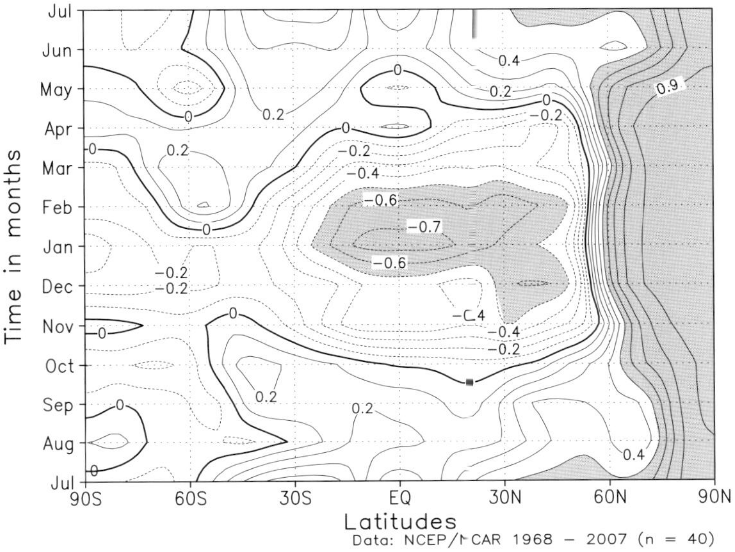

Fig. 11. Teleconnections (i.e., correlations) between the 100-hPa temperature at the North Pole and the zonal mean 70-hPa temperature (90°N–90°S) throughout the year. Shaded are correlations of more/less than +0.5/−0.5 (NCEP/NCAR re-analyses, period 1968–2007).

relations reach deep into the troposphere. Of great interest is the positive correlation in the tropical troposphere below the negative correlation in the tropopause region which appears to be a result from the warming over the Arctic.

This can be compared with the 5 strongest MMWs, Fig. 10(b). Here we are again dealing with the temperature anomalies, like in Fig. 9(b). The strong MMWs dominate the polar region not only in the stratosphere, but also in the troposphere where the positive anomalies reach to the ground, with more than 2 standard deviations in the upper troposphere over the Arctic. And outside the Arctic, south of about 60°N, the anomalies are significantly negative in the stratosphere over the Tropics and Subtropics. The structure of the anomalies indicates clearly a downward propagation of the stratospheric warming over the Arctic (e.g., Baldwin and Dunkerton, 2001).

In the stratosphere, the cooling outside the Arctic was to be expected from the teleconnections, see Fig. 10(a). The concurrent warming in the troposphere (most prominently over 30°S) is also consistent as regards the teleconnections (Salby and Callaghan, 2006).

4.2.2 The 5 coldest winters

These winters (1964, 1976, 1986, 1996, 2005) belong to the west phase winters in solar minimum and are among the coldest winters in the Arctic stratosphere since 1942, see Figs. 2 and 8. Accordingly, the deviations of the 70-hPa temperatures in (Jan+Feb)/2 from the long-term mean, Fig. 9(c), are negative in the Arctic stratosphere (more than one standard deviation) and positive from about 60°N till 40°S. The vertical section of the anomalies (Fig. 10(c)) is very similar to Fig. 10(b), but, ofcourse, with opposite sign.

4.2.3 Structure of the teleconnections through the year

In Fig. 10(a) we showed for (Jan+Feb)/2 a vertical meridional section of the teleconnections between the 100-hPa level over the Arctic and the rest of the atmosphere from the ground to 10 hPa. Here the link between high and low latitudes is strongest at the 70-hPa level. Therefore, we show in Fig. 11 the development of this link through the year. And the period with the strongest correlations coincides clearly with the period of the northern winter, when planetary scale waves are strongest.

5 Connections between the Stratosphere, the sun and the QBO during Summer

The stratosphere is dynamically least disturbed during the northern summer (Labitzke and van Loon, 1999) and the interannual variability is small, cf. Fig. 1. And if the different phases of the QBO are introduced a large solar signal is found. Figure 12 gives on the left the correlations of the 30-hPa temperatures with the 11-year SSC, with all data on top, the east phase data in the middle and the west phase data at the bottom.

The division into these two groups yields very clear results: the correlations for all years are a result of the strong correlations in the east phase, and the west phase contributes only very little. The very homogeneous pattern of the high correlations in the east phase is striking and convincing, as many different regions with different data problems give the same results. Accordingly, the temperature differences (on the right, middle panel) between solar max and min are over large areas higher than 2 standard deviations, see Fig. 1. To illustrate this strong correlation in summer, the time series of the monthly mean 30-hPa temperatures of July are given for a gridpoint near Nagasaki/Japan, Fig. 13. Again, the figure shows all data on top, the east phase in the middle and the west phase data at the bottom. The correlation of $r = 0.84$ (in the east phase) is clearly supported by the data of the single years. Only in July 1982 is the temperature a bit higher compared to the other solar maxima; but this is a response to the eruption of the Mexican volcanoe El Chichon (CH) in March/April 1982.

The correlations on the left of Fig. 14 show for July the vertical distribution of the relationship between the solar variability and the temperature, from the troposphere to the middle stratosphere. Again, the data are grouped in all years and in the east and west phases of the QBO. The corresponding temperature differences are given in the middle of the diagram, and the standardized temperature differences are on the right. Originally we made this division according to the phase of the QBO only in the northern winter data (Labitzke and van Loon, 1988); but it turns out that it is also a valid approach throughout the year (Labitzke, 2003, 2004a, b, 2005).

The results for the east phase (Fig. 14) are most striking: two centers with correlations above 0.7 are found over the Subtropics between 20 and 30 hPa. Further down, the double maximum in the east phase changes into one maximum, centered on the equatorial tropopause (Labitzke, 2003). The temperature differences between

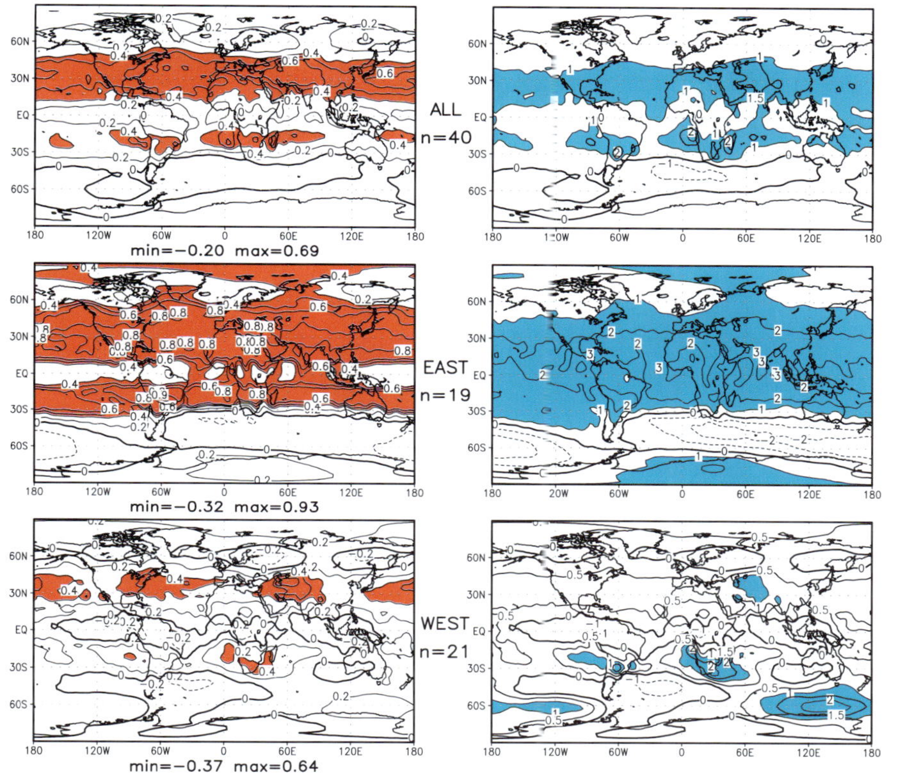

Fig. 12. Left: Correlations between the 10.7 cm solar flux and the detrended 30-hPa temperatures in July; shaded for emphasis where the correlations are above 0.4. Right: The respective temperature differences (K) between solar maxima and minima; shaded where the differences are above 1 K. Upper panels: all years; middle panels: only years in the east phase of the QBO; lower panels: only years in the west phase of the QBO. (NCEP/NCAR re-analyses, 1968–2007 (Labitzke, 2003, updated).)

solar maxima and minima are large, more than two standard deviations in some regions. This warming, i.e., positive anomalies, can only be explained by anomalous downwelling (i.e., reduced upwelling) over the Subtropics and Tropics (Kodera and Kuroda, 2002; Shepherd, 2002) which—in other words—means a weakening of the BDC for solar maxima/east phase of the QBO, as discussed above already for the northern winter.

In northern summer the solar variability signal is much weaker in the west phase years than in the east phase years and restricted to the northern hemisphere (lower panels in Fig. 14).

Figure 15 shows for the east and west phases of the QBO maps of the height differences (between solar maxima and minima) over the Subtropics and Tropics in

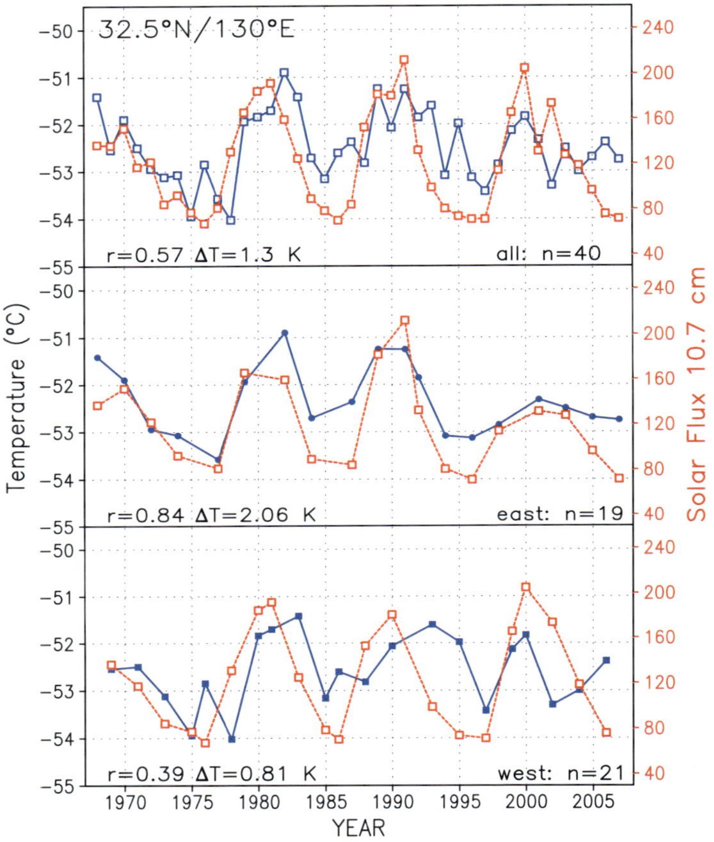

Fig. 13. Time series of monthly mean values of the 10.7 cm solar flux (dashed lines) and detrended 30-hPa temperatures in July at the grid point 32.5°N/130°E, (near Nagasaki/Japan). The upper panel shows all years ($n = 40$), the middle panel only years in the east phase of the QBO ($n = 19$), and the lower panel only years in the west phase of the QBO ($n = 21$). r = correlation coefficients; ΔT = temperature difference (K) between solar maxima and minima (NCEP/NCAR re-analyses, 1968–2007).

July for the 30-hPa level (about 24 km): again, the east phase dominates the solar signal (upper part of Fig. 15). In addition, the anomalous zonal (west-east) wind in the equatorial belt is affected by the solar variability on the decadal scale. At the top of the figure anomalously high values (more than two standard deviations, Fig. 1) are centered over the equator. This means that an anomalous ridge is centered on the equator in the solar maximum east years, connected with anomalous winds from the west. Therefore, during solar maxima in QBO/east years the low-latitude east wind is weakened, and conversely in the solar minimum years.

In the west phase of the QBO (bottom of Fig. 15), the pattern of the anomalies of

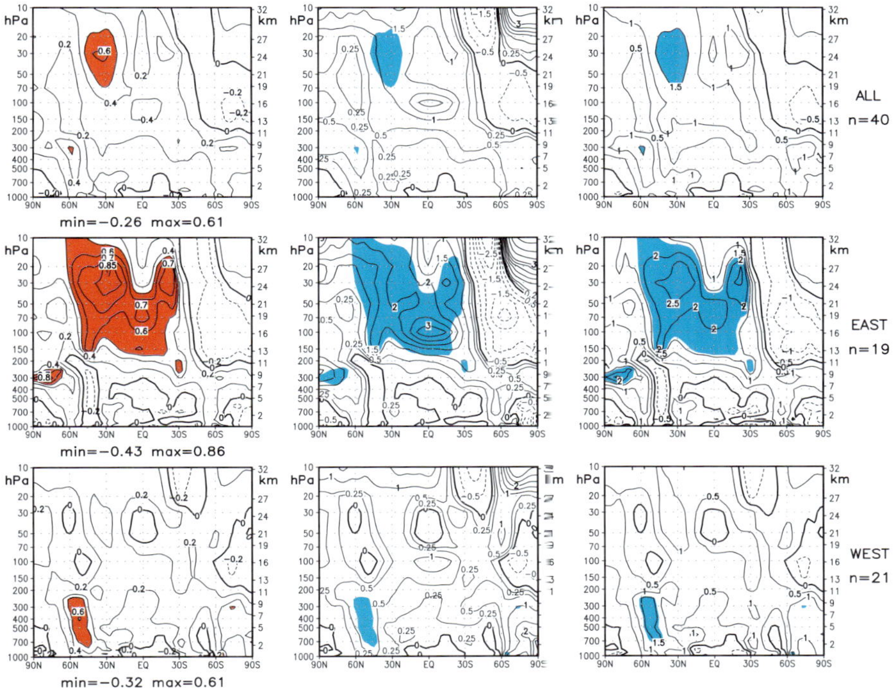

Fig. 14. Left: Vertical meridional sections of the correlations between the 10.7 cm solar flux and the detrended zonal mean temperatures in July; shaded for emphasis where correlations are above 0.5. Middle: The respective temperature differences (K) between solar maxima and minima, shaded where the corresponding correlations on the left hand side are above 0.5. Right: As for the middle row but standardized with the standard deviations. Upper panels: all years; middle panels: only years in the east phase of the QBO; lower panels: only years in the west phase of the QBO. (NCEP/NCAR re-analyses, 1968–2007.) (Labitzke (2003), updated.)

the geopotential heights is completely opposite, with the lowest values on the equator in solar maxima. The anomalous winds are from the east and the QBO is again weaker in solar maxima. The QBO thus not only modulates the solar signal on the decadal scale, but is itself modulated by the solar variability (Salby and Callaghan, 2000; Gray *et al.*, 2001; Soukharev and Hood, 2001; Labitzke, 2003, 2004b).

6 Summary

During solar minima the so-called "Holton-Tan Relationship" is valid for both phases of the QBO: In the QBO-eastphase the Arctic is relatively warm, with frequent MMWs, while during the westphase cold polar vortices dominate. However, during solar maxima the situation is reversed, especially during the westphase of the QBO.

New results based on an extended, 66-year long data set fully confirm our earlier

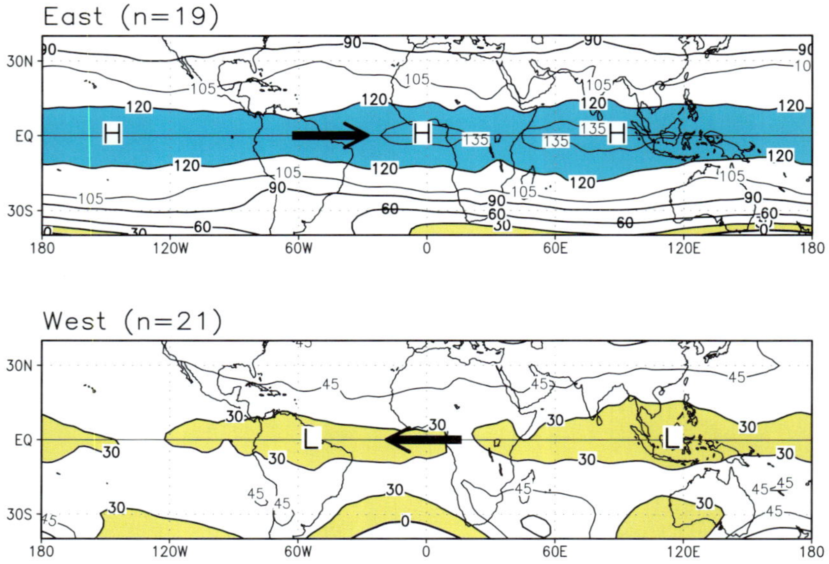

Fig. 15. Maps of the 30-hPa height differences (geopot.m) in July between solar maxima and minima. The upper map is for years in the east phase of the QBO, and the lower map for years in the west phase of the QBO. Arrows indicate the direction of the anomalous winds. (NCEP/NCAR re-analyses, 1968–2007.) (Labitzke (2003), adopted from her figure 5.)

findings and suggest a significant effect of the SSC on the occurrence of the Major Midwinter Warmings (MMWs) as well as on the strenght of the stratospheric polar vortex and on the mean meridional circulation.

Teleconnections show the dynamical interaction between the Arctic and the Tropics, and between the stratosphere and the troposphere. Their structure is very similar to the structure of the anomalies during strong MMWs in the stratosphere and of the anomalies during very cold stratospheric polar winters, respectively.

The results shown here suggest strongly that for the period from 1942–2007 ($n = 66$) the teleconnections between the Arctic and the Tropics and Subtropics during the northern winters were determined (in the stratosphere and in the troposphere) by the events in the stratosphere over the Arctic: i.e. by 27 stratospheric MMWs in January or February (44% of all winters) and by 29 COLD winters (about 50% of all). And these events are coupled to the 11-year SSC and to the QBO, as described above.

Outside the northern winter the eastphase of the QBO dominates with strong positive correlations over the Tropics and Subtropics, with the largest signal of the 11-year SSC (more than 2 standard deviations) during the northern summer.

Acknowledgments. We thank the members of the Stratospheric Research Group, FUB for professional support and Harry van Loon for many years of close cooperation. The 10.7 cm solar flux data are from the World Data Center A, Boulder, Colorado. NCEP Reanalysis Derived data provided by the NOAA/OAR/ESRL PSD, Boulder, Colorado, USA, from their Web site at http://www.cdc.noaa.gov/.

References

Baldwin, M. P. and T. J. Dunkerton, Stratospheric harbingers of anomalous weather regimes, *Science*, **294**, 581–584, 2001.

Baldwin, M. P. and Coauthors, The Quasi-Biennial Oscillation, *Rev. Geophys.*, **39**, 179–229, 2001.

Brönnimann, S., G. P. Compo, P. D. Sardeshmukh, R. Jenne, and A. Sterin, New approaches for extending the 20th century climate record, *EOS Trans AGU*, **86**, 2–7, 2005.

Brönnimann, S., J. L. Annis, C. Vogler, and P. D. Jones, Reconstructing the quasi-biennial oscillation back to the early 1900s, *Geophys. Res. Lett.*, **34**, L22805, doi:10.1029/2007GL031354, 2007.

Crooks, S. A. and L. J. Gray, Characterization of the 11-year solar signal using a multiple regression analysis of the ERA-40 dataset, *J. Clim.*, **18**(7), 996–1015, 2005.

Fritz, S. and S. D. Soules, Large-scale temperature changes in the stratosphere observed from Nimbus III, *J. Atmos. Sci.*, **27**, 1091–1097, 1970.

Gray, L., E. F. Drysdale, T. J. Dunkerton, and B. Lawrence, Model studies of the interannual variability of the northern hemisphere stratospheric winter circulation: The role of the Quasi-Biennial Oscillation, *Q. J. Roy. Met. Soc.*, **127**, 1985–2003, 2001.

Holton, J. and H. Tan, The influence of the equatorial Quasi-Biennial Oscillation on the global circulation at 50 mb, *J. Atmos. Sci.*, **37**, 2200–2208, 1980.

Hood, L. L., Thermal response of the tropical tropopause region to solar ultraviolet variations, *Geophys. Res. Lett.*, **30**(23), 2215, doi:10.1029/2003GL018364, 2003.

Hood, L. L., Effects of solar UV variability on the stratosphere, in *Solar variability and its effect on the Earth's atmosphere and climate system, AGU Monograph Series*, edited by J. Pap *et al.*, 283–304, American Geophysical Union, Washington D.C., 2004.

Hood, L. L. and S. Soukharev, Quasi-decadal variability of the tropical lower stratosphere: the role of extratropical wave forcing, *J. Atmos. Sci.*, **60**, 2389–2403, 2003.

Kalnay, E. and Co-authors, The NCEP/NCAR 40-year re-analysis project, *Bull. Am. Meteor. Soc.*, **77**, 437–471, 1996.

Kodera, K., Influence of stratospheric sudden warming on the equatorial troposphere, *Geophys. Res. Lett.*, **33**, L06804, doi:10.1029/2005GL024510, 2006.

Kodera, K. and Y. Kuroda, Dynamical response to the solar cycle, *J. Geophys. Res.*, **107**(D24), 4749, doi:10.1029/2002JD002224, 2002.

Kuroda, Y. and K. Kodera, Effect of solar activity on the polar-night jet oscillation in the northern and southern hemisphere winter, *J. Met. Soc. Jpn.*, **80**, 973–984, 2002.

Labitzke, K., The interaction between the stratosphere and mesosphere in winter, *J. Atmos. Sci.*, **29**, 1395–1399, 1972.

Labitzke, K., On the interannual variability of the middle stratosphere during the northern winters, *J. Met. Soc. Jpn.*, **60**, 124–139, 1982.

Labitzke, K., Sunspots, the QBO, and the stratospheric temperature in the north polar region, *Geophys. Res. Lett.*, **14**, 535–537, 1987.

Labitzke, K., The solar signal of the 11-year sunspot cycle in the stratosphere: Differences between the northern and southern summers, *J. Met. Soc. Jpn.*, **80**, 963–971, 2002.

Labitzke, K., The global signal of the 11-year sunspot cycle in the atmosphere: When do we need the QBO?, *Meteorolog. Z.*, **12**, 209–216, 2003.

Labitzke, K., On the signal of the 11-year sunspot cycle in the stratosphere over the Antarctic and its modulation by the Quasi-Biennial Oscillation (QBO), *Meteorolog. Z.*, **13**, 263–270, 2004a.

Labitzke, K., On the signal of the 11-year sunspot cycle in the stratosphere and its modulation by the Quasi-Biennial Oscillation (QBO), *J. Atmos. Sol.-Terr. Phys.*, **66**, 1151–1157, 2004b.

Labitzke, K., On the solar cycle-QBO relationship: a summary, *J. Atmos. Sol.-Terr. Phys.*, **67**, 45–54, 2005.

Labitzke, K. and J. J. Barnett, Global time and space changes of satellite radiances received from the stratosphere and lower mesosphere, *J. Geophys. Res.*, **78**, 483–496, 1973.

Labitzke, K. and Collaborators, *The Berlin Stratospheric Data Series*, CD from Meteorological Institute, Free University Berlin, 2002.

Labitzke, K. and K. Kunze, Stratospheric temperatures over the Arctic: Comparison of three data sets, *Meteorol. Z.*, **14**, 65–74, 2005.

Labitzke, K. and H. van Loon, The stratosphere in the Southern Hemisphere, in *Meteorology of the Southern Hemisphere*, edited by C. W. Newton, *Met. Monogr.*, **13**(35), Chapter 7, 113–138, 1972.

Labitzke, K. and H. van Loon, Associations between the 11-year solar cycle, the QBO and the atmosphere. Part I: The troposphere and stratosphere in the northern hemisphere winter, *J. Atmos. Terr. Phys.*, **50**, 197–206, 1988.

Labitzke, K. and H. van Loon, Associations between the 11-year solar cycle, the quasi-biennial oscillation and the atmosphere: a summary of recent work, *Phil. Trans. R. Soc. Lond.*, **A330**, 577–589, 1990.

Labitzke, K. and H. van Loon, The effect on the stratosphere of three tropical volcanic eruptions, in *The Mount Pinatubo eruption—Effects on the Atmosphere and Climate*, edited by Fiocco *et al.*, *Nato ASI Series I*, **42**, 113–125, Springer Berlin-Heidelberg, 1996.

Labitzke, K. and H. van Loon, *The Stratosphere (Phenomena, History, and Relevance)*, 179 pp., Springer, Berlin, Heidelberg, New York, 1999.

Labitzke, K., M. Kunze, and S. Brönnimann, Sunspots, the QBO, and the stratosphere in the North Polar region—20 years later, *Meteorolog. Z.*, **15**, 355–363, 2006.

Matthes, K., Y. Kuroda, K. Kodera, and U. Langematz, The transfer of the solar signal from the stratosphere to the troposphere: Northern winter, *J. Geophys. Res.*, **111**, D06108, doi:10.1029/2005JD006283, 2006.

Naujokat, B., An update of the observed Quasi-Biennial Oscillation of the stratospheric winds over the tropics, *J. Atmos. Sci.*, **43**, 1873–1877, 1986.

Pawson, S. and B. Naujokat, The cold winters of the middle 1990s in the northern middle stratosphere, *J. Geophys. Res.*, **104**, 14,209–14,222, 1999.

Salby, M. L. and P. Callaghan, Connection between the solar cycle and the QBO: The missing link, *J. Clim.*, **13**, 2652–2662, 2000.

Salby, M. L. and P. Callaghan, Evidence of the solar cycle in the general circulation of the stratosphere, *J. Clim.*, **17**, 34–46, 2004.

Salby, M. L. and P. Callaghan, Relationship of the quasi-biennial oscillation to the stratospheric signature of the solar cycle, *J. Geophys. Res.*, **111**, DO6110, doi:10.1029/2005JD006012, 2006.

Shea, D., H. van Loon, and K. Labitzke, Teleconections in the stratosphere, *Techn. Notes*, NCAR, Boulder/USA, 1991.

Shepherd, T. G., Issues in stratosphere-troposphere coupling, *J. Met. Soc. Jpn.*, **80**, 769–792, 2002.

Soukharev, B. and L. L. Hood, Possible solar modulation of the equatorial quasi-biennial oscillation: additional statistical evidence, *J. Geophys. Res.*, **106**, 14,855–14,868, 2001.

van Loon, H. and K. Labitzke, On the Southern Oscillation. Part V: The anomalies in the lower stratosphere of the Northern Hemisphere in winter and a comparison with the Quasi-Biennial Oscillation, *Monthly Weather Rev.*, **115**, 357–369, 1987.

van Loon, H. and K. Labitzke, Interannual variations in the stratosphere of the Northern Hemisphere: A description of some probable influences. Interactions between Global Climate Subsystems, The Legacy of Hann, *Geophys. Monogr.*, **15**, IUGG 15, 111–122, 1993.

van Loon, H. and K. Labitzke, The 10–12 year atmospheric oscillation, *Meteorolog. Z.*, **3**, 259–266, 1994.

van Loon, H. and K. Labitzke, The global range of the stratospheric decadal wave. Part I: Its association with the sunspot cycle in summer and in the annual mean, and with the troposphere, *J. Clim.*, **11**, 1529–1537, 1998.

van Loon, H. and K. Labitzke, The influence of the 11-year solar cycle on the stratosphere below 30 km: A review, *Space Sci. Rev.*, **94**, 259–278, 2000.

Climate and Weather of the Sun-Earth System (CAWSES): Selected Papers from the 2007 Kyoto Symposium,
Edited by T. Tsuda, R. Fujii, K. Shibata, and M. A. Geller, pp. 279–293.

Gravity wave coupling from below: A review

Robert A. Vincent

School of Chemistry and Physics, University of Adelaide, Adelaide, SA 5005, Australia
E-mail: robert.vincent@adelaide.edu.au

Gravity waves efficiently couple energy and momentum from source regions in the lower atmosphere to the middle and upper atmosphere. Wave dissipation results in a drag that drives meridional circulations that can profoundly affect the state of the middle atmosphere. New developments in various observing techniques, as well as a better appreciation of the limits imposed by observational selection, allow for a better understanding of the role of gravity waves in the atmosphere. Here we summarise recent advances in the measurement of wave parameters, especially those obtained by satellite and *in situ* techniques. Of particular importance is the development of techniques to measure momentum fluxes on a hemispheric or even global scale. Improved understanding of gravity wave coupling processes into the thermosphere-ionosphere has resulted from better formulations of dissipation effects. Allied with new rapid incoherent scatter measurements it may soon be possible to identify sources of small and medium scale ionospheric travelling disturbances that are not linked to auroral phenomena.

1 Introduction

Gravity waves, or more accurately buoyancy waves, are ubiquitous in the atmosphere. Wave periods range from minutes to hours and spatial scales extend up to thousands of km. They are important because they efficiently couple energy and momentum from their source regions in the lower atmosphere to the middle and upper atmosphere. Their presence in the lower atmosphere has been noted for over one hundred years, but is only relatively recently that their importance in the middle and upper atmosphere has been appreciated.

Wavelike perturbations in the ionospheric F-region are often referred to as travelling ionospheric disturbances (TIDs). Both early and more recent studies of TIDs recognized that they are often not generated by geomagnetic disturbances but have possible links to meteorological sources in the lower atmosphere (e.g. Munro, 1958; Waldock and Jones, 1987). It was the realization that TIDs were the manifestation of gravity waves that led in part to the seminal paper by Hines (1960) whose pioneering work also helped establish the importance of wave transfer of energy and momentum from the lower to the middle and upper atmosphere (Hines, 1974).

The last decade has seen new techniques brought to bear in the study of gravity waves and their effects in the middle atmosphere and thermosphere-ionosphere.

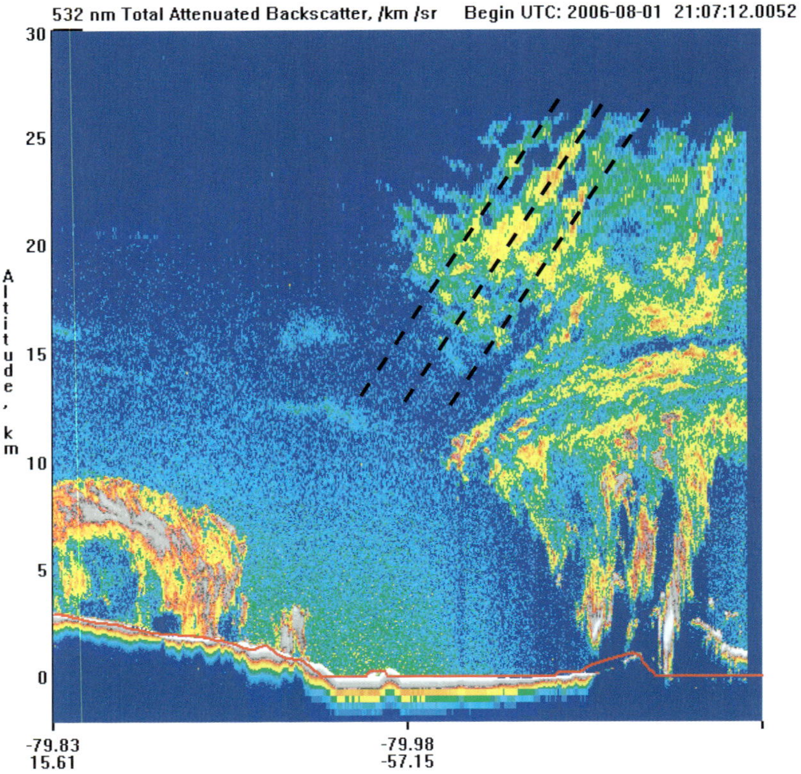

Fig. 1. Polar stratospheric clouds observed using a downward viewing lidar on the *Calipso* satellite as it passed over the Patagonian Andes and the Antarctic. The dashed lines show the phase fronts of a quasi-monochromatic gravity wave with an approximate vertical wavelength of about 10 km.

Satellite observations provide a global perspective and wave effects can be observed in somewhat unexpected ways. For example, Fig. 1 shows lidar signals transmitted from the *Calipso* satellite that are backscattered from polar stratospheric clouds during a pass over the Antarctic Peninsula. Gravity waves modulate the cloud density, as indicated by the dashed lines, with the tilt of the constant phase surfaces characteristic of upward propagating waves. These wave effects in the middle stratosphere are probably manifestations of mountain waves produced by the strong winter eastward flow over the Antarctic Peninsula.

Transfer of momentum by gravity waves is particularly important. The quantity $\rho_0 \overline{u'w'}$, where ρ_0 is atmospheric density and u' and w' are the horizontal and vertical perturbation amplitudes, is conserved in the absence of wave dissipation. Momentum deposition associated with wave breaking or dissipation exerts a body force on the atmosphere (Andrews *et al.*, 1987). The zonal-average zonal wind momentum

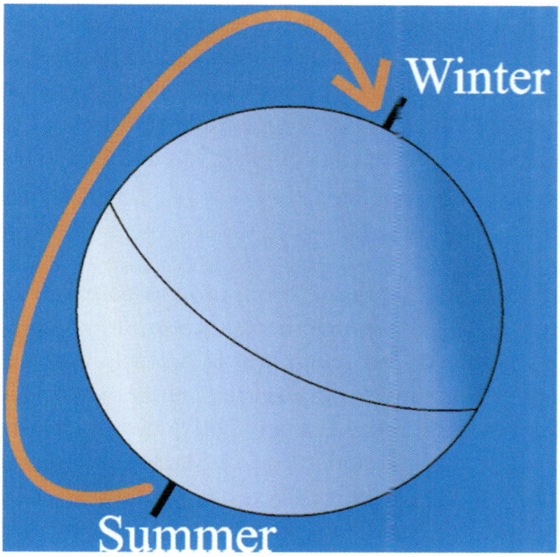

Fig. 2. Cartoon of the gravity wave driven summer-to-winter pole circulation at the mesopause.

equation is

$$\frac{\partial \overline{u}}{\partial t} - f\overline{v} = -\frac{1}{\rho_o}\frac{\partial\left(\rho_o\overline{u'w'}\right)}{\partial z} \tag{1}$$

At mid- to high-latitudes at the solstices the acceleration of the mean zonal wind $\overline{u}$ is small and there is a balance between the Coriolis term and flux divergence. At the solstices the wave drag drives a pole-to-pole circulation in the mesosphere that profoundly changes the thermal structure of this region. The upwelling over the summer pole (Fig. 2) and the associated adiabatic cooling produces temperatures that can be as low as 100 K at heights near 85 km. The cold temperatures lead to freezing of even the small quantities of water vapour present in the mesosphere and this is manifest in such phenomena as noctilucent clouds and polar mesospheric summer echoes observed with ground-based radars (e.g. Rapp and Lübken, 2004). The associated sinking of air over the winter pole leads to a warming whose effects can be felt as low as 20 km (Garcia and Boville, 1994).

In the tropics, where $f \sim 0$, the second term in Eq. (1) is negligible and so gravity waves can drive an acceleration of the zonal-mean zonal wind. Hence gravity waves probably play important roles in driving the quasi-biennial oscillation (QBO) in the stratosphere and semiannual oscillations near the stratopause and mesopause (Dunkerton, 1997; Baldwin *et al.*, 2001).

Full understanding of the role of gravity waves in the middle and upper atmosphere requires improved measurements on a global scale of wave parameters, especially momentum fluxes. Ideally these measurements should be made as a function of wave phase speed, so that their effects can be fully incorporated in numerical climate models and vertical coupling effects properly understood. Hamilton (1999) provides a background to early gravity wave observations and theoretical developments while Fritts and Alexander (2003) provide a recent and comprehensive review of gravity wave theory and effects. Here we review recent observations of gravity waves made using a variety of techniques as well as coupling effects.

2 Observations

2.1 Observational selection

A wide variety of techniques, ranging from ground-based to *in-situ* balloon-borne measurements and space-based measurements, are used to study gravity waves and their characteristics. However, no technique is perfect; it is not possible for one method to measure wave parameters across the whole wave spectrum (Alexander, 1998). For example, Alexander and Barnet (2007) discuss the response of various satellite viewing systems, as illustrated in Fig. 3. Satellite instruments that view along the limb, such as HIRDLS, have excellent vertical resolution, but relatively poor horizontal resolution. In contrast, a nadir-viewing instrument such as AIRS has excellent horizontal resolution compared with the vertical. Alexander *et al.* (2008b) made a detailed investigation of the application of the GPS radio occultation technique to gravity wave studies. Their analysis shows that best retrievals occur for waves with quasi-horizontal phase surfaces, i.e. for waves with quasi-inertial frequencies. They conclude that caution must be used in using occultation data to retrieve GW parame-

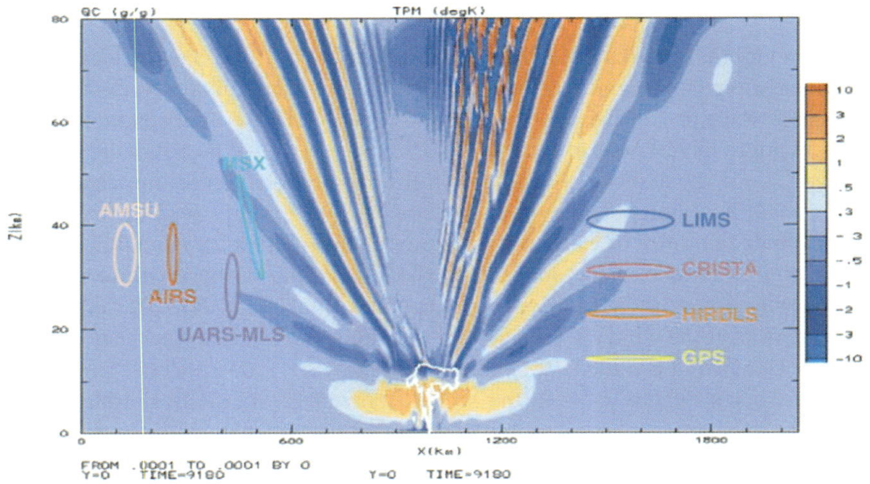

Fig. 3. Illustration of space-based observations of gravity waves. The background colours show temperature perturbations of modelled gravity waves, while the ovals illustrate the weighting function cross-sections for various satellite viewing modes. Those on the left are for downward viewing satellites, while the ovals on the right are for limb viewing geometries. (From figure 3, Alexander and Barnet (2007). See their paper for more details.)

ters.

Gravity wave dispersion is another selection factor that needs to be taken into account. Just because a gravity wave packet is in the viewing region of a particular instrument it does not necessarily mean that accurate measurements of wave amplitude and fluxes can be made. Short-period gravity waves tend to propagate vertically much faster than longer period waves. Consequently, a high frequency wave packet will propagate up through a region of observation more quickly than a longer period packet (Alexander and Barnet, 2007) so that observations of long-period waves are favoured.

2.2 Satellite observations

The advent of microsatellites in low earth orbit, such as GPS/MET, CHAMP and the COSMIC constellation, has allowed gravity wave activity to be studied globally using the GPS occultation technique (e.g. Tsuda *et al.*, 2000; de la Torre *et al.*, 2004, 2006; Ratnam *et al.*, 2004; Baumgaertner and McDonald, 2007; Hei *et al.*, 2008). As a low earth orbit satellite sets, the signals from an occulting GPS satellite are refracted by the intervening atmosphere and ionosphere. The resulting ray bending gives information on the vertical distributions of electron density, temperature and, low in the atmosphere, humidity. Gravity wave potential energy per unit mass information is derived from the temperature profiles

$$E_{\mathrm{p}} = \frac{1}{2}\frac{g^2}{N^2}\frac{T'^2}{T_{\mathrm{o}}^2}, \tag{2}$$

where N is the Brunt frequency, g is the acceleration due to gravity and T' and T_{o} are the temperature perturbation and mean temperature, respectively. As shown in Fig. 4, all the necessary information to compute E_{p} can be derived from the GPS

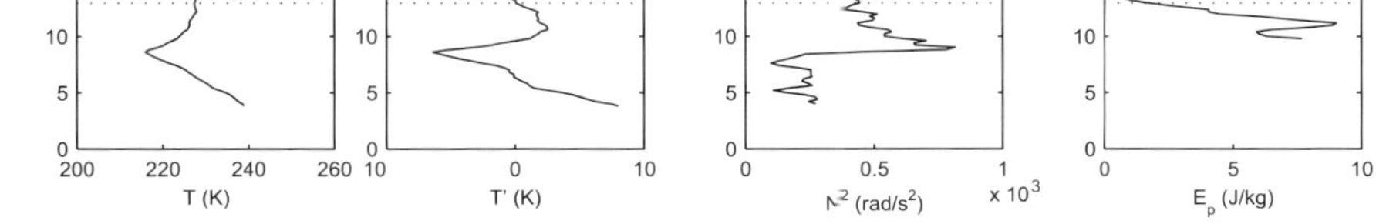

Fig. 4. GPS estimates of gravity wave perturbation amplitudes. (From Baumgaertner and McDonald, A gravity wave climatology for Antarctica compiled from Challenging Minisatellite Payload/Global Positioning System (CHAMP/GPS) radio occultations, *J. Geophys. Res.*, **112**, D05103, 2007. Copyright 2007 American Geophysical Union. Reproduced by permission of American Geophysical Union.)

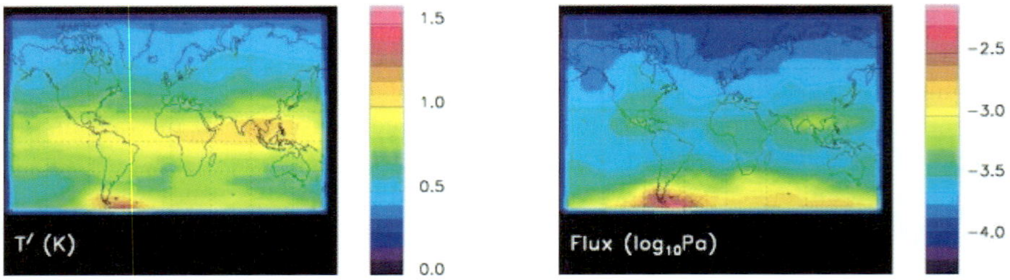

Fig. 5. HIRDLS observations at 25 km for May 2006. (Left) Temperature perturbations. (Right) Absolute momentum flux, $\rho_0 \left|\overline{u'w'}\right|$. (From Alexander *et al.*, Global estimates of gravity wave momentum flux from High Resolution Dynamics Limb Sounder (HIRDLS) observations, *J. Geophys. Res.*, **113**, D15S18, 2008. Copyright 2008 American Geophysical Union. Reproduced by permission of American Geophysical Union.)

measurements, although as noted by Alexander *et al.* (2008b), caution is required in deriving GW parameters from occultation data.

Using several years of CHAMP data, Baumgaertner and McDonald (2007) were able to study the seasonal cycle in GW E_p in the Antarctic lower stratosphere at heights between ~10 and 35 km. There is an annual cycle in E_p with largest amplitudes in winter, a result similar to that reported by Zink and Vincent (2001) for the SH mid-latitude site of Macquarie Island. Baumgaertner and McDonald (2007) also reported an interesting longitudinal variation of wave activity, as well as a strong enhancement of wave energy at the edge of the polar vortex. Their results show a strong relationship between gravity wave activity and geographic location, indicating that topography is a strong source for wave activity, especially over the Antarctic Peninsula and the Southern Andes. Mountain waves from this region have even been linked to disturbances in the lower ionosphere (Hocke *et al.*, 2002). On the other hand, an analysis by Hei *et al.* (2008) of 5 years of GPS data from the CHAMP satellite for both polar regions indicates the limited geographical extent of topographically forced waves and the importance of waves that they ascribe to gravity generation by planetary wave transience and/or breaking.

In the equatorial stratosphere, Tsuda *et al.* (2000), Alexander *et al.* (2002) and de la Torre *et al.* (2006) all report a maximum in GW E_p extending between about 1°S and 10°N. However, the E_p variations show significant geographic as well as interannual variability. de la Torre *et al.* (2006) found interannual enhancements in wave activity related to the QBO, similar to the findings of Vincent and Alexander (2000) made using radiosonde observations at Christmas Island (12°S) in the Indian Ocean. Geographic variability was discussed by Tsuda *et al.* (2000), who reported increases in E_p in regions with strong convection, such as the Maritime Continent in the Western Pacific.

Local enhancements in E_p, however, do not necessarily mean correspondingly large momentum fluxes. As discussed by Ern *et al.* (2004) and Alexander *et al.*

(2008a), the relation between absolute momentum flux and temperature variance is

$$\left|u'w'\right| = \frac{1}{2}\frac{\lambda_z}{\lambda_h}\frac{g^2}{N^2}\frac{T'^2}{T_o^2} \tag{3}$$

where λ_z and λ_h are the GW vertical and horizontal wavelengths, respectively. Comparison of vertical temperature profiles derived from adjacent scans through the atmosphere by the HIRDLS instrument, Alexander *et al.* (2008a) were able to estimate both λ_z and λ_h and so derive absolute fluxes. This is illustrated in Fig. 5, which shows average values of T' and $\rho_o\left|u'w'\right|$ for May 2006 (SH early winter). Note not only the relatively high values of T' over Southeast Asia, but also the correspondingly small flux values. In contrast, there are large temperature perturbations and large fluxes over the southern Andes. These geographic variations in flux can be ascribed to source-related differences in the ratio of λ_z and λ_h, which is an important factor in determining GW momentum via Eq. (3). Furthermore, using HIRDLS data Alexander *et al.* (2008a) were also able to follow waves generated over the southern Andes up to heights in the lower mesosphere.

2.3 *In-situ* observations

2.3.1 Radiosonde observations

A relatively inexpensive way to derive information on gravity wave activity in the lower stratosphere is to use conventional radiosonde soundings made by national weather agencies (Allen and Vincent, 1995). High vertical resolution (~20–50 m) soundings are now routinely made and archived by many weather agencies. Analysis of these data enables GW to be resolved, although selection of short vertical wavelength ($\lambda_z < 7$ km) waves is favoured (Alexander, 1998).

Wang and Geller (2003) and Wang *et al.* (2005) exploited routine high-resolution soundings of winds and temperatures made at more than 90 stations to study GW activity in the troposphere and lower stratosphere over the continental USA, Alaska and islands in the Pacific and the Caribbean. Analysis of a 4-year dataset that covers a wide variety of terrain ranging from oceanic to mountainous enabled them to distinguish source from propagation effects. Wang and Geller (2003) reported an overall decrease of stratospheric wave activity with increasing latitude, similar to previous studies (e.g. Allen and Vincent, 1995) in the Australian region. Seasonally, stratospheric wave amplitudes are larger in winter than in summer. Interestingly, tropospheric amplitudes are not always well correlated with stratospheric amplitudes. For example, in the troposphere there are clear GW energy maxima over the Rocky Mountains, whereas in the stratosphere maxima occur over the southeastern United States.

Combining wind and temperature profiles gives propagation directions in both the vertical and horizontal (Vincent *et al.*, 1997). Using the US radiosonde dataset Wang *et al.* (2005) showed that at least 75% of wave energy is up going in the lower stratosphere. They also report that the ratio of mean intrinsic frequency to inertial frequency is ~2.4–3, with a weak latitudinal dependence. These results are consistent with previous studies. The dominance of low intrinsic frequency waves is probably

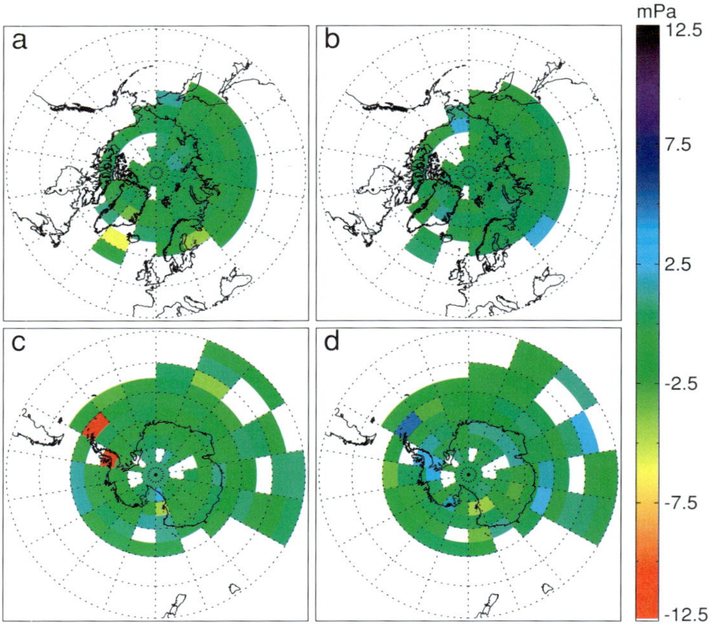

Fig. 6. Average values of momentum fluxes from SPB observations for the Arctic (top) and Antarctic (bottom). Panels a and c refer to zonal fluxes, while panels b and d show meridional values. Note the large fluxes over Greenland and the Antarctic Peninsula.

an observational effect, since the ~ 5 m s^{-1} ascent rate of radiosondes is inadequate for sampling long vertical wavelength, high phase speed waves—i.e. this is another manifestation of the "observational selection" issue discussed by Alexander (1998) and Alexander and Barnet (2007).

2.3.2 Superpressure balloon observations

Superpressure balloons (SPB) provide a powerful means for studying GW in the lower stratosphere (Hertzog and Vial, 2001). The envelope of an SPB is inextensible so when it is injected with a fixed amount of gas the pressure inside is greater than ambient pressure at the float level. Hence the balloon floats on a constant density (isopycnic) surface. The advantage of the SPB technique over ground or space-based techniques is that the balloons are advected by the background wind and hence measure the frequency of waves relative to the background flow, the so-called intrinsic frequency, $\hat{\omega}$. It is $\hat{\omega}$ that determines wave properties (e.g. Fritts and Alexander, 2003).

SPB developed by CNES, the French Space Agency, have 8 and 10-m diameters and float at pressure levels between 75 and 55 hPa, corresponding to altitudes of about 17 and 18 km, respectively. The balloons carry a small gondola to measure pressure p and temperature T as well as position using GPS techniques. Early results

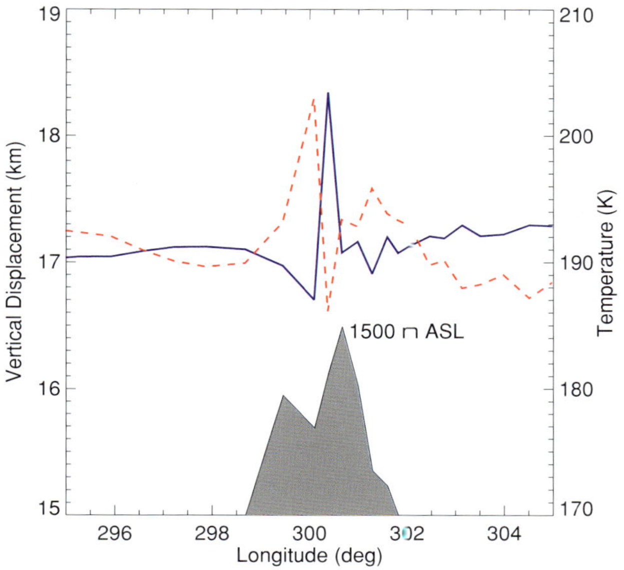

Fig. 7. Vertical displacement (blue, left scale) and temperature (red, right scale) recorded on an 8-m diameter SPB during its passage over the Antarctic Peninsula. The shaded region shows the cross-section through the Peninsula.

were limited by the accuracy of the GPS measurements and by the relatively slow 15-min sampling interval dictated by the rate at which data could be transmitted back to ground stations via the Argos satellite system. More information about the SPB technique and how the data are converted to GW momentum fluxes is given in Vincent *et al.* (2007), while Boccara *et al.* (2008) provide a detailed error analysis of the retrieved fluxes and phase speeds.

A major SPB campaign was made from McMurdo (78°S) in the SH spring of 2005 during which 27 balloons were launched to study the evolution of the vortex. In preparation for this campaign a number of test flights were made from Kiruna (68°N). Results from Vorcore and the Kiruna test flights enable GW momentum fluxes in the northern and southern polar lower stratospheres to be compared. Figure 6 shows density-weighted ($\rho_o\,\overline{u'w'}$, $\rho_o\,\overline{v'w'}$) momentum fluxes for flights in the Arctic and Antarctic (Vincent *et al.*, 2007). Note the large flux values in the vicinity of mountain ranges, such as eastern Greenland, the southern Andes and the Antarctic Peninsula. Overall, however, when the fluxes are zonally averaged over regions of high topography and oceanic regions it is found that the fluxes are comparable in magnitude (Hertzog *et al.*, 2008).

Wave amplitudes observed over mountainous regions appear to be significantly more intermittent than values measured over the oceans, with very large fluxes sometimes observed over mountains (Hertzog *et al.*, 2008). Figure 7 illustrates an extreme case where an 8-m diameter balloon suffered an approximate 1.5 km vertical displacement during a severe mountain wave event over the Antarctic Peninsula. The corresponding peak-to-peak temperature change was about 15 K. From the estimated 1 hr intrinsic period and approximate horizontal and vertical perturbation velocities of 10 m s^{-1} and 0.7 m s^{-1}, respectively, it is possible to estimate an average momentum flux of 0.7 Pa. This is a factor of about 70 times larger than the average flux in this region. Plougonven *et al.* (2008) provide a more detailed analysis of this event. Such large amplitude waves will break low in the stratosphere, whereas weaker amplitude waves can penetrate well into the middle atmosphere before breaking and depositing their momentum (e.g. Alexander *et al.*, 2008a).

3 Gravity Wave Coupling into the Thermosphere

Large-scale gravity waves associated with magnetic storms are a well-known feature of the ionosphere. After a major storm wave disturbances can produce ionospheric effects on a global scale (e.g. Karpachev *et al.*, 2007). Smaller scale waves propagating from sources such as tropical convection in the lower atmosphere have been suggested as the cause of equatorial spread-F (e.g. McClure *et al.*, 1998; Prakash, 1999), although Kudeki *et al.* (2007) recently questioned the need for gravity wave seeding of spread-F.

Identifying the sources of waves observed in the thermosphere-ionosphere is a challenging problem. To ray-trace backwards a wave observed in the ionosphere to a lower atmospheric source requires good knowledge of the intermediate wind and temperature fields. These cause the waves to be refracted and advected horizontally. The problem is particularly acute in the region near 100 km where there are often large amplitude tidal wind and temperature fluctuations and very large wind shears (Larsen, 2002). Molecular viscosity and thermal diffusive damping effects also become important in the region above 100 km. This complicates the application of GW ray-tracing techniques.

Recently, Vadas and Fritts (2005) provided an improved form of the GW dispersion relation that incorporates kinematic effects. Applying this new formulation Vadas and Fritts (2004) employed a linear model to study the thermospheric response to a GW spectrum generated by vertical motions induced in a tropospheric mesoscale convective complex. Molecular viscosity and thermal diffusivity act as selective filters on a GW spectrum, allowing only the high-frequency, largest vertical wavelength, waves to propagate to the highest altitudes (Vadas and Fritts, 2005). Strong body forces can be induced by the resulting momentum flux deposition at heights well into the thermosphere. There is also a significant solar cycle effect. GWs penetrate to much higher altitudes during active solar conditions when thermospheric temperatures are high compared with solar minimum conditions (Vadas and Fritts, 2006). This is in part because kinematic viscosity decreases more slowly with altitude when temperature increases, and in part due to the increase in λ_z associated with the higher temperatures.

Vadas (2007) recently explored the properties of GW propagating into the thermosphere-ionosphere in more detail, including the effects of both vertical and horizontal

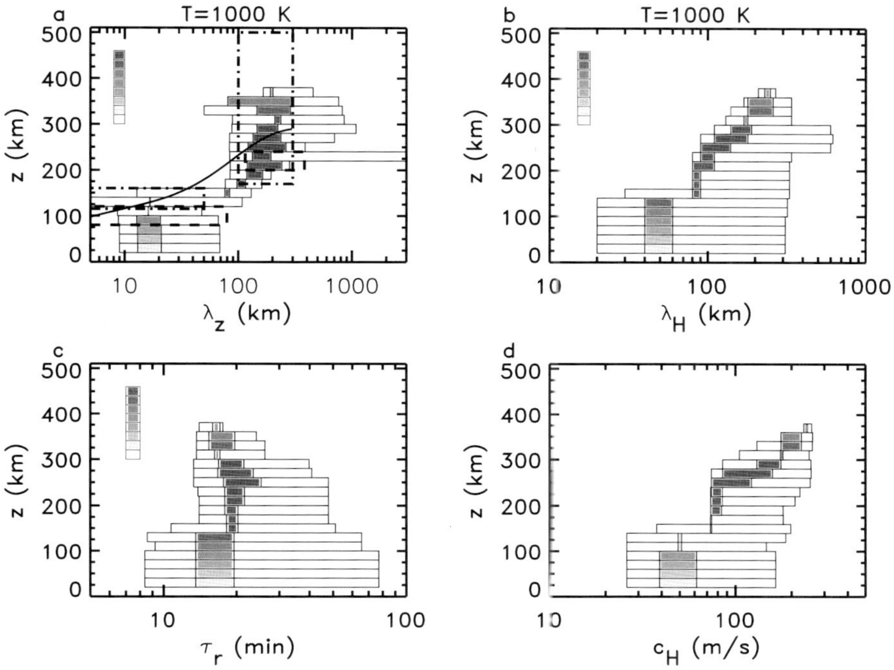

Fig. 8. Estimated GW spectra as a function of altitude for GWs generated by a deep convective plume and propagating into a thermosphere with an exospheric temperature of 1000 K. A sudden shear of -100 m s^{-1} is assumed at $z = 120$ km. (a) Vertical wavelength spectrum. (b) Horizontal wavelength spectrum. (c) Ground-based wave period spectrum. (d) Ground-based horizontal phase speed spectrum. Shading shows regions of strongest amplitude. (From figure 16, Vadas, Horizontal and vertical propagation and dissipation of gravity waves in the thermosphere from lower atmospheric and thermospheric sources, *J. Geophys. Res.*, **112**, A06305, 2007. Copyright 2007 American Geophysical Union. Reproduced by permission of American Geophysical Union.)

dispersion. Figure 8 shows the evolution of GW spectra as a function of altitude for waves generated by a deep convective plume in the lower atmosphere and with a background wind that is westward in the thermosphere. Note how the spectra are dominated by waves with periods ~15–20 min and that have horizontal scales of about 200–400 km. The largest amplitude waves are propagating eastward, against the mean flow. It should be noted that the effects of ion drag, wave breaking and eddy viscosity are neglected in this calculation (see Vadas (2007) for more details). Despite these limitations the results are in broad agreement with observations.

Sun *et al.* (2007) also explored the problem of GW propagation into the thermosphere using a three-dimensional transfer function technique. Similar to Vadas (2007) they found that GWs in a 15–30 min period band and horizontal wavelength ~200–400-km were most likely to propagate to the 300-km level.

While these studies have made significant improvements to our understanding of

GW propagation into the thermosphere-ionosphere system inclusion of the effects of the ionosphere itself are required, since wave motions move charge in the presence of the Earth's magnetic field. This leads in turn to selective damping, depending on the propagation direction of the waves relative to the magnetic field and the associated ion-drag effects (e.g. Hines, 1968; Tsugawa *et al.*, 2004).

4 Summary and Conclusions

The past decade has seen a great improvement in our understanding of GW coupling and their role in determining the state of the middle and upper atmosphere. Improvements result from the development and application of new observing techniques as well as better appreciation of observational limitations. New satellite and balloon measurements, in particular, provide better understanding of flux variability on both regional and global scales.

In summary:

1. The majority of the waves observed in the lower stratosphere couple energy and momentum upward into the middle and upper atmosphere.
2. Largest momentum fluxes are observed over regions of high topography, but these regions have the greatest wave variability.
3. On a zonally averaged basis, momentum fluxes over mountains and oceans are approximately equal.
4. Thermospheric GWs are selectively filtered by the kinematic dissipation. Only the high frequency, long vertical wavelength components to penetrate to the highest altitudes.
5. There is a strong solar cycle effect in GW propagation into the thermosphere. GWs propagate to higher altitudes during high sunspot conditions than during solar minimum conditions.

Despite the significant progress made in the past decade there are still a number of uncertainties. While it can be relatively straightforward to identify waves associated with topography, it is less easy to identify the GW sources over the ocean. The relative importance of different sources, such as shear and geostrophic adjustment, has still to be established. This is especially true for mid- to high-latitude sources. Convection is the main GW source in the tropics and there are now a number of models that attempt to specify wave fluxes above convection (e.g. Song *et al.*, 2003; Beres *et al.*, 2004; Lane and Moncrieff, 2008), although the model results have yet to be fully tested by observation.

With the deployment of new instruments we can expect good progress in understanding wave coupling into the thermosphere-ionosphere system. New radars, such as the Advanced Modular Incoherent Scatter Radar (AMISR), make three-dimensional simultaneous measurements of ionospheric parameters (Nicolls and Heinselman, 2007). This flexibility provides better insight into the relationship between TIDs and gravity waves and outcomes give hope for better identification of wave sources. In turn, improved appreciation of wave dissipation effects in the thermosphere can now be used to provide better understanding of atmospheric parameters in a region that is otherwise difficult to probe. For example, using GW observations made with AMISR, Vadas and Nicolls (2008) were able to infer the average background horizontal winds in the 160–240 km height range over Poker Flat, Alaska.

The challenge now is to apply these new instrumental developments, as well as advances in theory and modelling, to understand better upward coupling by gravity waves. This will be a major focus of research during CAWSES-II.

Acknowledgments. Discussions with M. J. Alexander, M. Geller, A. Hertzog and A. Klekociuk are gratefully acknowledged.

References

Alexander, M. J., Interpretations of observed climatological patterns in stratospheric gravity wave variance, *J. Geophys. Res.*, **103**, 8627–8640, 1998.

Alexander, M. J. and C. Barnet, Using satellite observations to constrain parameterizations of gravity wave effects for global models, *J. Atmos. Sci.*, **64**, 1652–1665, 2007.

Alexander, M. J., T. Tsuda, and R. A. Vincent, Latitudinal variations observed in gravity waves with short vertical wavelengths, *J. Atmos. Sci.*, **59**, 1394–1404, 2002.

Alexander, M. J., J. C. Gille, C. Cavanaugh, M. Coffey, C. Craig, V. Dean, T. Eden, G. Francis, C. Halvorson, T. Hannigan, R. Khosravi, D. Kinneson, H. Lee, S. Massie, N. Nardi, and A. Lambert, Global estimates of gravity wave momentum flux from High Resolution Dynamics Limb Sounder (HIRDLS) observations, *J. Geophys. Res.*, **113**, D15S18, doi:10.1029/2007JD008807, 2008a.

Alexander, P., A. de la Torre, and P. Llamedo, Interpretation of gravity wave signatures in GPS radio occultations, *J. Geophys. Res.*, **113**, D16117, doi:10.1029/2007JD009390, 2008b.

Allen, S. J. and R. A. Vincent, Gravity wave activity in the lower atmosphere: Seasonal and latitudinal variations, *J. Geophys. Res.*, **100**, 1327–1350, 1995.

Andrews, D. G., J. R. Holton, and C. B. Leovy, *Middle Atmosphere Dynamics*, 489 pp., Academic, San Diego, Calif, 1987.

Baldwin, M. P., L. J. Gray, T. J. Dunkerton, K. Hamilton, P. H. Haynes, W. J. Randel, J. R. Holton, M. J. Alexander, I. Hirota, T. Horinouchi, D. B. Jones, J. S. Kinnersley, C. Marquardt, K. Sato, and M. Takahashi, The quasi-biennial oscillation, *Rev. Geophys.*, **39**, 179–229, 2001.

Baumgaertner, A. J. G. and A. J. McDonald, A gravity wave climatology for Antarctica compiled from Challenging Minisatellite Payload/Global Positioning System (CHAMP/GPS) radio occultations, *J. Geophys. Res.*, **112**, D05103, doi:10.1029/2006JD007504, 2007.

Beres, J. H., M. J. Alexander, and J. R. Holton, A Method of specifying the gravity wave spectrum above convection based on latent heating properties and background wind, *J. Atmos. Sci.*, **61**, 324–337, 2004.

Boccara, G., A. Hertzog, R. A. Vincent, and F. Vial, Estimation of gravity-wave momentum fluxes and phase speeds from quasi-Lagrangian stratospheric balloon flights. 1: Theory and simulations, *J. Atmos. Sci.*, **65**, 3042–3055, 2008.

de la Torre, A., T. Tsuda, G. A. Hajj, and J. Wickert, A global distribution of the stratospheric gravity wave activity from GPS occultation profiles with SAC-C and CHAMP, *J. Meteor. Soc. Jpn.*, **82**, 407–417, 2004.

de la Torre, A., T. Schmidt, and J. Wickert, A global analysis of wave potential energy in the lower stratosphere derived from 5 years of GPS radio occultation data with CHAMP, *Geophys. Res. Lett.*, **33**, L24809, doi:10.1029/2006GL027696, 2006.

Dunkerton, T. J., The role of gravity waves in the quasi-biennial oscillation, *J. Geophys. Res.*, **102**, 26,053–26,076, 1997.

Ern, M., P. Preusse, M. J. Alexander, and C. D. Warner, Absolute values of gravity wave momentum flux derived from satellite data, *J. Geophys. Res.*, **109**, D20103, doi:10.1029/2004JD004752, 2004.

Fritts, D. C. and M. J. Alexander, Gravity wave dynamics and effects in the middle atmosphere, *Rev. Geophys.*, **41**, 1003, doi:10.1029/2001RG000106, 2003.

Garcia, R. R. and B. A. Boville, "Downward control" of the mean meridional circulation and temperature distribution of the polar winter stratosphere, *J. Atmos. Sci.*, **51**, 2238–2245, 1994.

Hamilton, K., Dynamical coupling of the lower and middle atmosphere: Historical background to current research, *J. Atmos. Sol.-Terr. Phys*, **61**, 73–84, 1999.

Hei, H., T. Tsuda, and T. Hirooka, Characteristics of atmospheric gravity wave activity in the polar regions revealed by GPS radio occultation data with CHAMP, *J. Geophys. Res.*, **113**, D04107, doi:10.1029/2007 JD008938, 2008.

Hertzog, A. and F. Vial, A study of the dynamics of the equatorial lower atmosphere by use of ultra-long-duration balloons 2. Gravity waves, *J. Geophys. Res.*, **106**, 22,745–722,761, 2001.

Hertzog, A., G. Boccara, R. A. Vincent, F. Vial, and P. Cocquerez, Estimation of gravity-wave momentum fluxes and phase speeds from quasi-Lagrangian stratospheric balloon flights. 2: Results from the Vorcore campaign in Antarctica, *J. Atmos. Sci.*, **65**, 3056–3070, 2008.

Hines, C. O., Internal atmospheric gravity waves at ionospheric heights, *Can. J. Phys.*, **38**, 1441–1481, 1960.

Hines, C. O., The effect of ohmic losses in upper atmospheric gravity waves, *J. Atmos. Sol.-Terr. Phys.*, **30**, 851–856, 1968.

Hines, C. O., *The Upper Atmosphere in Motion*, 1027 pp., AGU, Washington, D.C., 1974.

Hocke, K., T. Tsuda, and A. de la Torre, A study of stratospheric GW fluctuations and sporadic E at mid-latitudes with focus on possible orographic effect of Andes, *J. Geophys. Res.*, **107**, 4428, doi:10.1029/2001JD001330, 2002.

Karpachev, A. T., G. F. Deminova, N. Beloff, T. D. Carozzi, P. F. Denisenko, T. J. T. Karhunen, and M. Lester, Global pattern of the ionospheric response to large-scale internal gravity waves, *J. Atmos. Sol.-Terr. Phys.*, **69**, 906–924, 2007.

Kudeki, E., A. Akgiray, M. Milla, J. L. Chau, and D. L. Hysell, Equatorial spread-*F* initiation: Post-sunset vortex, thermospheric winds, gravity waves, *J. Atmos. Sol.-Terr. Phys.*, **69**, 2416–2427, 2007.

Lane, T. P. and M. W. Moncrieff, Stratospheric gravity waves generated by multiscale tropical convection, *J. Atmos. Sci.*, **65**, 2598–2614, 2008.

Larsen, M. F., Winds and shears in the mesosphere and lower thermosphere: Results from four decades of chemical release wind measurements, *J. Geophys. Res.*, **107**, doi:10.1029/2001JA0000218, 2002.

McClure, J. P., S. Singh, D. K. Bamgboye, F. S. Johnson, and H. Kil, Occurrence of equatorial *F* region irregularities: Evidence for tropospheric seeding, *J. Geophys. Res.*, **103**, 29,119–29,126, 1998.

Munro, G. H., Travelling ionospheric disturbances in the *F*-region, *Aust. J. Phys.*, **11**, 91–112, 1958.

Nicolls, M. J. and C. J. Heinselman, Three-dimensional measurements of travelling ionospheric disturbances with the Poker Flat incoherent scatter radar, *Geophs. Res. Lett.*, **34**, L21104, doi:10.1029/2007GL031506, 2007.

Plougonven, R., A. Hertzog, and H. Teitelbaum, Observations and simulations of a large-amplitude mountain wave breaking over the Antarctic Peninsula, *J. Geophys. Res.*, **113**, D16113, doi:10.1029/2007JD009739, 2008.

Prakash, S., Production of electric field perturbations by gravity wave winds in the *E* region suitable for initiating equatorial spread *F*, *J. Geophys. Res.*, **104**, 10,051–10,069, 1999.

Rapp, M. and F.-J. Lübken, Polar mesosphere summer echoes (PMSE): Review of observations and current understanding, *Atmos. Chem. Phys*, **4**, 2601–2633, 2004.

Ratnam, M. V., G. Tetzlaff, and C. Jacobi, Global and seasonal variations of stratospheric gravity wave activity deduced from the CHAMP/GPS satellite, *J. Atmos. Sci.*, **61**, 1610–1620, 2004.

Song, I.-S., H.-Y. Chun, and T. P. Lane, Generation mechanisms of convectively forced internal gravity waves and their propagation to the stratosphere, *J. Atmos. Sci.*, **60**, 1960–1980, 2003.

Sun, L., W. Wan, F. Ding, and T. Ma, Gravity wave propagation in the realistic atmosphere based on a three-dimensional transfer function model, *Ann. Geophys.*, **25**, 1979–1986, 2007.

Tsuda, T., M. Nishida, C. Rocken, and R. Ware, Global morphology of gravity wave activity in the stratosphere revealed by the GPS occultation Data (GPS/MET), *J. Geophys. Res.*, **105**, 7257–7274, 2000.

Tsugawa, T., A. Saito, and Y. Otsuka, A statistical study of large-scale travelling ionospheric disturbances using the GPS network in Japan, *J. Geophys. Res.*, **109**, A06302, doi:10.1029/2003JA010302, 2004.

Vadas, S. L., Horizontal and vertical propagation and dissipation of gravity waves in the thermosphere from lower atmospheric and thermospheric sources, *J. Geophys. Res.*, **112**, A06305, doi:10.1029/2006JA011845, 2007.

Vadas, S. L. and D. C. Fritts, Thermospheric responses to gravity waves arising from mesoscale convective complexes, *J. Atmos. Sol.-Terr. Phys.*, **66**, 781–804, 2004.

Vadas, S. L. and D. C. Fritts, Thermospheric responses to gravity waves: Influences of increasing viscosity and thermal diffusivity, *J. Geophys. Res.*, **110**, D15103, doi:10.1029/2004JD005574, 2005.

Vadas, S. and D. C. Fritts, Influence of solar variability on gravity wave structure and dissipation in the thermosphere from tropospheric convection, *J. Geophys. Res.*, **111**, A10S12, doi:10.1029/2005JA011510, 2006.

Vadas, S. L. and M. J. Nicolls, Using PFISR measurements and gravity wave dissipative theory to determine the neutral, background thermospheric winds, *Geophys. Res. Lett.*, **35**, L02105, doi:10.1029/2007GL031522, 2008.

Vincent, R. A. and M. J. Alexander, Gravity waves in the tropical lower stratosphere: An observational study of seasonal and interannual variability, *J. Geophys. Res.*, **105**, 17,971–17,982, 2000.

Vincent, R. A., S. J. Allen, and S. D. Eckermann, Gravity-wave parameters in the lower stratosphere, in *Gravity wave processes: Their parameterization in global climate models*, edited by K. Hamilton, 7–25, Springer-Verlag, Berlin, 1997.

Vincent, R. A., A. Hertzog, G. Boccara, and F. Vial, Balloon-borne measurements of gravity-wave momentum fluxes in the polar stratosphere, *Geophys. Res. Lett.*, **34**, L19804, doi:10.1029/2007GL031072, 2007.

Waldock, J. A. and T. B. Jones, Source regions of medium scale travelling ionospheric disturbances observed at midlatitudes, *J. Atmos. Sol.-Terr. Phys.*, **49**, 105–114, 1987.

Wang, L. and M. A. Geller, Morphology of gravity-wave energy as observed from 4 years (1998–2001) of high vertical resolution U.S. radiosonde data, *J. Geophys. Res.*, **108**, 4489, doi:10.1029/2002JD002786, 2003.

Wang, L., M. A. Geller, and M. J. Alexander, Spatial and temporal variations of gravity wave parameters. Part I: Intrinsic frequency, wavelength, and vertical propagation direction, *J. Atmos. Sci.*, **62**, 125–142, 2005.

Zink, F. and R. A. Vincent, Wavelet analysis of stratospheric gravity wave packets over Macquarie Island, 1, Wave parameters, *J. Geophys. Res.*, **106**, 10,275–10,287, 2001.

Climate and Weather of the Sun-Earth System (CAWSES): Selected Papers from the 2007 Kyoto Symposium,
Edited by T. Tsuda, R. Fujii, K. Shibata, and M. A. Geller, pp. 295–336.

What we have learnt from CPEA (Coupling Processes in the Equatorial Atmosphere): A review

Shoichiro Fukao[1,2]*

[1]*Research Institute for Sustainable Humanosphere, Kyoto University, Uji, Kyoto 611-0011, Japan*
E-mail: fukao@rish.kyoto-u.ac.jp
[2]*Research Institute of Science and Technology, Tokai University, Tomigaya, Shibuya, Tokyo 151-0063, Japan*

It is well recognized that active cumulus convection in the equatorial region is the source of many unique atmospheric processes that couples the entire atmosphere vertically from bottom to top. In particular, it generates a variety of atmospheric waves, which propagate upward and interact with the background mean flow at high altitudes. The atmospheric waves cause typical global-scale oscillations, such as the quasi-biennial oscillation (QBO) and semiannual oscillation (SAO). It is suspected that equatorial plasma irregularities in the ionosphere are seeded by small-scale atmospheric waves. The mechanisms of these coupling processes, however, have not yet been made clear due, primarily, to the sparseness of observational data. The Indonesian Maritime Continent (IMC; or the Indonesian Archipelago) in particular is known for its convection, the most active and strongest over the globe, and consequently the IMC is considered as the region where vertical coupling takes place most conspicuously. Coupling Processes in the Equatorial Atmosphere (CPEA) is a research project of Japan developed to study dynamical and electrodynamical coupling processes in the equatorial atmosphere by conducting various observations in the Indonesian equatorial region. The experimental campaigns of CPEA covered the six-year period September 2001 to March 2007. The present review outlines the CPEA project and presents new results first demonstrated by CPEA.

1 Introduction

In the equatorial region, atmospheric convection is driven primarily by an imbalance of net solar radiation: a net radiative gain near the Earth's surface and radiative cooling at upper levels. Such an imbalance, if static, would lead to a continuous increase in temperature in the equatorial region. In the real atmosphere, however, convection causes the rise of warm air in the equatorial region followed by sinking of cold air at higher latitudes, forming a meridional overturning cell (Hadley circulation) (e.g., Holton, 1972). This circulation acts to transfer heat from low to higher

*Now at Department of Space Communication Engineering, Fukui University of Technology, Gakuen, Fukui 910-8505, Japan.

latitudes. Thus, convection characterizes the dynamics and determines the structure of the equatorial atmosphere, spanning from synoptic to cumulus scales, with each scale playing an important role in maintaining the balance of heat, moisture, and momentum in the atmosphere. However, it is not easy in practice to observe simultaneously both the cumulus scale motions and the synoptic scale disturbances in which the cumulus motions are embedded.

Convection in the equatorial region is thought to be the source of many unique atmospheric processes that couples the entire atmosphere vertically from bottom to top and horizontally from equator to pole as well. Nevertheless, the equatorial atmosphere is not yet well understood and the coupling processes are poorly understood. For instance, the active convection in the equatorial region generates a variety of atmospheric waves, such as equatorial waves, atmospheric tides, and internal atmospheric gravity waves in a manner yet poorly understood (e.g., Andrews *et al.*, 1987).

The minimal Coriolis force effect present in the equatorial region allows the widest range of wave periods for their propagation over the globe. Their amplitudes grow with altitude while propagating upward in response to exponential decrease of the atmospheric density with altitude. The energy and momentum of these waves are transported upward and deposited at high altitudes. The waves interact with each other or with the background mean flow at high altitudes, generating typical global-scale oscillations, such as the quasi-biennial oscillation (QBO) in the stratosphere (e.g., Baldwin *et al.*, 2001) and semiannual oscillation (SAO) in the stratosphere and mesosphere (e.g., Garcia *et al.*, 1997). These interactions are quite significant because it is the small-scale processes of atmospheric waves generated in the lower atmosphere that control the large-scale process of global circulation in the middle atmosphere, a long distance from their source (Garcia and Solomon, 1985). However, the wave species allegedly responsible for individual processes have not been well identified due to a sparsity of observations.

On the other hand, the geomagnetic field lines are parallel to the Earth's surface or perpendicular to the gravitational force in the equatorial ionosphere. The consequence of this geometry is the unique electrodynamical features that occur above 100 km altitude in the equatorial ionosphere, such as the equatorial electrojet (EEJ) in the E region and the equatorial ionization anomaly (EIA) and equatorial plasma bubbles in the F region (e.g., Kelley, 1989 and references therein).

Atmospheric tides and internal gravity waves are either generated in situ or propagate up to the ionosphere, and play an important role in the dynamics and energetics of the equatorial ionosphere, particularly in the altitude range of 100 to 300 km. For instance, mesoscale gravity waves are suspected to seed equatorial plasma bubbles (e.g., Kelley *et al.*, 1981; Hysell *et al.*, 1990). However, our current understanding falls far short of full elucidation of these phenomena owing, again, to the sparsity of proper observations.

Consequently, it would be of great significance to observe various dynamical and electrodynamical processes in the whole equatorial atmosphere from a unified viewpoint of vertical couplings due to the convection-induced atmospheric waves. A research project called Coupling Processes in the Equatorial Atmosphere (CPEA) was conducted, primarily by ground-based observations, to elucidate basic processes of the vertical couplings on various spatial and temporal scales occurring in the equatorial low, middle, and upper atmosphere and ionosphere (Fukao, 2006). CPEA was funded as a Grant-in-Aid for Scientific Research on Priority Areas project by the Ministry of Education, Culture, Sports, Science and Technology of Japan during the six-year period September 2001 to March 2007.

The general outline of CPEA and its major outcomes from two extensive observational campaigns are briefly reviewed in this article.

2 Prelude to CPEA

The land and sea distribution and the associated east-west difference of sea surface temperature are especially conspicuous in the equatorial region, which leads to the non-uniform distribution of convective clouds. OLR (Outgoing Longwave Radiation) satellite data is a good proxy for convective cloud activity in the tropics. Nishida *et al.* (2000) demonstrated that the tallest clouds (deepest convection) tend to appear over the equatorial region, especially, the Indonesian Maritime Continent (IMC; or the Indonesian Archipelago). The activity of mesoscale internal gravity waves with vertical wavelengths shorter than 10 km was estimated from their potential energy at lower stratospheric heights (Tsuda *et al.*, 2000). The potential energy was particularly enhanced over the IMC where the cumulonimbus clouds were supposed to be tallest as revealed by the least OLR intensity observed by satellites in the equatorial region. The good correlation between these two types of data well demonstrates that atmospheric waves are most actively generated by convection there. Therefore, the vertical coupling is expected to occur most significantly there, and consequently, the Indonesian Maritime Continent was selected as the main observational site for the CPEA research.

Recognizing the importance of vertical coupling in the equatorial atmosphere, the Scientific Committee on Solar-Terrestrial Physics (SCOSTEP) sponsored a series of international scientific programs that were closely concerned with this subject, among them the Solar Terrestrial Energy Program (STEP; 1990–97). The Energy Processes Including Coupling (EPIC; 1998–2002) project was sponsored subsequently by SCOSTEP as one of the four post-STEP mini projects.

On the other hand, the TOGA/COARE (Tropical Ocean Global Atmospheres/Coupled Ocean Atmosphere Response Experiment) was an international research program promoted by the World Climate Research Programme (WCRP) that studied the interaction of the ocean and atmosphere in the western Pacific. Several scientific groups pioneered observational and theoretical investigations of coupling in the equatorial atmosphere. Among them, extensive field observations were conducted with deployment of various instruments including Doppler radars (ship, land, and air) in the IMC (Webster and Lukas, 1992).

Meanwhile endeavors have been made for establishing core observatories and/or facilities in the equatorial region to integrate these investigations. For instance, a chain of atmospheric radars (wind profilers), called the Trans Pacific Profiler Network (TPPN), was installed at Piura, Peru (6°S, 81°W), Christmas Island, Kiribati (2°N,

157°E), Pohnpei, Federal States of Micronesia (7°N, 158°E) and Biak, Indonesia (1°S, 136°E) by the US National Oceanic and Atmospheric Administration (NOAA), and successfully operated for almost one decade from the mid 1980s (e.g., Gage *et al.*, 1991). Subsequently another wind profiler started operation at Darwin, Australia (12°S, 131°E) in 1990 (Schafer *et al.*, 2003). On the other hand, a meteor radar was installed in Jakarta, Indonesia (6.4°N, 106.7°E) in 1992 (Tsuda *et al.*, 1995), while three MF radars were deployed in Christmas Island, Pontianak, Indonesia (0°N, 109°E), and Tirunelveli, India (9°N, 78°E) in early 1990s.

The Equatorial Atmosphere Radar (EAR) was installed through collaboration between Japan and Indonesia right at the equator in Kototabang, West Sumatra, Indonesia (0.20°S, 100.32°E) in March 2001 (Fukao *et al.*, 2003a). It started observations in July 2001 in an institutional collaboration between the Radio Science Center for Space and Atmosphere (now Research Institute for Sustainable Humanosphere), Kyoto University of Japan and the National Institute of Aeronautics and Space (LAPAN) of Indonesia. CPEA is based on these developments, with the EAR site serving as a focus for its activities, and will encourage an extension of the network to incorporate as many techniques as possible to ensure that the equatorial atmosphere is characterized as broadly as possible.

3 Studies in the CPEA

No doubt, it is short-period internal gravity waves and equatorial waves that hold the key to the vertical coupling of the equatorial atmosphere. Data on these atmospheric waves have been considerably accumulated in the last decades, owing to the novel observation techniques like atmospheric radars and satellite radio occultation (RO). For instance, it is considered that momentum transport by gravity waves and equatorial waves is particularly important in producing the QBO and SAO in the middle atmosphere. Also, tides and short-period gravity waves directly bring their signatures to upper atmospheric phenomena such as airglow, electric fields and the equatorial electrojet (EEJ) (e.g., Burrage *et al.*, 1995 for tidal-origin airglow; Taylor *et al.*, 1987 for gravity-wave-origin airglow; Gelinas *et al.*, 2002 for the EEJ). The wave-mean flow interaction plays a major role in generating these oscillations. However, identification of the wave species responsible for generating them is still controversial; the wave species concerned with the QBO and SAO have not yet been well understood (e.g., Mayr *et al.*, 1998; Baldwin *et al.*, 2001; Horinouchi *et al.*, 2003), although the theoretical principle for the wave-mean flow interactions has been already established (Dunkerton, 1997; Fritts and Alexander, 2003).

Spatial distributions and seasonal variability of atmospheric waves depend not only upon cumulus convection as the wave sources but also wave propagation characteristics. Gravity waves that propagate upward from tropospheric sources experience a manifold of interactions on their way to the middle atmosphere, which include wave-mean flow interaction, wave-wave interaction, wave ducting by mean flow, wave generation by shear flow, and wave absorption at critical levels. Consequently, the relationship between the wave species and the global-scale oscillations are, in general, not so straightforward, and even one of the interactions makes their one-to-one correlation quite difficult.

It is well known that the zonal winds in the equatorial mesosphere and lower thermosphere (MLT; 60–100 km in altitude) are characterized by a semiannual oscillation (SAO) with westward winds during equinox and eastward winds during solstice (Vincent, 1993; Garcia *et al.*, 1997). Burrage *et al.* (1996) found that the SAO is modulated by the quasi-biennial oscillation (QBO) with the first westward winds becoming quite stronger in alternate years. This suggests a dynamical coupling between the stratosphere and the mesosphere caused, probably, by atmospheric waves, but the way of participation of atmospheric waves in generating this coupling process seems very complicated.

Besides the SAO and QBO, the MLT zonal winds undergo variabilities with periods of a few tens of days, which resemble those of intraseasonal oscillations (ISO; or intraseasonal variations, ISV) in the troposphere (Eckermann *et al.*, 1997). This correlation demonstrates that signals from cumulus convection in the troposphere reach as high as the MLT region. However, this raises the further interesting question to be solved of how the ISO, a large-scale wave disturbance, which slowly propagates in the zonal direction, can propagate upward into the MLT region. One plausible explanation is that gravity waves and nonmigrating (not Sun-synchronous) tides whose intensity is modulated by the tropospheric ISO directly propagate to the MLT region and induce similar ISO in the zonal winds through wave-mean flow interactions (Eckermann *et al.*, 1997). No doubt, putting this working hypothesis to the test may require a comprehensive lineup of various observational instruments.

In the thermosphere/ionosphere, it was only recently that the neutral atmosphere or the thermosphere was directly sensed by novel optical techniques from ground and space. Day-to-day variability of the winds and electric fields, associated with variability in the tides, planetary waves, and gravity waves, causes variabilities in ionospheric structure and dynamics, which still remain poorly understood. Recent concurrent airglow, radar, and ionospheric sounding observations have provided various evidences of significant planetary wave signatures in the thermosphere/ionosphere over equatorial latitudes. For instance, planetary scale oscillations of different periodicities with episodic nature occurred simultaneously in the mesospheric winds and EEJ intensity over India (e.g., Abdu *et al.*, 2006). Also, nighttime F-region height oscillations at planetary-wave periodicities were demonstrated to be associated with similar oscillations in the mesosphere over Brazil (e.g., Abdu *et al.*, 2006; Takahashi *et al.*, 2006).

The ionospheric wind dynamo generates large-scale electric fields that cause plas ma drifts, forming the equatorial ionization anomaly (EIA) in electron density (e.g., Kelley, 1989). Recent analyses of the NASA Imager for Magnetopause-to-Aurora Global Exploration (IMAGE) satellite airglow observations (at the far-ultraviolet (FU V) 135.6-nm wavelength) have revealed four longitudinal enhancements of the EIA (Sagawa *et al.*, 2005; Immel *et al.*, 2006). Sagawa *et al.* and Immel *et al.* considered that longitudinal variations in nonmigrating tides that were excited by latent heat release associated with raindrop formation in the tropical troposphere could propagate upward into the lower thermosphere, affect the dynamo process in the E region, and produce the four-peaked longitudinal EIA variations in the F region. This hypothesis has been tested with the National Center for Atmospheric Research (NCAR) Thermosphere-Ionosphere-Mesosphere-Electrodynamics General Circulation Model (TIME-GCM), confirming that the four-peaked feature is caused by an eastward propagating zonal wavenumber -3 diurnal tide of tropospheric origin (Hagan *et al.*, 2007). These results provide marked evidence that deep convection in the troposphere indirectly modifies the top-side ionosphere, a few 100 km in altitude, by modulating the E-region dynamo in the lower thermosphere.

Strong electric fields and plasma density gradients, probably produced in part by gravity waves, are considered to cause instabilities that give rise to plasma irregularities on a wide range of spatial scales. However, day-to-day quiet time variability of occurrence and spatial distributions of equatorial spread F (ESF) or plasma bubbles still remains poorly understood, and it is possible that some physics is missing in the prevailing theory of ESF generation (e.g., Tsunoda, 2006).

The conventional but still refreshing questions will be concerned with transport and sink mechanisms for metallic atoms and ions in the equatorial ionosphere, which should be investigated from the new unified point of view of CPEA.

In the CPEA, spatial and temporal variabilities of various atmospheric waves in the equatorial atmosphere are investigated in the wide altitude range from the troposphere to the thermosphere/ionosphere. The main dissipation mechanisms through which atmospheric waves interact with the mean flow are of an interesting subject for CPEA. Cumulus convection, as the most important wave source in the tropics, is investigated over the IMC and contrasted with that over the Indian Ocean. CPEA seeks comprehensive solutions to some of the above-mentioned important scientific questions relevant to dynamical and electrodynamical coupling in the equatorial atmosphere, and is expected to give important clues to some of them by continuous well-coordinated observation with various ground-based instruments, including radars and lidars, radiosondes and satellites. Readers should refer other scientific questions to Fukao (2006).

4 Project Outline of CPEA

CPEA is an observational project conducted in the Indonesian Maritime Continent (IMC) for elucidation of the vertical coupling of the equatorial atmosphere, primarily by means of ground-based observations (Fukao, 2006). During the six-year experimental period of CPEA, the instruments shown in Fig. 1 were developed and installed at the EAR site and its periphery. Although the altitude coverage of individual instruments is limited, the whole atmosphere up to several hundred km altitude is complementally covered by operating all instruments at the same time.

The vertical coupling of the equatorial atmosphere has been investigated from the following six approaches (Fukao *et al.*, 2006): The first three approaches are physics-oriented, being primarily concerned with the dynamical and electrodynamical coupling processes in the lower, middle and upper atmosphere, respectively, while the remaining three are instrumentation-oriented, seeking new processes that are expected to occur over the whole equatorial atmosphere. They constitute warp and weft

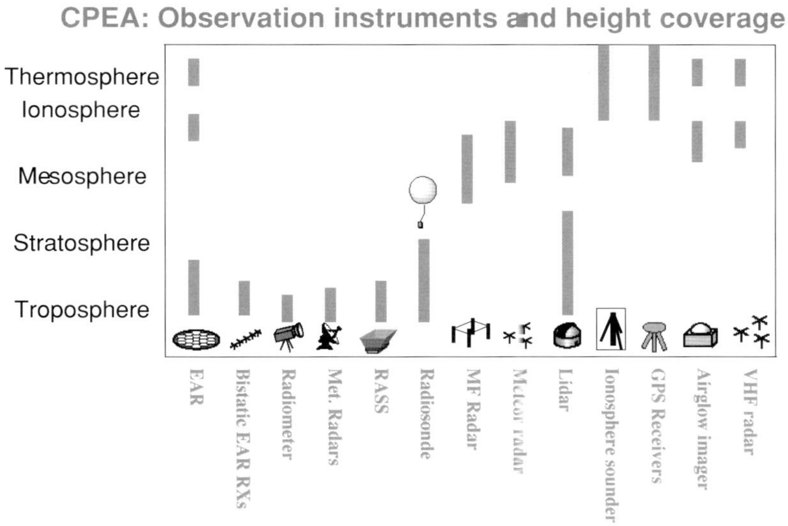

Fig. 1. Main instruments of CPEA installed at the Equatorial Atmosphere Observatory (EAO)/Equatorial Atmosphere Radar (EAR) site and its periphery in Kototabang, West Sumatra, Indonesia (0.20°S, 100.32°E). The vertical bars show height coverage. (From Fukao (2006).)

woven, respectively, for elucidation of the coupling processes in the equatorial atmosphere.

1. Studies on evolution of cumulus convection and its coupling with meso- to synoptic-scale precipitation systems: An extensive radiosonde sounding network is established in Indonesia, Singapore, and Malaysia to understand the coupling processes of convective clouds from meso- to synoptic-scale and generation mechanisms of atmospheric waves by cumulus convection.

2. Studies on the four-dimensional structure of atmospheric waves and transport of momentum and energy: The radio sounding network is utilized to determine the four-dimensional structure of atmospheric waves in the troposphere and stratosphere. A regional MF/meteor radar network is established in Indonesia to determine the wave structure in the MLT region. Satellite data are also utilized in order to supplement the ground-based and radiosonde observations.

3. Studies on the variations of the ionosphere and thermosphere owing to the atmospheric waves: Optical and radio techniques including airglow imaging and GPS are utilized to investigate variations of the ionosphere and thermosphere above 80 km altitude and the seeding of plasma instabilities featured in the geomagnetic equatorial region due to internal gravity waves penetrating through the middle atmosphere from below.

4. Studies on basic features of the equatorial atmospheric and ionospheric processes with EAR: Long-term observations with the EAR and other instruments are conducted to investigate various atmospheric and ionospheric processes, including convection, tropospheric vertical wind, generation of small-scale gravity waves, the tropical tropopause layer (TTL), and ionospheric irregularities.

5. Studies on small-scale wind and turbulence structure: An innovative multi-static radar system that utilizes EAR as the transmitter-receiver station is composed by adding two digital receiver sub-arrays as its part to investigate the extremely small-scale three-dimensional air motion inside the volume illumined by the EAR antenna beam.

6. Studies on the vertical structure of the equatorial atmosphere with a multi-lidar system: A sophisticated multi-lidar system consisting of several kinds of lidar has been developed to investigate the vertical structures of temperature, clouds, and water vapor in the equatorial atmosphere, and sporadic metallic layers, mesospheric inversion layers, and gravity waves near the mesopause.

5 Main Instrumentation and Networks

The Equatorial Atmosphere Observatory (EAO), located close to the geographic equator at Kototabang in West Sumatra, Indonesia (0.20°S, 100.32°E), is the base for the various CPEA activities. Its longitude is unique, being on the border between

Fig. 2. The Equatorial Atmosphere Radar (EAR) in Kototabang, West Sumatra, Indonesia (top) and its 560 three-element Yagi antennas, each with a transmitter-receiver module mounted near the ground (bottom). The operational frequency is 47.0 MHz. The antenna is a near-circular array with a diameter of 110 m, and the peak transmit power is 100 kW. (From Fukao *et al.*, Equatorial Atmosphere Radar (EAR): System description and first results, *Radio Sci.*, **38**, 1053, 2003. Copyright 2003 American Geophysical Union. Reproduced by permission of American Geophysical Union.)

the Indian Ocean and the Indonesian Maritime Continent. The core instrument of the EAO is the Equatorial Atmosphere Radar (EAR), shown in Fig. 2. EAR is a 47.0-MHz Doppler radar with a near-circular active phased array antenna system that enables fast beam steering to any direction (Fukao *et al.*, 2003a). The antenna aperture is 110 m in diameter while the peak power is 100 kW, one tenth of that of the MU radar (Fukao *et al.*, 1985a, b), and accordingly the height coverage is more limited than that of the MU radar. EAR can observe atmospheric echoes up to 20 km in altitude and plasma irregularity echoes from the ionospheric *E* and *F* regions. See Fukao *et al.* (2003a) for more details. Observational data from the EAR are available in real time at the following web site: http://www.rish.kyoto-u.ac.jp/ear/data/.

In general, observation of the wind velocity vector with single atmospheric radar is performed under an assumption that the velocity field is uniform over the horizontal space illuminated by at least three antenna beams. In order to investigate fine-scale wind perturbations and turbulence in the lower troposphere, Nishimura *et al.* (2006) have developed a multistatic radar capability for EAR, with which the scattered echoes are simultaneously received at two receiving sites as well as the EAR itself. The major technical difficulty was the existence of strong ground clutter echoes, which has been successfully overcome by applying the adaptive sidelobe control algorithm developed by Kamio *et al.* (2004).

For precipitation observations, an X-band weather radar and a boundary-layer radar (BLR) were installed at Kototabang, and an additional X-band Doppler weather radar at Sungai Puar in West Sumatra, Indonesia (0.36°S, 100.41°E). RASS (Radio Acoustic Sounding System) experiments were conducted with EAR to measure fine altitude profiles of temperature and humidity over the EAO. Surface meteorological instruments, such as a radiometer, rain gauge, and disdrometer, were installed at EAO too (Kozu *et al.*, 2006).

A radiosonde-sounding network was established for the campaigns through collaboration between Indonesia, Singapore and Malaysia (see Fig. 3). CPEA-I was joined by seven radiosonde stations; Kototabang (EAO/EAR site; 0.20°S, 100.32°E), Bandung (6.90°S, 107.60°E), Padang (0.88°S, 100.35°E), Jambi (1.60°S, 103.65°E), Kuala Lumpur (2.73°N, 101.70°E), Kuching (1.48°N, 110.33°E), and Singapore (1.22°N, 103.59°E). On the other hand, the launching was made at three stations, Kototabang, Padang, and Jambi, during CPEA-II. Intensive launching of radiosondes was conducted in close collaboration with Frontier Observational Research System for Global Change (FORSGC) of Japan and with Kyoto University Active Geosphere Investigations for the 21st Century COE (KAGI21) (Kozu *et al.*, 2006; Tsuda *et al.*, 2006).

The regional radar network for MLT observations is a quasi-equilateral triangle with one side 700 km, consisting of an MF radar at Pontianak (0.0°N, 109.3°E), Indonesia, a meteor radar at Kototabang (0.20°S, 100.32°E), and another MF radar at Pameungpeuk (7.5°S, 107.5°E), Indonesia (e.g., Sridharan *et al.*, 2006b). The first radar has been operational since 1995 while the latter two were deployed in March 2004. MLT winds have been measured continuously with these three radars afterward, with occasional interruptions due to commercial power line failure, light-

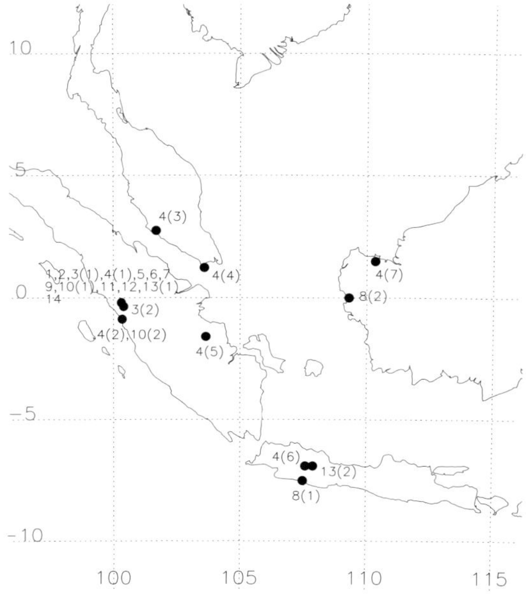

Fig. 3. Locations of major observational instruments for CPEA. The numbers in the map correspond to those in Table 1. (From Fukao (2006).)

ning damage, etc. Also, an international equatorial radar network was composed, by collaborating with the MF radars at Christmas Island in the Central Pacific and Tirunelveli in India.

The multi-lidar system consists of four lidars; Mie and Raman lidars for measurements of tropospheric aerosol/cirrus clouds and water vapor, respectively, Rayleigh lidar for stratospheric and mesospheric temperature and aerosol measurements, and resonance lidar for measurements of density and temperature of metallic species, such as Na, Fe, K, and Ca, in the mesopause region. Most of this multi-lidar system is remotely controlled via the Internet from Japan (Shibata *et al.*, 2006).

For observation of the thermosphere and ionosphere, GPS receivers, an all-sky CCD airglow imager (ASI), and an airglow temperature photometer were put in operation. Also, simultaneous mid-latitude ASI observations were conducted at Sata, Japan (31.02°N, 130.68°E; geomagnetic latitude: 21.47°N) and Darwin, Australia (12.44°S, 130.96°E; geomagnetic latitude: 21.86°S), which are nearly geomagnetically conjugate (Shiokawa *et al.*, 1999, 2004; Otsuka *et al.*, 2002). In addition, VHF (30.8 MHz) radar to detect ionospheric irregularities has been operational at EAO since February 2006 (Otsuka *et al.*, 2009).

FM-CW ionospheric sounders were constructed at three places along the meridian of the EAO/EAR site; Kototabang (dip latitude: 10.36°S), Chumphon in Thailand (10.72°N, 99.37°E; dip latitude: 3.28°N; close to the geomagnetic equator) and

Table 1. Sites of the major instruments.

No.	Instruments	Location
1	EAR	Kototabang (0.20°S, 100.32°E)
2	Boundary layer radar (BLR)	Kototabang (0.20°S, 100.32°E)
3	Weather radar	(1) Kototabang (0.20°S, 100.32°E)
	Doppler weather radar	(2) Sungai Puar (0.36°S, 100.41°E)
4	Radiosondes	(1) Kototabang (0.20°S, 100.32°E) (2) Padang (0.88°S, 100.35°E) (3) Kuala Lumpur (2.73°N, 101.70°E) (4) Singapore (1.22°N, 103.59°E) (5) Jambi (1.60°S, 103.65°E) (6) Bandung (6.90°S, 107.60°E) (7) Kuching (1.48°N, 110.33°E)
5	Radiometer and other surface instruments	Kototabang (0.20°S, 100.32°E)
6	Multi-lidar system	Kototabang (0.20°S, 100.32°E)
7	Meteor Radar	Kototabang (0.20°S, 100.32°E)
8	MF Radars	Tirunelveli (8.7°N, 77.8°E) (1) Pameungpeuk (7.5°S, 107.5°E) (2) Pontianak (0.0°N, 109.3°E)
9	FM-CW ionospheric sounders (Ionosondes)	Kototabang (0.20°S, 100.32°E; dip latitude: 10.36°S) Chumphon (10.72°N, 99.37°E; dip latitude: 3.28°N) Chiang Mai (18.76°N, 98.93°E; dip latitude: 13.03°N)
10	GPS receivers	(1) Kototabang (0.20°S, 100.32°E) (2) Padang (0.9[illegible]°S, 100.46°E)

Table 1. (continued).

No.	Instruments	Location
11	VHF radar (30.8 MHz)	Kototabang (0.20°S, 100.32°E)
12	Magnetometer	Kototabang (0.20°S, 100.32°E)
13	Airglow imagers	Kototabang (0.20°S, 100.32°E) Tanjungsari (6.9°S, 107.9°E)
14	Airglow temperature photometer	Kototabang (0.20°S, 100.32°E)

Chiang Mai in Thailand (18.76°N, 98.93°E; dip latitude: 13.03°N; close to the geomagnetic conjugate point of Kototabang) by the National Institute of Information and Communication Technology (NICT) of Japan (Maruyama *et al.*, 2007).

6 The Observations and CPEA-I and -II Campaigns

Observations with individual instruments were initiated immediately after installation at EAO. The EAR, the MF radar at Pontianak, and the BLR at Kototabang, which had been already operational when CPEA started, were operated continuously during the whole CPEA period. Using all instruments available, two extensive observation campaigns were conducted; the first one, CPEA-I, in the period from March 10 to May 9, 2004, and the second, CPEA-II, in the period from November 18 to December 23, 2005. The major results to be reviewed in the following sections are primarily the outcomes from these campaigns.

The CPEA-I was conducted as part of the CAWSES (Climate and Weather of the Sun-Earth System) campaign of SCOSTEP. Various instruments, shown in Fig. 1, participated in the campaign in order to cover a wide altitude range in the equatorial atmosphere, from the troposphere to the thermosphere/ionosphere. The sites of these instruments are presented in Fig. 3 and Table 1. See Fukao (2006) for more details regarding the setup of CPEA-I and -II and the activity of cumulus convection during the periods.

7 Principal Research Highlights

7.1 Unique features of cumulus convection over Sumatra

Tropical cloud systems associated with the 30–60-day period intraseasonal oscillations (ISOs) play an important role in driving the global circulation through latent heat release. They are known as organized convective systems on wide spatial and temporal scales based upon the satellite observations (Nakazawa, 1988; Lau *et al.*, 1991). Synoptic-scale convective systems are recognized as eastward-propagating envelope of super cloud clusters (SCCs), which are composed of westward-propagating cloud clusters (CCs) on mesoscale (Nakazawa, 1988).

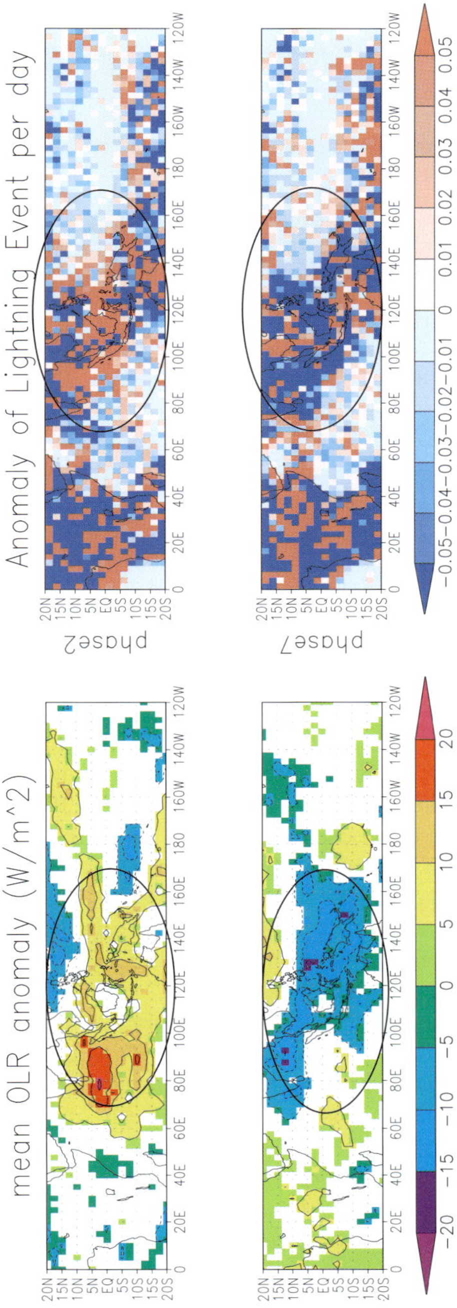

Fig. 4. (Left) OLR anomalies when cumulus convection is inactive (top) and active (bottom) over the Indonesian Maritime Continent, encircled with solid line. (Right) Anomalies of lightning events (flash rate) observed by TRMM during the same period as that shown in the left. (Reprinted from *Dyn. Atmos. Oceans*, **42**, Morita *et al.*, Analysis of rainfall characteristics of the Madden-Julian oscillation using TRMM satellite data, 107–126, Copyright 2006, with permission from Elsevier.)

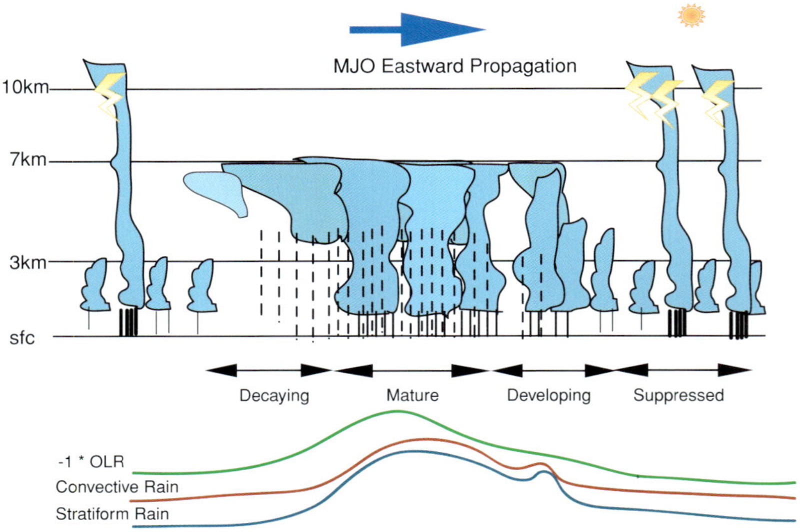

Fig. 5. A conceptual model of an ISO life cycle. Vertical solid lines represent convective rainfall and vertical dashed lines stratiform rainfall. Time proceeds from right to left. Schematics for anomalies of the inverse of OLR, the convective rain amount, and the stratiform rain amount are shown in the bottom. (Reprinted from *Dyn. Atmos. Oceans*, **42**, Morita *et al.*, Analysis of rainfall characteristics of the Madden-Julian oscillation using TRMM satellite data, 107–126, Copyright 2006, with permission from Elsevier.)

It is well-known that SCCs are significantly modified by the topography over the Indonesian Maritime Continent (IMC) (e.g., Nitta *et al.*, 1992). Shibagaki *et al.* (2006a) studied behaviors of westerly wind bursts (WWBs) associated with an SCC over the IMC, using data from radiosondes launched from multiple sites during CPEA-I. They showed a marked structural change of WWBs due to interception by the elevated topography of Sumatra. It was also reported that the migration speed of WWBs over Sumatra was approximately half that over the sea between Sumatra and Kalimantan due to the orographic effect.

Shibagaki *et al.* (2006b) also studied the internal structure of CCs developed in other SCCs over the mountainous region in Sumatra, based on Equatorial Atmosphere Radar (EAR) and weather radar observations. They showed vertical circulation of a mesoscale convective precipitation system (horizontal scale of ~40 km), which is composed of multiple precipitation cells (horizontal scale of <10 km), generated by low-level wind intrusion in relation to a valley-mountain breeze.

It used to be considered from OLR (Outgoing Longwave Radiation) anomalies shown in Fig. 4 that convective activity over the IMC is more intense during active ISO periods than inactive periods (Morita *et al.*, 2006). On the other hand, lightning flash rate anomalies observed by the TRMM (Tropical Rainfall Measurement Mission) satellite are more frequent when the ISO is inactive and less frequent when the

ISO is active. Also, the X-band Doppler weather radar at Sungai Puar demonstrated that deep convection occurred in inactive ISO periods, while shallow convection was dominant in active ISO periods (Kawashima *et al.*, 2006). Kozu *et al.* (2005, 2006) discussed the intraseasonal variation of raindrop size distribution (DSD) over EAR, demonstrating that the DSD was broad in the inactive phase of an ISO, but narrowed as the phase of an ISO shifted to active.

Using rawinsonde data from multiple stations, Kodama *et al.* (2006) quantified the latent heat and eddy transport term Q1 as well as the moisture sink term Q2, supporting this result that the height of the positive peak of Q1 was much higher than that of Q2 during inactive ISO periods. In this stage, not well organized but very tall (more than $\sim$10 km) convective rain (afternoon shower) seemed to dominate with more frequent lightning flashes. On the other hand, both terms were comparable during active ISO periods, and well-organized mesoscale convective systems with stratiform rain dominated with fewer lightning flashes. Consequently, upward heat transport was locally much stronger during inactive ISO periods than active periods.

This feature can be summarized in the schematic of Fig. 5 (Morita *et al.*, 2006). In the IMC, islands and inland seas are stationary heat sources, and forcing due to the sensible heating is significant when an ISO is inactive. Convective cloud height should be taller than that of matured clouds but its horizontal size seems small compared with the resolution of satellite observations; that is the reason why cloud information from satellites does not show a similar feature.

All of these features seem to oppose our conventional understanding that cumulus convection is more active during the active phase of ISOs as classified by OLR (or TBB). This comes from the fact that local convective processes are significant in the inactive ISO phase due to the topographic effect of Sumatra.

7.2 Excitation of internal gravity waves by convection

Internal gravity waves are known to play an important role in dynamics and energy and momentum balance in the Earth's atmosphere (Fritts and Alexander, 2003 and references therein). They propagate upward from sources in the lower atmosphere, growing in amplitude until they break and deposit energy and momentum into the background atmosphere. As mentioned in Section 1, this is significant since small-scale gravity waves generated in the lower atmosphere control large-scale oscillations in the middle atmosphere, such as the quasi-biennial oscillation (QBO) and the semiannual oscillation (SAO), through the wave-mean flow interaction process (Dunkerton, 1997).

A variety of sources have been identified for generation of gravity waves for different geographical and meteorological conditions (Fritts and Alexander, 2003). Among them, cumulus convection is recognized as the major source for generation of gravity waves in the equatorial region (e.g., Salby and Garcia, 1987; Pfister *et al.*, 1993; Alexander *et al.*, 1995; Ricciardulli and Garcia, 2000). Since they are not constant in time and space, our understanding of gravity wave characteristics is far from complete. Consequently, continuing observations of gravity waves are expected for different times and regions.

By means of an intensive radiosonde campaign conducted at five sites during

CPEA-I, Ratnam *et al.* (2006a) obtained clear evidence for dominant inertia-gravity waves with periods of 2–3 days and vertical wavelengths of ~3–5 km in the upper troposphere and lower stratosphere (UTLS) region. The wave energy was found to maximize near 20 km altitude, consistent with the profile of the background Brunt-Väisälä frequency squared. The horizontal wavelength was estimated to be approximately 2,000 to 3,000 km. They suggested that the plausible source of the inertia-gravity waves was SCCs with an envelope comparable to the horizontal wavelength of gravity waves, which were located far west of the EAR in the Indian Ocean and slowly advected eastward toward the EAR site. No inertia-gravity wave was generated from stationary type convective clouds. Thus, they conjectured that the inertia-gravity waves were temporally enhanced in response to the eastward-advection of SCCs (Ratnam *et al.*, 2006a).

It has been well recognized that strong localized convection generates gravity waves with a broad emission spectrum at low latitudes (known as convectively-induced gravity waves). Short-period gravity waves propagate nearly vertically, while long-period gravity waves propagate at increasing zenith angles. The short-period components (periods ranging from a few minutes to a few hours) are visible directly above convection. They play a major role in transporting wave energy and momentum vertically to the middle and upper atmosphere (e.g., Garcia and Solomon, 1985; Fritts and Alexander, 2003).

Alexander *et al.* (2008) examined five years of tropospheric data below 12 km from the EAR for seasonal and interannual gravity wave activity. A clear semiannual cycle of vertical wind variances with maxima occurring at the equinoxes and minima at the solstices was found inside and above convection. This semiannual cycle was consistently correlated with TRMM storm height and surface rainfall rate, and consequently it was mainly due to convection. According to them, half of the variances were due to short-period gravity waves with periods less than 2 hours, with the rest coming from waves with periods of less than 24 hours. The average momentum fluxes estimated above the convection cells in the UTLS were 0.7–0.9 $m^2\ s^{-2}$ for both zonal and meridional components (Alexander *et al.*, 2006). The magnitudes are generally consistent with the aircraft measurements of Alexander and Pfister (1995) and the model results of Lane and Reeder (2001).

Ratnam *et al.* (2009) detected short-period gravity waves with a dominant period of ~3 hours and vertical wavelength of ~10 km that were generated by convection localized around the EAR site, by hourly launching of radiosondes during the CPEA-II period. In addition, direct evidence of convectively-induced gravity waves with a period of ~1 hour, a vertical wavelength as short as ~2 km, and an amplitude of 3–10 mPa was provided by EAR in the UTLS region (Dhaka *et al.*, 2005).

Small-scale gravity waves in the MLT region, which are supposed to dominate the mesospheric momentum flux (e.g., Fritts and Vincent, 1987), were observed by an OH airglow imager at Tanjungsari, West Jawa (6.9°S, 107.9°E), during the period September 2000 to September 2001, and their horizontal propagation characteristics were analyzed in detail. The observed period, horizontal wavelength, and horizontal phase speeds were typically 5–13 min, 13–45 km and 37–75 m/s. This analysis suggests that active convective clouds in GMS images should be the source of the observed gravity waves, and the horizontal propagation characteristics over the equatorial region are affected mainly by the distribution of cumulus convection in the troposphere. The effect of the background mean wind in the middle atmosphere is not significant in the equatorial region where the mean wind is comparatively weak (~20 m/s) (Nakamura *et al.*, 2003).

7.3 Vertical wind in the middle-to-upper troposphere

Vertical wind is one of the most important factors that control the energy, momentum, and air mass transports in the upper troposphere, and is determined for phenomena that range from synoptic scale to mesoscale. Tropospheric vertical wind (W), which is difficult to observe directly by other techniques, has been determined for the first time over the IMC by using EAR (Nishi *et al.*, 2007; Yamamoto *et al.*, 2007). In the middle-to-upper troposphere (8–14 km), three-hour averaged W showed upward motion up to 0.09 m/s with a 2-day average value of 0.05 m/s in the vicinity of deep convective clouds within synoptic-scale SCCs. On the other hand, the upward motion almost disappeared and the 2-day averaged W was only 0.01 m/s when the SCCs moved to the east of Sumatra (Seto *et al.*, 2006). This result quantitatively indicates that cumulus activity is an important factor that determines vertical transport from the middle to upper troposphere (e.g., Balsley *et al.*, 1988).

Yamamoto *et al.* (2007) further investigated dominance of downward motion of ~0.11 m/s observed just above the enhanced cumulus convection in the altitude range from 14 km to the tropopause. Based upon the European Center for Medium-Range Weather Forecasts (ECMWF) analysis, they showed that a major portion of the downward motions was caused by leeward wind and leeward downward tilt of isentropes, both of which prevailed over western Sumatra. This result is qualitatively consistent with the long-term average obtained in the Central Pacific (Balsley *et al.*, 1988). In addition, Yamamoto *et al.* (2008) accurately determined particle fall velocity in cirriform clouds with high time resolution, by operating a 95-GHz cloud profiling radar simultaneously with EAR.

In the altitude range where tilted scattering layers exist, Doppler velocity observed in the vertically-pointed beam of atmospheric radars suffers contamination from horizontal wind (e.g., Fukao, 2007). This fact indicates that vertical wind is difficult to observe directly in such regions. However, using the bias component of zonal wind observed in the vertically-pointed radar beam and the echo power data obtained by two beams symmetrically offset from the zenith in a vertical plane, Yamamoto *et al.* (2003) showed frequent occurrence of shear instability (Kelvin-Helmholtz instability) around the tropical tropopause, when easterly wind in the upper troposphere indicated a significant vertical decrease around the tropopause. It may be interesting to examine the possible contribution of this instability to the troposphere-stratosphere exchange process (Holton *et al.*, 1995).

7.4 Identification of 5–8-day waves in the equatorial middle atmosphere

Equatorial Kelvin waves are eastward propagating modes with periods ranging from a few days to a few tens of days, which are trapped within ~15° of the equator. Slow and Fast Kelvin waves are considered to play a major role in driving the east-

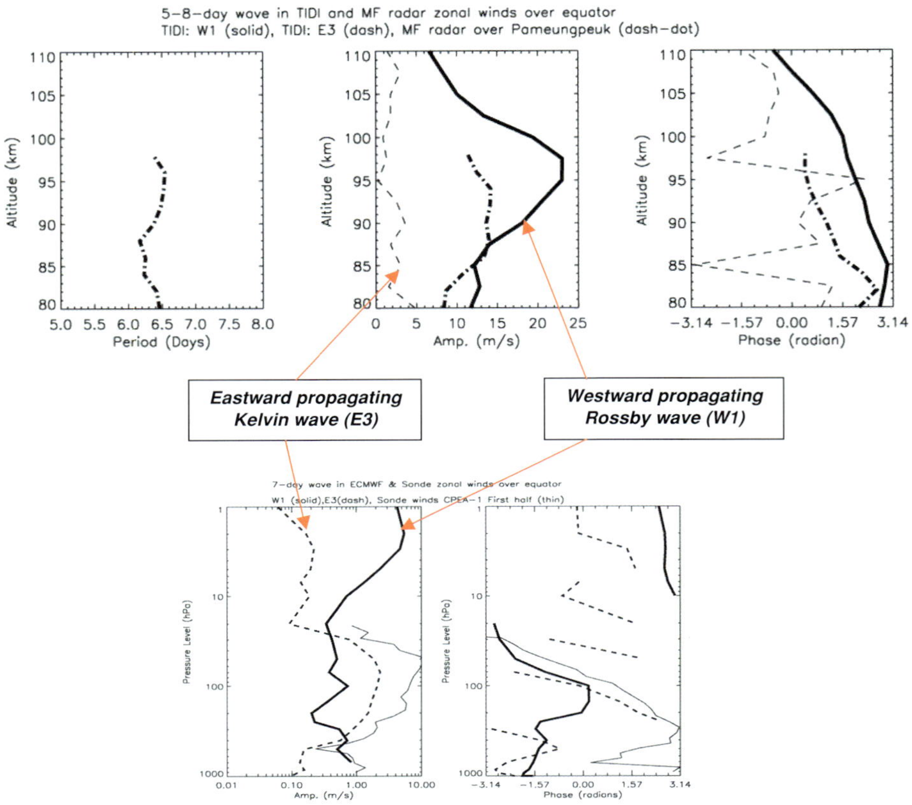

Fig. 6. (Top panels) Altitude profiles of dominant wave period (left), 6.5-day wave amplitude (middle) and phase (right) in the zonal winds over Pameungpeuk (dash-dot). The profiles for the W1 and E3 of 5–8 day wave in the TIDI winds are shown by solid and dashed lines, respectively (courtesy of S. Sridharan). (Bottom panels) Amplitude and phase of 7-day wave in the radiosonde winds are shown by thin solid curve. Those of W1 (thick solid curves) and E3 (dashed curves) components in the ECMWF winds during CPEA-I are shown for comparison. (From Sridharan *et al.* (2008).)

ward phase of the QBO in the lower stratosphere (e.g., Holton and Lindzen, 1972; Plumb, 1982) and the SAO in the mesosphere (e.g., Hirota, 1978). On the other hand, the presence of a wave with a period of ~6.5 days has been often indicated in the MLT region, and its resemblance to and difference from the normal mode of ~5-day periodicity with symmetric meridional structure and zonal wavenumber 1 have been discussed (Wu *et al.*, 1994; Meyer and Forbes, 1997; Talaat *et al.*, 2001).

Sridharan *et al.* (2006a) detected an eastward propagating Kelvin wave of period ~7 days in the winds and temperature in the UTLS region using radiosonde and satellite data obtained during the CPEA-I period. It dominates in the UTLS region as shown by the thin solid lines in the bottom panel of Fig. 6. The presence of 5–8 day eastward propagating Kelvin wave with zonal wavenumber 3 was confirmed at 21 km by the Sounding of Atmosphere using Broadband Emission Radiometry (SABER) on board TIMED satellite. This wave became damped above 23 km, where a wave of period of ~6.5 days observed simultaneously in the MLT region by meteor radar at Kototabang, and by MF radars at Pameungpeuk and Pontianak began to increase in amplitude as shown in the top panel of Fig. 6 (Sridharan *et al.*, 2006b). From the horizontal wave characteristics inferred from the TIMED Doppler Interferometer (TIDI) and SABER, the wave is identified to be the symmetric wavenumber 1 Rossby wave.

Altitude profiles of a westward propagating wave with zonal wavenumber 1 (W1) and an eastward propagating wave with zonal wavenumber 3 (E3) inferred from the ECMWF wind analysis are shown for comparison in the bottom panel of Fig. 6 (Sridharan *et al.*, 2008). The predominance of W1 in the MLT region and of E3 in the UTLS region is consistent with the observations. The downward phase propagation from the MLT region to the tropopause indicates that the wave may have its source in the troposphere. During the observation period, the OLR distribution in the tropics showed similar periodicities, propagating eastward with zonal wave number 3 and westward with zonal wave number 1. This observational evidence clearly indicates that tropical convective heating is a common source for both waves.

7.5 Equatorial Kelvin waves and coupling processes in the UTLS region

It has been recognized in earlier studies that equatorial Kelvin waves significantly affect thermal structure around the tropical tropopause and dehydration and cirrus formation processes (e.g., Fujiwara *et al.*, 1998, 2001). Among these effects, breaking of the Kelvin waves near the tropopause is considered to play an important role in stratosphere-troposphere exchange, enhancing turbulence transport of stratospheric ozone into the troposphere. However, details of the impact of Kelvin waves on the tropical tropopause have not yet been made clear.

The Global Positioning System (GPS)/ Radio Occultation (RO) measurement technique is capable of providing global measurements with high accuracy and resolution (e.g., Tsuda *et al.*, 2000; Ratnam *et al.*, 2004). Figure 7(a) shows one such result obtained with the German CHAllenging Mini satellite Payload (CHAMP) GPS at 16 km altitude during November 2002 (Ratnam *et al.*, 2006b). Temperature perturbations (dotted thin line) clearly show eastward movement from the Indian Ocean to the Indonesian Maritime Continent, particularly after 7 November 2002. The thin smooth curves indicate fitted curves with only wavenumber 1 and 2 components. Panel (b) shows the residue between the original and fitted curves. It consists mainly of higher wavenumber components (>2), which propagate to the east with increasing amplitude, particularly from the Indian Ocean to the Indonesian Maritime Continent (IMC). Superimposed in the same panel is the OLR averaged between 10°N and 10°S. It also showed clear concurrent eastward movement and reached Kototabang on 18–19 November 2002. Therefore, convection is supposed as the source of the higher wavenumber components. The top and bottom panels in (c) show tropopause height and temperature measured by CHAMP GPS within ±10° latitude and ±20° longitude of Kototabang in comparison with the radiosonde observations from the

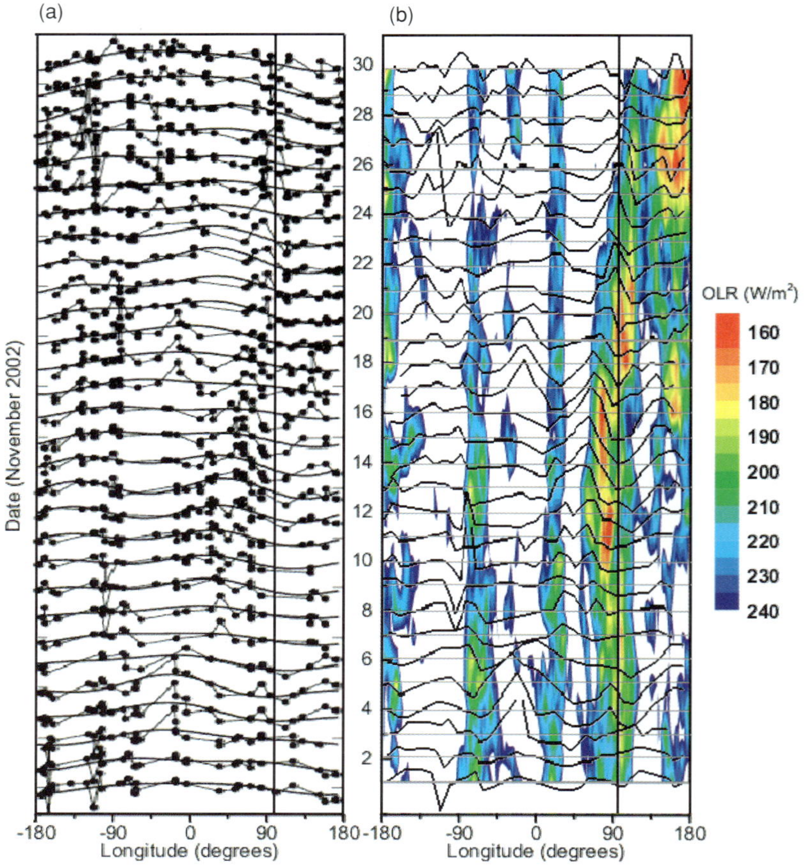

Fig. 7. (a) Time-longitude section of temperature anomaly at 16 km altitude observed during November 2002 (dotted lines). The thin smooth curves indicate fitted curves with wavenumber 1 and 2 components. The location of Kototabang is indicated by the vertical line near 100°E longitude. (b) The residue between the original and fitted curves of the temperature anomaly (solid curves) and OLR (background color shade) observed during the same period. (c) The time variation of the tropopause height and temperature obtained by radiosondes (open circle) and CHAMP GPS RO technique (cross) are shown in the top and bottom panels, respectively. (From Ratnam *et al.* (2006b).)

same site. The temperature and height were considerably modulated peak-to-peak by 5–10 K and ~3 km, both starting on November 19–20, 2002. Consequently, Kelvin waves are thought to be responsible for the modulation of the tropopause, and higher wavenumber components are likely dominant below the tropopause (Tsuda *et al.*, 2006).

Breaking of the equatorial Kelvin waves and the associated increase of turbulence were, for the first time, directly detected by the EAR in the tropopause region (Fuji-

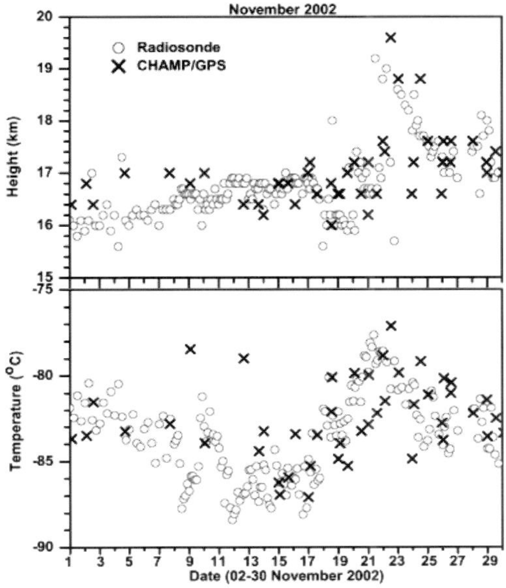

Fig. 7. (continued).

wara *et al.*, 2003). The top panel of Fig. 8 shows the zonal wind anomaly in the UTLS region observed by the EAR. A significant zonal wind oscillation due to Kelvin waves was observed. The jump of the tropopause was caused by the downward progression of the maximum eastward phase of the Kelvin wave. It was confirmed that the Kelvin wave was saturated and broken, generating convective turbulence there. The bottom panel shows the spectral width of the scattered spectrum of the EAR, which is proportional to the kinetic energy of turbulence. Turbulence intensity became ~5 times larger in association with the Kelvin wave breaking than that in other periods. Wave breaking turbulence in the tropopause region merged into the turbulence below, and caused effective and irreversible transport of lower stratospheric air mass deeply into the troposphere. Similar prominent events occurred over the EAR at least three more times between July and December 2001. It is hoped that a possible relation of this Kelvin wave mode with the above-mentioned higher wavenumber components will be investigated.

Using stratospheric temperature profiles derived from CHAMP GPS RO measurements during June 2001 to May 2006, Tsuda *et al.* (2009) studied the climatological behavior of atmospheric wave activities in the tropics. They found that spatial and seasonal variations of the potential energy for waves (primarily gravity waves and partly Kelvin waves) with vertical scale less than 7 km were, in general, closely related to the distribution of convective clouds below. They point out that one additionally needs to take into account wave propagation characteristics for precise

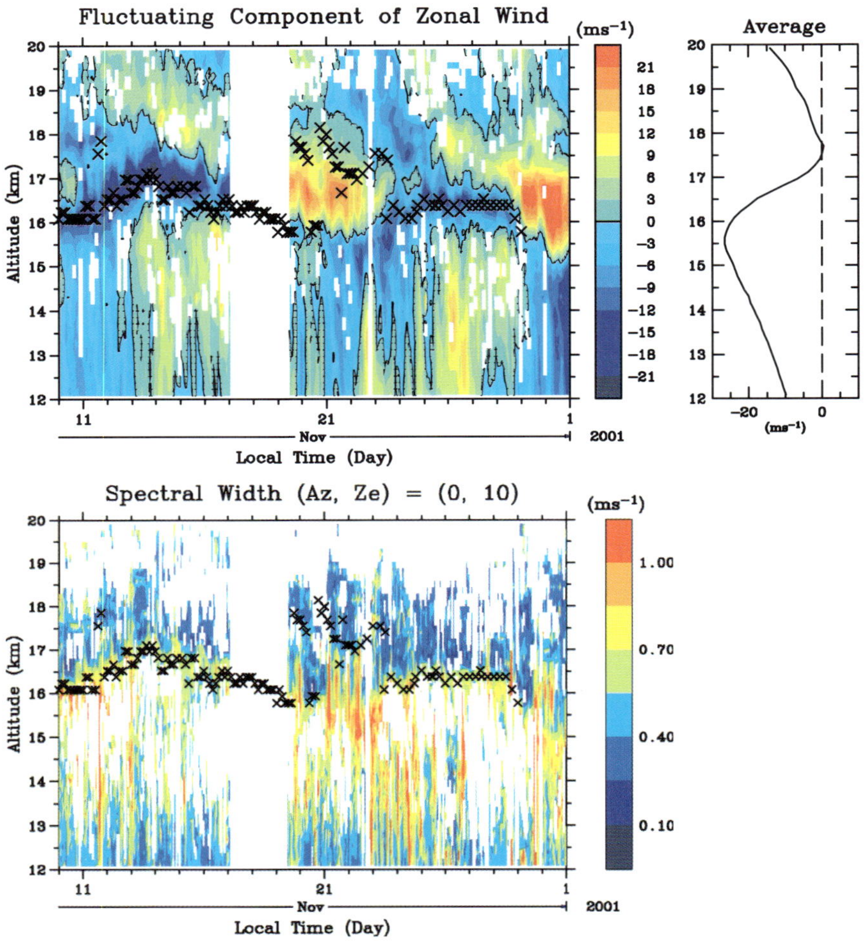

Fig. 8. Time-altitude sections of zonal wind anomaly measured by the EAR during November 10 to 30, 2001 (eastward positive; top left) with the average wind profile (top right), and the turbulence component of the spectral width obtained by EAR during the same period. The tropopause height defined by the temperature minimum measured by radiosondes is indicated by crosses. (From Fujiwara *et al.*, Turbulence at the tropopause due to breaking Kelvin waves observed by the Equatorial Atmosphere Radar, *Geophys. Res. Lett.*, **30**, 1171, 2003. Copyright 2003 American Geophysical Union. Reproduced by permission of American Geophysical Union.)

description of the wave climatology in the tropical stratosphere.

7.6 Aerosols and metallic atoms in the equatorial atmosphere

The multi-lidar system is fully operational and dedicated to revealing the dynamical and compositional structure with high altitude resolution over a wide altitude range (Nagasawa *et al.*, 2006; Abo *et al.*, 2006). For instance, preceding resonance

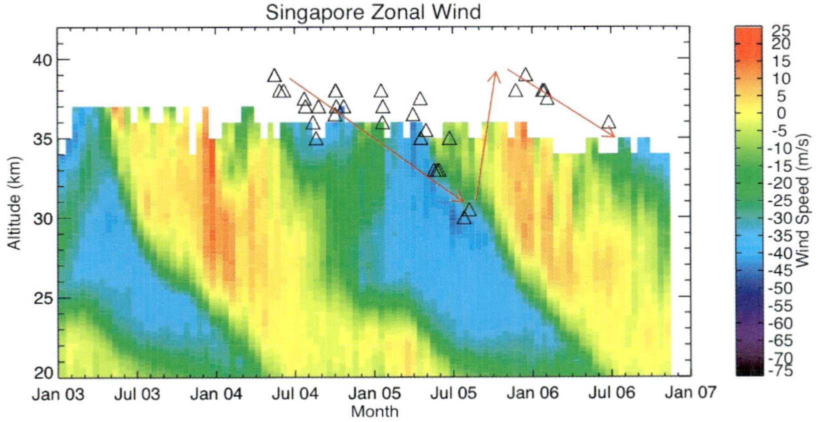

Fig. 9. Altitude-time section of the top height of stratospheric aerosol layer observed by the Rayleigh lidar at Kototabang (triangles) and the zonal wind over the equator observed by radiosondes at Singapore (color shades). (Courtesy of M. Abo.)

lidar observations have been conducted primarily at mid-latitudes or near the poles, but Shibata *et al.* (2006) successfully observed sodium (Na) and iron (Fe) layers in the MLT region over Kototabang, and found more sporadic-Na layers than at mid-latitudes after midnight.

The Rayleigh lidar offers high-resolution profiles of temperature fluctuations in the 30 to 80 km altitude range (Nagasawa *et al.*, 2006). The Mie lidar started observations of subvisible cirrus near the tropical tropopause layer. The cirrus is related to the freeze drying caused by sedimentation and lofting of air parcels affected through enhanced IR radiative heating of cloud particles that is thought to control water vapor entering the stratosphere.

A stratospheric aerosol layer now remains in steady state after the eruption of Mt. Pinatubo in 1991. In general, the layer is distributed in the altitude range from the tropopause to ~30 km at mid latitudes. However, Rayleigh lidar observations at Kototabang have revealed that the top height of the stratospheric aerosol layer extends to ~40 km, as shown by the triangle in Fig. 9 (M. Abo, private communication). The more important outcome is that the top height of the layer descends with time, synchronized with the QBO in the zonal wind observed over Singapore. The QBO signals of the aerosol layer are noticed in the altitude range from 40 to 30 km (M. Abo, private communication). In addition, the tropospheric aerosol amount observed around the tropopause over Kototabang is much more than at mid-latitudes. They suspect that this is evidence of active material exchange between the troposphere and the stratosphere occurring over the equatorial region.

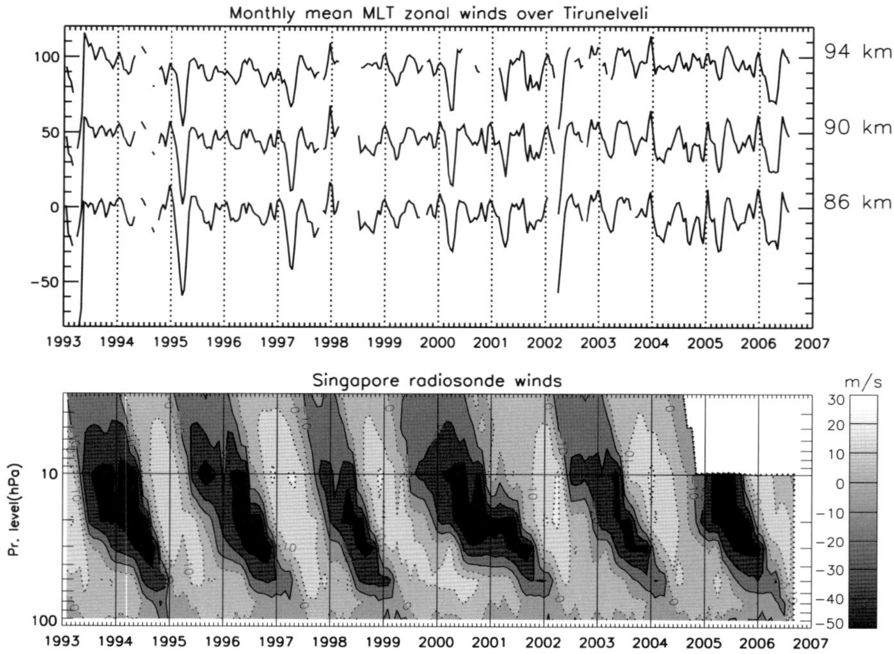

Fig. 10. Monthly mean zonal winds (top) over Tirunelveli for the altitudes 86, 90 and 94 km and (bottom) over Singapore for the pressure levels 90–3 hPa. (From Sridharan *et al.*, Radar observations of long-term variability of mesosphere and lower thermosphere winds over Tirunelveli (8.7°N, 77.8°E), *J. Geophys. Res.*, **112**, D23105, 2007. Copyright 2007 American Geophysical Union. Reproduced by permission of American Geophysical Union.)

7.7 Tropospheric signals in the mesosphere

Equatorial zonal winds in the mesosphere and lower thermosphere (MLT) region are characterized by a predominant semiannual variability with westward winds during equinox and eastward winds during solstice (Hirota, 1978; Vincent, 1993; Garcia *et al.*, 1997). The first westward phase undergoes interannual variability with large winds occurring in alternate years. This variability has been interpreted as the quasi-biennial oscillation (QBO) in the MLT region, and its possible relation with the stratospheric QBO (SQBO), which usually has nearly two year periodicity, has been noted; a large westward MLT wind phase coincides with the eastward wind phase of the SQBO at all pressure levels (e.g., Burrage *et al.*, 1996). Sridharan *et al.* (2003, 2007) analyzed data from the MF radar at Tirunelveli, India, for the years 1993–2006, and found a period of nearly 3 years during 1998–2001, as shown in Fig. 10. Simultaneously, the period of the SQBO was extended to 3 years during 1998–2002, with large westward winds in the MLT region coinciding with the eastward phase of the stratospheric QBO.

Based upon the same data set, it has been indicated that the annual mean north-

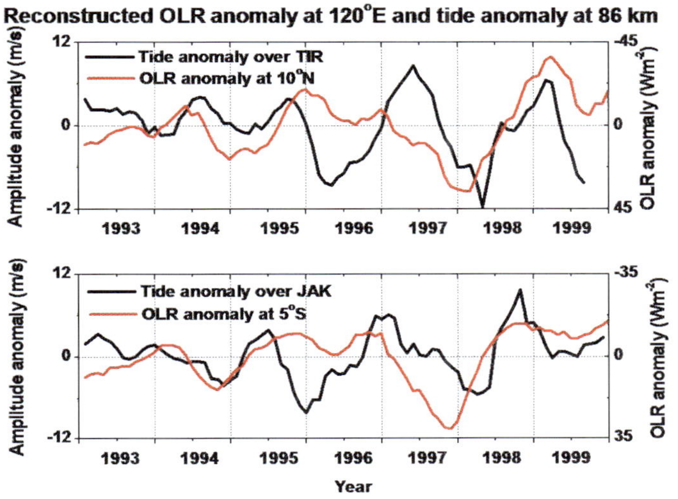

Fig. 11. Anomalies of diurnal tidal amplitude at 86 km in altitude over Tirunelveli, India (8.7°N, 77.8°E; top) and Jakarta; Indonesia (6.4°S, 106.7°E; bottom) shown in black line. OLR anomalies at 120°E are shown in red line at 10°N and 5°S for Tirunelveli (top) and Jakarta (bottom), respectively. (From Gurubaran *et al.*, Interannual variability of diurnal tide in the tropical mesopause region: A signature of the El Nino-Southern Oscillation (ENSO), *Geophys. Res. Lett.*, **32**, L13805, 2005. Copyright 2005 American Geophysical Union. Reproduced by permission of American Geophysical Union.)

ward winds showed monotonic change from northward to southward from 1993 to 2005 (Sridharan *et al.*, 2007). It is interesting to contrast this result near the equator with the decreasing trend in the annual mean southward winds reported from midlatitudes (Jacobi *et al.*, 2005). The solar cycle variations, cooling of the middle atmosphere owing to increase of CO_2 and/or decrease of O_3, and other effects may be suspected as the cause of the long-term MLT wind decrease.

It has been demonstrated that the El Nino-Southern Oscillation (ENSO) affects tidal fields at altitudes as high as the mesopause region. Figure 11 compares OLR anomalies over the IMC with anomalies of diurnal atmospheric tides in the meridional wind near the mesopause as observed with radars at Jakarta, Indonesia (6.4°S, 106.7°E) and Tirunelveli, India (8.7°N, 77.8°E) during 1993–1999 (Gurubaran *et al.*, 2005). Anomalies are defined for both OLR and tidal amplitudes by removing the respective composite seasonal cycles. OLR is used as a proxy for deep tropical convection. Based upon the good correlation between the two, the interannual variability of the diurnal tide is likely caused by cumulus convection over the IMC (~120°E) in response to El Nino-Southern Oscillation (ENSO). It has been suggested that nonmigrating tides excited by large-scale convective systems in the troposphere propagate upward to the mesopause level and compete with the dominant migrating tide in that region (Gurubaran *et al.*, 2005). Recently, longitudinal variabilities of diurnal tides have been investigated in detail based upon data from five MF radars distributed from

$\sim$80°E to $\sim$160°W in the tropical region (Gurubaran *et al.*, 2009). An interesting feature to be noticed is that the long-term tidal fields are not uniform over longitude but differ significantly. For instance, the tidal amplitudes reveal a strong QBO signature over Tirunelveli in India but not over Kauai in the Central Pacific, these sites being separated by more than 120°.

As mentioned in Section 3, the MLT zonal winds also undergo intraseasonal oscillations (ISO) with periods of tens of days (Eckermann and Vincent, 1994). It has been known that there was a strong ISO signal in the amplitude of the diurnal tide and a weak signal in that of high-frequency gravity waves (Eckermann *et al.*, 1997; Isoda *et al.*, 2004). However, the tropospheric ISO is unable to propagate upward directly into the MLT region. Eckermann *et al.* (1997) proposed a hypothesis that tropospheric convection associated with an ISO modulates the intensity of upward propagating gravity waves and nonmigrating tides, and that they induce similar intraseasonal oscillation in the zonal winds through the wave-mean flow interaction in the MLT region. More extensive observations on gravity waves and nonmigrating tides in the middle atmosphere are required to test this working hypothesis.

Even ionospheric irregularities are well correlated with convective activity in the tropics. Based upon the GPS/MET (GPS Meteorology) experiment and ECMWF reanalysis, Hocke and Tsuda (2001) demonstrated that gravity waves were excited by cumulus convection over the continents (including the IMC), and enhanced the activity of the sporadic-E and other ionospheric irregularities aloft in the tropics (with vertical scale less than 7 km).

7.8 Equatorial plasma bubbles

No doubt, equatorial plasma bubbles or huge depletions of the plasma density that are produced by equatorial spread F (ESF) are one of the most prominent features caused by the electrodynamical coupling in the equatorial ionosphere (e.g., Woodman and LaHoz, 1976; Tsunoda *et al.*, 1979; Tsunoda and White, 1981; Hysell and Burcham, 2002). Geomagnetically conjugate structures of the plasma bubbles, which extend to an apex altitude of $\sim$1800 km over the geomagnetic equator, have been observed by all-sky CCD airglow imagers (ASIs) located at Sata and Darwin (Otsuka *et al.*, 2002; Shiokawa *et al.*, 2004). Their detailed features and temporal evolution are quite similar at the two stations, suggesting that the small- to 100-km-scale electron density structures associated with the equatorial plasma bubbles are elongated along the geomagnetic field lines. The occurrence of the bubbles is well correlated with large-scale (about 1000 km in longitude) electron density structures in the equatorial ionization anomaly (EIA), as imaged by satellite (Ogawa *et al.*, 2005).

The Equatorial Atmosphere Radar (EAR), which has a unique active phased-array antenna system, is capable of steering its antenna beam as much as $\pm 50°$ in azimuth on the plane perpendicular to the geomagnetic field lines (Fukao *et al.*, 2003a). The EAR can take a snapshot of radar backscatter echoes (backscatter plumes) from ESF with zonal and altitude extents of several hundred kilometers within a few minutes, a resolution that has been heretofore unachievable (Fukao *et al.*, 2003a, b, 2004). Figure 12 shows a typical example of signal-to-noise ratio (SNR) maps of backscatter plumes from ESF (Yokoyama and Fukao, 2006). Since the first operation of the EAR

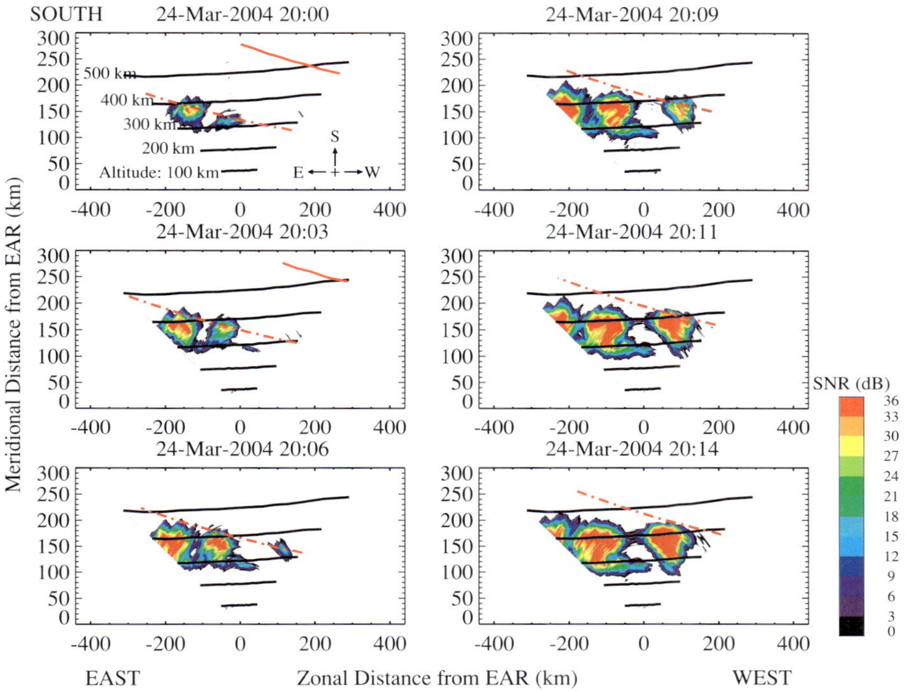

Fig. 12. Fan sector maps of signal-to-noise ratio (SNR) for growing backscatter plumes (equatorial plasma bubbles) observed on March 24, 2004. The five quasi-horizontal lines in each panel indicate altitudes every 100 km from 100 to 500 km, at which the perpendicularity between the radar beam and the geomagnetic field lines is achieved. The inclined solid and dash-dotted lines represent the local and apex sunset terminators, respectively. (From Yokoyama and Fukao, Upwelling backscatter plumes in growth phase of equatorial spread *F* observed with the Equatorial Atmosphere Radar, *Geophys. Res. Lett.*, **33**, L08104, 2006. Copyright 2006 American Geophysical Union. Reproduced by permission of American Geophysical Union.)

in 2001, many important aspects of the equatorial plasma bubbles, for example, onset time, growth rate, spatial distribution, and so forth, have been revealed.

Backscatter plumes observed by the EAR and the associated airglow depletion observed by the ASI installed at the EAO/EAR site are compared in Fig. 13 (Otsuka *et al.*, 2004a). Both airglow depletion and backscatter plume appeared as band-like structures elongated in the meridional direction with a zonal width of about 100 km. The plumes were projected onto a horizontal plane at 250-km altitude, where the 630-nm airglow layer generally exists. It has been found that the backscatter plume (3-m scale irregularities) occur within the entire airglow depletion region and the strongest echoes come from the most depleted region of plasma density (Otsuka *et al.*, 2004a). This demonstrates that the backscatter plumes are a good proxy for equatorial plasma bubbles.

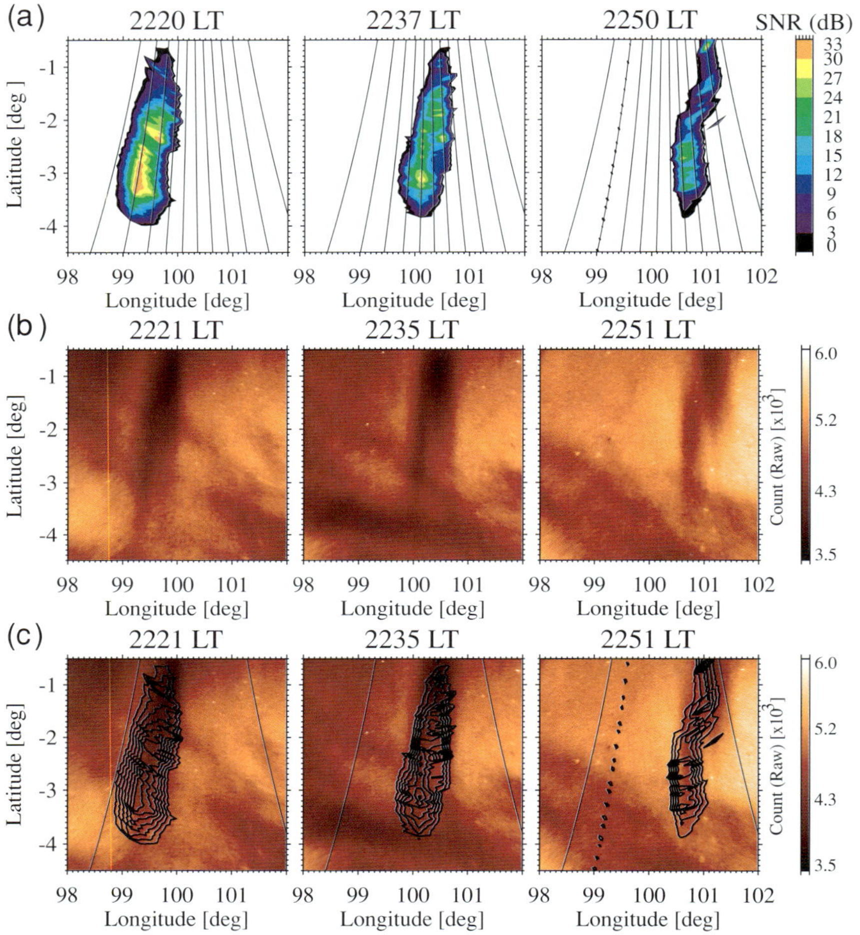

Fig. 13. (a) Two-dimensional distribution of backscatter plume observed with the EAR between 2220 and 2250 LT on April 1, 2003. The plume echo intensity is projected along the geomagnetic field lines onto the surface of 250 km altitude. (b) 630-nm airglow images observed with the all-sky imager at the EAR site. (c) FAI echo intensity shown in (a) is superposed on each airglow image shown in (b). (From Otsuka *et al.*, Spatial relationship of equatorial plasma bubbles and field-aligned irregularities observed with an all-sky airglow imager and the Equatorial Atmosphere Radar, *Geophys. Res. Lett.*, **31**, L20802, 2004. Copyright 2004 American Geophysical Union. Reproduced by permission of American Geophysical Union.)

Yokoyama *et al.* (2004) have shown, for the first time, by direct observation that all backscatter plumes (and accordingly, plasma bubbles) appear at or immediately before sunset at the altitude of the apex of the geomagnetic field line connected with the observed area in magnetically quiet conditions. The snapshots of plasma bubbles

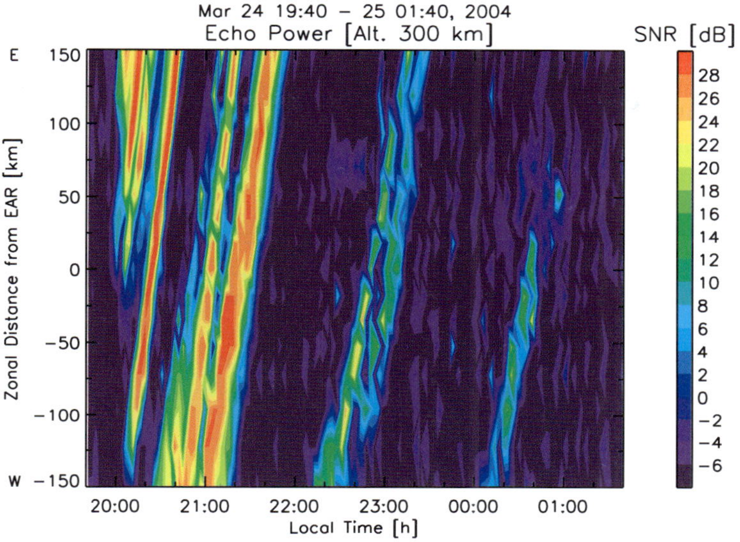

Fig. 14. Temporal variation of plasma bubbles along zonal distance on a plane at an altitude of 300 km from 19:40 LT on March 24 to 01:40 LT on March 25, 2004. Vertical axis indicates zonal distance (positive eastward) from the EAR. (From Fukao *et al.* (2006).)

as shown in Fig. 12 are taken every 2–3 minutes so that the onset and growth of the bubbles can be identified with high time and space resolution. Most notably, the EAR can discriminate fresh plasma bubbles from fossil ones that are generated far to the west of the EAR and drift eastward over it. Almost all bubbles begin to appear at or before sunset at the altitude of the apex of the geomagnetic field line connected with the observed area, and the onset time of more than half of the events corresponds exactly to apex sunset time (Yokoyama *et al.*, 2004).

Nonlinear evolution of plasma bubbles, which could not be understood by conventional single fixed-beam radar observations, has been studied with the EAR for the first time. The rise velocity of the observed plasma bubbles, ranging from a few tens to a few hundreds of meter per second, is comparable to the nonlinear evolution of plasma bubbles simulated by the previous numerical studies (Yokoyama and Fukao, 2006).

Occurrence location and zonal distance of plasma bubbles have been determined by multi-beam observation with the EAR. Figure 14 shows the temporal variation of plasma bubbles along zonal distance on a plane at the altitude of 300 km. The eastward traverse velocity of plasma bubble clusters can be estimated from their slopes, and then the zonal distance between the bubble clusters can be estimated from the velocity. This method of estimation has been applied to 16 nights when distances between clusters of plasma bubbles can be estimated, resulting an average distance of 690 km ranging from 370 to 1000 km. Comparing the occurrence of the equatorial plasma bubbles and the F-layer uplift measured by FM-CW sounders, it has been shown that the F-layer altitude rapidly increases only when plasma bubbles are initiated within the scanned area around sunset. This result suggests that the prereversal enhancement and the ESF occurrence have a characteristic structure with a scale of several hundreds of kilometers (Fukao *et al.*, 2006).

Collaboration with satellite observations is also significant. While EAR observed plasma bubbles (plasma density depletion), plasma density enhancements (plasma blobs) were detected by ROCSAT-1 in the northern hemisphere. The plasma bubbles grew upward with large Doppler velocity away from the EAR late in the pre-midnight sector, in association with the appearance of the plasma blobs. The zonal structure and upward drift velocity of the blobs corresponded to those of the bubbles on the common magnetic flux tube. Localized eastward polarization electric fields probably play an important role in the generation of plasma blobs as well as the resurgence of the plasma bubbles (Yokoyama *et al.*, 2007).

7.9 Impacts on the equatorial ionosphere from below

The unique location of the EAR at the dip latitude of 10.36°S enables us to study an interesting electrodynamical coupling process because the E region illuminated by the EAR is connected to the F region over the dip equator by the geomagnetic field line. The E-region irregularities observed with the EAR are disrupted during the

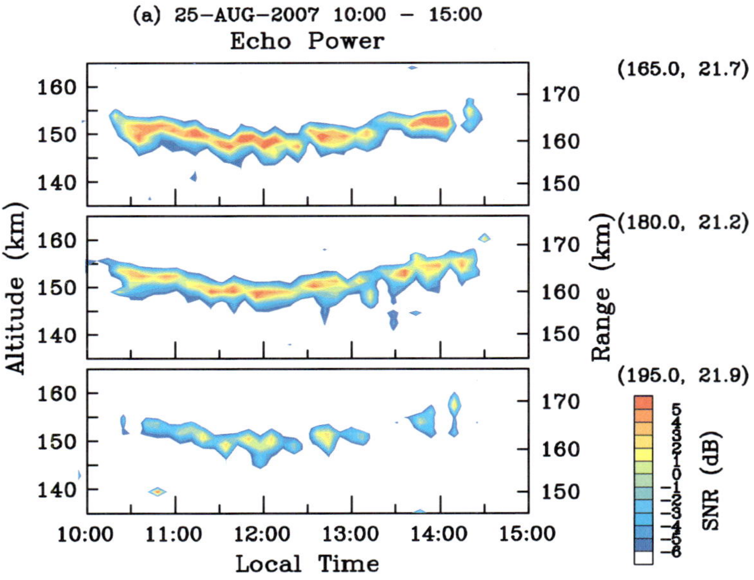

Fig. 15. Altitude-time-intensity plots of 150-km echoes observed in three azimuthal directions on 25 August 2007. (From Patra *et al.*, Daytime 150-km echoes observed with the Equatorial Atmosphere Radar in Indonesia: First results, *Geophys. Res. Lett.*, **35**, L06101, 2008. Copyright 2008 American Geophysical Union. Reproduced by permission of American Geophysical Union.)

growth phase of a plasma bubble simultaneously observed in the *F* region. A relatively weak vertical downward electric field and a large eastward polarization electric field associated with plasma density depletion in the evening equatorial ionosphere, which are mapped to the low latitude *E* region over the EAR, have been proposed to be responsible for the inhibition of growth of the irregularities (Patra *et al.*, 2005). On the other hand, weak irregularities are observed in the valley region (120–160 km altitudes) during the post-sunset period. These echoes have horizontal structure similar to that of *F*-region plasma bubbles and propagate eastward with the same velocity (Yokoyama *et al.*, 2005). These two results demonstrate the strong electrodynamical coupling between the *E* and *F* regions.

Daytime 150-km echoes have been detected also with the EAR as shown in Fig. 15. Major characteristics, such as a necklace shape, a gravity wave-like modulation, and outward and westward drift velocities are common to those observed previously in the equatorial regions. On the other hand, the echo intensity in the east-west plane is not uniform but has a strong asymmetry that has not been noticed in preceding equatorial observations (Patra *et al.*, 2008).

Two-year observations of the ASI at the EAO/EAR site have revealed existence of poleward-moving, quasi-periodic waves in the equatorial anomaly region (Shiokawa *et al.*, 2006). The waves have east-west phase fronts, propagate toward higher latitudes repeatedly at a speed of $\sim$310 m/s with a period of $\sim$40 min, and appear most frequently in June. These features may suggest upward propagation of internal gravity waves from below as their source.

The huge earthquake that occurred in West Sumatra at 0058 UT on December 26, 2004, induced big disturbances in ionospheric total electron content (1.6 to 6.9 TECU), primarily, to the north of the epicenter. A model calculation has shown that such disturbances can be caused by an anisotropic response of the ionosphere to acoustic waves launched from the earthquake (Otsuka *et al.*, 2006). The above results are manifestations that the equatorial ionosphere and thermosphere are disturbed in a measureable way by forcing from below.

It has been conjectured that small-scale gravity waves may seed the Rayleigh-Taylor instability in the nighttime *F* region, whose growth rate is considered too small to account for the rapid growth of ESF without any seeding mechanism (Kelley *et al.*, 1981; Hysell *et al.*, 1990). The relationship between the mechanism that seeds the plasma instability and large-scale plasma structures is still unknown, although gravity waves propagating from below are thought to play an important role as discussed by Ogawa *et al.* (2009).

7.10 Coupling with the opposite hemisphere and midlatitudes

Medium-scale travelling ionospheric disturbances (MSTIDs), which have horizontal wavelengths on the order of several hundred kilometers, have been considered to be generated by either internal gravity waves that propagate upward from the lower atmosphere or heating due to precipitating particles associated with auroral activity (Hines, 1960; Hunsucker, 1982). However, recent observations of nighttime airglow images over Japan have demonstrated that most of nighttime MSTIDs propagate southwestward irrespective of season (Shiokawa *et al.*, 2003a). This preference

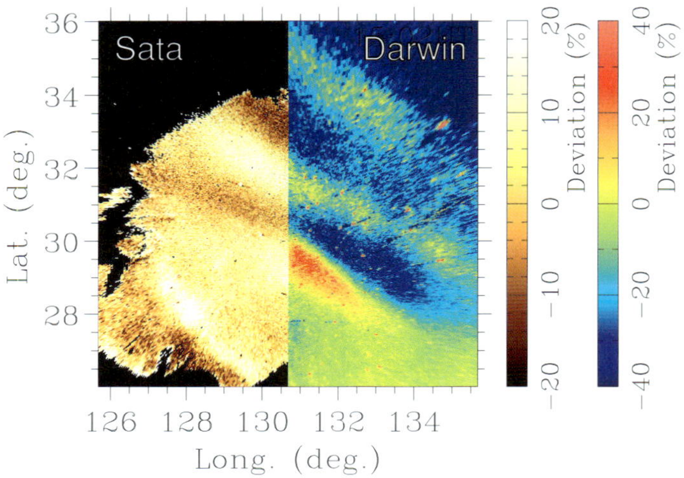

Fig. 16. 630-nm all sky airglow images of MSTIDs in geographic coordinates simultaneously observed at (left) Sata, Japan and (right) Darwin, Australia. The Darwin image is mapped to the northern hemisphere along the geomagnetic field line. MSTID structures extending in the NW-SE direction are matched in detail on both images. (From Otsuka *et al.*, Geomagnetic conjugate observations of medium-scale traveling ionospheric disturbances at midlatitude using all-sky airglow imagers, *Geophys. Res. Lett.*, **31**, L15803, 2004. Copyright 2004 American Geophysical Union. Reproduced by permission of American Geophysical Union.)

of propagation direction is not straightforwardly explained by the classical theory of gravity waves (Kelley and Miller, 1997; Miller *et al.*, 1997), and is suggestive of involvement of electric field perturbations, which are caused by plasma instabilities (Perkins, 1973; Saito *et al.*, 1998; Kelley and Makela, 2001; Shiokawa *et al.*, 2003b).

This hypothesis was tested by Otsuka *et al.* (2004b) by imaging spatial structures of nighttime MSTIDs with the ASIs installed at Sata and Darwin. As shown in Fig. 16, the MSTID structure extending in the northeast-to-southwest direction in the southern hemisphere is matched well on both images, exhibiting excellent geomagnetic conjugacy in great details (Otsuka *et al.*, 2004b; see also Shiokawa *et al.*, 2005). This mirrored structure of MSTIDs strongly suggests that their origin and temporal development are controlled by electric field perturbations transported between the two hemispheres via the geomagnetic field lines. However, simultaneous observations of neutral winds in the conjugate hemispheres are required to clarify the electrodynamic coupling processes between the hemispheres and between the *F* and *E* regions as well.

Multi-year monitoring of GPS scintillations at the EAO/EAR site tells that GPS scintillations (S_4) occur in the pre-midnight sector (between sunset and midnight) in equinoctial months (i.e., maxima in March and September, minima in June and December) (Ogawa *et al.*, 2009; Otsuka *et al.*, 2009); that is the well known characteristic of equatorial plasma bubble occurrence over the Indonesian Maritime Continent

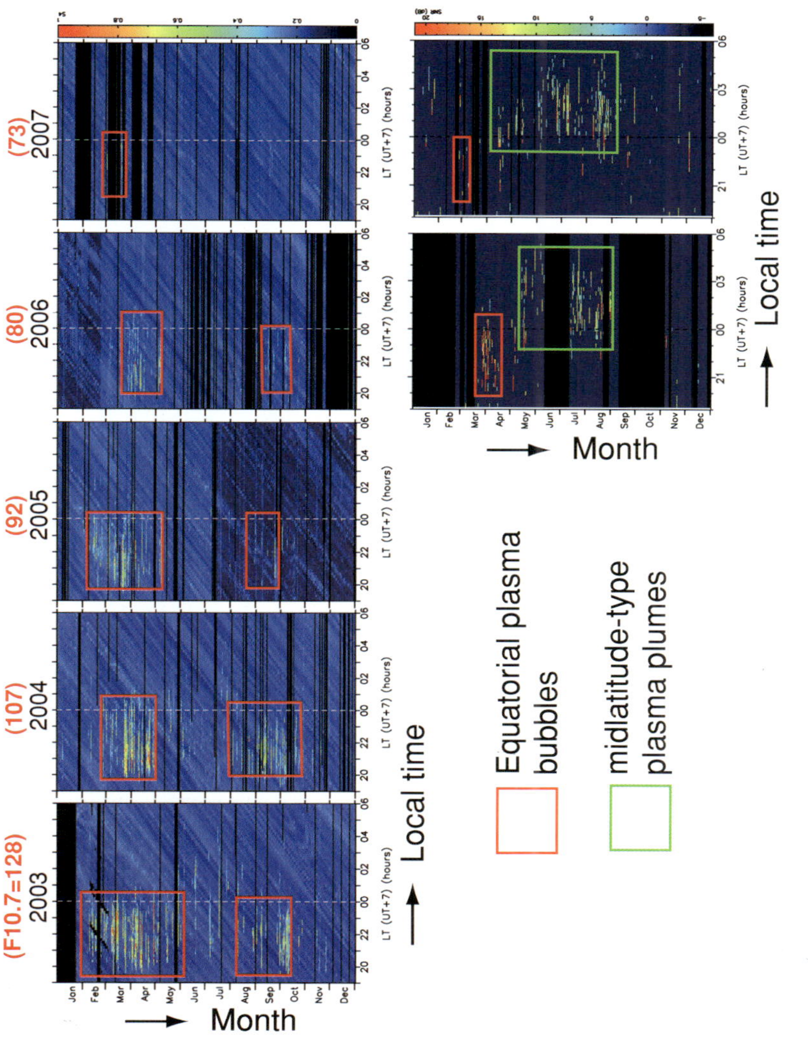

Fig. 17. Seasonal and local time variations of GPS scintillation index (S_4) observed in the period 2003–2007 (top) and radar backscatter plumes observed with the on-site 30.8-MHz radar between 2006 and 2007 (bottom) at Kototabang, Indonesia (0.2°S, 100.3°E; dip latitude 10.4°S). (Courtesy of Y. Otsuka.)

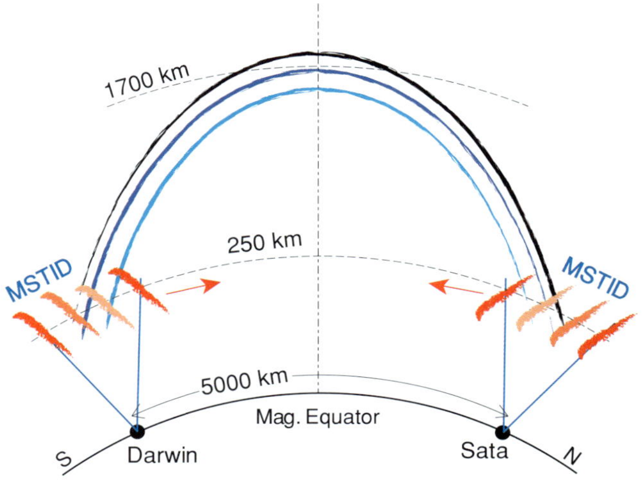

Fig. 18. This cartoon schematically shows that the mirrored MSTID images are obtained at Darwin, Australia and Sata, Japan at the same time on the same day. The distance between the two sites is ~5000 km. (Courtesy of T. Ogawa.)

(e.g., Aarons, 1993). The scintillation occurrence rapidly decreases with decreasing solar activity, as shown in the top panels of Fig. 17. The corresponding backscatter echoes (pre-midnight ESF echoes) obtained by the on-site 30.8-MHz Doppler radar with a peak power of 20 kW are indicated in the bottom panels of the same figure. They propagate primarily to the east and their seasonal and local time variations of occurrence are consistent with those of equatorial plasma bubbles (Otsuka *et al.*, 2009).

While the occurrence of pre-midnight ESF echoes decreases in the low solar activity period, post-midnight ESF echoes are frequently observed with the EAR. However, they are not accompanied by GPS scintillations. Their seasonal and local time variations are not consistent with those of the equatorial plasma bubbles. They often propagate westward (not eastward) with the wave front aligned from northeast to southwest in the southern hemisphere. These features are quite similar to those of the mid-latitude spread *F* observed with the MU radar in Japan rather than ESF (Fukao *et al.*, 2004; Otsuka *et al.*, 2009). Mid-latitude backscatter plumes are supposed to be accompanied by MSTIDs (Fukao *et al.*, 1991). As shown in the schematic of Fig. 18, the MSTIDs are closely coupled between the hemispheres, and propagate down to considerably low latitudes, producing mid-latitude-type plumes on their way.

8 Conclusions

The Coupling Processes in the Equatorial Atmosphere (CPEA) project has been based on a unified viewpoint which integrates lower-, middle- and upper-atmosphere processes, and combines dynamical and electrodynamical approaches that used to be studied independently. The Indonesian Maritime Continent (IMC), the center of the most intense cumulus convection over the globe, is considered as the region where the vertical coupling is most conspicuous, and consequently is set as the focus of the CPEA research activities. The CPEA sought pioneering research by means of ground-based observations with supplemental satellite measurements in the IMC. After six years (2001–2007) of endeavor, a new Equatorial Atmosphere Observatory (EAO) was established for equatorial atmosphere research with EAR as the core facility in Kototabang, West Sumatra, Indonesia.

Many observational studies on various types of dynamical and electrodynamical coupling in the equatorial atmosphere over the IMC were conducted as a part of CPEA. Among them, two campaigns of CPEA have provided numerous new results on the vertical coupling of the equatorial atmosphere, which had not been previously made clear due to the sparseness of observational data in that region. Some examples from our noteworthy results are peculiar features of cumulus convection over Sumatra, short-period gravity wave excitation, ENSO and ISO signals in the MLT region, the spatial structure and onset of equatorial plasma bubbles, and the conjugacy of MSTIDs between two hemispheres. It has been clearly demonstrated by much of this observational evidence that the equatorial atmosphere is much more closely coupled on multiple temporal and spatial scales in the vertical direction than previously expected. More details on the individual topics and other new areas of achieved progress of CPEA should be referred to the more than 200 papers that have been published in international peer-reviewed journals. Especially, a bunch of papers regarding the CPEA-I and -II campaigns are contained in the special issues edited by the *Journal of Meteorological Society of Japan* in 2006 and by *Earth, Planets and Space* in 2009.

Finally, the principal scientific objective of CPEA was to promote and coordinate interdisciplinary research on the dynamical and electrodynamical processes, which vertically couple the equatorial atmosphere and thermosphere/ionosphere, but it was also intended, through the scientific activities of CPEA, to promote and enhance the links between research groups in the countries of the equatorial region, which goals, it may be concluded, have been successfully achieved.

Acknowledgments. The CPEA project was conducted through collaboration between the Research Institute for Sustainable Humanosphere (RISH), Kyoto University, the National Institute of Aeronautics and Space (LAPAN) of Indonesia, the Indonesian Agency for the Assessment and Application of Technology (BPPT), and the Indonesian Meteorological and Geophysical Agency (BMG) during 2001–07. The author deeply appreciates the staff of LAPAN, BPPT and BMG for their extensive collaborations. The CPEA was supported by Grant-in-Aid for Scientific Research on Priority Area—764 of the Ministry of Education, Culture, Sports, Science and Technology of Japan. The author expresses his appreciation to the principal investigators of the six study groups of CPEA for their full collaborations; Profs. Mamoru Yamamoto (Kyoto University), Toru Sato (Kyoto University), Toshiaki Kozu (Shimane University), Toshitaka Tsuda (Kyoto University), Chikao Nagasawa (Tokyo Metropolitan University), and Tadahiko Ogawa (Nagoya University), and all of their co-investigators. He thanks Drs. Masayuki Yamamoto (Kyoto University), Tatsuhiro Yokoyama (Cornell University), Yoshiaki Shibagaki (Osaka Electro-Communication University), and Prof. Makoto Abo (Tokyo Metropolitan University) for valuable comments for writing this review article. He is also grateful to Profs. S. Gurubaran (Indian Institute of Geomagnetism, Tirunelveli, India), Kazuo Shiokawa (Nagoya University), Drs. Sundararajan Sridharan (National Atmospheric Research Laboratory, Pakala, India), M. Venkat Ratnam (National Atmospheric research Laboratory, Gadanki, India), Hiroyuki Hashiguchi (Kyoto University), Yuichi Otsuka (Nagoya University) for useful suggestions, and to Dr. William L. Oliver (Boston University) for careful editing of the original manuscript. Finally, the author wishes to thank Prof. Yamamoto for his dedicated efforts as Secretary General to the CPEA project. The operation of EAR is based upon the Agreement between RISH and LAPAN signed on September 8, 2000.

References

Aarons, J., The longitudinal morphology of equatorial F-layer irregularities relative to their occurrence, *Space Sci. Rev.*, **63**, 209–243, 1993.

Abdu, M. A., T. K. Ramkumar, I. S. Batista, C. G. M. Brum, H. Takahashi, B. W. Reinisch, and J. H. A. Sobral, Planetary wave signatures in the equatorial atmosphere-ionosphere system, and mesosphere—E- and F-region coupling, *J. Atmos. Sol.-Terr. Phys.*, **68**, 509–522, 2006.

Abo, M., C. Nagasawa, and Y. Shibata, Cirrus clouds climatology over the equatorial region, *Reviewed and revised papers presented at the 23rd International Laser Radar Conference*, 415–418, 2006.

Alexander, M. J. and L. Pfister, Gravity wave momentum flux in the lower stratosphere over convection, *Geophys. Res. Lett.*, **22**, 2029–2032, 1995.

Alexander, M. J., J. R. Holton, and D. R. Durran, The gravity-wave response above deep convection in a squall line simulation, *J. Atmos. Sci.*, **52**, 2212–2226, 1995.

Alexander, S., T. Tsuda, J. Furumoto, T. Shimomai, T. Kozu, and M. Kawashima, A statistical overview of convection during CPEA campaign, *J. Meteor. Soc. Jpn.*, **84A**, 57–93, 2006.

Alexander, S. P., T. Tsuda, Y. Shibagaki, and T. Kozu, Seasonal gravity wave activity observed with the Equatorial Atmosphere Radar and its relation to rainfall information from the Tropical Rainfall Measuring Mission, *J. Geophys. Res.*, **113**, D02104, doi:1029/2007JD008777, 2008.

Andrews, D. G., J. R. Holton, and C. B. Leovy, *Middle Atmosphere Dynamics*, Academic Press, Inc., San Diego, Calif, 1987.

Baldwin, M. P., L. J. Gray, T. J. Dunkerton, K. Hamilton, P. H. Haynes, W. J. Randel, J. R. Holton, M. J. Alexander, I. Hirota T. Horinouchi, D. B. A. Jones, J. S. Kinnersley, C. Marquardt, K. Sato, and M. Takahashi, The quasi-biennial oscillation, *Rev. Geophys.*, **39**, 179–229, 2001.

Balsley, B. B., W. L. Ecklund, D. A. Carter, A. C. Riddle, and K. S. Gage, Average vertical motions in the tropical atmosphere observed by a radar wind profiler on Pohnpei (7°N latitude, 157°E longitude), *J. Atmos. Sci.*, **45**, 396–405, 1988.

Burrage, M. D., D. L. Wu, W. R. Skinner, D. A. Ortland, and P. B. Hays, Latitude and seasonal dependence of the semidiurnal tide observed by the high-resolution Doppler imager, *J. Geophys. Res.*, **100**, 11313–11321, 1995.

Burrage, M. D., R. A. Vincent, H. G. Mayr, W. R. Skinner, N. F. Arnold, and P. B. Hays, Long-term variability in the equatorial middle atmosphere zonal wind, *J. Geophys. Res.*, **101**, 12847–12854, 1996.

Dhaka, S. K., M. K. Yamamoto, Y. Shibagaki, H. Hashiguchi, M. Yamamoto, and S. Fukao, Convection-induced gravity wave observed by the Equatorial Atmosphere Radar (0.20°S., 100.32°E) in Indonesia, *Geophys. Res. Lett.*, **12**, L14820, doi:10.1029/2005GL022907, 2005.

Dunkerton, T. J., The role of gravity waves in the quasi-biennial oscillation, *J. Geophys. Res.*, **102**, 26053–26076, 1997.

Eckermann, S. D. and R. A. Vincent, First observations of intraseasonal oscillations in the equatorial mesosphere and lower thermosphere, *Geophys. Res. Lett.*, **21**, 265–268, 1994.

Eckermann, S. D., D. K. Rajopadhyaya, and R. A. Vincent, Intraseasonal wind variability in the equatorial mesosphere and lower thermosphere: Long-term observations from the central Pacific, *J. Atmos. Sol.-Terr. Phys.*, **59**, 603–627, 1997.

Fritts, D. C. and R. A. Vincent, Mesospheric momentum flux studies at Adelaide, Australia: Observations and a gravity wave—tidal interaction model, *J. Atmos. Sci.*, **44**, 605–619, 1987.

Fritts, D. C. and M. J. Alexander, Gravity wave dynamics and effects in the middle atmosphere, *Rev. Geophys.*, **41**, 1003, doi:10.1029/2001RG000106, 2003.

Fujiwara, M., K. Kita, and T. Ogawa, Stratosphere-troposphere exchange of ozone associated with the equatorial Kelvin wave as observed with ozonesondes and rawinsondes, *J. Geophys. Res.*, **103**, 19173–19182, 1998.

Fujiwara, M., F. Hasebe, M. Shiotani, N. Nishi, H. Voemel, and S. J. Oltmans, Water vapor control at the tropopause by equatorial Kelvin waves observed over the Galapagos, *Geophys. Res. Lett.*, **28**, 3143–3146, 2001.

Fujiwara, M., M. K. Yamamoto, H. Hashiguchi, T. Horinouchi, and S. Fukao, Turbulence at the tropopause due to breaking Kelvin waves observed by the Equatorial Atmosphere Radar, *Geophys. Res. Lett.*, **30**, 1171, doi:10.1029/2002GL016278, 2003.

Fukao, S., Coupling Processes in the Equatorial Atmosphere (CPEA): A Project Overview, *J. Meteor. Soc. Jpn.*, **84A**, 1–18, 2006.

Fukao, S., Recent Advances in Atmospheric Radar Study, *J. Meteor. Soc. Jpn.*, **85B**, 215–239, 2007.

Fukao, S., T. Sato, T. Tsuda, S. Kato, K. Wakasugi, and T. Makihira, The MU radar with an active phased array system, 1. Antenna and power amplifiers, *Radio Sci.*, **20**, 1155–1168, 1985a.

Fukao, S., T. Tsuda, T. Sato, S. Kato, K. Wakasugi, and T. Makihira, The MU radar with an active phased array system, 2. In-house equipment, *Radio Sci.*, **20**, 1169–1176, 1985b.

Fukao, S., M. C. Kelley, T. Shirakawa, T, Takami, M. Yamamoto, T. Tsuda, and S. Kato, Turbulent upwelling of the mid-latitude ionosphere, 1. Observational results, *J. Geophys. Res.*, **96**, 3725–3746, 1991.

Fukao, S., H. Hashiguchi, M. Yamamoto, T. Tsuda, T. Nakamura, M. K. Yamamoto, T. Sato, M. Hagio, and Y. Yabugaki, Equatorial Atmosphere Radar (EAR): System Description and First Results, *Radio Sci.*, **38**, 1053, doi:10.1029/2002RS002767, 2003a.

Fukao, S., Y. Ozawa, M. Yamamoto, and R. T. Tsunoda, Altitude-extended equatorial spread F observed near sunrise terminator over Indonesia, *Geophys. Res. Lett.*, **30**, 2137, doi:10.1029/2003GL018383, 2003b.

Fukao, S., Y. Ozawa, T. Yokoyama, M. Yamamoto, and R. T. Tsunoda, First observations of the spatial structure of F region 3-m-scale field-aligned irregularities with the Equatorial Atmosphere Radar in Indonesia, *J. Geophys. Res.*, **109**, A02304, doi:10.1029/2003JA010096, 2004.

Fukao, S., T. Yokoyama, T. Tayama, M. Yamamoto, T. Maruyama, and S. Saito, Eastward traverse of equatorial plasma plumes observed with the Equatorial Atmosphere Radar in Indonesia, *Ann. Geophys.*, **24**, 1411–1418, 2006.

Gage, K. S., B. B. Balsley, W. L. Ecklund, D. A. Carter, and J. R. McAfee, Wind profiler-related research in the tropical Pacific, *J. Geophys. Res.*, **96**, 3209–3220, 1991.

Garcia, R. R. and S. Solomon, The effect of breaking gravity-waves on the dynamics and chemical-composition of the mesosphere and lower thermosphere, *J. Geophys. Res.*, **90**, 3850–3868, 1985.

Garcia, R. R., T. J. Dunkerton, R. S. Lieberman, and R. A. Vincent, Climatology of the semiannual oscillation of the tropical middle atmosphere, *J. Geophys. Res.*, **102**, 26019–26032, 1997.

Gelinas, L. J., M. C. Kelley, and M. F. Larsen, Large-scale E-region electric field structure due to gravity wave winds, *J. Atmos. Sol.-Terr. Phys.*, **64**, 1465–1469, 2002.

Gurubaran, S., R. Rajaram, T. Nakamura, and T. Tsuda, Interannual variability of diurnal tide in the tropical mesopause region: A signature of the El Nino-Southern Oscillation (ENSO), *Geophys. Res. Lett.*, **32**, L13805, doi:10.1029/2005GL022928, 2005.

Gurubaran, S., R. Rajaram, T. Nakamura, T. Tsuda, D. Riggin, and R. A. Vincent, Radar observations of the diurnal tide in the tropical mesosphere-lower thermosphere region: Longitudinal variabilities, *Earth Planets Space*, 2009 (in press).

Hagan, M. E., A. Maute, R. G. Roble, A. D. Richmond, T. J. Immel, and S. L. England, Connections between deep tropical clouds and the Earth's ionosphere, *Geophys. Res. Lett.*, **34**, L20109, doi:10.1029/2007GL030142, 2007.

Hines, C. O., Internal atmospheric gravity waves at ionospheric heights, *Can. J. Phys.*, **38**, 1441–1481, 1960.

Hirota, I., Equatorial waves in the upper stratosphere and mesosphere in relation to the semiannual oscillation of the zonal wind, *J. Atmos. Sci.*, **35**, 714–722, 1978.

Hocke, K. and T. Tsuda, Gravity waves and ionospheric irregularities over tropical convection zones observed by GPS/MET radio occultation, *Geophys. Res. Lett.*, **28**, 2815–2818, 2001.

Holton, J. R., *An Introduction to Dynamic Meteorology*, Academic Press, New York and London, 319 pp., 1972.

Holton, J. R. and R. S. Lindzen, An updated theory for the quasi-biennial cycle of the tropical stratosphere, *J. Atmos. Sci.*, **29**, 1076–1080, 1972.

Holton, J. R., P. H. Haynes, M. E. McIntyre, A. R. Douglass, P. B. Rood, and L. Pfister, Stratosphere-troposphere exchange, *Rev. Geophys.*, **33**, 403–439, 1995.

Horinouchi, T, S. Pawson, K. Shibata, U. Langematz, E. Manzini, M. A. Giorgetta, F. Sassi, R. J. Wilson, K. Hamilton, J. de Grandpre, and A. A. Scaifel, Tropical cumulus convection and upward-propagating waves in middle-atmospheric GCMs, *J. Atmos. Sci.*, **60**, 2765–2782, 2003.

Hunsucker, R. D., Atmospheric gravity waves generated in the high altitude ionosphere: A review, *Rev. Geophys.*, **20**, 293–315, 1982.

Hysell, D. L. and J. Burcham, Long term studies of equatorial spread *F* using the JULIA radar at Jicamarca, *J. Atmos. Terr. Phys.*, **64**, 1531–1543, 2002.

Hysell, D., M. C. Kelley, W. E. Swartz, and R. F. Woodman, Seeding and layering of equatorial spread *F* by gravity waves, *J. Geophys. Res.*, **95**, 17253–17260, 10.1029/ 90JA01213, 1990.

Immel, T. J., E. Sagawa, S. L. England, S. B. Henderson, M. E. Hagan, S. B. Mende, H. U. Frey, C. M. Swenson, and L. J. Paxton, Control of equatorial ionospheric morphology by atmospheric tides, *Geophys. Res. Lett.*, **33**, L15108, doi:10.1029/ 2006GL026161, 2006.

Isoda, F., T. Tsuda, T. Nakamura, R. Vincent, I. Reid, E. Acmad, A. Sadewo, A. Nuryanto, Intraseasonal oscillations of the zonal wind near the mesopause observed with MF and meteor radars in the tropics, *J. Geophys. Res.*, **109**, D21108, doi:10.1029/ 2003JD003378, 2004.

Jacobi, C., Y. I. Portnyagin, E. C. Merzlyakov, T. V. Solovjova, N. A. Makarov, and D. Kurschner, A long-term comparison of mesopause region wind measurements over eastern and central Europe, *J. Atmos. Sol.-Terr. Phys.*, **67**, 229–240, 2005.

Kamio, K., K. Nishimura, and T. Sato, Adaptive sidelobe control for clutter rejection of atmospheric radars, *Ann. Geophys.*, **22**, 4005–4012, 2004.

Kawashima, M., Y. Fujiyoshi, M. Ohi, T. Honda, T. Kozu, T. Shimomai, and H. Hashiguchi, Overview of Doppler radar observations of precipitating cloud systems in Sumatera Island during CPEA-IOP '04, *J. Meteor. Soc. Jpn.*, **84A**, 33–56, 2006.

Kelley, M. C., *The Earth's Ionosphere: Plasma Physics and Electrodynamics*, Academic Press, Inc., San Diego, Calif, 1989.

Kelley, M. C. and C. A. Miller, Electrodynamics of midlatitude spread *F*: 3. Electrodynamic waves? A new look at the role of electric field in thermospheric wave dynamics, *J. Geophys. Res.*, **102**, 11539–11547, 1997.

Kelley, M. C. and J. J. Makela, Resolution of the discrepancy between experiment and theory of midlatitude *F*-region structures, *Geophys. Res. Lett.*, **28**, 2589–2592, 2001.

Kelley, M. C., M. F. Larsen, C. A. LaHoz, and J. P. McClure, Gravity wave initiation of equatorial spread *F*: A case study, *J. Geophys. Res.*, **86**, 9087–9100, 1981.

Kodama, Y.-M., M. Tokuda, and F. Murata, Convective activity over the Indonesian Maritime Continent during CPEA-I as evaluated by lightning activity and Q1 and Q2 profiles, *J. Meteor. Soc. Jpn.*, **84A**, 133–149, 2006.

Kozu, T., T. Shimomai, Zainul Akramin, Marzuki, Y. Shibagaki, and H. Hashiguchi, Intraseasonal variation of raindrop size distribution at Koto Tabang, West Sumatra, Indonesia, *Geophys. Res. Lett.*, **32**, L07803, doi:10.1029/2004GL022340, 2005.

Kozu, T., K. K. Reddy, S. Mori, M. Thurai, J. T. Ong, D. N. Rao, and T. Shimomai, Seasonal and diurnal variations of raindrop size distribution in Asian Monsoon region, *J. Meteor. Soc. Jpn.*, **84A**, 195–209, 2006.

Lane, T. P. and M. J. Reeder, Convectively generated gravity waves and their effect on the cloud environment, *J. Atmos. Sci.*, **58**, 2427–2440, 2001.

Lau, K., T. Nakazawa, and C. H. Sui, Observations of cloud cluster hierarchies over the tropical western Pacific, *J. Geophys. Res.*, **96**, 3197–3208, 1991.

Maruyama, T., M. Kawamura, S. Saito, K. Nozaki, H. Kato, H. Hemmakorn, T. Boonchuk, T. Komolmis, and C. Ha Duyen, Low latitude ionosonde network for ionosphere-thermosphere coupling studies, *Ann. Geophys.*, **25**, 1569–1577, 2007.

Mayr, H. G., J. G. Mengel, C. A. Reddy, K. L. Chan, and H. S. Porter, Variability of the equatorial oscillations induced by gravity wave filtering, *Geophys. Res. Lett.*, **25**, 2629–2632, 1998.

Meyer, C. K. and J. M. Forbes, A 6.5-day westward propagating wave: Origin and characteristics, *J. Geophys. Res.*, **102**, 26173–26178, 1997.

Miller, C. A., W. E. Swartz, M. C. Kelley, M. Mendillo, D. Nottingham, J. Scali, and B. Reinisch, Electrodynamics of midlatitude spread F: 1. Observations of unstable, gravity wave-induced ionospheric electric fields at tropical latitudes, *J. Geophys. Res.*, **102**, 11521–11532, 1997.

Morita, J., Y.-N. Takayabu, S. Shige, and Y.-M. Kodama, Analysis of rainfall characteristics of the Madden-Julian oscillation using TRMM satellite data, *Dyn. Atmos. Oceans*, **42**, 107–126, 2006.

Nagasawa, C., M. Abo, and Y. Shibata, Lidar system for observations of equatorial lower and upper atmosphere, *Reviewed and revised papers presented at the 23rd International Laser Radar Conference*, 43–46, 2006.

Nakamura, T., T. Aono, T. Tsuda, A. G. Admiranto, E. Achmad, and Suranto, Mesospheric gravity waves over a tropical convective region observed by OH airglow imaging in Indonesia, *Geophys. Res. Lett.*, **30**, 1882, doi:10.1029/2003GL017619, 2003.

Nakazawa, T., Tropical super clusters within intraseasonal variations over the western Pacific, *J. Meteor. Soc. Jpn.*, **66**, 823–839, 1988.

Nishi, N., M. K. Yamamoto, T. Shimomai, A. Hamada, and S. Fukao, Fine structure of vertical motion in the stratiform precipitation region observed by a VHF Doppler radar installed in Sumatra, Indonesia, *J. Appl. Meteor. Climatol.*, **46**, 522–537, 2007.

Nishida, M., A. Shimizu, T. Tsuda, C. Rocken, and R. H. Ware, Seasonal and longitudinal variations in the tropical tropopause observed with the GPS occultation technique (GPS/MET), *J. Meteor. Soc. Jpn.*, **78**, 691–700, 2000.

Nishimura, K., E. Gotoh, and T. Sato, Fine scale 3D wind field observed with multistatic equatorial atmosphere radar, *J. Meteor. Soc. Jpn.*, **84A**, 227–238, 2006.

Nitta, Ts., T. Mizuno, and K. Takahashi, Multi-scale convective systems during the initial phase of the 1986/87 EI Nino, *J. Meteor. Soc. Jpn.*, **70**, 447–466, 1992.

Ogawa, T., E. Sagawa, Y. Otsuka, K. Shiokawa, T. J. Immel, S. B. Mende, and P. Wilkinson, Simultaneous ground-and satellite-based airglow observations of geomagnetic conjugate plasma bubbles in the equatorial anomaly, *Earth Planets Space*, **57**, 385–392, 2005.

Ogawa, T., Y. Miyoshi, Y. Otsuka, T. Nakamura, and K. Shiokawa, Equatorial GPS ionospheric scintillations over Kototabang, Indonesia and their relation to atmospheric waves from below, *Earth Planets Space*, 2009 (in press).

Otsuka, Y., K. Shiokawa, T. Ogawa, and P. Wilkinson, Geomagnetic conjugate observations of equatorial airglow depletions, *Geophys. Res. Lett.*, **29**, doi:10.1029/2002GL015347, 2002.

Otsuka, Y., K. Shiokawa, T. Ogawa, T. Yokoyama, M. Yamamoto, and S. Fukao, Spatial relationship of equatorial plasma bubbles and field-aligned irregularities observed with an all-sky airglow imager and the Equatorial Atmosphere Radar, *Geophys. Res. Lett.*, **31**, L20802, doi:10.1029/2004GL020869, 2004a.

Otsuka, Y., K. Shiokawa, T. Ogawa, and P. Wilkinson, Geomagnetic conjugate observations of medium-scale traveling ionospheric disturbances at midlatitude using all-sky airglow imagers, *Geophys. Res. Lett.*, **31**, L15803, doi:10.1029/2004GL020262, 2004b.

Otsuka, Y., N. Kotake, T. Tsugawa, K. Shiokawa, T. Ogawa, Effendy, S. Saito, M. Kawamura, and T. Maruyama, GPS detection of total electron content variations over Indonesia and Thailand following the 26 December 2004 earthquake, *Earth Planets Space*, **58**, 159–165, 2006.

Otsuka, Y., T. Ogawa, and Effendy, VHF radar observations of nighttime F-region field-aligned irregularities over Kototabang, Indonesia, *Earth Planets Space*, 2009 (in press).

Patra, A. K., T. Yokoyama, M. Yamamoto, S. Saito, T. Maruyama, and S. Fukao, Disruption of E region echoes observed by the EAR during the development phase of equatorial spread F: a manifestation of electrostatic field coupling, *Geophys. Res. Lett.*, **32**, L17104, doi:10.1029/2005GL022868, 2005.

Patra, A. K., T. Yokoyama, Y. Otsuka, and M. Yamamoto, Daytime 150-km echoes observed with the Equatorial Atmosphere Radar in Indonesia: First results, *Geophys. Res. Lett.*, **35**, L06101, doi:10.1029/2007GL033130, 2008.

Perkins, F., Spread F and ionospheric currents, *J. Geophys. Res.*, **78**, 218–226, 1973.

Pfister, L., K. R. Chan, T. B. Bui, S. Bowen, M. Legg, B. Gary, K. Kelly, M. Proffitt, and W. Starr, Gravity waves generated by a tropical cyclone during the STEP tropical field program, *J. Geophys. Res.*, **98**, 8611–8638, 1993.

Plumb, A., *The quasi-biennial oscillation in Dynamics of the Middle Atmosphere*, 217–251, D. Reidel, Norwell, Mass, 1982.

Ratnam, M. V., T. Tsuda, C. Jacobi, and Y. Aoyama, Enhancement of gravity wave activity observed during a major Southern Hemisphere stratospheric warming by CHAMP/GPS measurements, *Geophys. Res. Lett.*, **31**, L16101, doi:10.1029/2004GL019789, 2004.

Ratnam, M. V., T. Tsuda, Y. Shibagaki, T. Kozu, and S. Mori, Gravity wave characteristics over the equator observed during CPEA campaign using simultaneous data from multiple stations, *J. Meteor. Soc. Jpn.*, **84A**, 239–257, 2006a.

Ratnam, M. V., T. Tsuda, S. Mori, and T. Kozu, Modulation of Tropopause Temperature Structure Revealed by Simultaneous Radiosonde and CHAMP GPS Measurements, *J. Meteor. Soc. Jpn.*, **84**, 989–1003, 2006b.

Ratnam, M. V., S. P. Alexander, T. Kozu, and T. Tsuda, Characteristics of gravity waves observed with intensive radiosonde campaign during November-December 2005, *Earth Planets Space*, 2009 (in press).

Ricciardulli, L. and R. R. Garcia, The excitation of equatorial waves by deep convection in the NCAR Community Climate Model (CCM3), *J. Atmos. Sci.*, **57**, 3461–3487, 2000.

Sagawa, E., T. J. Immel, H. U. Frey, and S. B. Mende, Longitudinal structure of the equatorial anomaly in the nighttime ionosphere observed by IMAGE/FUV, *J. Geophys. Res.*, **110**, A11302, doi:10.1029/2004JA010848, 2005.

Saito, A., T. Iyemori, L. G. Blomberg, M. Yamamoto, and M. Takeda, Conjugate observations of the mid-latitude electric field fluctuations with the MU radar and the Freja satellite, *J. Atmos. Sol.-Terr. Phys.*, **60**, 129–140, 1998.

Salby, M. L. and R. R. Garcia, Transient-response to localized episodic heating in the tropics. 1. Excitation and short-time near-field behavior, *J. Atmos. Sci.*, **44**, 458–498, 1987.

Schafer, R., S. K. Avery, and K. S. Gage, A comparison of VHF wind profiler observations and the NCEP-NCAR reanalysis over the tropical Pacific, *J. Appl. Meteor.*, **42**, 873–889, 2003.

Seto, T. H., M. K. Yamamoto, H. Hashiguchi, S. Fukao, M. Abo, T. Kozu, and M. Kudsy, Observational study on westerly wind burst over Sumatra, Indonesia by the Equatorial Atmosphere Radar—A case study during the first CPEA campaign—, *J. Meteor. Soc. Jpn.*, **84A**, 95–112, 2006.

Shibagaki, Y., T. Kozu, T. Shimomai, S. Mori, F. Murata, Y. Fujiyoshi, H. Hashiguchi, and S. Fukao, Evolution of a super cloud cluster and the associated wind fields observed over the Indonesian Maritime Continent during the first CPEA campaign, *J. Meteor. Soc. Jpn.*, **84A**, 19–31, 2006a.

Shibagaki, Y., T. Shimomai, T. Kozu, S. Mori, Y. Fujiyoshi, H. Hashiguchi, M. K. Yamamoto, S. Fukao, and M. D. Yamanaka, Multiscale aspects of convective systems associated with an intraseasonal oscillation over the Indonesian Maritime Continent, *Mon. Wea. Rev.*, **134**, 1682–1696, 2006b.

Shibata, Y., C. Nagasawa, M. Abo, T. Maruyama, S. Saito, and T. Nakamura, Lidar observations of sporadic Fe and Na layers in the mesopause region over Equator, *J. Meteor. Soc. Jpn.*, **84A**, 317–325, 2006.

Shiokawa, K., Y. Katoh, M. Satoh, M. K. Ejiri, T. Ogawa, T. Nakamura, T. Tsuda, and R. H. Wiens, Development of optical mesosphere thermosphere imagers (OMTI), *Earth Planets Space*, **51**, 887–896, 1999.

Shiokawa, K., C. Ihara, Y. Otsuka, and T. Ogawa, Statistical study of nighttime medium-scale traveling ionospheric disturbances using midlatitude airglow images, *J. Geophys. Res.*, **108**, 1052, doi:10.1029/2002JA009491, 2003a.

Shiokawa, K., Y. Otsuka, C. Ihara, T. Ogawa, and F. J. Rich, Ground and satellite observations of nighttime medium-scale traveling ionospheric disturbance at midlatitude, *J. Geophys. Res.*, **108**, 1145, doi:10.1029/2002JA009639, 2003b.

Shiokawa, K., Y. Otsuka, T. Ogawa, and P. Wilkinson, Time evolution of high-altitude plasma bubbles imaged at geomagnetic conjugate points, *Ann. Geophys.*, **22**, 3137–3143, 2004.

Shiokawa, K., Y. Otsuka, T. Tsugawa, T. Ogawa, A. Saito, K. Ohshima, M. Kubota, T. Maruyama, T. Nakamura, M. Yamamoto, and P. Wilkinson, Geomagnetic conjugate observation of nighttime medium- and large-scale traveling ionospheric disturbances: FRONT3 campaign, *J. Geophys. Res.*, **110**, A05303, doi:10.1029/2004JA010845, 2005.

Shiokawa, K., Y. Otsuka, and T. Ogawa, Quasi-periodic southward-moving waves in 630-nm airglow images in the equatorial thermosphere, *J. Geophys. Res.*, **111**, A06301, doi:10.1029/2005JA011406, 2006.

Sridharan, S., S. Gurubaran, and R. Rajaram, QBO influences on the variability of planetary waves in the equatorial mesopause region, *Earth Planets Space*, **55**, 687–696, 2003.

Sridharan, S., T. Tsuda, T. Nakamura, T. Kozu, S. Mori, and J. M. Russell, Observations of the 7-day Kelvin wave in the tropical atmosphere during the CPEA campaign, *J. Meteor. Soc. Jpn.*, **84A**, 259–275, 2006a.

Sridharan, S., T. Tsuda, R. A. Vincent, T. Nakamura, and Effendy, A Report on radar observations of 5–8-day waves in the equatorial MLT region, *J. Meteor. Soc. Jpn.*, **84A**, 295–304, 2006b.

Sridharan, S., T. Tsuda, and S. Gurubaran, Radar observations of long-term variability of mesosphere and lower thermosphere winds over Tirunelveli (8.7°N, 77.8°E), *J. Geophys. Res.*, **112**, D23105, doi:10.1029/2007JD008669, 2007.

Sridharan, S., T. Tsuda, T. Nakamura, and T. Horinouchi, The 5–8 day Kelvin and Rossby waves in the tropics as revealed by ground and satellite-based observations, *J. Meteor. Soc. Jpn.*, **86**, 43–55, 2008.

Takahashi, H., C. M. Wrasse, D. Pancheva , M. A. Abdu , I. S. Batista , L. M. Lima , P. P. Batista , B. R. Clemesha, and K. Shiokawa, Signatures of 3–6 day planetary waves in the equatorial mesosphere and ionosphere, *Ann. Geophys.*, **24**, 3343–3350, 2006.

Talaat, E. R., J. H. Yee, and X. Zhu, Observations of the 6.5 day wave in the mesosphere and lower thermosphere, *J. Geophys. Res.*, **106**, 20715–20723, 2001.

Taylor, M. J., M. A. Hapgood, and P. Rothwell, Observations of gravity-wave propagation in the OI (557.7 nm), Na (589.2 nm) and the near-infrared OH nightglow emissions, *Planet. Space Sci.*, **35**, 413–427, 1987.

Tsuda, T., S. Fukao, M. Yamamoto, T. Nakamura, M. D. Yamanaka, T. Adachi, H. Hashiguchi, N. Fujioka, M. Tsutsumi, S. Kato, S. W. B. Harijono, T. Sribimawati, B. P. Sitorus, R. B. Yahya, M. Karmini, F. Renggono, B. L. Parapat, W. Djojonegoro, P. Mardio, N. Adikusumah, H. T. Endi, and H. Wiryosumarto, A preliminary report on observations of Equatorial Atmosphere Dynamics in Indonesia with radars and rawinsondes, *J. Meteor. Soc. Jpn.*, **73**, 393–406, 1995.

Tsuda, T., M. Nishida, C. Rocken, and R. H. Ware, A global morphology of gravity wave activity in the stratosphere revealed by the GPS occultation data (GPS/MET), *J. Geophys. Res.*, **105**, 7257–7273, 2000.

Tsuda, T., M. V. Ratnam, T. Kozu, and S. Mori, Characteristics of 10-day Kelvin wave observed with radiosondes and CHAMP/GPS occultation during the CPEA campaign (April–May, 2004), *J. Meteor. Soc. Jpn.*, **84A**, 277–293, 2006.

Tsuda, T., M. V. Ratnam, S. P. Alexander, T. Kozu, and Y. Takayabu, Temporal and spatial distributions of atmospheric wave energy in the equatorial stratosphere revealed by GPS radio occultation temperature data obtained with the CHAMP satellite during 2001–2006, *Earth Planets Space*, 2009 (in press).

Tsunoda, R. T., Day-to-day variability in equatorial spread *F*: is there some physics missing?, *Geophys. Res. Lett.*, **33**, L16106, doi:10.1029/2006GL025956, 2006.

Tsunoda, R. T. and B. R. White, On the generation and growth of equatorial backscatter plumes, 1. Wave structure in the bottomside *F* layer, *J. Geophys. Res.*, **86**, 3610–3616, 1981.

Tsunoda, R. T., M. J. Baron, J. Owen, and D. M. Towle, Altair: An incoherent scatter radar for equatorial spread *F* studies, *Radio Sci.*, **14**, 1111–1119, 1979.

Vincent, R. A., Long-period motions in the equatorial mesosphere, *J. Atmos. Sol.-Terr. Phys.*, **55**, 1067–1080, 1993.

Webster, P. J. and R. Lukas, TOGA COARE: The Coupled Ocean-Atmosphere Response Experiment, *Bull. Amer. Meteor. Soc.*, **73**, 1377–1416, 1992.

Woodman, R. F. and C. LaHoz, Radar observations of *F*-region equatorial irregularities, *J. Geophys. Res.*, **81**, 5447–5466, 1976.

Wu, D. L., P. B. Hays, and W. R. Skinner, Observations of the 5-day wave in the mesosphere and lower thermosphere, *Geophys. Res. Lett.*, **21**, 2733–2736, 1994.

Yamamoto, M. K., M. Fujiwara, T. Horinouchi, H. Hashiguchi, and S. Fukao, Kelvin-Helmholtz instability around the tropical tropopause observed with the Equatorial Atmosphere Radar, *Geophys. Res. Lett.*, **30**, 1476, doi:10.1029/2002GL016685, 2003.

Yamamoto, M. K., N. Nishi, T. Horinouchi, M. Niwano, and S. Fukao, ertical wind observation in the tropical upper troposphere by VHF wind profiler: A case study, *Radio Sci.*, **42**, RS3005, doi:10.1029/2006RS003538, 2007.

Yamamoto, M. K., Y. Ohno, H. Horie, N. Nishi, H. Okamoto, K. Sato, H. Kumagai, M. Yamamoto, H. Hashiguchi, S. Mori, N. O. Hashiguchi, H. Nagata, and S. Fukao, Observation of particle fall velocity in cirriform cloud by VHF and millimeter-wave Doppler radars, *J. Geophys. Res.*, **113**, D12210, doi:10.1029/2007JD009125, 2008.

Yokoyama, T. and S. Fukao, Upwelling backscatter plumes in growth phase of equatorial spread *F* observed with the Equatorial Atmosphere Radar, *Geophys. Res. Lett.*, **33**, L08104, doi:10.1029/2006GL025680, 2006.

Yokoyama, T., S. Fukao, and M. Yamamoto, Relationship of the onset of equatorial *F*-region irregularities with the sunset terminator observed with the Equatorial Atmosphere Radar, *Geophys. Res. Lett.*, **31**, L24804, doi:10.1029/2004GL021529, 2004.

Yokoyama, T., A. K. Patra, S. Fukao, and M. Yamamoto, Ionospheric irregularities in the low-latitude valley region observed with the Equatorial Atmosphere Radar, *J. Geophys. Res.*, **110**, A10207, doi:10.1029/2005JA011208, 2005.

Yokoyama, T., S.-Y. Su, and S. Fukao, Plasma blobs and irregularities concurrently observed by ROCSAT-1 and Equatorial Atmosphere Radar, *J. Geophys. Res.*, **112**, A05311, doi:10.1029/2006JA012044, 2007.

Climate and Weather of the Sun-Earth System (CAWSES): Selected Papers from the 2007 Kyoto Symposium,
Edited by T. Tsuda, R. Fujii, K. Shibata, and M. A. Geller, pp. 337–348.

Vertical coupling by the semidiurnal tide in Earth's atmosphere

Jeffrey M. Forbes

*Department of Aerospace Engineering Sciences, University of Colorado,
Boulder, CO 80309, USA
E-mail: forbes@colorado.edu*

In this paper I provide a perspective on the global manifestations and consequences of the migrating (Sun-synchronous) semidiurnal tide in Earth's atmosphere. Topics include tide/mean-wind interactions; production of non-Sun-synchronous tides due to interaction between sun-synchronous tides and stationary planetary waves; modification of the zonal mean circulation due to momentum deposition by dissipating tides; and tidal modulation by the quasi-2-day wave and related effects in the thermosphere-ionosphere system. Experimental and modeling evidence for the above phenomena are presented with a view towards providing an appreciation for the pervasiveness of semidiurnal tidal effects from the surface to the ionosphere-thermosphere system.

1 Excitation of the Semidiurnal Tide

Because the Earth rotates, any point in the atmosphere is subject to solar radiation that varies with the diurnal cycle. If this radiation is absorbed locally by various atmospheric constituents and soon afterwards liberated in the form of heat, then the solar heating varies similarly with time. If we imagine a heating distribution $J(t)$ that is non-zero during the day and zero at night, and a Fourier decomposition is performed where $\Omega = \dfrac{2\pi}{24}$ and t is in hours:

$$J(t) = \sum_{n=0}^{N} A_n(\theta, z, \lambda) \cos(n\Omega t - \phi_n(\theta, z)) \tag{1}$$

then the heating is separated into so-called diurnal ($n = 1$), semidiurnal ($n = 2$), terdiurnal ($n = 3$) components, etc. In a linear system, which is often applicable to atmospheric tides, the atmospheric response to each of the above wave components can be calculated separately and then summed *a posteriori* to get the total tidal response. Generally the diurnal amplitude is the largest in the expansion by at least a factor of two. However, it is the global response to the semidiurnal component of heating that concerns us in this paper. Note that if Eq. (1) is applied throughout the atmosphere, the A_n and ϕ_n represent the amplitude and phase, respectively, of the semidiurnal tidal heating as a function of height (z), latitude (θ) and longitude (λ).

One can further imagine, again due to rotation of the Earth, that the heating distribution ought to be periodic in longitude, too. An extension to Eq. (1) may therefore be written:

$$J(t) = \sum_{s=-\infty}^{s=+\infty} \sum_{n=0}^{N} A_{n,s}(\theta, z) \cos(n\Omega t + s\lambda - \phi_{n,s}(\theta, z)) \tag{2}$$

where s = zonal wavenumber and A_n and ϕ_n are functions of height and latitude. Thus, the semidiurnal ($n = 2$) tide can in principle consist of a number of zonal wavenumber components depending on the zonal distribution of heating. Moreover, if we calculate the zonal phase speed of these components by setting the quantity in square brackets in Eq. (2) equal to a constant and differentiating with respect to time:

$$C_{\text{ph}} = \frac{d\lambda}{dt} = -\frac{2\Omega}{s} \tag{3}$$

then we see that, depending on s, a semidiurnal tide can be propagating westward with the phase speed of the Sun ($s = 2$; $C_{\text{ph}} = -\Omega$); westward with a phase speed twice that of the Sun ($s = 1$; $C_{\text{ph}} = -2\Omega$); westward with a phase speed slower than the Sun ($s > 2$); or eastward ($s < 0$). The zonally-symmetric oscillation does not propagate zonally, but instead oscillates with semidiurnal period at all longitudes simultaneously. In the context of a linear perturbation model (e.g., the Global Scale Wave Model (GSWM); Hagan and Forbes, 2002, 2003), the total response to a longitudinal distribution of heating can be obtained by calculating the response to each (n, s) component of heating individually, and then all of the solutions summed to get the global tidal response.

The westward-propagating $s = 2$ component of the semidiurnal tide (hereafter SW2) is Sun-synchronous ($s = 2$; $C_{\text{ph}} = -\Omega$) and is said to *migrate* with the apparent motion of the Sun to a ground-based observer. This tide is generated by interaction of solar radiation with a distribution of atmospheric constituents (and the subsequent liberation of heat) that is zonally-symmetric. On the other hand, in the troposphere, the semidiurnal tide excitation due to latent heating associated with deep tropical convection and insolation absorption by water vapor are longitude-dependent and these sources excite a spectrum of semidiurnal tides consisting of various zonal wavenumbers. Hagan and Forbes (2002, 2003) address this problem, and calculate the spectrum of semidiurnal tides propagating into the lower thermosphere due to these sources. While this is an extremely interesting and timely area of study, in keeping with the need to focus the work at hand, attention here is devoted to SW2, and specifically that component of SW2 generated by insolation absorption by ozone between about 30 and 50 km. The ozone distribution at these altitudes is smoothly-varying with latitude and zonally-symmetric (to a good approximation) with longitude between $\pm 50^\circ$ latitude. At higher latitudes the contribution to global excitation of the semidiurnal tide is of secondary importance. Therefore, the latitude distribution of semidiurnal tidal heating due to O_3 insolation absorption is very broad and quasi-symmetric about the equator, with some shift of the maximum heating with the subsolar point. This latitude distribution projects very well onto the fundamental semidiurnal mode ("Hough function") of tidal theory, which is symmetric about the equator and extends from pole to pole with a full width at half-maximum of 60° latitude. This zonally-symmetric distribution of heating between 30 and 60 km, and peaking near 45 km with a near-Gaussian distribution (Chapman and Lindzen, 1970; Forbes and Garrett, 1979), represents the starting point for the remainder of the paper.

2 Interaction between SW2 and the Zonal Mean Wind

As noted above, the broad latitudinal distribution of heating depicted in Fig. 1 primarily excites a fundamental semidiurnal mode that is symmetric about the equator and moreover has a vertical wavelength in excess of 200 km. This wave propagates to the surface, and the semidiurnal signal can be seen in tropical surface pressure records (Chapman and Lindzen, 1970). As this wave propagates upward and encounters latitudinal and vertical shears and Doppler-shifting effects of the middle atmosphere zonal mean wind jets, the horizontal structure of SW2 becomes distorted and non-symmetric. In the context of linear tidal theory (e.g., Chapman and Lindzen, 1970) this is manifested in the form of high-order orthogonal modes which, when superimposed according to the relative amplitudes demanded by orthogonal decomposition, reproduce the distorted shape. Based on experience with numerical models that properly account for tide-mean wind interactions (Lindzen and Hong, 1974; Walterscheid and Venkateswaran, 1979a, b; Forbes, 1982) these higher-order modes are generated as though the above 'mode coupling' process (Lindzen and Hong, 1974) serves as a secondary tidal source with each mode propagating upwards as an in-

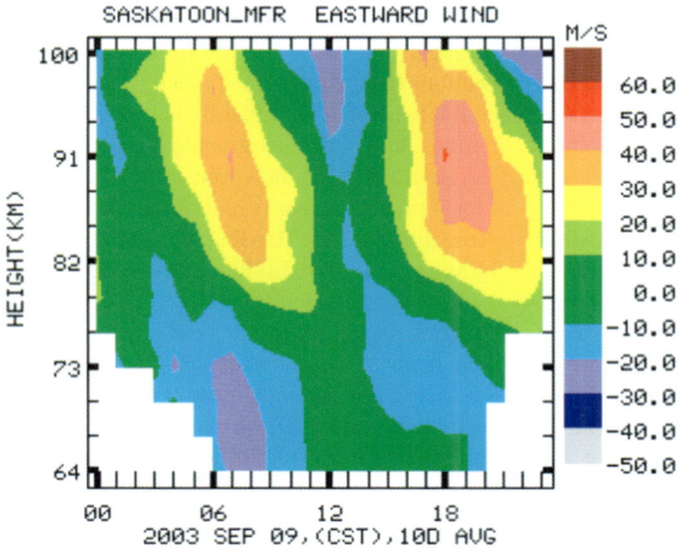

Fig. 1. Zonal wind measurements by the Saskatoon, Canada (52°N, 107°W) MF radar on September 9, 2003, illustrating the solar semidiurnal tide. (Figure courtesy of Prof. Alan Manson.)

dependent global oscillation. Further, these higher-order modes tend to have their maximum amplitudes at middle to high latitudes, with progressively shorter vertical wavelengths (e.g., 82, 54, 41, 34 km) and increasing numbers of 180° shifts in phase (nodes) between equator and pole (e.g., 1, 2, 3, 4) as the order of the Hough function increases.

3 Observations of SW2

An example of a semidiurnal signature in the zonal wind field in the mesosphere-lower thermosphere region is provided in Fig. 1. These data (courtesy of Prof. Alan Manson) originate from the University of Saskatchewan MF radar in Saskatoon, Canada (52°N, 107°W), and clearly illustrate a semidiurnal oscillation of order ±50 m s^{-1}. Moreover, the phase of the oscillation indicates downward phase progression corresponding to a vertical wavelength of about 50 km. Downward phase progression for a gravity wave or propagating tide correspond to upward group velocity, therefore implying a source region at lower altitudes. Note that the vertical wavelength is much shorter than that of the first fundamental mode excited by ozone heating. The oscillation revealed in Fig. 1 is primarily forced by 'mode coupling' due to distortion of the upward-propagating semidiurnal tide by the zonal mean mesospheric jets just below the height region covered by the MF radar in Fig. 1. Tidal waves forced by higher-order components (beyond the fundamental mode) of the ozone heating and troposphere heating sources also make some contributions to Fig. 1 tidal structure.

Observations at a single site like those depicted in Fig. 1 cannot distinguish contributions from migrating (longitude-independent) versus non-migrating (longitude-dependent) tides. Observations from space have the potential to do this, provided that the satellite orbit precesses in local time such that the tidal structures are sampled sufficiently fast that other variations (e.g., in the zonal mean, such as the semiannual variation) do not alias into the tidal signal. The difficulty is that the faster precessing orbits have low orbital inclinations, and hence geographical coverage that is limited in latitude. Study of the semidiurnal tide does, however, offer some advantage since in principle only 12 hours of local time coverage is required to define its structure.

The Upper Atmosphere Research Satellite (UARS) was the first mission to fully delineate the structure of migrating semidiurnal tidal winds from equatorial to middle latitudes (McLandress *et al.*, 1996) between 90 and 110 km. Due to constraints imposed by yaw maneuvers, instrument sampling and local time precession, this tidal analysis was restricted to 60-day mean values between ±40° latitude and 90–110 km altitude. More recently the SABER instrument on the TIMED satellite provided semidiurnal temperature measurements from 20–120 km and ±50° latitude (Forbes *et al.*, 2008). These data were obtained from full 24-hour local time sampling every 60 days, and processed such that a moving 60-day window is used to determine the background 60-day mean temperature each day, and the residuals from this mean are analyzed for the tidal perturbations. Due to yaw maneuvers conducted by the satellite every ≈60 days, continuous sampling and the tides derived according to this method are restricted to ±50° latitude. An example is shown in Fig. 2, and provides some sense of intra-seasonal and inter-annual variability. Note that the latitudinal structures

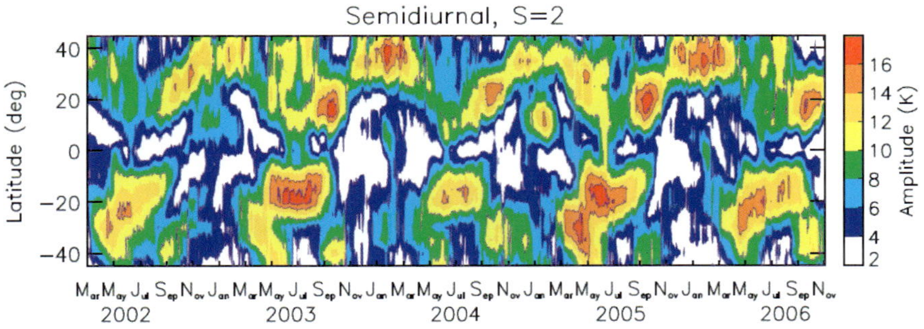

Fig. 2. Latitude versus time contours of semidiurnal migrating tidal temperature amplitudes (K) at 100 km altitude, covering March 2002 through December 2006. (From Forbes *et al.*, Tidal variability in the ionospheric dynamo region, *J. Geophys. Res.*, **113**, A02310, 2008. Copyright 2008 American Geophysical Union. Reproduced by permission of American Geophysical Union.)

do not conform to the broad structure corresponding to ozone insolation absorption described previously, again reflecting the presence of higher-order tidal modes generated by the mode coupling process referred to earlier.

Referring to Fig. 2, and as noted by Forbes *et al.* (2008), in the Southern Hemisphere the semidiurnal component is generally characterized by elevated amplitudes during April–September between -5° and -40°, whereas the semidiurnal tide in the Northern Hemisphere is characterized by a band of elevated amplitudes that oscillates sinusoidally from about 5–30° latitude near June to around 30–45° around January. Along this band, 2–4 K enhancements are seen around September–October and December–March during 2003–2004 and 2005–2006, following a 4 K elevated maximum in the Southern Hemisphere during 2003 and 2005. These features are suggestive of a quasi-biennial oscillation (QBO) effect for the semidiurnal tide extending to 40° latitude in the Northern Hemisphere and -20° to -30° latitude in the Southern Hemisphere. QBO modulation of the semidiurnal tide is likely a tide-zonal mean wind interaction effect, with the modulating zonal mean winds in the stratosphere and lower mesosphere varying at the QBO period ($\approx$26 months).

4 Interactions between SW2 and Stationary Planetary Waves

The QBO is not the only quasi-periodic modulation of SW2. Consider now a situation similar to the tide-zonal mean wind interaction described in Section 3, except that the mean winds vary with longitude. For simplicity consider first the stationary planetary wave (SPW) with $s = 1$ (SPW1). Through nonlinear terms in the momentum equation, momentum forcing terms involving the background and perturbation winds appear:

$$\underset{\text{SW2}}{\cos(2\Omega t + 2\lambda)} \times \underset{\text{SPW1}}{\cos\lambda} \rightarrow \underset{\text{SW1}}{\cos(2\Omega t + \lambda)} + \underset{\text{SW3}}{\cos(2\Omega t + 3\lambda)} \tag{4}$$

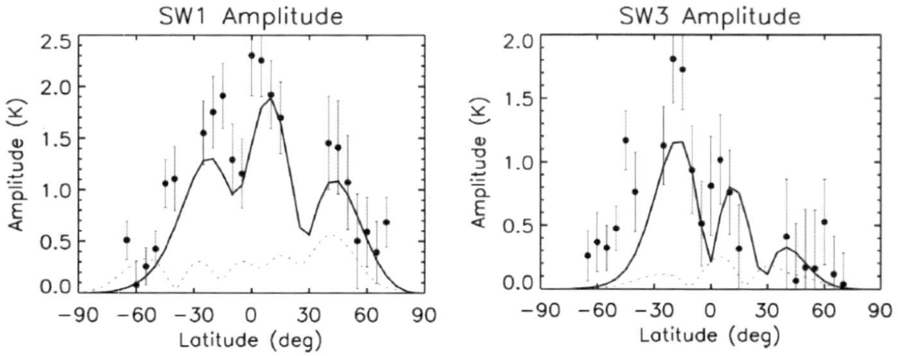

Fig. 3. Latitude structures of January semidiurnal temperature amplitudes for SW1 (left) and SW3 (right) at 86 km altitude. The solid lines represent Hough mode fits to these data taking into account the first two symmetric and antisymmetric propagating and trapped modes. The vertical bars represent 1-σ uncertainty estimates from the least squares fitting algorithm. The dashed lines are values from the GSWM taking into account forcing only by latent heat release in the Tropics (Hagan and Forbes, 2003), suggesting that this source is much less important than SW2-SPW1 nonlinear interaction.

In other words, the westward-propagating non-migrating semidiurnal tides with $s = 1$ and $s = 3$ (SW1 and SW3) are generated through a process that looks very much like amplitude modulation of a periodic signal. A theoretical basis for the above where the "sum" and "difference" waves (in both frequency and wavenumber) are generated can be found in Teitelbaum *et al.* (1989) and Teitelbaum and Vial (1991). The interaction of SW2 with a stationary wave-1 structure was first put forth by Forbes *et al.* (1995) to explain observation of a large-amplitude SW1 oscillation propagating zonally around the South Pole at 92 km altitude. Following the derivation of Eq. (3) for SW1 instead of SW2, it is clear that the zonal phase speed of SW1 is -2Ω, and thus propagates around the pole at a speed *twice as fast as the Sun.* In a manner similar to the mode coupling process described in Section 3, SW1 and SW3 combine with SW2 to produce the longitudinal distortion in the total semidiurnal tide that results from the longitudinally-varying background wind structure through which the tide is propagating.

The generation of SW1 and SW3 by the SW2-SPW1 interaction has been investigated numerically by Angelats i Coll and Forbes (2002) and Yamashita *et al.* (2002), and shown to be a viable explanation for the South Pole observations, although the computed SW1 maxima occur about 10–20 km higher than 92 km. Note that the appearance of strong SW1 over South Pole occurs during local summer and the SW2-SPW1 interaction mainly occurs in middle to high latitudes of the winter hemisphere, so that inter-hemispheric propagation is involved. These models also verify that SW1 and SW3 propagate away from the source region as independent oscillations, and extend into the thermosphere. In fact, molecular dissipation of SW1 and SW3 alone

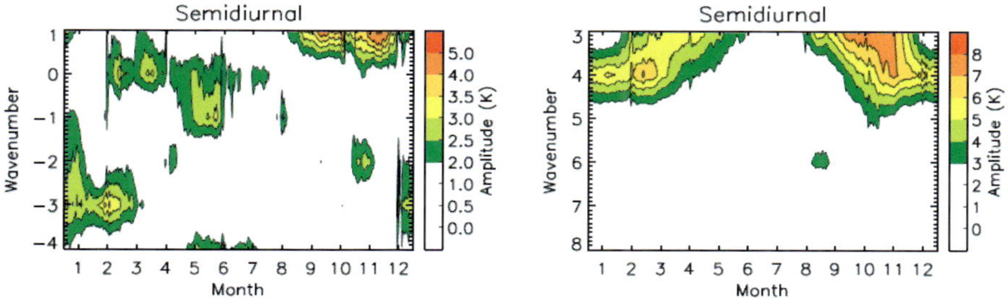

Fig. 4. Nonmigrating semidiurnal tidal temperature amplitudes at 116 km over the equator versus zonal wave number (y-axis) and day of year (x-axis), averaged over 2002–2006. Zonal wave numbers are negative for eastward propagating waves and positive for westward propagating waves. The white areas denote the "noise floor" that can be considered an approximate measure of uncertainty in the displayed amplitudes.

are shown by Angelats i Coll and Forbes (2002) to produce mean zonal and meridional wind amplitudes of order 10–20 m s^{-1} and 5–10 m s^{-1}, respectively, in the 110–200 km altitude region.

Recent observations at middle and low latitudes also reveal the presence of SW1 and SW3 in the 80–120 km height region (Forbes and Wu, 2006; Forbes *et al.*, 2008). Examples are shown in Figs. 3 and 4. Figure 3 (Forbes and Wu, 2006) illustrates latitudinal structures of SW1 and SW3 in the temperature field at 86 km. The latitudinal shapes are generally similar to classical Hough functions of classical tidal theory, and support the first-order interpretation that these waves are excited and propagate as independent oscillations. The semidiurnal tidal spectrum at 116 km over the equator (Fig. 4; Forbes *et al.*, 2008) is derived from TIMED/SABER temperatures in the same manner as described in connection with Fig. 2. This figure is derived from temperature residuals averaged over 2001–2006, and thus represent tidal oscillations that survive averaging over this period, i.e., climatologically persistent features. It is noteworthy that SW1 and SW3 remain at amplitudes of order 5–8 K, and that they co-exist during September–December. It is not known exactly why SW3 appears during February–April, but SW1 does not. However, it must be remembered that Fig. 4 only depicts the equator at 116 km altitude, and thus provides limited insight into what is going on globally. Evidence for co-existence of S0 and SW4 are also apparent during February–March, which would be consistent with the wave-wave interaction (4), with SPW2 replacing SPW1:

$$\underset{\text{SW2}}{\cos(2\Omega t + 2\lambda)} \times \underset{\text{SPW2}}{\cos 2\lambda} \rightarrow \underset{\text{S0}}{\cos(2\Omega t)} + \underset{\text{SW4}}{\cos(2\Omega t + 4\lambda)} \tag{5}$$

The SE1 and SE3 non-migrating tides in Fig. 4 likely arise from latent heating due to deep tropical convection, but further discussion is out of the scope of the current paper.

5 Interactions between SW2 and Traveling Planetary Waves

By extension of Eq. (5) into the time domain, we can anticipate that interaction of SW2 with prominent traveling planetary waves (e.g., with 2-day, 5-day, 10-day, 16-day periods) might lead to other secondary waves. Interaction of SW2 with the quasi-2-day wave (QTDW) is the only one that has been studied extensively (i.e., in terms of both comprehensive modeling and observation) to date. The QTDW is a mesospheric oscillation common to the summer season of each hemisphere, and is traditionally thought to be a westward-propagating Rossby normal mode (Salby, 1981a, b) with $s = 3$, although more recent work indicates that other zonal wavenumbers (e.g., $s = 4$) are present as well (Salby and Callaghan, 2001, 2003). For illustration purposes we will assume $s = 3$. The SW2-QTDW interaction

$$\underset{\text{SW2}}{\cos(2\Omega t + 2\lambda)} \times \underset{\text{QTDW}}{\cos(0.5\Omega t + 3\lambda)} \rightarrow \underset{\text{W5, 9.6 h}}{\cos(2.5\Omega t + 5\lambda)} + \underset{\text{E1, 16.0 h}}{\cos(1.5\Omega t - \lambda)} \quad (6)$$

thus consists of 2-day modulation of SW2, or equivalently two sideband or secondary ('sum and difference') oscillations: a westward-propagating wave with 9.6-hour period and $s = 5$, and a 16-hour period eastward-propagating wave with $s = -1$. These sidebands have been observed in wind spectra of the mesosphere and lower thermosphere (Cevolani and Kingsley, 1992) and interpreted as resulting from Eq. (6). Figure 5 provides an example of day-to-day variability in the semidiurnal tide of the high-latitude E-region during March, 1988, that appears to be the result of modulation by the QTDW (Huuskonen *et al.*, 1991), although one primarily expects to see such effects during local summer.

A numerical modeling study of the SW2-QTDW interaction was performed by

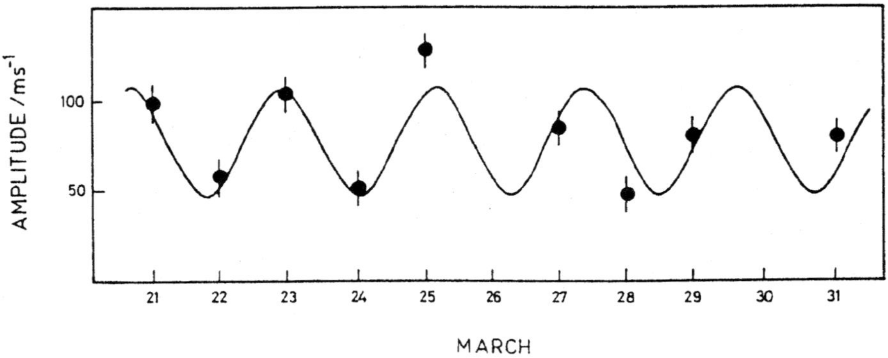

Fig. 5. Amplitude of the meridional wind amplitude of the semidiurnal tide (dots) at 70°N averaged over 105–130 km during March 1988 from EISCAT observations. The solid line represents a modulating wave with period of 2.2 days. (With kind permission from Springer Science+Business Media: *Ann. Geophys.*, Observations of day-to-day variability in the meridional semi-diurnal tide at 70°N, **9**, 1991, 407–415, Huuskonen *et al.*, figure 7.)

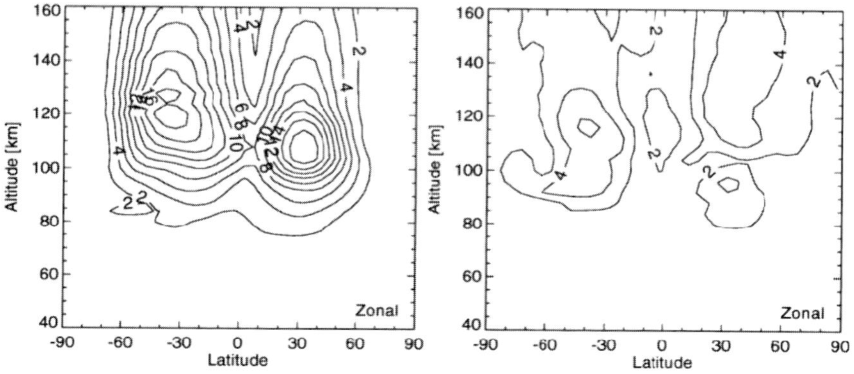

Fig. 6. Height versus latitude zonal wind amplitude structures of the westward-propagating 9.6-h $s = 5$ (left) and eastward-propagating 16-h $s = -1$ (right) waves generated via nonlinear interaction between SW2 and the westward-propagating 2-day wave with $s = 3$ (Palo *et al.*, 1999).

Palo *et al.* (1999) that provided further insight into this process and its extension into the thermosphere. These authors introduced the QTDW at the lower boundary of their model (ca. 30 km) at an amplitude that resulted in a realistic QTDW structure in the mesosphere. SW2 was generated internally by insolation absorption by ozone. They then found that the above W5 and E1 waves were generated within the model by nonlinear wave-wave interactions, and determined their height versus latitude structures as shown in Fig. 6. Again, it appears as though these secondary waves, in a manner similar to the secondary waves generated in connection with the interactions of SW2 with zonal mean winds and with stationary planetary waves (see Sections 3 and 4), propagate upward as independent oscillations. As shown in Fig. 6, these oscillations appear to maximize between about 100 and 160 km, which coincides with the ionospheric dynamo region. Here, it is possible that the sideband waves modulate the semidiurnal tide (both excited in-situ and propagating upwards from below) at a quasi-2-day period, and that this 2-day periodicity is imposed upon the dynamo-generated electric fields that map into the F-region along equipotential field lines and redistribute plasma (Forbes, 1996). A possible manifestation of this process is shown in Fig. 7, which illustrates a quasi-2-day variation in f_oF_2 over Okinawa during the same period of time when a QTDW was measured in the mesosphere and lower thermosphere wind field. Okinawa is near the crest latitude of the so-called ionospheric fountain effect (or equatorial ionization anomaly, EIA), and is thus a repository for plasma that is uplifted at the equator by the daytime eastward electric field that arises primarily through the E-region wind dynamo. If the semidiurnal tidal wind in the dynamo region is varying with a 2-day period, then it is likely that the equatorial electric field and the fountain effect will be similarly modulated. However, this process needs much more numerical modeling work and observations before it is fully understood

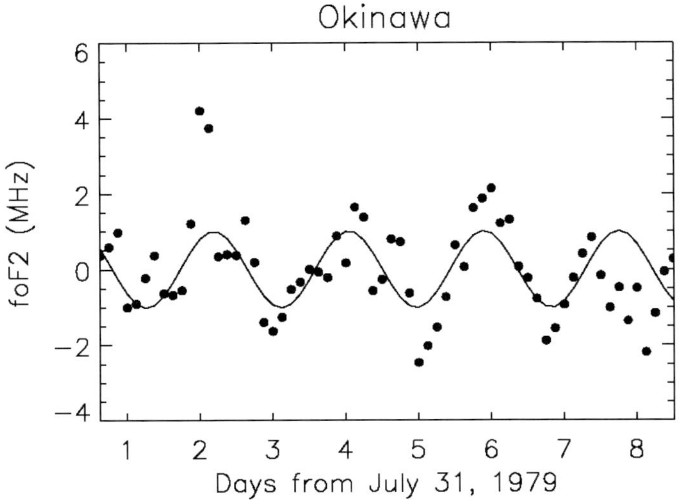

Fig. 7. Quasi-2-day oscillation in f_oF_2 observed over Okinawa, August 1–8, 1979. This figure illustrates observations from the database of ionospheric 2-day wave events analyzed by Forbes and Zhang (1997).

and established to be the primary mechanism. For instance, composition variations or 2-day modulation of the semidiurnal tide at F-region altitudes could also be involved.

The literature contains much additional observational evidence for planetary wave modulations of tides (e.g., Kamalabadi *et al.*, 1997; Beard *et al.*, 1999; Pancheva, 2000; Palo *et al.*, 2007); unfortunately, limitations on the scope and magnitude of the present contribution prevent further elucidation of these fascinating results.

6 Conclusions

The semidiurnal solar thermal tide is ubiquitous in Earth's atmosphere, and serves to globally couple the troposphere, stratosphere, mesosphere, thermosphere and ionosphere. Dissipation of SW2 and its offspring induce significant modifications of the lower thermosphere (100–200 km) mean circulation. Modulation of SW2 by planetary waves in the 2–20 day period range provides a means of modulating the E-region dynamo at these periods, and generating E-fields that impose these periodicities on the F-region ionosphere aloft. Direct vertical penetration of SW2 and/or sideband waves produced as a result of modulation by planetary waves is an additional means of similarly affecting the neutral thermosphere, and the solar semidiurnal tide is the vehicle through which this can happen. The semidiurnal tide plays similar roles in other planetary atmospheres, particularly Mars. At Mars, insolation absorption by dust plays a similar role in exciting SW2 that ozone does in Earth's atmosphere, and the consequences (Forbes and Miyahara, 2006) are even more dramatic than in Earth's atmosphere.

Acknowledgments. This work was supported under Grant ATM-0346218 from the National Science Foundation to the University of Colorado.

References

Angelats i Coll, M. and J. M. Forbes, Nonlinear interactions in the upper atmosphere: The $s = 1$ and $s = 3$ nonmigrating semidiurnal tides, *J. Geophys. Res.*, **107**, 1157, doi:10.1029/2001JA900179, 2002.

Beard, A. G., N. J. Mitchell, P. J. S. Williams, and M. Kunitake, Non-linear interactions between tides and planetary waves resulting in periodic tidal variability, *J. Atmos. Sol.-Terr. Phys.*, **61**, 363–376, 1999.

Cevolani, G. and S. P. Kingsley, Non-Linear Effects on tidal and planetary waves in the lower thermosphere, *Adv. Space Res.*, **12**, 77–80, 1992.

Chapman, S. and R. S. Lindzen, *Atmospheric Tides: Thermal and Gravitational*, 200 pp., Gordon and Breach, 1970.

Forbes, J. M., Atmospheric tides 2. The solar and lunar semidiurnal components, *J. Geophys. Res.*, **87**, 5241–5252, 1982.

Forbes, J. M., Planetary waves in the thermosphere-ionosphere system, *J. Geomag. Geoelectr.*, **48**, 91–98, 1996.

Forbes, J. M. and H. B. Garrett, Thermal excitation of atmospheric tides due to insolation absorption by O_3 and H_2O, *Geophys. Res. Lett.*, **5**, 1013–1016, 1979.

Forbes, J. M. and X. Zhang, The quasi 2-day ionospheric oscillation: A statistical study, *J. Atmos. Sol.-Terr. Phys.*, **59**, 1025–1034, 1997.

Forbes, J. M. and S. Miyahara, Solar semidiurnal tide in the dusty mars atmosphere, *J. Atmos. Sci.*, **63**, 1798–1817, 2006.

Forbes, J. M. and D. Wu, Solar tides as revealed by measurements of mesosphere temperature by the MLS experiment on UARS, *J. Atmos. Sci.*, **63**, 1776–1797, 2006.

Forbes, J. M., N. A. Makarov, and Yu. I. Portnyagin, First results from the meteor radar at South Pole: A large 12-hour oscillation with zonal wavenumber one, *Geophys. Res. Lett.*, **22**, 3247–3250, 1995.

Forbes, J. M., X. Zhang, S. Palo, J. Russell, C. J. Mertens, and M. Mlynczak, Tidal variability in the ionospheric dynamo region, *J. Geophys. Res.*, **113**, A02310, doi:10.1029/2007JA012737, 2008.

Hagan, M. E. and J. M. Forbes, Migrating and nonmigrating diurnal tides in the middle and upper atmosphere excited by tropospheric latent heat release, *J. Geophys. Res.*, **107**, 4754, doi:10.1029/2001JD001236, 2002.

Hagan, M. E. and J. M. Forbes, Migrating and nonmigrating semidiurnal tides in the middle and upper atmosphere excited by tropospheric latent heat release, *J. Geophys. Res.*, **108**, 1062, doi:10.1029/2002JA009466, 2003.

Huuskonen, A., T. S. Virdi, G. O. L. Jones, and P. J. S. Williams, Observations of day-to-day variability in the meridional semi-diurnal tide at 70°N, *Ann. Geophys.*, **9**, 407–415, 1991.

Kamalabadi, F., J. M. Forbes, N. A. Makarov, and Yu. I. Portnyagin, Evidence for non-linear coupling of planetary waves and tides in the Antarctic mesopause, *J. Geophys. Res.*, **102**, 4437–4446, 1997.

Lindzen, R. S. and S.-S. Hong, Effects of mean winds and meridional temperature gradients on solar and lunar semidiurnal tides in the atmosphere, *J. Atmos. Sci.*, **31**, 1421–1466, 1974.

McLandress, C., G. G. Shepherd, and B. H. Solheim, Satellite observations of thermospheric tides: Results from the Wind Imaging Interferometer on UARS, *J. Geophys. Res.*, **101**, 4093–4114, 1996.

Palo, S. E., R. G. Roble, and M. E. Hagan, Middle atmosphere effects of the quasi two-day wave determined from a General Circulation Model, *Earth Planets Space*, **51**, 629–647, 1999.

Palo, S. E., J. M. Forbes, X. Zhang, J. M. Russell III, and M. G. Mlynczak, An eastward propagating two-day wave: Evidence for nonlinear planetary wave and tidal coupling in the mesosphere and lower thermosphere, *Geophys. Res. Lett.*, **34**, L07807, doi:10.1029/2006GL027728, 2007.

Pancheva, D., Evidence for nonlinear coupling of planetary waves and tides in the lower thermosphere over Bulgaria, *J. Atmos. Sol.-Terr. Phys.*, **62**, 115–132, 2000.

Salby, M. L., Rossby normal modes in nonuniform background configurations I, Simple fields, *J. Atmos. Sci.*, **38**, 1803–1826, 1981a.

Salby, M. L., Rossby normal modes in nonuniform background configurations II, Equinox and solstice conditions, *J. Atmos. Sci.*, **38**, 1827–1840, 1981b.

Salby, M. L. and P. F. Callaghan, Seasonal amplification of the 2-day wave: Relationship between normal mode and instability, *J. Atmos. Sci.*, **58**, 1858–1869, 2001.

Salby, M. L. and P. F. Callaghan, Dynamics of the 2-day wave in a nonlinear model of the middle and upper atmosphere, *J. Geophys. Res.*, **108**(D23), 4713, doi:10.1029/2003JD003648, 2003.

Teitelbaum, H. and F. Vial, On tidal variability induced by nonlinear interaction with planetary waves, *J. Geophys. Res.*, **96**, 14,169–14,178, 1991.

Teitelbaum, H., F. Vial, A. H. Manson, R. Giraldez, and M. Massbeuf, Non-linear interaction between the diurnal and semidiurnal tides: terdiurnal and diurnal secondary waves, *J. Atmos. Terr. Phys.*, **51**, 627–634, 1989.

Walterscheid, R. L. and S. V. Venkateswaran, Influence of mean zonal motion and meridional termperature gradients on the solar semidiurnal atmospheric tide: A spectral study, Part 1, Theory, *J. Atmos. Sci.*, **36**, 1623–1635, 1979a.

Walterscheid, R. L. and S. V. Venkateswaran, Influence of mean zonal motion and meridional termperature gradients on the solar semidiurnal atmospheric tide: A spectral study, Part 2, Numerical results, *J. Atmos. Sci.*, **36**, 1636–1662, 1979b.

Yamashita, K., S. Miyahara, Y. Miyoshi, K. Kawano, and J. Ninomiya, Seasonal variation of non-migrating semidiurnal tide in the polar MLT region in a general circulation model, *J. Atmos. Sol.-Terr. Phys.*, **64**, 1083–1094, 2002.

Index